AF901312

Springer
Berlin
Heidelberg
New York
Barcelona
Hong Kong
London
Milan
Paris
Tokyo

中文大學出版社
THE CHINESE UNIVERSITY PRESS
www.chineseupress.com

Physics and Astronomy ONLINE LIBRARY
http://www.springer.de/phys/

SPRINGER SERIES IN PHOTONICS

Series Editors: T. Kamiya B. Monemar H. Venghaus Y. Yamamoto

The Springer Series in Photonics covers the entire field of photonics, including theory, experiment, and the technology of photonic devices. The books published in this series give a careful survey of the state-of-the-art in photonic science and technology for all the relevant classes of active and passive photonic components and materials. This series will appeal to researchers, engineers, and advanced students.

1 **Advanced Optoelectronic Devices**
By D. Dragoman and M. Dragoman

2 **Femtosecond Technology**
Editors: T. Kamiya, F. Saito, O. Wada, H. Yajima

3 **Integrated Silicon Optoelectronics**
By H. Zimmermann

4 **Fibre Optic Communication Devices**
Editors: N. Grote and H. Venghaus

5 **Nonclassical Light from Semiconductor Lasers and LEDs**
By J. Kim, S. Somani, and Y. Yamamoto

6 **Vertical-Cavity Surface-Emitting Laser Devices**
By H. Li and K. Iga

7 **Active Glass for Photonic Devices**
Photoinduced Structures and Their Application
Editors: K. Hirao, T. Mitsuyu, J. Si, and J. Qiu

8 **Nonlinear Photonics**
Nonlinearities in Optics, Optoelectronics and Fiber Communications
By Y. Guo, C.K. Kao, E.H. Li, and K.S. Chiang

Series homepage – http://www.springer.de/phys/books/ssp/

Y. Guo C.K. Kao E.H. Li K.S. Chiang

Nonlinear Photonics

Nonlinearities in Optics, Optoelectronics and Fiber Communications

With 152 Figures, 29 Tables,
Numerous Examples and Problems

The Chinese University Press

Springer

Professor Yili Guo
Department of Electronic Engineering
Tsinghua University
100084 Beijing
People's Republic of China

Professor E. Herbert Li (1957-2001)

Professor Charles K. Kao
ITx Services Ltd.
c/o S.K. Yee Medical Foundation
Unit 1708, Office Tower, Convention Plaza
1 Harbour Road, Wan Chai
Hong Kong

Professor Kin S. Chiang
Optoelectronics Research Centre
and Department of Electronic Engineering
City University of Hong Kong
83 Tat Chee Avenue, Kowloon
Hong Kong

Series Editors:

Professor Takeshi Kamiya
Ministry of Education, Culture, Sports,
Science and Technology
National Institution for Academic Degrees
3-29-1 Otsuka, Bunkyo-ku, Tokyo, 112-0012
Japan

Professor Bo Monemar
Department of Physics
and Measurement Technology
Materials Science Division
Linköping University
58183 Linköping
Sweden

Dr. Herbert Venghaus
Heinrich-Hertz-Institut
für Nachrichtentechnik Berlin GmbH
Einsteinufer 37
10587 Berlin
Germany

Professor Yoshihisa Yamamoto
Stanford University
Edward L. Ginzton Laboratory
Stanford, CA 94305
USA

ISSN 1437-0379
ISBN 962-201-861-0 The Chinese University Press Hong Kong
ISBN 3-540-43123-3 Springer-Verlag Berlin Heidelberg New York

Library of Congress Cataloging-in-Publication Data applied for.
Die Deutsche Bibliothek - CIP-Einheitsaufnahme
Nonlinear photonics : nonlinearities in optics, optoelectronics and fiber communications ; with 29 tables, numerous examples and problems / Y. Guo ... - Berlin ; Heidelberg ; New York ; Barcelona ; Hong Kong ; London ; Milan ; Paris ; Tokyo : Springer, 2002 (Springer series in photonics ; 8) (Physics and astronomy online library) ISBN 3-540-43123-3

This work is subject to copyright. All rights are reserved, whether the whole or part of the material is concerned, specifically the rights of translation, reprinting, reuse of illustrations, recitation, broadcasting, reproduction on microfilm or in any other way, and storage in data banks. Duplication of this publication or parts thereof is permitted only under the provisions of the German Copyright Law of September 9, 1965, in its current version, and permission for use must always be obtained from Springer-Verlag. Violations are liable for prosecution under the German Copyright Law.

Springer-Verlag Berlin Heidelberg New York
a member of BertelsmannSpringer Science+Business Media GmbH

http://www.springer.de

© The Chinese University Press and Springer-Verlag Berlin Heidelberg 2002
Printed in Hong Kong

The use of general descriptive names, registered names, trademarks, etc. in this publication does not imply, even in the absence of a specific statement, that such names are exempt from the relevant protective laws and regulations and therefore free for general use.

Cover concept: eStudio Calamar Steinen; Cover production: *design & production* GmbH, Heidelberg
SPIN: 10791124 57/3141/di 5 4 3 2 1 0

In memory of
the late Professor E. Herbert Li,
who wished to dedicate this book
to his two beloved daughters,
Victoria and Elizabeth.

Preface

In the last decade of the twentieth century, 1991 to 2000, we witnessed an enormous growth in the information market, which gave rise to huge advancement in information science and technology. Over those years, mobile communications, personal computers, the Internet, optical fiber communications, WDM networks, etc., have become indispensable in our daily lives. Their rapid developments owe much to the technological advances in microelectronics, optolectronics, optical fibers, and fiber devices in the past forty years.

Entering the twenty-first century, optical fibers, semiconductor devices, and optical communications are still under active development. Optical fibers with new dopants and structural designs are emerging and finding important applications. Semiconductor devices using new compositions and operation principles, in particular, multiple-quantum-well based opto-electronic devices, are being developed. These developments have laid an important foundation for future optical fiber communication systems.

In 1990, single-wavelength transmission was the major form of optical communication, which provided a typical capacity of 622 Mbit/s. By 2000, multi-wavelength transmission had become the dominant technology. A single fiber with 64 wavelength channels can readily achieve a capacity of 640 Gbit/s. Transmission distance had also increased by more than 100 times through the use of cascaded optical fiber amplifiers.

The tremendous growth in bit rate, number of wavelength channels, and transmission distance, has set much more stringent requirements for information transmission in the time, frequency, and spatial domains. The fundamental issue is that many undesirable linear optical effects (dispersion, dispersion slope, polarization mode dispersion, chirp, Rayleigh scattering) and nonlinear optical effects (SPM, XPM, FWM, SBS, SRS), as well as their interactions, can cause significant signal deterioration in high-capacity systems and thus seriously limit further development of such systems into fiber networks with much higher capacity. Therefore, much effort has been made, from principle to practice, to suppress these undesirable nonlinear effects or their undesirable interactions with linear effects. At the same time, significant progress has been made in the exploration of new techniques and applications based on nonlinear interactions. It is the purpose of this book to describe these endeavors.

This book starts with a traditional analysis of the nonlinear interaction between matter and field, with emphasis on the discussion of optical absorption and nonlinear refractive index and their relations with the interacting field, followed by a brief review of some typical nonlinear optical effects. We move quickly into discussions of various nonlinear effects in optical fibers, fiber devices, multiple quantum wells, and waveguide devices, as well as in dense WDM, high-speed TDM, and long-distance fiber communication systems. Particular attention is given to the description of the techniques developed for the suppression or the utilization of these effects. To distinguish this book from traditional books on nonlinear optics, its title is chosen to be *Nonlinear Photonics*. It is hoped that the book can provide a comprehensive coverage of the subject, as well as working knowledge, that will be useful to researchers, engineers, and postgraduate students, who are involved in the research and development of the optical fiber communication technology.

The writing of this book started in 1995, when Professor Charles K. Kao was Vice-Chancellor of the Chinese University of Hong Kong (CUHK). Through the arrangement of Professor Kao, Professor Yili Guo of Tsinghua University, Beijing, visited the CUHK for nine months. During that period of time, Professor Guo, together with Professor Kin S. Chiang of the City University of Hong Kong and the late Professor E. Herbert Li of the University of Hong Kong, planned this book. This book is the result of their joint effort, with Chapter 1 by Professor Charles K. Kao, Chapters 2–8 by Professor Yili Guo, Chapters 9 by Professsor Kin S. Chiang and Chapters 10 and 11 by the late Professor E. Herbert Li.

Yili Guo
Charles K. Kao
Kin S. Chiang

Beijing and Hong Kong
January 2002

Acknowledgments

The authors are very grateful to the publishers for their great amount of very professional and thorough editorial work in publishing this book, in particular, to Dr. Steven K. Luk and Mr. Yat-Kong Fung of the Chinese University Press and to Dr. Claus E. Ascheron and Dr. Angela M. Lahee of Springer-Verlag.

It is very sad that one of the authors, Professor E. Herbert Li, and the former Director of the Chinese University Press, Mr. Paul S. Wong, passed away in the course of publication of this book. This book represents the warmest tributes to them.

Professor Yili Guo has taught the subject of nonlinear photonics at Tsinghua University, Beijing, China, for many years and has been engaged in a large number of relevant research projects. He would like to express his sincere gratitude to his colleagues in China, who have collaborated with him in the research on nonlinear optics, for their strong support and encouragement to his work since 1980's, in particular, to his colleagues at the Physics Institute of Academia Sinica, the Semiconductor Institute of Academia Sinica, Fudan University, and Xidian University. The contributions from the research students of Tsinghua University in the last 20 years are also gratefully acknowledged.

Special thanks are also given to Professor G. P. Agrawal of the University of Rochester, Dr. Tingye Li and Dr. Tien-Pei Lee of Bell Laboratories, and Professor W. A. Gambling of the City University of Hong Kong for their kind support, and especially to Dr. Aleksandra B. Djurisic of the University of Hong Kong for her assistance in revising the late Professor E. Herbert Li's work.

Contents

1. Optical Communication

Electromagnetic energy at optical wavelengths is increasingly being harnessed as the means to enhance our communication systems. The 1.3 to 1.55 micrometer optical window is being intensely exploited for optical fiber transmission purposes. Other wavelengths from long IR to soft X-rays are being used to fabricate new materials and devices needed for building sensors, displays, and other I/O equipment. Together with the electromagnetic energy at radio and microwave frequencies, we shall achieve affordable communication for everyone, at anytime and at any place. The communication capabilities will allow, not only, human to human communication and contact, but also human to machine and machine to machine interaction. The communication will allow our visual, audio, and touch senses to be contacted as a virtual 3-D presence. The volume and rate of information to be handled will be enormous and will have to be handled by using both electronic and optic technologies together. The progress made so far is impressive. Information rate at 1 terabit/s can be handled by a single fiber. To push this frontier further, the optoelectronic devices will be necessary and the nonlinear photonic characteristics will be indispensable for the creation of novel high speed devices. In this chapter, a short history of the development of optical fiber communication, and the roles of optoelectronics are presented. Without this ideal transmission medium which allowed low cost, wide bandwidth global transport of information possible, and advances in electronics, computers, and optoelectronics, the communication revolution heralding the new information age will not be possible.

1.1 A Brief History of the Development of Optical Fiber Technology

Optical fibers were first made by drawing molten glass from a crucible filled with glass. Due to the viscous nature of glass and the natural strong surface tension forces, the resultant fiber has a perfect circular cross-section. With a steady rate of draw, long lengths of fiber with uniform dimension can be made. The diameter is controlled by the rate of draw, and can be as fine as a few micrometers. Glass fiber is highly flexible and traps light within it. As long as the bending radii are not too small, light is reflected through internal total reflection as it travels through the fiber. O'Brian in the USA and Hopkins in the UK were working on the fiber image bundles almost simultaneously. O'Brian had the idea of cladding a single fiber filament with an additional outer layer of glass with a lower refractive index. This technique ensured that light in one fiber would not leak into another one when they are in touch with each other. It also meant that the fiber can be touched by other objects without loosing light. Hopkins also suggested for image bundles the fibers can be isolated with a dielectric material as a cladding. The jacket, however, reduces the angular aperture of the fiber, since the internal reflection critical angle is reduced to a smaller value. Even though less light can be carried through the fiber, the protective layer is essential for the fiber to perform with greater consistency.

The thinly cladded fiber enabled several new applications to be explored. When fibers are laid in a regular closely packed manner forming an arrayed fiber-bundle, an image presented at one end of the bundle will emerge at the other end with the image staying spatially coherent. Such a bundle is referred to as a coherent bundle. This was the start of endoscopy as it is known today. The coherent image transmission not only allows optical image transfer but also is a useful way of presenting an image with a narrow viewing angle to the viewer. This property can be explored as a means to view a transmitted image without it being obscured by bright ambient illumination. Even in the extreme case when sunlight is shining directly on the end of the bundle the image from the bundle is still discernable. This is possible since the light from the bundle is highly directed while the ambient light is usually omni-directional. When the arrayed fiber-bundle is cut perpendicular to the fiber bundle axis into thin slices and then polished, the slices are referred to as fiber face-plates. These face plates have the same directional properties as the bundle. They are used as cover glasses on instrument panels in aircrafts such that the instrument readings can be read even with sunlight shining on the cover glasses. For this type of application, the fiber bundle must be pre-fused to form a consolidated rod before cutting so that the slices are hermetically sealed. For fine resolution the fiber core size is reduced and the cladding is kept as thin as possible without loosing the isolation property. The core size of a typical fiber for image bundle or face-plate applications are around 10 micrometers with cladding at around 1 micrometer.

This technology was later developed for another important application, namely as the key component for a low light level viewing instrument. In this application the fiber cover glass is transformed into an image-preserving photomultiplier without the use of complicated lens optics. The core glass is selected to be soluble to a chosen solvent which dissolves differentially the core glass but not the cladding glass. The cladding glass is selected such that the surface can be coated by a cathodic material which, when suitably processed, would emit secondary electrons when bombarded by incoming photon-induced electrons under an accelerating voltage. The cover glass now acts as a compact image-forming photo-multiplier. These optical viewing instruments, called night vision scopes or star light scopes, allow surveillance using only illumination from the stars in the night sky alone.

Glass fibers for light conduit applications are made with a high-index core glass, usually a lead glass. The cladding is a soda-lime silicate glass with a melting point somewhat higher than the core glass. These fibers are made by first sleeving a rod of core glass with a tube of cladding glass collapsed over the rod. By heating the end of the combination rod to a temperature when viscous flow can be achieved an elongated fiber can be pulled out. Surprisingly the geometrical precision is preserved through this elongation process, and a cross-sectional dimension reduction of several orders of magnitude achieved. A start rod of 15mm diameter can be reduced to a fiber with an overall diameter of 0.15mm very readily. This 100-fold reduction means that a rod of 1 meter can be transformed to 10km of fiber.

For light conduit applications the attenuation of light propagating through such fiber structures should be relatively low. Such levels of attenuation can easily be met by optical quality glass for optical instrumentation purposes. Although fiber is a known electromagnetic wave-guide for optical waves, little attention was paid to the use of the fiber as a transmission line for communication applications, not until lasers were invented. Even then, improvement in the transparency of optical glass by many orders of magnitude was not expected to be possible. Also there was no reason to search for very low attenuation glasses. Nevertheless, towards the end of 1950s, the prospect for light waves to be used as a carrier for communication was increasing. Based on the theory of the dielectric waveguide, established by Hondros and Debye in 1908, the modes of propagation of optical waves in the optical fiber were first evaluated by Snitzer. Burke and Kapany later were able to demonstrate, using specially prepared waveguide samples, optical wave propagation of the low order waveguide modes.

Despite of this knowledge on fibers as a optical waveguide, it was not until 1960s, when Kao and Hockham were the first to do in-depth studies on the practical feasibility of fiber as a light guide for communication applications, before wider thought was given to this aspect of usage. Their paper in 1966 clearly defined the structure, the physical and material requirements, for optical fiber to be suitable for use as a transmission medium for broadband communication. Three further papers were published by Kao and his co-workers in the following years, 1967 and 1968, providing experimental verification that the transparency of the glassy materials with the required low losses could be realized physically. The invention of the optical fiber lightguide for communication application was completed. Much work and several diversionary steps were taken by a large number of people worldwide before the proposed single-mode structure was became the standard.

The first major issue was the problem of lowering the attenuation of glass to a level that would allow light to travel a reasonable distance before amplification must be provided. This was the key question which had to be answered before fiber could be regarded as a viable contender as the basic transmission line for broad optical communications. Kao and his co-workers showed unequivocally that the theoretically predicted low loss can be achieved by removing impurities in the glass. The series of papers published in 1967/68, mentioned earlier, convinced the scientific world to commence a concerted effort into research and development of optical fiber for communications applications.

Due principally to the interest for the military applications of optical fiber as a light weight, small size, convenient-to-handle, strong, electro-magnetic immune, non-interferable, broadband and low loss lightguide, the requisite funding for early development work was made available. The basic work carried out for these applications proved the practical feasibility of fiber for use in very stringent operational environments. This work laid the foundation for optical fiberguides to be further developed for commercial applications. In the course of the development, it was discovered that the fiber lived up to all the expectations. By 1980s two methods for producing fibers were achieved by Corning and ATT, known as outside vapor deposition (OVD) and modified chemical vapor deposition (MCVD) respectively. A third alternative method, vapor axial deposition (VAD) was achieved in Japan. These techniques pointed to a successful realization of a low cost fiber that could be produced in large quantities.

During the first commercialization phase, a graded index fiber with a 50 micrometer diameter core was used. This was a diversion since this type of the fiber was intrinsically more costly than the signal mode fiber and it had inferior information bandwidth characteristics. By 1990, single mode fiber was the undisputed standard. The single mode fiber made with a simple germanium-doped silica glass as the core and pure fused silica as the cladding was found to be close to the optimum for operation at 1.3 micrometer wavelength. At that wavelength, the dispersion of the wave-guide is cancelled by the material dispersion resulting in a fiber with an extremely broad information-carrying capacity. However, the loss was not at a minimum at 1.3μm wavelength. At 1.55 micrometer wavelength, where the fiber loss is at a minimum, the dispersion is finite. By tailoring the geometry of the waveguide, a so-called dispersion-compensated fiber, operating at 1.55μm, was successfully demonstrated. Still, for ultra-long distance communication application, the selection of the best waveguide is not obvious and this will be discussed a little later.

With a loss of less than 0.5dB/km at 1.3μm and less than 0.2dB/km at 1.55μm, the fiber as a transmission line is unparallelled in performance compared with any other form of communication transmission lines. In fact, it can be shown that for almost all future communication applications little advantage will be accrued as a result of further efforts to find an alternative transmission line. Effects to reduce scattering loss, however, are still worthwhile. One way is to reduce one scattering loss through changing the glass and portion the other is a possibility of making a fiber for an operating optical wavelengths longer than 1.55μm. Since loss is inversely proportional to the 4th power of wavelength, a theoretical loss of less than 0.2dB/km is possible. At 4μm extrapolating for the scattering of loss silica loses a loss as low as 0.001dB/km possible. This would allow repeaterless transmission anywhere around the globe.

The recent success in making optical fiber amplifiers, at 1.3μm, opens the way for fiber to stretch to any distance without the use of electronic amplifiers. This was another major milestone for fiber systems. The broadband performance of the amplifier permits the fiber to operate with many discrete wavelengths simultaneously. The 1.3μm amplifier is widely used but the development of 1.55μm is still continuing at the experimental phase. A broadband amplifier covering 1.3 to 1.55μm is not yet available. The ultimate bandwidth capacity of a single physical single-mode fiber enables a system bandwidth in excess of 1 terahertz to operate over any distance. It is interesting to note that zero dispersion at any wavelength would allow carriers of different wavelengths to couple with each other. For DWDM operation a non-zero dispersion fiber is preferred.

At this juncture in time, fibers have been installed at a rate of over 50,000km per year. The world will be wired with sufficient superhighways for information transmission so that with wireless task, any-time, any-place, any-person communication will be an economical possibility. We are in an age where information is central to our lives. The all-capable communication network is the infrastructure that would permit us to increase productivity/ capita and to adopt an altogether different life-style. This will be an evolutionary process, and we will work and play very differently in the future.

Throughout the period of the development of the low loss fiber for communication purposes, the assumption has always been made that the fiber has insignificant nonlinear characteristics. As we pushed for higher system performances, such as longer span lengths, by increasing the power level of the optical carrier and by reducing the loss and increasing the bandwidth, we started to encounter nonlinear phenomena. First it was the second and third order nonlinearity, such as Brilluion and Raman scattering, causing self-focussing due to power-density-induced refractive index change. Such nonlinear effects were first encountered when the maximum input power of the optical carrier launched into the fiber was investigated. It was later realized that even at much lower carrier power level, the nonlinearity provides a coupling mechanism for different wavelength carriers present in the same waveguide. The coupling coefficients are small but, as the transmission length is increased and light source coherence is improved, the coupling over simultaneous propagation distance of kilometers for two or more carriers of different wavelengths can be very significant. The nonlinearity characteristics become central to limit system performances.

The choice of the operating wavelength of 1.3 or 1.55μm depends on more than just loss and bandwidth considerations. For systems with span lengths less than 50km the choice can be arbitrary. For long systems, the distance, the repeater spacing, the ease of powering the repeaters, and the bandwidth capacity must all be taken into consideration for the choices that would realize the most economic system solution. For major trunk systems that require continuous upgrade of system capacity during their lifetime, the consideration of using wavelength multiplex will be high on the agenda. In these cases the choice of 1.3μm operation and the ease of using optical fiber amplifiers was an important advantage. The convenience of upgrading using multi-wavelengths was the deciding factor. Even with the advantage of having the lower loss at 1.55μm, the difficulty in implementing amplification and wavelength multiplexing due to nonlinearity effects would make this option less attractive, but the non-zero dispersion fiber mentioned earlier have altered this situation.

Despite of the problems caused by the nonlinear effects, the understanding of nonlinearity opens up new application areas for fiber. This is a very exciting area since nonlinear coupling can be maintained over very large distances in fiber; a situation hitherto not available for experimental physics as well as for yet-to-be discovered new application areas.

In other application areas for low loss fibers, one exciting example is the fiber gyro. If polarization of the propagating mode can be maintained over a long distance, the prospect of making a fiber gyro with performance matching or exceeding that of the magnetic compass can be achieved. This instrument will be robust and reliable and can be conveniently carried since it has no moving parts. It will allow every moving thing to sense its position anywhere on earth and, if linkages are provided through communication channels, relative to each other. This is indeed another break through in our desire to be auto-piloted to any destination. The fiber gyro is close to practical realization for commercial use. It is a very important fiber based component.

A number of other fiber based-components have already been envisaged to be of great importance: these include the Erbium-doped fiber amplifier mentioned above, fiber resonators, fiber delay lines, and fiber couplers offering 1 to n and n to m coupling. With nonlinear properties included, the four-wave mixing and other nonlinear coupling mechanisms will allow passive shifting of wavelengths and parametric amplification. A host of new powerful components will be available to enhance system performances.

The importance of studying fiber nonlinearity is the underlying reason why this book is written. It gives the reader the tools for understanding nonlinear phenomena in general and nonlinear phenomena in fiber. Besides fibers, nonlinearity in bulk optics and guided optics will be explored for achieving both active and passive devices for optical signal processing purposes. The theoretical methods contained in this book are also applicable to device applications. The future opportunities for exploring nonlinearity in fibers and in devices are almost unlimited. Fibers together with the optical signal processing devices will find their way into a combined role of transmission, sensing and control. It is the basis for an integrated infrastructure for our information age.

1.2 The Roles of Optoelectronics

The operating wavelengths of optical fiber systems are chosen to match the characteristics of the silica-based fibers. Even though it is not the ultimate transmission system, its performance is not a limiting factor in achieving the information carrying capacity for our future networks. The current carrying capacity is over 1 terabit/s over 10s of km in a single fiber. This should satisfy the foreseeable requirement of our fixed networks. The challenge is to make the necessary optoelectronic components at economically attractive cost and with performances meeting the application needs. Fortunately, the mechanism of interaction between electrons and photons are sufficiently well understood and the fabrication technologies are available for the making the required components. The current optoelectronic devices are expected to work at up to 100Gb/s. Beyond that speed, pure optical devices are needed. Further extension of speed is expected to be achievable when nano technology allows atomic size devices to be made. Optical methods based on the use of nonlinearity in materials are expected to create large storage devices with convenient signal processing capabilities. Current activities can be grouped into seven different areas:

Information communication;
Information recording;
Display;
I/O equipment;
Energy supply;

Manufacturing technology; and
Sensing and measurement.

Much work are in progress towards discovering new materials and processing technologies needed for the realization of new wavelength conversion, optical access, optical polymer and interconnect devices. Much of these activities are exploiting the nonlinear photonic effects in polymers, glasses and semiconductors in order to achieve desired characteristics in devices and systems.

Important theoretical tools and design methods in nonlinear photonics are needed to improve the understanding of how nonlinear effects can be harnessed as the means to achieve optoelectronic components and systems. The theory and design tools in optics, optoeletronics and fiber communications are discussed systematically in the succeeding chapters. The aim is to extend the discussion of nonlinear effects in crystalline optical material, optical fiber, and quantum-well structures in semiconductors by creating first a unified theoretical basis. Emphasis is placed on explaining how optical field influences the nonlinear and absorption coefficients of material systems. Analysis of the performance of optoelectronic devices and the propagation characteristics of signals along the optical fiber are then presented as illustrations. For future developments of ultra long fiber systems operating with DWDM, and quantum well devices, the analysis provided in this book is of fundamental importance.

In particular Chaps. 2, 3, and 4 establish the fundamental unified theory applicable to nonlinear photonics. Materials equations, nonlinear electronic transitions, wave propagation formulae, nonlinear wave coupling equations, conservation of energy and momentum are addressed. Chapter 5 deals with classical nonlinear problems. Chapters 6 and 7 treat the application of the unified theory to determine the nonlinear phenomena in optical fiber and semiconductors including quantum well structures. Chapter 8 discusses the photonic nonlinearity and its influence on the performance of optical fiber systems. Chapter 9 presents the photonic nonlinearity and its influence on planar optical waveguide components. Chapters 10 and 11 deal with bandwidth engineering through modifying the optical characteristics of the material. This effect is known as QW intermixing. Strictly speaking this topic is outside the scope of the main topics under discussion.

The semi classical theoretical formulation of the physics of nonlinear photonics enables numerical solutions to be computed for applications such as the ultra long fiber systems and wavelength conversion devices. It is a needed step towards realizing many new components needed for the seven application areas listed earlier.

1.3 The Information Age

Through the use of optical fiber, wireless and computer technologies, universal provision of information services on anytime and anywhere basis for anyone are being realized. The Internet enables a myriad of information services to be accessed on a global basis. Currently computer access for the Internet is generally along a phone line. This ease of access is going to be broadened within a couple of years when the third generation mobile phones and other equipment such as the PCS become widespread.

Linux, Java, and Jini languages have released the source codes of their operating systems. This would allow users and software developers to jointly improve the development of services to meet the exact demands of the users.

The states and large organizations have the ability to access and to use exclusive printed records. This situation gives them a greater influence over the public than otherwise. By controlling broadcasting, television, and telecommunication facilities the state becomes even more powerful. However, as the channels for the distribution of information and knowledge spread into more private hands, the power of the people increases. Even before the advent of the new information technology which enabled anyone to access information, anytime, anywhere, and at affordable costs, the election of the President, for example, in the United States, is significantly influenced by the candidate's ability to project the right image on the TV.

The new term, "Information Age", is coined to describe the age we are now entering, where information and knowledge become a form of capital asset, in the same manner that physical resources are a capital asset. The equal opportunity to access information at an affordable cost is changing our life fundamentally. It is the beginning of a new epoch in which the individual may be as powerful as a state.

All of us are, as the scribes of yore, empowered when pertinent information and knowledge enable us to perform our tasks more easily and more effectively. The social and political structures, the institutions and economic pillars, the proud and the meek, all are being impacted. However, such monumental changes are not likely to occur overnight. The transition will be an evolutionary one.

As we get more familiar with the World Wide Web on the Internet, we certainly have a wealth of information for the picking. In addition, we find the convenience of writing emails for rapid personal and business transactions. We visit the chat rooms for person to person and group contacts. We visit the on-line stock market for convenient trading. We even find pleasure in just surfing the Internet for undefined purposes. However, only a few companies engaged in the gathering, sorting, processing and then presenting the information have been very successful financially. So far, e-commerce is, with only a very small number of exceptions, an alternate form of normal business processes. Yahoo, Amazon and AOL are exceptions. They make money through the high valuation given by the investors who believe their potential future earnings. For most companies, their presence on the Web is hardly noticed by most people. Information overload becomes the significant problem facing us. This situation should encourage the emergence of agents or intermediaries, who would search, efficiently and quickly, for specific pertinent information for a fee. Indeed they are already appearing. They are the new form of consumer council or honest advertising agents to help the consumers. We are beginning to create a world just for me, I-World, and, just for us, We-World.

As public and private information becomes less clearly distinguishable, the invasion of privacy starts in earnest. The credit cards, the cash cards, the personal identity cards, the electronically stored forms and records, and the health and medical information of all the patients cannot be too tightly and securely locked away. In an IT world, each person need to have his/her own identity and personal records certified in order to protect oneself from being falsely accused of having opened a bank account or not having a valid insurance policy.

Human rights is of paramount importance. However, in the IT world our personal space within which we can exercise our rights are shrinking. The reason is simple. IT makes our actions global. We must be aware that our rights must not infringe on other persons' rights. Therein lies the rub. We need to have a very tolerant society to allow: any form of freedom of speech, the practice of religion, the respect of other peoples' cultural heritage, etc. The Chinese Diaspora spreads far and wide around the world, and so are the Irish Diaspora and the Jewish Diaspora. Are they infecting the purity of the local culture? Should they be banned on the WWW?

The technological advances in health sciences are even more a cause for the social paradigm to shift. Why should the pro-life and pro-choice groups claim that their opposing believes are absolutely right. The belief was so right that killing was permitted, just as the Christians did against the infidels in the Middle Ages. At the same time doctors have to take life and death decisions when technology has advanced to the point when a brain dead person can be kept alive for years.

The IT age has begun. Even though the Haves and Have-nots will live with cultural gaps, they all have vital roles to play. They will be living in a global village, and the growth of the world economy will be increasingly interdependent as each group is engaged in a vital aspect within the economic processes. They will be playing on a level playing field in a stable world. Their prospect of attaining a better life is excellent.

The impact of having pertinent information, in multimedia form at anytime, any place and for everyone, is transforming the roles and functions of the individuals and organizations and even inanimate objects, large and small. It is a new epoch in the making. It is a situation rightfully termed as the beginning of the Information Age. The current social paradigm will be altered, as our life style will be changing. The economy of the world will be increasingly dependent on the use of intellectual assets besides capital assets.

This major change is brought about through the advances in fiber optics, optoelectronics, electronics, computers, wireless and wired communication technologies. During the past twenty years, the cost of the transmission, storage, and processing of data per unit time per unit distance decreased at an exponential rate. This trend is expected to continue over the next twenty years. It is obvious by now that the future network would allow information to be used at an ever-increasing rate and at an affordable cost by all users. Pertinent and useful information and knowledge allow us to add value to our productive efforts.

The prospect of having unlimited bandwidth for transport, storage, and processing of information is real. What does this situation mean to the functioning of a community? What are the characteristics of an IT-City? Is it going to be an I-World or a We-World? These are the questions to be answered. This situation also presents opportunities for us, the human race, to conduct our affairs with greater efficiency and with improved productivity. Our lives will be significantly altered.

The timely maturation of the optoelectronics and communication technologies enables the world to take an important step in addressing the issue of using IT to improve our chance of achieving a sustainable development. We see that a stable economic growth can be achieved for the Haves and Have-nots. The interdependence would allow the gradual narrowing of the economic gap. It is a triumph for technology.

References

[1.1] C.K. Kao, G.A. Hockham: Dielecric-fibre surface waveguides for optical frequencies, *Proc. I.E.E.* **113**(7), 1151–1158 (July 1966)

[1.2] Charles K. Kao: *Optical Fibers Systems: Technology, Design and Applications*, McGraw-Hill, New York (1982)

[1.3] C.K. Kao: *Optical Fiber*, Peter Peregrinus for Institution of Electrical Engineers, London (1988)

2. Medium Equations and Electrical Susceptibilities in Nonlinear Photonics

Theory treats and solves the electromagnetic field equations and medium equations simultaneously, to find both the lightwave fields propagating in the photonic medium and the medium's nonlinear response to the radiation fields. The most convenient approach to the interaction of lightwave fields and media is the semiclassical, which treats the field as an electromagnetic field described by Maxwell's equations, and the medium as a quantum ensemble described by quantum mechanics.

The goal of the theory is to get the macroscopic average value of a certain physical quantity, such as the polarization intensity P of four-wave mixing, as an expected result of interaction between radiation fields and medium. The relation between the expected P, the susceptibility χ of medium, and the known fields E_i ($i = 1, 2, ...r$) is natural, so a large part of this chapter concentrates on the representation and properties of χ.

The semiclassical approach is based on the density matrix and the dipole transitions, by which the exact relations between χ and the micro-structures of the medium can be obtained. Resonant or nearly resonant operation is always used to strengthen χ hence to enhance the nonlinear effects, since the higher orders of $\chi^{(n)}$ are usually rather weak. The processes and resultant formulas of χs solution are all too complicated to memorize. A diagrammatic representation, the Feynman diagram for steady-state response, is discussed in this chapter.

Optical fibers and semiconductor materials are the principal media for lightwave technologies and optoelectronics. Fundamentally and theoretically, the polarization $P^{(n)}(\omega)$, susceptibility $\chi^{(n)}(\omega)$, and Kramers-Kronig relations of these kinds of media should be discussed in the unified medium equations developed in this chapter.

2.1 Medium Equations in Nonlinear Photonics

2.1.1 Electromagnetic Field Equations with Nonlinear Sources

Classical electromagnetic fields obey Maxwell's equation:

$$\nabla \times \boldsymbol{E} = -\frac{\partial \boldsymbol{B}}{\partial \mathrm{t}},$$

$$\nabla \times \boldsymbol{H} = +\frac{\partial \boldsymbol{D}}{\partial \mathrm{t}} + \boldsymbol{J}, \tag{2.1}$$

$$\nabla \cdot \boldsymbol{D} = \rho,$$

$$\nabla \cdot \boldsymbol{B} = 0.$$

Considering only the induced polarization, consequently $\rho = 0$ and $\boldsymbol{J} = 0$, since there are neither current nor charge. The equations become

$$\nabla \times \boldsymbol{E} = -\frac{\partial \boldsymbol{B}}{\partial t},$$

$$\nabla \times \boldsymbol{H} = +\frac{\partial \boldsymbol{D}}{\partial t}, \tag{2.2}$$

$$\nabla \cdot \boldsymbol{D} = 0,$$

$$\nabla \cdot \boldsymbol{B} = 0.$$

This group of equations expresses the electromagnetic fields in the medium when interaction is taken into account. The electromagnetic properties of the medium should be considered,

$$\boldsymbol{B} = \mu \boldsymbol{H}, \tag{2.3}$$

$$\boldsymbol{D} = \varepsilon \boldsymbol{E}.$$

In most nonlinear cases, the electric field $\boldsymbol{E}$ and the electric polarization $\boldsymbol{P}$ are dominant, hence the electric-displacement vector $\boldsymbol{D}$ in the last equation becomes

$$\boldsymbol{D} = \varepsilon \boldsymbol{E} = \varepsilon_0 \boldsymbol{E} + \boldsymbol{P}. \tag{2.4}$$

When the interaction between the field and medium has nonlinear properties, the total polarization $\boldsymbol{P}$ should be

$$\boldsymbol{P} = \boldsymbol{P}_{\mathrm{L}} + \boldsymbol{P}_{\mathrm{NL}} = \varepsilon_0 \chi^{(1)} \cdot \boldsymbol{E} + \boldsymbol{P}_{\mathrm{NL}}, \tag{2.5}$$

where $\boldsymbol{P}_{\mathrm{NL}}$ is a key factor in nonlinear photonics. When the electric field is strong enough but not extremely so, the resultant $\boldsymbol{P}$ may be expanded as a power series of the field $\boldsymbol{E}$. This series is perturbed and convergent.

The wave-propagation equation in a medium can be obtained from Maxwell's equations (2.2) and the medium equations (2.4) and (2.5). Apply curl to both sides of the first equation in (2.2):

$$\nabla\times\nabla\times\boldsymbol{E}=-\frac{\partial}{\partial t}(\nabla\times\boldsymbol{B})=\mu_0\frac{\partial^2\boldsymbol{D}}{\partial t^2}. \tag{2.6}$$

Substituting (2.4) and (2.5) into (2.6), we get

$$\nabla\times\nabla\times\boldsymbol{E}=-\mu_0\varepsilon_0\frac{\partial^2\boldsymbol{E}}{\partial t^2}-\mu_0\varepsilon_0\chi^{(1)}\frac{\partial^2\boldsymbol{E}}{\partial t^2}-\mu_0\frac{\partial^2\boldsymbol{P}_{\mathrm{NL}}}{\partial t^2}.$$

With $\varepsilon=\varepsilon_0[1+\chi^{(1)}]$, the nonlinear wave equation is

$$\nabla\times\nabla\times\boldsymbol{E}+\mu_0\varepsilon\frac{\partial^2\boldsymbol{E}}{\partial t^2}=-\mu_0\frac{\partial^2\boldsymbol{P}_{\mathrm{NL}}}{\partial t^2}. \tag{2.7}$$

Thus in nonlinear photonics, the electromagnetic field is described by a second-order nonlinear differential equation with a secondary source contributed by nonlinear polarization $\boldsymbol{P}_{\mathrm{NL}}$, which causes the equation's nonlinearity. In most cases, there is no analytical solution for (2.7). When perturbation, slow-varying amplitude, and nearly planar wave approximations are used, the analytic solutions of some coupling-wave equations can be obtained, as will be discussed in Chap. 3. First we must find the response of the medium with which the polarization P can be found from the medium equation.

2.1.2 Medium Equation in the Time-Space Domain

In the linear case, $\boldsymbol{P}$ can be expressed as a simple linear function

$$\boldsymbol{P}(r,t)=\int_{-\infty}^{+\infty}\varepsilon_0\chi^{(1)}(r-r',t-t')\cdot\boldsymbol{E}(r',t')\,dr'dt'. \tag{2.8}$$

In the nonlinear case, where the field interacts more strongly than in the linear case, the nonlinear polarization $\boldsymbol{P}$ is usually expanded as a power series of $\boldsymbol{E}(r,t)$:

$$
\begin{aligned}
P(r,t) = & \int_{-\infty}^{+\infty} \varepsilon_0 \chi^{(1)}(r-r_1, t-t_1) \cdot E(r_1,t_1) dr_1 dt_1 \\
& + \int_{-\infty}^{+\infty} \varepsilon_0 \chi^{(2)}(r-r_1, t-t_1; r-r_2, t-t_2) \\
& : E(r_1,t_1) E(r_2,t_2) dr_1 dr_2 dt_1 dt_2 \\
& + \int_{-\infty}^{+\infty} \varepsilon_0 \chi^{(3)}(r-r_1, t-t_1; r-r_2, t-t_2; r-r_3, t-t_3) \\
& \vdots\, E(r_1,t_1) E(r_2,t_2) E(r_3,t_3) dr_1 dr_2 dr_3 dt_1 dt_2 dt_3 \\
& + \cdots
\end{aligned}
\tag{2.9}
$$

Equation (2.9) equates two specifications. First, the polarization $P(r,t)$ is constructed in two parts:

$$
\begin{aligned}
P_{\mathrm{L}} \equiv P^{(1)} \equiv & \int_{-\infty}^{+\infty} \varepsilon_0 \chi^{(1)}(r-r', t-t') \cdot E(r',t') dr' dt' \\
\equiv & \varepsilon_0 \chi^{(1)} E(r,t)
\end{aligned}
\tag{2.10}
$$

and

$$
\begin{aligned}
P_{\mathrm{NL}} \equiv & P^{(2)} + P^{(3)} + \ldots \\
\equiv & \int_{-\infty}^{+\infty} \varepsilon_0 \chi^{(2)}(r-r_1, t-t_1; r-r_2, t-t_2) : E(r_1,t_1) E(r_2,t_2) dr_1 dr_2 dt_1 dt_2 \\
& + \int_{-\infty}^{+\infty} \varepsilon_0 \chi^{(3)}(r-r_1, t-t_1; r-r_2, t-t_2; r-r_3, t-t_3) \\
& \vdots\, E(r_1,t_1) E(r_2,t_2) E(r_3,t_3) dr_1 dr_2 dr_3 dt_1 dt_2 dt_3 + \ldots \\
\equiv & \varepsilon_0 \chi^{(2)} : E(r,t) E(r,t) + \varepsilon_0 \chi^{(3)} \vdots\, E(r,t) E(r,t) E(r,t) + \ldots
\end{aligned}
\tag{2.11}
$$

where $\chi^{(1)}$ is the linear electric susceptibility of the medium, $\chi^{(2)}$, $\chi^{(3)}$, ... , $\chi^{(n)}$ are the nonlinear susceptibilities, and χ is a macroscopic susceptibility.

The nonlinear polarization $P_{\mathrm{NL}} = P^{(2)} + P^{(3)} + \ldots$ is a perturbation term in the entire polarization, and each term $P^{(n)}$ is always smaller than the previous term, $P^{(n-1)}$. In this case, the power series is always convergent.

2.1.3 Medium Equation in the Frequency Domain

If the field is expressed as a group of monochromatic planar waves:

$$
E(\boldsymbol{r},t) = \sum_i E(\boldsymbol{k}_i, \omega) \tag{2.12}
$$

where k_i is the wave-vector of wave component E_i, $k = 2\pi/\lambda = \omega/v$, then the Fourier transform of (2.9) is

$$\boldsymbol{P}(k,\omega) = \boldsymbol{P}^{(1)}(k,\omega) + \boldsymbol{P}^{(2)}(k,\omega) + \boldsymbol{P}^{(3)}(k,\omega) + \ldots \tag{2.13}$$

with

$$\boldsymbol{P}^{(1)}(k,\omega) = \varepsilon_0\chi^{(1)}(k,\omega)\cdot\boldsymbol{E}(k,\omega) \tag{2.14}$$

$$\boldsymbol{P}^{(2)}(k,\omega) = \varepsilon_0\chi^{(2)}(k=k_i+k_j;\omega=\omega_i+\omega_j) : \boldsymbol{E}(k_i,\omega_i)\boldsymbol{E}(k_j,\omega_j) \tag{2.15}$$

$$\boldsymbol{P}^{(3)}(k,\omega) = \varepsilon_0\chi^{(3)}(k=k_i+k_j+k_l;\omega=\omega_i+\omega_j+\omega_l) \vdots\ \boldsymbol{E}(k_i,\omega_i)\boldsymbol{E}(k_j,\omega_j)\boldsymbol{E}(k_l,\omega_l) \tag{2.16}$$

The Fourier element of susceptibility is:

$$\begin{aligned}&\chi^{(n)}(k = k_1 + k_2 \ldots + k_n; \omega = \omega_1 + \omega_2 + \ldots + \omega_n)\\ &\quad \equiv \int_{-\infty}^{+\infty} \chi^{(n)}(r - r_1, t - t_1; \ldots r - r_n, t - t_n)\\ &\quad \times e^{i[k_1(r-r_1)-\omega_1(t-t_1)\ldots+k_n(r-r_n)-\omega_n(t-t_n)]}\cdot dr_1 dr_2 \ldots dr_n dt_1 dt_2 \ldots dt_n\end{aligned} \tag{2.17}$$

In the electric dipole approximation, $\chi^{(n)}(r,t)$ is independent of r, it is always expressed as $\chi^{(n)}(t)$. Moreover, $\chi^{(n)}(k,\omega)$ is independent of k, thus it is presented by $\chi^{(n)}(\omega)$. Hence (2.13) becomes the nonlinear medium equation (2.18) shown below:

$$\begin{aligned}\boldsymbol{P}(\omega) &= \varepsilon_0\chi^{(1)}(\omega)\cdot\boldsymbol{E} + \varepsilon_0\chi^{(2)}(\omega;\omega_1,\omega_2) : \boldsymbol{E}(\omega_1)\boldsymbol{E}(\omega_2)\\ &\quad + \varepsilon_0\chi^{(3)}(\omega;\omega_1,\omega_2,\omega_3) \vdots\ \boldsymbol{E}(\omega_1)\boldsymbol{E}(\omega_2)\boldsymbol{E}(\omega_3) + \cdots.\end{aligned} \tag{2.18}$$

Here $\chi^{(1)}(\omega)$, the first order of susceptibility, is a second-rank tensor (a scalar); $\chi^{(2)}(\omega)$, the second order of susceptibility, is a third-rank tensor (a vector); and $\chi^{(3)}(\omega)$, the third order of susceptibility, is a fourth-rank tensor (a quadratic vector); and all the susceptibility elements depend on the microscopic structure of the medium and the features of interactive fields.

The Fourier elements of polarization P are

$$\boldsymbol{P}^{(2)}(\omega) = \int_{-\infty}^{+\infty} \varepsilon_0\chi^{(2)}(\omega_3;\omega_1,\omega_2) : \boldsymbol{E}_1(\omega_1)\boldsymbol{E}_2(\omega_2)\mathrm{d}\omega_1\mathrm{d}\omega_2 \tag{2.19}$$

$$\begin{aligned}\boldsymbol{P}^{(3)}(\omega) &= \int_{-\infty}^{+\infty} \varepsilon_0\chi^{(3)}(\omega_4;\omega_1,\omega_2,\omega_3)\\ &\quad \vdots\ \boldsymbol{E}_1(\omega_1)\boldsymbol{E}_2(\omega_2)\boldsymbol{E}_3(\omega_3)d\omega_1 d\omega_2 d\omega_3\end{aligned} \tag{2.20}$$

$$\boldsymbol{P}^{(n)}(\omega)$$
$$= \int_{-\infty}^{+\infty} \varepsilon_0 \chi^{(n)}(\omega_{n+1};\omega_1,\omega_2 \ldots \omega_n)(\cdot)^n \boldsymbol{E}_1(\omega_1)\boldsymbol{E}_2(\omega_2)\ldots\boldsymbol{E}_n(\omega_n)\,d\omega_1 d\omega_2 \ldots d\omega_n. \tag{2.21}$$

The polarization can be expanded as a power series composed of the electric field $\boldsymbol{E}$, the magnetic field B, the spatial gradient $\nabla_\beta \boldsymbol{E}_\alpha$, etc. Electro-optic, magneto-optic, multiwave mixing, and stimulated Brillouin and Raman scattering can all be described by the series

$$\begin{aligned} \boldsymbol{P}_\mu &= \boldsymbol{P}_\mu^0 + \varepsilon_0\, \chi_{\mu\alpha} \boldsymbol{E}_\alpha^{\omega} + \varepsilon_0\, \chi_{\mu\alpha\beta} \nabla_\beta \boldsymbol{E}_\alpha^{\omega} \\ &\quad + \varepsilon_0\, \chi_{\mu\alpha\beta} \boldsymbol{E}_\alpha^{\omega_1} \boldsymbol{E}_\alpha^{\omega_2} + \varepsilon_0\, \chi_{\mu\alpha\beta\gamma} \boldsymbol{E}_\alpha^{\omega_1} \boldsymbol{E}_\beta^{\omega_2} \boldsymbol{E}_\gamma^{\omega_3} \\ &\quad + \varepsilon_0\, \chi_{\mu\alpha\beta} \boldsymbol{E}_\alpha^{\omega_1} \boldsymbol{E}_\alpha^{\omega_2} + \cdots . \end{aligned} \tag{2.22}$$

Here the superscript ω_i is the frequency of the field $\boldsymbol{E}_j$, the subscripts are the Cartesian coordinates, $\boldsymbol{P}_\mu{}^0$ is the intrinsic polarization in the absence of any extra field, $\varepsilon_0\chi_{\mu\alpha}\boldsymbol{E}_\alpha{}^\omega$ is the linear polarization, and $\varepsilon_0\chi_{\mu\alpha\beta}\nabla_\beta \boldsymbol{E}_\alpha{}^\omega$ is the optical rotation. Distinct components of the electric polarization $\boldsymbol{P}$ correspond to the distinct physical effects.

From this discussion, we see that the perturbation power series, expanded for the interaction between fields and medium, is essential for nonlinear optics. This representation was proposed by Bloembergen [2.1]. In principle, the macroscopic susceptibility $\chi(\omega)$ can be calculated by quantum-mechanical or semiclassical methods, but real media are very complicated in structure and energy level states. Only rarely, as for alkaline metal vapor, is it possible to find $\chi(\omega)$ by calculation, but the theoretical representation of $\chi(\omega)$ is still very important for understanding and implementing the different kinds of nonlinear effects. The susceptibility $\chi(\omega)$ can also be derived by a classical approach, the anharmonic oscillator model, which, though simple, can offer some fundamental insight.

2.2 The Classical Approach of Electrical Susceptibility

In the classical oscillator model, the medium is treated as a group of electrons and nuclei with resonant characteristics. The interactions between the fields and medium are considered as a vibration forced by the field.

2.2.1 Classical Linear Oscillators and Linear Susceptibility

The linear oscillator is a first-order approximation. The potential energy, in which the oscillator moves, is a second-power function of displacement ($U \propto r^2$), and the restoring force is $-kr$. According to the classical model, the oscillator has an intrinsic frequency ω_0, a damping term ς, and a linear restoring force $-\omega_0^2 r$. The motion equation is

$$\frac{d^2 r_1}{dt^2} + 2\varsigma \frac{dr}{dt} = -\frac{e}{m} E - \omega_0{}^2 r_1, \tag{2.23}$$

where r_1 is the displacement from the balance position. Using the trial solution $r_1 = r_1 e^{-i\omega t}$, (2.23) becomes $-\omega^2 r_1 - 2i\omega\zeta r_1 + \omega_0{}^2 r_1 = eE/m$, the solution of which is

$$r_1 = -\frac{e}{m} E(\omega) \cdot \frac{1}{\omega_0{}^2 - \omega^2 - 2i\varsigma\omega},$$

where $E = E(\omega)\exp(-i\omega t)$ + c. c is substituted into the above equation.

From the definition on electric polarization of an oscillator, $P = Ner$, E, P and r are treated as scalars in this section, then

$$P(\omega) = -Ner_1 = \frac{Ne^2}{m} E(\omega) \frac{1}{\omega_0{}^2 - \omega^2 - 2i\varsigma\omega},$$

where N is the number of oscillators in a unit volume. Inserting

$P(\omega) = \varepsilon_0 \chi^{(1)}(\omega) E(\omega)$

we have

$$\begin{aligned} \chi^{(1)}(\omega) &= \frac{P}{\varepsilon_0 E} = \frac{Ne^2}{\varepsilon_0 m} \cdot \frac{1}{\omega_0{}^2 - \omega^2 - 2i\varsigma\omega} \\ &= \chi'(\omega) + i\chi''(\omega). \end{aligned} \tag{2.24}$$

The susceptibility $\chi(\omega)$ is always a complex quantity. Thus,

$$\chi'(\omega) = \frac{Ne^2}{\varepsilon_0 m} \frac{\omega_0^2 - \omega^2}{(\omega_0^2 - \omega^2)^2 + 4\varsigma^2\omega^2} \tag{2.25}$$

$$\chi''(\omega) = \frac{Ne^2}{\varepsilon_0 m} \frac{2\varsigma\omega}{(\omega_0^2 - \omega^2)^2 + 4\varsigma^2\omega^2}. \tag{2.26}$$

Combining (2.2), (2.3), (2.4) and (2.17) yields

$$\varepsilon_r(\omega) = 1 + \chi(\omega) = 1 + \frac{Ne^2}{\varepsilon_0 m} \cdot \frac{1}{[(\omega_0{}^2 - \omega^2) - 2i\varsigma\omega]}. \tag{2.27}$$

The relation between dielectric ε_r and refractive index is

$$\varepsilon_r(\omega) = \tilde{n}^2(\omega), \tag{2.28}$$

$\tilde{n}$ is the complex refractive index,

$$\tilde{n}(\omega) = n(\omega) + i\kappa(\omega), \tag{2.29}$$

where $\kappa(\omega)$ is the extinction coefficient. Equations (2.27), (2.28) and (2.29) imply that

$$n^2(\omega) - \kappa^2(\omega) = 1 + \frac{Ne^2}{\varepsilon_0 m} \cdot \frac{(\omega_0^2 - \omega^2)}{[(\omega_0^2 - \omega^2) + 4\varsigma^2\omega^2]} = 1 + \chi'(\omega) \tag{2.30}$$

$$\kappa(\omega)n(\omega) = \frac{Ne^2}{2\varepsilon_0 m} \cdot \frac{\varsigma\omega}{[(\omega_0^2 - \omega^2) + 4\varsigma^2\omega^2]} = \frac{\chi''(\omega)}{2}. \tag{2.31}$$

Since $\kappa(\omega)$ is much smaller than the refractive index $n(\omega)$, (2.30) and (2.31) become

$$n^2(\omega) \cong 1 + \frac{Ne^2}{\varepsilon_0 m} \cdot \frac{(\omega_0^2 - \omega^2)}{[(\omega_0^2 - \omega^2) + 4\varsigma^2\omega^2]} = 1 + \chi'(\omega), \tag{2.32}$$

$$\kappa(\omega) = \frac{\chi''(\omega)}{2n(\omega)} = \frac{Ne^2}{2n(\omega)\varepsilon_0 m} \cdot \frac{\varsigma\omega}{[(\omega_0^2 - \omega^2) + 4\varsigma^2\omega^2]}. \tag{2.33}$$

Equation (2.32) is the refractive index dispersion and (2.33) is the absorption characteristics of the medium.

Conventionally, the contours of (2.25) and (2.26) are shown in Fig. 2.1 to indicate the refractive index and absorption characteristics. Near the resonant frequency ω_0, the absorption χ'' has a sharp peak and the absorption curve has the Lorentz line shape, whose linewidth ζ is always $<< \omega_0$. Here the linewidth ζ is just the earlier damping coefficient ζ in (2.23) (this is to be proved in Problem 2.1 of this chapter). The absorption changes very rapidly near ω_0. Consequently the absorption can be neglected after only a few multiples of ζ and hence the medium is effectively transparent.

The curve $\chi'(\omega)$, here $\chi' = n^2 - 1$, expresses the refractive index dispersion. The regions in which the frequencies are lower than ω_1 (the positive peak of χ') and higher than ω_2 (the negative peak of χ') are the normal-dispersion regions. In these, the refractive index n increases with frequency ω. The region in which $n(\omega)$ varies between these two peaks is the anomalous-dispersion region. In the anomalous-dispersion region, n

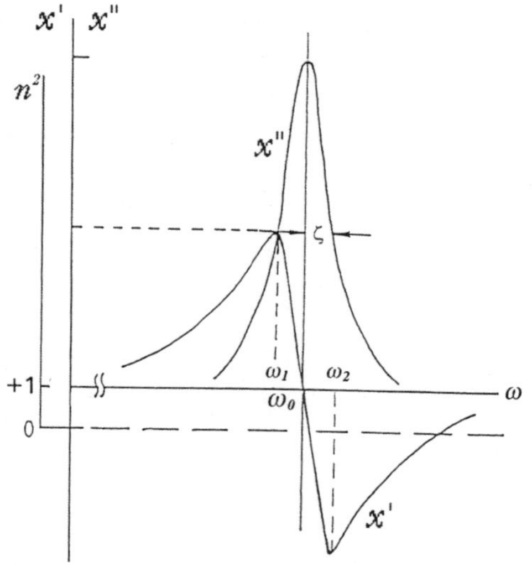

Fig. 2.1 The real and imaginary parts of the linear susceptibility which are associated with the refractive index $n(\omega)$ and absorption $\kappa(\omega)$ or $\alpha(\omega)$ of the medium.

decreases with ω. The linewidth ζ is just the frequency-difference Δf between the peak of χ' and the resonant point of χ''.

In the transparent regions a few linewidths from ω_0, $\chi'' \ll \chi'$, and so the absorption may be neglected. Most nonlinear crystals and undoped silica fibers operate in these regions for communications. Energy transfer occursonly between original and retransmitted lightwaves. These transfers are reversible, hence $\omega_1 + \omega_2 \leftrightarrow \omega_3 + \omega_4$. However, no energy transfer occurs between lightwaves and the medium.

2.2.2 Classical Anharmonic Oscillators

In the classical approach, the anharmonic oscillator is a second- or higher-order approximation. In such a case, the fields are more intense than in the linear case. The oscillators' motions become much larger such that the potential energy distribution is described by the third- or higher-order terms. The anharmonic restoring force $-\omega_0{}^2 r + \alpha r^2 + \beta r^3$ can be introduced into (2.23) by substitution for $-\omega_0{}^2 r$, thus

$$\frac{d^2 r}{dt^2} + 2\varsigma \frac{dr}{dt} + \omega_0{}^2 r - \alpha r^2 - \beta r^3 = -\frac{eE}{m}. \tag{2.34}$$

This is a nonlinear differential equation. The analytical solution for r can be obtained by expanding it to a polynomial of r, which is actually a power series of E:

$$r_1 + r_2 + r_3 + \ldots + r_n + \ldots \tag{2.35}$$

Substituting (2.35) into (2.34), we have

$$\frac{d^2 r_1}{dt^2} + 2\varsigma \frac{dr_1}{dt} + \omega_0{}^2 r_1 = -\frac{e}{m} E \,, \tag{2.36}$$

$$\frac{d^2 r_2}{dt^2} + 2\varsigma \frac{dr_2}{dt} + \omega_0{}^2 r_2 = \alpha r_1{}^2 \,, \tag{2.37}$$

$$\frac{d^2 r_3}{dt^2} + 2\varsigma \frac{dr_3}{dt} + \omega_0{}^2 r_3 = 2\alpha r_1 r_2 + \beta r_1{}^3 \,, \tag{2.38}$$

$$\frac{d^2 r_n}{dt^2} + 2\varsigma \frac{dr_n}{dt} + \omega_0{}^2 r_n = f(r_1, r_2, \ldots r_{n-1}) \,. \tag{2.39}$$

The solution r_1 of (2.36) is the same as that of (2.23). Substituting it into (2.37), r_2 is obtained as

$$\begin{aligned} r_2 &= \alpha \frac{e^2}{m^2} E^2(\omega) \cdot \frac{\exp(-2i\omega t)}{(\omega_0{}^2 - 2i\varsigma\omega - \omega^2)^2 (\omega_0{}^2 - 4i\varsigma\omega - 4\omega^2)} \\ &+ \alpha \frac{e^2}{m^2} E(\omega) E^*(\omega) \cdot \frac{1}{\omega_0{}^2 (\omega_0{}^2 - 2i\varsigma\omega - \omega^2)(\omega_0{}^2 + 2i\varsigma\omega - \omega^2)} \\ &+ \text{c.c.} \end{aligned} \tag{2.40}$$

where $E(\omega)$ is the complex amplitude, and $E^*(\omega)$ is the complex conjugate in which ω is replaced by $-\omega$. There is a second harmonic term in (2.40), since a 2ω appeared in the term related with time t.

Similarly, the resulting r_1, r_2, and r_3 are permuted into (2.38) to yield

$$\begin{aligned} r_3 &= -\frac{e^3}{m^3} E^3 \cdot \left[\frac{2\alpha^2}{(\omega_0{}^2 - 4i\varsigma\omega - 4\omega^2)} + \beta \right] \\ &\times \frac{\exp(-3i\omega t)}{(\omega_0{}^2 - 6i\varsigma\omega - 9\omega^2)^2 (\omega_0{}^2 - 2i\varsigma\omega - \omega^2)} \\ &- \frac{e^2}{m^2} E^2 E^* \left[\frac{2\alpha^2}{(\omega_0{}^2 - 4i\varsigma\omega - 4\omega^2)} + \frac{4\alpha^3}{\omega_0{}^2} + 3\beta \right] \\ &\times \frac{\exp(-i\omega t)}{(\omega_0{}^2 - 2i\varsigma\omega - \omega^2)(\omega_0{}^2 + 2i\varsigma\omega - \omega^2)} + \text{c.c.} \end{aligned} \tag{2.41}$$

Finally, the expected second-order nonlinear electric polarization $P^{(2)}$ and the third-order nonlinear electric polarization $P^{(3)}$ are

$$P^{(2)} = -Ner_2$$
$$= \varepsilon_0\chi^{(2)}(\omega,\omega)E^2(\omega)\exp(-2i\omega t) + \varepsilon_0\chi^{(2)}(\omega,-\omega)\cdot E(\omega)E^*(\omega) + \text{c.c.} \quad (2.42)$$

$$P^{(3)} = -Ner_3 = \varepsilon_0\chi^{(3)}(\omega,\omega,\omega)E^3(\omega)\exp(-3i\omega t)$$
$$+ 3\varepsilon_0\chi^{(3)}(\omega,\omega,-\omega)\cdot E^2(\omega)E^*(\omega)\exp(-i\omega t) + \text{c.c.} \quad (2.43)$$

When we find (2.42) and (2.43), the resultant polarization P has components not only of the frequency ω, but also of the SHG frequency 2ω in $P^{(2)}$, and the THG frequency 3ω in $P^{(3)}$. These specify the nonlinear optical processes.

When we substitute (2.40) into (2.42), $\chi^{(2)}(\omega_1,\omega_2)$ becomes

$$\chi^{(2)}(\omega_1,\omega_2) = -\frac{N\alpha e^3}{\varepsilon_0 m^2} \times \left[\frac{1}{(\omega_0^2-\omega_1^2-2i\varsigma\omega_1)}\cdot\frac{1}{(\omega_0^2-\omega_2^2-2i\varsigma\omega_2)} \cdot\frac{1}{[\omega_0^2-(\omega_1+\omega_2)^2-2i\varsigma(\omega_1+\omega_2)]}\right] \quad (2.44)$$

When (2.41) is substituted into (2.43), $\chi^{(3)}(\omega_1,\omega_2,\omega_3)$ becomes

$$\chi^{(3)}(\omega_1,\omega_2,\omega_3) = -\frac{N\alpha e^4}{\varepsilon_0 m^3}$$
$$\times\left\{\beta+\frac{2}{3}\alpha^2\left[\frac{1}{[\omega_0^2-(\omega_1+\omega_2)^2-2i\varsigma(\omega_1+\omega_2)}\right.\right.$$
$$+\frac{1}{[\omega_0^2-(\omega_2+\omega_3)^2-2i\varsigma(\omega_2+\omega_3)]}$$
$$\left.\left.+\frac{1}{[\omega_0^2-(\omega_{31}+\omega_1)^2-2i\varsigma(\omega_{31}+\omega_{12})]}\right]\right\} \quad (2.45)$$
$$\times\left[\frac{1}{(\omega_0^2-\omega_1^2-2i\varsigma\omega_1)}\cdot\frac{1}{(\omega_0^2-\omega_2^2-2i\varsigma\omega_2)}\cdot\frac{1}{(\omega_0^2-\omega_3^2-2i\varsigma\omega_3)}\right.$$
$$\left.\times\frac{1}{[\omega_0^2-(\omega_1+\omega_2+\omega_3)^2-2i\varsigma(\omega_1+\omega_2+\omega_3)]}\right]$$

In (2.42), if all the frequencies are different, sixteen combinations exist in second-order nonlinear effects. This is because in $\chi^{(2)}$ (ω_m, ω_n), four possibilities exist both for m: 1, 2, –1, –2, and for n: 1, 2, –1, –2 (Table 2.1).

Table 2.1 Combinations of second-order nonlinear effects.

	Susceptibility	$\omega_{\text{out-put}}$	Patterns in physics	Direction of waves
1	$\chi^{(2)}(\omega_1,\omega_2)$	$(\omega_1+\omega_2)$	Sum-frequency	Same direction
2	$\chi^{(2)}(\omega_1,-\omega_2)$	$(\omega_1-\omega_2)$	Difference-frequency	Same direction
3	$\chi^{(2)}(\omega_1,-\omega_1)$	0	Rectification	
4	$\chi^{(2)}(\omega_1,\omega_1)$	$-2\omega_1$	SHG	Same direction
5	$\chi^{(2)}(-\omega_1,\omega_2)$	$-(\omega_2-\omega_1)$	Difference-frequency	Same direction
6	$\chi^{(2)}(-\omega_1,-\omega_2)$	$(\omega_1+\omega_2)$	Sum-frequency	Opposite direction
7	$\chi^{(2)}(-\omega_1,-\omega_1)$	$2\omega_1$	SHG	Opposite direction
8	$\chi^{(2)}(-\omega_1,\omega_1)$	0	Rectification	
9	$\chi^{(2)}(\omega_2,\omega_2)$	$-2\omega_2$	SHG	Same direction
10	$\chi^{(2)}(\omega_2,-\omega_2)$	0	Rectification	
11	$\chi^{(2)}(\omega_2,-\omega_1)$	$-(\omega_2-\omega_1)$	Difference-frequency	Same direction
12	$\chi^{(2)}(\omega_2,\omega_1)$	$-(\omega_2+\omega_1)$	Sum-frequency	Same direction
13	$\chi^{(2)}(-\omega_2,\omega_1)$	$-(\omega_1-\omega_2)$	Difference-frequency	Same direction
14	$\chi^{(2)}(-\omega_2,-\omega_1)$	$(\omega_1+\omega_2)$	Sum-frequency	Opposite direction
15	$\chi^{(2)}(-\omega_2,-\omega_2)$	$2\omega_2$	SHG	Opposite direction
16	$\chi^{(2)}(-\omega_2,\omega_2)$	0	Rectification	

2.3 The Semiclassical Approach of Electrical Susceptibility and the Equations of Motion of the Density Operator

2.3.1 $\chi^{(n)}$ and Microscopic Quantum Processes

The classical oscillator model offers some primary results, but these are qualitative. It is impossible to quantitatively calculate $\chi^{(n)}$. Since $\chi^{(n)}$ stems from the distortion of electron clouds, or ion displacements, such microscopic effects should be described by quantum mechanics. In the theory of molecular structure, the behavior of atomic and molecular ensembles interacting with fields can be derived by variations of their energy states. With the semiclassical approach, we can understand $\chi^{(n)}$ and its quantitative relations to the medium more precisely.

If we have a system of a very large number of particles, N per unit volume, the system has $3N$ degrees of freedom. To determine its state, we need to measure $3N$ independent parameters. But N may be at least 10^{23}cm^{-3} for a gas, so the only practical way is to use statistical mechanics.

Statistics has the concept of "ensemble" to describe a physical system. We suppose a large number of unique systems are in the same macroscopic condition, but different microscopic motion states, thus forming a statistical ensemble.

In quantum mechanics, the systems in an ensemble are represented by wavefunctions, from which observable values can be obtained as average values in a probability distribution domain. To find these macroscopic values, we need to introduce ensemble averages. When dealing with averages from statistical and quantum mechanics, we use the density-matrix approach.

The density matrix is the counterpart to the distribution function in classical statistics, and is the most general description of ensembles constituted by independent quantum-mechanical systems. Suppose there are N systems in an ensemble, each represented by a normalized wavefunction $\Psi^k(q)$. In this system, we consider an observable physical quantity P, and evaluate this quantity in this ensemble. The average value of P will be calculated by the operator $\tilde{P}$ acting on the wavefunction Ψ^k which is a quantum average and an intrinsic feature of quantum mechanics. Now, we find the average of the whole ensemble expressed in the matrix-trace form as

$$\langle P\rangle = Tr(\rho P)\,, \tag{2.46}$$

which can also be written as

$$\langle P\rangle = \sum_{m,n} \rho_{nm} P_{mn}\,. \tag{2.47}$$

The quantum statistical average of a physical quantity is expressed as the matrix-trace of the product of the density matrix and this physical quantity. The advantage of using a density matrix is that an ensemble can be described even though its wavefunction is not yet well understood. This is quite enough to find the average physical quantities if the occupied probabilities of each level or state are known. Another advantage is that the coherent relations between the different states can be shown clearly. Thus one can solve problems related to quantum coherence, such as collisions, which can be described directly by off-diagonal elements of the density matrix.

2.3.2 The Equations of Motion of the Density Matrix ρ

When the medium and radiation interact, the ensemble is verified experimentally. Thus the states of a quantum ensemble vary with the time t, the position of the particles r and their velocities v. Such an ensemble should be described by the equation of motion of ρ. In this case, the action of the extra fields is described by a perturbed Hamilton H' to find the nonlinear electric susceptibility of a quantum ensemble included in a perturbed electromagnetic field. Ordinarily, the equation of motion of ρ with damping is

$$i\hbar\frac{d\rho(t)}{dt} = [H, \rho(t)] + i\hbar\left(\frac{d\rho(t)}{dt}\right)_{\text{relax}} . \tag{2.48}$$

The second term on the right side introduced in a phenomenological fashion is the relaxation term of the density matrix caused by damping. The Hamiltonian of the ensemble is

$$H = H_0 + H'(t), \tag{2.49}$$

where H_0 is the nonperturbed term of an isolated ensemble in the absence of any extra electromagnetic fields, and H' is the interaction Hamiltonian of a medium interacting with an extra field, dependent on t. Expanding the density matrix operator with perturbation as a power series,

$$\rho(t) = \rho_0 + \rho_1(t) + \ldots + \rho_r(t), \tag{2.50}$$

where the rth term $\rho_r(t)$ is proportional to the rth power of the perturbation Hamiltonian H', as long as ρ_0 is the initial density matrix operator of an unperturbed ensemble.

Initially, if the medium is in thermal equilibrium, then

$$\rho_0 = z \cdot \mathrm{e}^{-\frac{H_0}{kT}} \tag{2.51}$$

where z is a normalization constant, k is the Boltzmann constant, T is the absolute temperature of the medium, ρ_0 is a diagonal matrix, and its matrix element $\rho_0(n,n)$ is the quantum-statistical weight of energy level n at thermal equilibrium. Substituting (2.49), (2.50) and (2.51) into (2.48), we have a group of equations that is satisfied by different ranks of the expanded density matrix (2.52):

$$\begin{aligned} i\hbar \frac{d}{dt}\rho_1 &= [H_0, \rho_1] + [H', \rho_0] + i\hbar\Gamma\rho_1 \\ i\hbar \frac{d}{dt}\rho_2 &= [H_0, \rho_2] + [H', \rho_1] + i\hbar\Gamma\rho_2 \\ i\hbar \frac{d}{dt}\rho_3 &= [H_0, \rho_3] + [H', \rho_2] + i\hbar\Gamma\rho_3 \\ i\hbar \frac{d}{dt}\rho_r &= [H_0, \rho_r] + [H', \rho_{r-1}] + i\hbar\Gamma\rho_r \end{aligned} \tag{2.52}$$

where the term $i\hbar\Gamma\rho_r$ relates to the damping term $i\hbar(d\rho/dt)_r$. As the damping is supposed to be quite weak, Γ is the damping constant introduced empirically. It will be shown later that Γ has a definite meaning connected with the relaxation of nondiagonal elements of the density matrix.

Generally, since the population probability in thermal equilibrium ρ_0 and the interactive Hamiltonian are known, from (2.52) ρ_1, ρ_2, ... ρ_r can be found by permutation, and the total density matrix operator ρ is then determined.

Lightwave fields can be indicated as a combination of a definite number of monochromatic wave terms by the Fourier series

$$E(t) = \sum_n E(\omega_n) \cdot e^{i\omega_n t} . \tag{2.53}$$

We suppose here that the particle's dimension a_0 is much smaller than the wavelength λ, so the field magnitude E is essentially constant across a_0 and only varies with time t. Hence we can take the electric dipole approximation. The interaction Hamiltonian will be

$$\begin{aligned} H_1(t) &= -E(t) \cdot R = -\sum_n R \cdot E(\omega_n) \cdot e^{i\omega_n t} \\ &= \sum_n H_1(\omega_n) \cdot e^{i\omega_n t}, \end{aligned} \tag{2.54}$$

where R is the electric dipole moment of a particle. Similarly, the elements of the various density matrix operators can be expressed as

$$\rho_r(t) = \sum_n \rho_r(\omega_n) \cdot e^{i\omega_n t} . \tag{2.55}$$

2.4 The Elements of the Density Matrix and of the Susceptibility Tensor

2.4.1 Polarization in An Independent Molecular System

By the definition of polarization $P(t)$, the expected value of the dipole moment per unit volume is

$$\boldsymbol{P}(t) \equiv \frac{1}{V}\langle R\rangle = \frac{1}{V} Tr(\rho\ \boldsymbol{R}) \tag{2.56}$$

where the matrix ($\rho\boldsymbol{R}$) is used to express an operator, $Tr(\rho\boldsymbol{R})$ is the matrix trace, and $\rho = \rho(t)$ is the density operator of an ensemble. This macroscopic electric polarization intensity is the quantum-statistical ensemble average of the electric dipole moment per unit volume.

We consider here an independent molecular ensemble of identical, independent, unidirectional molecules or particles. This simple model is easily used for a practical medium. In such an ensemble the particles are identical, and the polarization $\boldsymbol{P}(t)$ in (2.56) and the ensemble operator can both be expressed as those of a single molecule or particle in the ensemble. If N is the number of molecules or particles per unit volume in a uniform thin medium (such as the gas), then the polarization intensity in an independent molecular ensemble is

$$\boldsymbol{P}(t) = NTr\{\rho\,(t)\boldsymbol{R}_m\}\,. \tag{2.57}$$

Substituting (2.55) into (2.57) yields

$$\boldsymbol{P}(t) = N\sum_n Tr\{\rho(\omega_n)\cdot \boldsymbol{R}\}\cdot e^{i\omega_n t}\,, \tag{2.58}$$

then

$$\boldsymbol{P}_\mu(\omega_\sigma) = N\sum_{ab\cdots}(\boldsymbol{R})^\mu{}_{ab\cdots}\rho^{(n)}{}_{ab}(\omega_1\omega_2\cdots\omega_n) \tag{2.59}$$

where a, b, c, ... symbolize the energy-levels in the medium which are involved in the interactions. In this section, we will find first the density matrix elements $\rho_{ab}{}^{(1)}(\omega_1)$, $\rho_{ab}{}^{(2)}(\omega_1, \omega_2)$, $\rho_{ab}{}^{(n)}(\omega_1, \omega_2, \omega_3,\ldots\omega_n)$ and the dipole moment matrix elements of transition between energy levels $(R^\mu)_{ca}$, $(R^\alpha)_{ab}$, $(R^\beta)_{bc}$, ..., and then substitute into (2.59) to find the $P_\mu(\omega_\sigma)$. Using (2.18), the polarization projection element on the μ axis is (2.60):

$$P_\mu{}^{(n)}(\omega_\sigma) = \varepsilon_0\chi^{(n)}{}_{\mu\alpha\beta\gamma\ldots\xi}\,(\omega_\sigma;\ \omega_1,\omega_2,\omega_3,\ldots\omega_n)E_\alpha(\omega_1)E_\beta(\omega_2)E_\gamma(\omega_3)\ldots E_\xi(\omega_n) \tag{2.60}$$

where $\alpha, \beta, \gamma, \ldots\xi$ are the oscillating directions of the field vector, such as x, y, z in Cartesian coordinates, and the pairs of (μ,ω_σ), (α,ω_1), (β,ω_2),..., (ξ,ω_n) symbolize the different field elements. Comparing (2.60) and (2.59), the susceptibility tensor elements $\chi_{\mu\alpha\beta\gamma\ldots\xi}{}^{(n)}(\omega_\sigma;\omega_1,\omega_2,\omega_3,\ldots\ \omega_n)$ can be obtained. We will find the density-matrix elements $\rho_{ab}{}^{(n)}(\omega_1,\ \omega_2,\ \omega_3,\ldots\omega_n)$, next the susceptibility tensor elements $\chi_{\mu\alpha\beta\gamma\ldots\xi}{}^{(n)}\ (\omega_\sigma;\ \omega_1,\ \omega_2,\ \omega_3,\ldots\ \omega_n)$, and then the polarization P of the resultant nonlinear effect will be obtained.

2.4.2 Orders of the Density Matrix Elements

2.4.2.1 First-Order Elements

Starting from (2.55) to find the density matrix element, substitute (2.54) and

$$\rho_1(t) = \sum_n \rho_1(\omega_n)\cdot e^{i\omega_n t}$$

into the first equation in (2.52). We have the nonperturbed intrinsic function-set φ_m as a picture and use it on both the left and right sides of the terms in (2.52) to complete the matrix-element operation. Consequently the leading term on the right side of the first equation in (2.52) is

$$\begin{aligned}[H_0,\ \rho_1(\omega_n)]_{ab} &= \int \varphi_a^*[H_0,\ \rho_1(\omega_n)]\varphi_b d\tau = (E_a - E_b)[\rho_1(\omega_n)]_{ab} \\ &= \hbar\omega_{ab}[\rho_1(\omega_n)]_{ab}\end{aligned}$$

where the particle transitions from state a to state b, and φ_a and φ_b, are the state functions before and after the transition respectively. The next term on the right side of first equation in (2.52) is

$$[H'(\omega_n),\ \rho_0]_{ab} = (\rho_{bb}{}^{(0)} - \rho_{aa}{}^{(0)})\ [H'(\omega_n)]_{ab}.$$

The third term in (2.52) becomes

$$[\Gamma\rho_1(\omega_n)]_{ab} = \Gamma_{ab}[\rho_1(\omega_n)]_{ab}$$

where E_a and E_b are nonperturbed intrinsic energies of the ensemble, $\omega_{ab} = (E_a - E_b)/\hbar$ is Bohr's resonant frequency, and $\rho_{bb}{}^{(0)} - \rho_{aa}{}^{(0)}$ are diagonal elements of the zero-order density matrix ρ_1. $\Gamma_{ab} = \Gamma_{ba}$ is the damping factor corresponding to a nondiagonal element of the density matrix. In that case, the equation satisfied by the first-order density-matrix Fourier element is

$$\begin{aligned}-\hbar\omega_n[\rho_1(\omega_n)]_{ab} &= \hbar\omega_{ab}[\rho_1(\omega_n)]_{ab} + (\rho_{bb}{}^{(0)} - \rho_{aa}{}^{(0)})\ [H'(\omega_n)]_{ab} \\ &\quad + i\hbar\Gamma_{ab}[\rho_1(\omega_n)]_{ab}.\end{aligned}$$

Moving all three terms, the $[\rho_1(\omega_n)]_{ab}$ included, to the left side results in the first-order density matrix elements

$$\rho_{ab}^{(1)}(\omega_n) = \frac{H_{ab}^{(1)}(\omega_n)(\rho_{bb}^{(0)} - \rho_{aa}^{(0)})}{\hbar(\omega_{ba} - \omega_n - i\Gamma_{ab})} = \frac{H_{ab}^{(1)}(\omega_n)(\rho_{bb}^{(0)} - \rho_{aa}^{(0)})}{D_{ab}(\omega_n)}, \tag{2.61}$$

where $\rho_{ab}{}^{(1)}(\omega_n)$ is the simplified form of $[\rho_1(\omega_n)]_{ab}$, $H'_{ab}(\omega_n)$ is the simplified form of $[H'(\omega_n)]_{ab}$, $D_{ab}(\omega_n)$ is the simplified form of $\hbar(\omega_{ba} - \omega_n - i\Gamma_{ab})$, and $\omega_{ba} = -\omega_{ab}$.

2.4.2.2 Second-Order (Nonlinear) Elements

Substituting $\rho_2(t) = \sum_n \rho_2(\omega_n) \cdot e^{i\omega_n t}$ into the second equation in (2.52), then applying the matrix-element operation to every term in that equation yields

$$\sum_n (-\hbar\omega_n - \hbar\omega_{ab} - i\hbar\Gamma_{ab})[\rho_2(\omega_n)]_{ab} \cdot e^{i\omega_n t} = [H'(t), \rho_1(t)]_{ab}. \tag{2.62}$$

For the sum-frequency $(\omega_1 + \omega_2)$ effect generated by injecting two fields ω_1 and ω_2, $H'(t)$ can be written as

$$\begin{aligned} H'(t) &= [\boldsymbol{E}(\omega_1)e^{i\omega_1 t} + \boldsymbol{E}(\omega_2)e^{i\omega_2 t}] \cdot \boldsymbol{P} \\ &= H(\omega_1)e^{i\omega_1 t} + H'(\omega_2)e^{i\omega_2 t}. \end{aligned}$$

Putting this term and $\rho_1(t) = \Sigma\, \rho_1(\omega_n{}')\, e^{-i\omega_n' t}$ into the right side of (2.62) yields

$$\begin{aligned} \sum_n D_{ab}(\omega_n)[\rho_2(\omega_n)]_{ab} \cdot e^{i\omega_n t} &= \sum_n [H'(\omega_1), \rho_1(\omega_n')]_{ab} \cdot e^{i(\omega_l + \omega_n')t} \\ &= \sum_n [H'(\omega_2), \rho_1(\omega_n')]_{ab} \cdot e^{i(\omega_2 + \omega_n')t} \end{aligned}$$

where we use $\omega_n{}'$ in the $\rho_1(\omega_n{}')$ to distinguish it from ω_n in the $\rho_2(\omega_n)$. Equating the Fourier elements on both sides that involve the same frequency, we find

$$D_{ab}(\omega_1 + \omega_2)[\rho_2(\omega_1 + \omega_2)]_{ab} = [H'(\omega_1), \rho_1(\omega_2)]_{ab} + [H'(\omega_2), \rho_1(\omega_1)]_{ab}. \tag{2.63}$$

We search the matrix-elements by multiplexing operators $[AB]_{ab} = \Sigma(A)_{ac}(B)_{cb}$, and substitute into (2.62) for $\rho_1(\omega_1)$ and $\rho_2(\omega_2)$. Finally, we have the second-order density matrix element for the sum-frequency

$$\rho_{ab}^{(2)}(\omega_1+\omega_2)=\frac{1}{D_{ab}(\omega_1+\omega_2)}\sum_c\left\{\frac{H'_{ac}(\omega_1)\cdot H'_{cb}(\omega_2)[\rho_{bb}^{(0)}-\rho_{cc}^{(0)}]}{D_{cb}(\omega_2)}-\frac{H'_{ac}(\omega_2)\cdot H'_{cb}(\omega_1)[\rho_{cc}^{(0)}-\rho_{aa}^{(0)}]}{D_{ac}(\omega_2)}+\frac{H'_{ac}(\omega_2)\cdot H'_{cb}(\omega_1)[\rho_{bb}^{(0)}-\rho_{aa}^{(0)}]}{D_{cb}(\omega_1)}-\frac{H'_{ac}(\omega_1)\cdot H'_{cb}(\omega_2)[\rho_{cc}^{(0)}-\rho_{aa}^{(0)}]}{D_{ac}(\omega_1)}\right\} \tag{2.64}$$

where a, b, c, are different intrinsic states in the ensemble, $\rho ab^{(2)}(\omega_1+\omega_2)$ is the simplified form of $[\rho_2(\omega_1+\omega_2)]ab$, $Dab(\omega_1+\omega_2)=\hbar(\omega ba-\omega_1-\omega_2-i\Gamma ab)$, and $Dcb(\omega_2)=\hbar(\omega bc-\omega_2-i\Gamma ab)$. Equation (2.64) can be further simplified to

$$\rho_{ab}^{(2)}(\omega_1+\omega_2)=\frac{1}{D_{ab}(\omega_1+\omega_2)}\sum_{(\omega_1,\omega_2)}\sum_c\left\{H'_{ac}(\omega_1)\cdot H'_{cb}(\omega_2)\cdot\left[\frac{\rho_{bb}^{(0)}-\rho_{cc}^{(0)}}{D_{cb}(\omega_2)}+\frac{\rho_{aa}^{(0)}-\rho_{cc}^{(0)}}{D_{ac}(\omega_1)}\right]\right\} \tag{2.65}$$

where we introduce the symbol $\sum_{(\omega_1\omega_2)}$ to express the sum over all distinguishable permutations. It means exchanging the positions of ω_1 and ω_2 in each term of (2.65), then summing all the terms.

2.4.2.3 Third-Order and Higher-Order Elements

Very like the procedures above, the third-order density matrix element of the sum-frequency $(\omega_1+\omega_2+\omega_3)$ is

$$\rho_{ab}^{(3)}(\omega_1+\omega_2+\omega_3)=\frac{1}{D_{ab}(\omega_1+\omega_2+\omega_3)}\times\sum_{(\omega_1,\omega_2,\omega_3)}\sum_{cd}\left\{\frac{H'_{ac}(\omega_1)\cdot H'_{cb}(\omega_2)\cdot H'_{ab}(\omega_3)}{D_{cb}(\omega_2+\omega_3)}\cdot\left[\frac{\rho_{bb}^{(0)}-\rho_{dd}^{(0)}}{D_{db}(\omega_3)}+\frac{\rho_{cc}^{(0)}-\rho_{dd}^{(0)}}{D_{cd}(\omega_2)}\right]+\frac{H'_{ab}(\omega_2)\cdot H'_{dc}(\omega_3)\cdot H'_{cb}(\omega_1)}{D_{ac}(\omega_2+\omega_3)}\cdot\left[\frac{\rho_{dd}^{(0)}-\rho_{cc}^{(0)}}{D_{cd}(\omega_3)}+\frac{\rho_{dd}^{(0)}-\rho_{aa}^{(0)}}{D_{da}(\omega_2)}\right]\right\} \tag{2.66}$$

where the first summation symbol $\sum\limits_{(\omega_1,\omega_2,\omega_3)}$ depicts summing all the terms after changing mutually the positions of ω_1, ω_2 and ω_3 in each term in (2.66). With the same procedures, we can get the higher-order density matrix elements.

2.4.3 Orders of Susceptibility Tensor Elements

2.4.3.1 First-Order Elements

If there are n particles in a unit volume, the first order of polarization intensity $P^{(1)}(t)$ is

$$P^{(1)}(t) = NTr\{\rho_1(t)R\} \tag{2.67}$$

where R is the electric dipole momentum of a single particle in the independent system. Putting (2.55) into (2.67) yields

$$\boldsymbol{P}^{(1)}(t) = N\Sigma Tr\{\rho_1(\omega_n)\boldsymbol{R}\}\ e^{i\omega_n t}$$

and the complex amplitude of a Fourier element (a certain frequency ω in ω_n) in the first-order polarization vector is

$$\begin{aligned}\boldsymbol{P}_\mu^{(1)}(\omega) &= NTr\{\rho_1(\omega)\boldsymbol{R}\}\\ &= N\sum_{a,b}(\boldsymbol{R})_{ab}\cdot\rho_{ba}^{(1)}(\omega)\\ &= N\sum_{a,b}(\boldsymbol{R})_{ab}^{\mu}\cdot\rho_{ba}^{(1)}(\omega)\cdot\frac{\rho_{aa}^{(0)}-\rho_{bb}^{(0)}}{D_{ba}(\omega)}\end{aligned} \tag{2.68}$$

where the perturbed Hamiltonian matrix element $H'_{ba}(\omega)$ with dipole approximation can be written as

$$\begin{aligned}H'_{ba}(\omega) &= -[R\cdot E(\omega)]_{ba}\\ &= -\sum_\alpha [R^\alpha E_\alpha(\omega)]_{ba}\\ &= -\sum_\alpha (R^\alpha)_{ba}E_\alpha(\omega)\end{aligned}$$

where $\alpha = x, y, z$ are subscripts on axes, from (2.67) and (2.68). The direction of the polarization vector P is determined by that of the dipole operator R. We express $P_\mu(\omega_\sigma)$ as a projection element on a certain axis

$$
\begin{aligned}
\rho_{\mu}^{(1)}(\omega) &= -N\sum_{a,b}\frac{(R^{\mu})_{ab}(R^{\alpha})_{ba}}{D_{ba}}[\rho_{aa}^{(0)} - \rho_{bb}^{(0)}]E_{\alpha}(\omega) \\
&\equiv \varepsilon_0\chi_{\mu\alpha}^{(1)}(-\omega;\omega)E_{\alpha}(\omega)
\end{aligned}
\tag{2.69}
$$

where $(R^{\mu})_{ab}$ is from the polarization element $P_{\mu}^{(1)}(\omega)$, and $(R^{\alpha})_{ba}$ is from the interactive Hamiltonian. Hence the first-order linear susceptibility tensor element in the frequency domain is

$$
\begin{aligned}
\chi_{\mu\alpha}^{(1)}(-\omega;\omega) &= -\frac{N}{\varepsilon_0}\sum_{a,b}\frac{(R^{\mu})_{ab}\cdot(R^{\alpha})_{ba}}{D_{ba}(\omega)}\left[\rho_{aa}^{(0)} - \rho_{bb}^{(0)}\right] \\
&= +\frac{N}{\varepsilon_0\hbar}\sum_{a,b}\rho_{aa}^{(0)}\left[\frac{(R^{\mu})_{ab}(R^{\alpha})_{ba}}{\omega_{ba}-\omega}+\frac{(R^{\alpha})_{ab}(R^{\mu})_{ba}}{\omega_{ba}+\omega}\right].
\end{aligned}
\tag{2.70}
$$

2.4.3.2 Second-Order (Nonlinear) Elements

From (2.57)

$$
\boldsymbol{P}^{(2)}(t) = NTr\{\rho_2(t)\boldsymbol{R}\}. \tag{2.71}
$$

Substitute (2.55) into it and then express the Fourier element $P_{\mu}{}^{(2)}(\omega_3)$ as in (2.72):

$$
\begin{aligned}
P_{\mu}^{(2)}(\omega_1+\omega_2) &= N\sum_{a,b}(R^{\mu})_{ab}\rho_{ba}^{(2)}(\omega_1+\omega_2) \\
&= N\sum_{a,b}(R^{\mu})_{ab}\sum_{(\omega_1,\omega_2)}\sum_{c}\left\{\frac{H'_{bc}(\omega_1)\cdot H'_{ca}(\omega_2)}{D_{ba}(\omega_1+\omega_2)}\cdot\left[\frac{\rho_{aa}^{(0)}-\rho_{cc}^{(0)}}{D_{ca}(\omega_2)}+\frac{\rho_{bb}^{(0)}-\rho_{cc}^{(0)}}{D_{bc}(\omega_1)}\right]\right\} \\
&= N\sum_{(\omega_1,\omega_2)}\sum_{a,b,c}\frac{(R^{\mu})_{ab}(R^{\alpha})_{ab}(R^{\beta})_{ab}}{D_{ba}(\omega_1+\omega_2)}\left[\frac{\rho_{aa}^{(0)}-\rho_{cc}^{(0)}}{D_{ca}(\omega_2)}+\frac{\rho_{bb}^{(0)}-\rho_{cc}^{(0)}}{D_{bc}(\omega_1)}\right]E_{\alpha}(\omega_1)E_{\beta}(\omega_2)
\end{aligned}
\tag{2.72}
$$

where the second-order density matrix element (2.65) and the electric dipole approximation for interactive H' are used in this derivation. According to the definition of polarization (2.60), we have

$$
P_{\mu}{}^{(2)}(\omega_3) \equiv \varepsilon_0\chi_{\mu\alpha\beta}{}^{(2)}(-\omega_3\,;\,\omega_1,\omega_2)E_{\alpha}(\omega_1)E_{\beta}(\omega_2) \tag{2.73}
$$

where $(-\omega_3;\omega_1,\omega_2)$ means that in this process, two photons with frequencies ω_1 and ω_2 are absorbed by the medium and one photon with frequency ω_3 is emitted from the polarization in the medium. Comparing (2.73) with (2.72) yields

$$\begin{aligned}
\chi^{(2)}_{\mu\alpha\beta}(-\omega_3;\omega_1,\omega_2) &= \frac{1}{2}\left[\chi^{(2)}_{\mu\alpha\beta}(-\omega_3;\omega_1,\omega_2)+\chi^{(2)}_{\mu\alpha\beta}(-\omega_3;\omega_1,\omega_2)\right] \\
&= \frac{N}{2\varepsilon_0}\sum_{\alpha,\beta}^{(\omega_1,\omega_2)}\sum_{a,b,c}\frac{(R^\mu)_{ab}(R^\alpha)_{bc}(R^\beta)_{ca}}{D_{ba}(\omega_1+\omega_2)}\left[\frac{\rho^{(0)}_{aa}-\rho^{(0)}_{cc}}{D_{ca}(\omega_2)}+\frac{\rho^{(0)}_{bb}-\rho^{(0)}_{cc}}{D_{bc}(\omega_1)}\right] \\
&= \frac{N}{2\varepsilon_0}\sum_{\alpha,\beta}^{(\omega_1,\omega_2)}\sum_{a,b,c}\rho^{(0)}_{aa}\left[\frac{(R^\mu)_{ab}(R^\alpha)_{bc}(R^\beta)_{ca}}{D_{ba}(\omega_1+\omega_2)D_{ca}(\omega_2)}-\frac{(R^\mu)_{cb}(R^\alpha)_{ba}(R^\beta)_{ac}}{D_{bc}(\omega_1+\omega_2)D_{ac}(\omega_2)}\right. \\
&\qquad\left.+\frac{(R^\mu)_{ba}(R^\alpha)_{ac}(R^\beta)_{cb}}{D_{ab}(\omega_1+\omega_2)D_{ac}(\omega_1)}-\frac{(R^\mu)_{cb}(R^\alpha)_{bc}(R^\beta)_{ac}}{D_{bc}(\omega_1+\omega_2)D_{bc}(\omega_1)}\right]
\end{aligned} \tag{2.74}$$

where the first summation symbol $\sum\limits_{\alpha,\beta}^{(\omega_1\omega_2)}$ depicts changing mutually the positions of pairs $(\alpha\omega_1)$ and $(\beta\omega_2)$ in each term of (2.74), and then summing all the terms. A symmetric operator $\widetilde{S}$ is introduced to sum over the possible permutations of the pairs $(\alpha\omega_1)$ and $(\beta\omega_2)$. The second summation symbol $\sum\limits_{abc}$ represents the contributions of all the intrinsic energy states.

In the square bracket of (2.74), when the pairs $(\alpha\omega_1)$ and $(\beta\omega_2)$ exchange, the total summation will stay invariant. Consequently (2.74) can be written in the general form as (2.75):

$$\begin{aligned}
\chi^{(2)}_{\mu\alpha\beta}(-\omega_3;\omega_1,\omega_2) = \frac{\widetilde{S}}{2}\frac{N}{\varepsilon_0\hbar}\sum_{a,b,c}\rho^{(0)}_{aa}&\left[\frac{(R^\mu)_{ab}(R^\alpha)_{bc}(R^\beta)_{ca}}{(\omega_2+\omega_{ba})(\omega_1+\omega_2+\omega_{bc})}\right. \\
&-\frac{(R^\alpha)_{cb}(R^\mu)_{bc}(R^\beta)_{ca}}{(\omega_2+\omega_{ac})(\omega_1+\omega_2+\omega_{bc})} \\
&-\frac{(R^\beta)_{ab}(R^\mu)_{bc}(R^\alpha)_{ca}}{(\omega_2+\omega_{ba})(\omega_1+\omega_2+\omega_{bc})} \\
&\left.+\frac{(R^\beta)_{ab}(R^\alpha)_{bc}(R^\mu)_{ca}}{(\omega_2+\omega_{ac})(\omega_1+\omega_2+\omega_{bc})}\right].
\end{aligned} \tag{2.75}$$

Here, the symmetry operator $\tilde{S}$ implies that the final result of (2.75) is invariant when pairs $(\alpha\omega_1)$ and $(\beta\omega_2)$ are exchanged. When we exchange $(\alpha\omega_1)$ and $(\beta\omega_2)$ in the third term and add the second term, the second-order nonlinear susceptibility tensor element in the frequency domain $\chi_{\mu\alpha\beta}{}^{(2)}$ $(-\omega_3\,;\,\omega_1,\omega_2)$ can be expressed in three independent terms as

$$\chi^{(2)}_{\mu\alpha\beta}(-\omega_1-\omega_2;\omega_1,\omega_2)$$
$$=\frac{\tilde{S}}{2}\frac{N}{\varepsilon_0\hbar^2}\sum_{abc}\rho^{(0)}_{aa}\left\{\frac{R^{\mu}_{ab}R^{\alpha}_{bc}R^{\beta}_{ca}}{(\omega_{ab}+\omega_1+\omega_2)(\omega_{ac}+\omega_2)}+\frac{R^{\alpha}_{ab}R^{\mu}_{bc}R^{\beta}_{ca}}{(\omega_{ab}-\omega_1)(\omega_{ac}+\omega_2)}+\frac{R^{\alpha}_{ab}R^{\beta}_{bc}R^{\mu}_{ca}}{(\omega_{ab}-\omega_1)(\omega_{ac}-\omega_1-\omega_2)}\right\} \tag{2.76}$$

The three terms are all independent when the symbol $\tilde{S}$ is involved. Since $\tilde{S}$ implies two exchangeable terms, there are actually six (3×2) terms in all.

2.4.3.3 Third- and Higher-Order Elements

When the optical medium interacts simultaneously with the three monochromatic lightwave fields $E_1(\alpha\omega_1)$, $E_2(\beta\omega_2)$ and $E_3(\gamma\omega_3)$ with frequencies ω_1, ω_2, ω_3, and polarization directions α, β, γ respectively, the sum-frequency $(\omega_1+\omega_2+\omega_3)$ is generated. The third-order nonlinear susceptibility tensor element in the frequency domain is (2.77)

$$\chi^{(3)}_{\mu\alpha\beta\gamma}(-\omega_1-\omega_2-\omega_3;\omega_1,\omega_2,\omega_3)$$
$$=\frac{\tilde{S}}{3!}\frac{N}{\varepsilon_0\hbar^3}\sum_{abcd}\rho^{(0)}_{aa}\left\{\frac{R^{\mu}_{ab}R^{\alpha}_{bc}R^{\beta}_{cd}R^{\gamma}_{da}}{(\omega_{ba}-\omega_1-\omega_2-\omega_3)(\omega_{ca}-\omega_2-\omega_3)(\omega_{da}-\omega_3)}+\frac{R^{\alpha}_{ab}R^{\mu}_{bc}R^{\beta}_{cd}R^{\gamma}_{da}}{(\omega_{ba}+\omega_1)(\omega_{ca}-\omega_2-\omega_3)(\omega_{da}-\omega_3)}+\frac{R^{\alpha}_{ab}R^{\beta}_{bc}R^{\mu}_{cd}R^{\gamma}_{da}}{(\omega_{ba}+\omega_1)(\omega_{ca}+\omega_1+\omega_2)(\omega_{da}-\omega_3)}+\frac{R^{\alpha}_{ab}R^{\beta}_{bc}R^{\gamma}_{cd}R^{\mu}_{da}}{(\omega_{ba}+\omega_1)(\omega_{ca}+\omega_1+\omega_2)(\omega_{da}+\omega_1+\omega_2+\omega_3)}\right\} \tag{2.77}$$

where the four terms are all independent just as $\tilde{S}$ included. The symbol $\tilde{S}$ implicates six terms exchanged between $(\alpha\omega_1)$, $(\beta\omega_2)$ and $(\gamma\omega_3)$, hence if take off it, there are twenty four (4×6) terms in total.

In a like manner, the arbitrary r-order nonlinear susceptibility tensor element is expressed by the independent $(r+1)$ terms as:

$$\chi^{(r)}_{\mu\alpha\beta\gamma\ldots\xi}(\omega_1,\omega_2,\omega_3\ldots\omega_r)=\frac{\widetilde{S}}{r!}\frac{N}{\varepsilon_0\hbar^r}\sum_{abcd\ldots z}\rho^{(0)}_{aa}$$
$$\times\left\{\frac{R^{\mu}_{ab}R^{\alpha}_{bc}R^{\beta}_{cd}\ldots R^{\xi}_{za}}{(\omega_{ba}-\omega_1-\omega_2-\ldots-\omega_r)(\omega_{ca}-\omega_2-\omega_3\ldots-\omega_r)\ldots(\omega_{za}-\omega_r)}\right.$$
$$+\frac{R^{\alpha}_{ab}R^{\mu}_{bc}R^{\beta}_{cd}\ldots R^{\xi}_{za}}{(\omega_{ba}+\omega_1)(\omega_{ca}-\omega_2-\omega_3\ldots-\omega_r)\ldots(\omega_{za}-\omega_r)}$$
$$\left.+\ldots+\frac{R^{\alpha}_{ab}R^{\beta}_{bc}R^{\gamma}_{cd}\ldots R^{\xi}_{za}}{(\omega_{ba}+\omega_1)(\omega_{ca}+\omega_1+\omega_2)\ldots(\omega_{za}+\omega_1+\omega_2+\ldots+\omega_r)}\right\}. \tag{2.78}$$

2.5 The General Properties of Nonlinear Susceptibility and of the Feynman Representations

2.5.1 Symmetry of Nonlinear Susceptibility

2.5.1.1 Frequency-Permutation Symmetry

There are two kinds of frequency-sequence permutation symmetries, the inherent-permutation symmetry, and the total-permutation symmetry, which have different meanings in physics. By these symmetries, the susceptibility representation can be further simplified.

2.5.1.2 Inherent-Permutation Symmetry

The inherent-permutation symmetry implies that, in the susceptibility tensor element $\chi_{\mu\alpha\beta\gamma\ldots\xi}{}^{(n)}(-\omega_{n+1};\ \omega_1,\omega_2,\omega_3,\ \ldots\ \omega_n)$, when permutations are implemented between the pairs of indices $(\alpha\omega_1)$, $(\beta\omega_2)$, $(\gamma\omega_3)$ ... $(\xi\omega_n)$, the resultant susceptibility tensor is unchanged. The inherent-permutation symmetry of the susceptibilities is

$$\chi^{(n)}_{\mu\alpha\beta\gamma\cdots\xi}(-\omega_{n+1};\omega_1,\omega_2,\omega_3,\cdots,\omega_n)$$
$$=\chi^{(n)}_{\mu\xi\beta\gamma\cdots\alpha}(-\omega_{n+1};\omega_n,\omega_2,\omega_3,\cdots,\omega_1) \tag{2.79}$$
$$\vdots$$
$$=\chi^{(n)}_{\mu\beta\alpha\gamma\cdots\xi}(-\omega_{n+1};\omega_2,\omega_1,\omega_3,\cdots,\omega_n).$$

The essence of (2.79) is apparent physically. Since these field elements interact at the same time and place, permuting the sequence of these fields changes the resultant nonlinear effects. For (2.79), there are $n!$ inherent permutations for frequencies ω_1, ω_2, ω_3, ..., ω_n. For that reason, in (2.76), (2.77) and (2.78), 2!, 3! and $n!$ occur in the denominators. This inherent permutation symmetry is valid in both the resonant and nonresonant conditions.

2.5.1.3 Total-Permutation Symmetry

In total-permutation symmetry, the permutation occurs amongst not only the injected fields $E_1(\alpha\omega_1)$, $E_2(\beta\omega_2)$, $E_3(\gamma\omega_3)$, ..., $E_n(\xi\omega_n)$, but also the resultant field $E_{n+1}(\mu\omega_{n+1})$, which takes part in the permutation as well. Thus the overall permutation symmetry of the susceptibilities is

$$
\begin{aligned}
&\chi^{(n)}_{\mu\alpha\beta\gamma\cdots\xi}(-\omega_{n+1};\omega_1,\omega_2,\omega_3,\cdots,\omega_n)\\
&=\chi^{(n)}_{\mu\xi\beta\gamma\cdots\alpha}(-\omega_{n+1};\omega_n,\omega_2,\omega_3,\cdots,\omega_1)\\
&\vdots\\
&=\chi^{(n)}_{\xi\alpha\beta\gamma\cdots\mu}(\omega_n;\omega_1,\omega_2,\omega_3,\cdots,-\omega_{n+1}).
\end{aligned}
\tag{2.80}
$$

This symmetry is true only when all frequencies are far from resonances and thus the medium is lossless, so that the optical wave mixing processes are reversible. The total permutation symmetry (2.80) has profound physical implications. When the frequencies of injected fields ω_1, ω_2, ω_3, ..., ω_n and their combination ω_{n+1} are all far from the intrinsic resonant frequency ω_0 of the medium, the imaginary part of susceptibility $\chi''(\omega) = 0$. Hence the susceptibility is a real quantity and the medium is lossless in this case. As long as there is no energy-exchange between the fields, including the resultant field $E_{n+1}(\mu\omega_{n+1})$, the interactions can be inversed, such as $(\omega_1+\omega_2+\omega_3+...+\omega_n \leftrightarrow \omega_{n+1})$ or $(\omega_1-\omega_{n+1}+\omega_3+...+\omega_n \leftrightarrow -\omega_2)$. These can all exist simultaneously. Since they have the same representation, the representations can be further simplified by introducing the total symmetric permutation symbol $\tilde{S}_{TOTAL}$. The equations for $\chi^{(1)}(\omega)$, $\chi^{(2)}(\omega)$, $\chi^{(3)}(\omega)$, and $\chi^{(r)}(\omega)$ simplify to (2.81–2.84):

$$
\chi^{(1)}_{\mu\alpha}(-\omega;\omega)=+\frac{N}{\varepsilon_0\hbar}\tilde{S}_{TOTAL}\sum_{a,b}\rho^{(0)}_{aa}\left[\frac{(R^{\mu})_{ab}(R^{\alpha})_{ba}}{\omega_{ba}-\omega}\right] \tag{2.81}
$$

$$\chi^{(2)}_{\mu\alpha\beta}(-\omega_1-\omega_2;\omega_1,\omega_2)$$
$$=\widetilde{S}_{\text{TOTAL}}\cdot\frac{N}{2!\varepsilon_0\hbar^2}\sum_{abc}\rho^{(0)}_{aa}\cdot\frac{R^{\mu}_{ab}R^{\alpha}_{bc}R^{\beta}_{ca}}{(\omega_{ab}+\omega_1+\omega_2)(\omega_{ac}+\omega_2)} \tag{2.82}$$

$$\chi^{(3)}_{\mu\alpha\beta\gamma}(-\omega_1-\omega_2-\omega_3;\omega_1,\omega_2,\omega_3)$$
$$=\widetilde{S}_{\text{TOTAL}}\cdot\frac{N}{3!\varepsilon_0\hbar^3}\sum_{abcd}\rho^{(0)}_{aa}\cdot\left[\frac{R^{\mu}_{ab}R^{\alpha}_{bc}R^{\beta}_{cd}R^{\gamma}_{da}}{(\omega_{ba}-\omega_1-\omega_2-\omega_3)(\omega_{ca}-\omega_2-\omega_3)(\omega_{da}-\omega_3)}\right] \tag{2.83}$$

$$\chi^{(r)}_{\mu\alpha\beta\gamma\ldots\xi}(-\omega_{r+1};\omega_1,\omega_2\ldots\omega_r)$$
$$=\widetilde{S}_{\text{TOTAL}}\frac{(-1)^r N}{r!\varepsilon_0\hbar^r}\cdot\sum_{abc\ldots m}\rho^{(0)}_{aa}\frac{R^{\mu}_{ab}R^{\alpha}_{bc}\cdots R^{\xi}_{ma}}{D_{abc\ldots m}(\omega_1,\omega_2\ldots\omega_r)}. \tag{2.84}$$

2.5.1.4 Spatial Structural Symmetry

The susceptibility tensor is defined as the response of the anisotropic media when optical fields act on it. Thus the spatial symmetries expressed by 32 point groups are described by the susceptibilities, and when some of the symmetric operations transform according to these point group symmetries, the nonlinear susceptibility of the medium is unchanged. That means some of the tensor elements are zero and some relations exist amongst the non-zero elements. This reduces the number of independent tensor elements.

Some media have inversed symmetry, and are called centrosymmetric media or simply as symmetric media. In such media, if the inverse symmetry operation($x\to -x$, $y\to -y$, $z\to -z$) is taken, the resultant physical quantities, such as the amplitudes of $\boldsymbol{P}^{(1)}(r,t)$, $\boldsymbol{P}^{(2)}(r,t)$, $\boldsymbol{P}^{(3)}(r,t)$, are unchanged physically.

Referring to (2.10) and (2.11), when inverse symmetry operation is taken in centrosymmetric media, $\boldsymbol{P}^{(n)}(r,t)$ and $\boldsymbol{E}(r,t)$ are both changing their signs. When n is an odd number, such as $n = 1$ and 3, the signs on the left and right sides of the equations are both changed from "+" to "–" simultaneously. Then $\boldsymbol{P}^{(\text{odd})}(r,t)$ are unchanged both physically and mathematically in centrosymmetric media. But when n is an even number, the sign of $\boldsymbol{P}^{(\text{even})}(r,t)$ on the left is changed from "+" to "–" physically, while mathematically, the right sides of the equations are unchanged. Absolutely, the $\boldsymbol{P}^{(n)}(r,t)$ and $\boldsymbol{E}(r,t)$ being all zero for every (r,t) is impossible and senseless, then $\chi^{(\text{even})} = 0$ will be the only solution for centrosymmetric media.

The physical pattern of zero $\chi^{(2)}$ in the symmetric media could be explicated as follows. A symmetry operation transforms the medium into an indistinguishable orientation, and also transforms the elements according to the standard representations. If the medium is centro-symmetry, then the transformed elements must equal to the untrans-formed ones, and this places constraints on the elements. On the one hand, the product of field components $E(r,t)\cdot E(r,t)$ remains unchanged, whereas the polarization $P^{(2)}(r,t)$ changes sign. On the other hand, as the property of a symmetric medium, the susceptibility element should not change sign under inversion, and these criteria are consistent only if all tensor elements of $\chi^{(2)}$ are zero. It is extremely essential for fussed silica optical fiber nonlinearity, which will be mentioned later in Sect. 2.7.1.

Thus we have

$$\chi^{(2r)}_{\mu\alpha\beta\ldots\xi}(-\omega_\sigma;\omega_1,\omega_2\ldots\omega_{2r})=0 \tag{2.85}$$

for all even order of susceptibilities, or it is equivalent to

$$\chi^{(r)}_{\mu\alpha\beta\ldots\xi}(-\omega_\sigma;\omega_1,\omega_2\ldots\omega_r)=(-1)^{r+1}\chi^{(r)}_{\mu\alpha\beta\ldots\xi}(-\omega_\sigma;\omega_1,\omega_2\ldots\omega_r) \tag{2.86}$$

for the centrosymmetric media.

2.5.1.5 Time-Inversion Symmetry

In classical mechanics, time inversion means the time-measuring direction inverses, or that $+t$ is replaced by $-t$. Some of the dynamic parameters, for example, kinetic energy and location coordinates, their functions, and the events linked with the lossless energy-pulse shape distortion in fibers, are independent of time inversion. Furthermore, in quantum mechanics these dynamic parameters can be depicted by the real operator in the Schrödinger representation. Consequently, even most of the optical isolators are spin dependent, for simplicity, the mentioned energy-coordinate events can be treated as spin-free systems. The time inversion equals to a complex conjugation operation, and is significant for fiber dispersion compensation by FWM phase conjugations.

Since the Hamiltonian H_0, energy eigenfunction $u_m(r)$, and the dipole momentum $R^i_{m,n}$ are all real. From (2.84), $\chi^{(r)}\propto\left(R^i_{m,n},u_m(r),H_0\right)$, and therefore is time-inversed. Then if complex conjugation is done on (2.84), the frequencies ω_1, ω_2, ... ω_m are replaced by their complex conjugate frequencies as

$$\begin{aligned}\chi^{(r)}*\mu\alpha\beta\ldots\xi(-\omega_{r+1};\omega_1,\ldots\omega_r)&=\chi^{(r)}\mu\alpha\beta\ldots\xi(-\omega_{r+1}*;\omega_1*,\ldots\omega_r*)\\&=\chi^{(n)}\mu\alpha\beta\ldots\xi(+\omega_{r+1};-\omega_1,\ldots-\omega_r).\end{aligned}$$

From the time-inversion properties, we have

$$\chi^{(r)*}\ \mu\alpha\beta\ldots\xi(-\omega_{r+1};\omega_1,\ldots\omega_r) = \chi^{(r)}\ \mu\alpha\beta\ldots\xi(-\omega_{r+1};\omega_1,\ldots\omega_r).$$

Thus

$$\chi^{(r)}\ \mu\alpha\beta\ldots\xi(-\omega_{r+1};\omega_1,\ldots\omega_r) = \chi^{(r)}\ \mu\alpha\beta\ldots\xi(+\omega_{r+1};-\omega_1,\ldots-\omega_r)\,. \tag{2.87}$$

Equation (2.87) shows that when all the frequencies change their signs, the susceptibility is unchanged. This is the time-inversed symmetry of susceptibilities. A more detailed discussion can be found, for example, in Butcher and Cotter [2.3].

2.5.2 Intermolecular Interactions

In Sect. 2.4, (2.76), (2.77) and (2.78) are all considered in an independent molecular system, and the interaction between molecules was neglected. However, in the case of resonance, some terms in the denominators of these equations become zero, and hence the susceptibility tensor elements become infinite. This is physically impossible, but it is caused by the interactions neglected amongst molecules. In the semiclassical approach, there is the lifetime of particle stayed on an energy level. In other words, there are relaxations between different levels due to the interaction between molecules. Therefore, we must modify the susceptibilities.

2.5.2.1 Removing the Singularity from the Real Frequency Axis

We first discuss (2.70),

$$\chi^{(1)}_{\mu\alpha}(-\omega_i) = -\frac{N}{\varepsilon_0}\sum_{a,b}\frac{(R^\mu)_{ab}\cdot(R^\alpha)_{ba}}{D_{ba}(\omega)}\left[\rho^{(0)}_{aa} - \rho^{(0)}_{bb}\right]$$

$$= +\frac{N}{\varepsilon_0\hbar}\sum_{a,b}\rho^{(0)}_{aa}\left[\frac{(R^\mu)_{ab}(R^\alpha)_{ba}}{\omega_{ba}-\omega} + \frac{(R^\alpha)_{ab}(R^\mu)_{ba}}{\omega_{ba}+\omega}\right]$$

where $\omega_{ab} = (E_a - E_b)/\hbar = \omega_a - \omega_b = -\omega_{ba}$ is a definite frequency. If the weak interactions between molecules are considered, the energy-level will have an uncertainty of the same magnitude as the interaction between molecules.

Suppose the levels $E_a = \hbar\omega_a$ and $E_b = \hbar\omega_b$ have uncertainties $\hbar\Gamma_a$ and $\hbar\Gamma_b$, then the transition frequency's uncertainty between levels a and b is $\Gamma_{ab} = \Gamma_a + \Gamma_b = \Gamma_{ba}$. The resonant transition frequency between levels a and b has a linewidth Γ_{ab} which equals the linewidth ς. From the

Uncertainty Principle, if the energy level has an uncertainty ΔE, the particle has a lifetime Δt on its energy level,

$$\Delta E \Delta t \geq \hbar. \tag{2.88}$$

Then in $\chi_{\mu\alpha}{}^{(1)}(\omega)$, the intrinsic resonant real frequency ω_{ab} is replaced by a complex frequency $\omega_{ab} \pm i\Gamma_{ab}$, and (2.70) becomes (2.89)

$$\chi_{\mu\alpha}^{(1)}(\omega) = -\frac{N}{\varepsilon_0 \hbar} \sum_{a,b} \rho_{aa}^{(0)} \left[\frac{(R^\mu)_{ab}(R^\alpha)_{ba}}{\omega_{ab} + \omega + i\Gamma_{ab}} + \frac{(R^\alpha)_{ab}(R^\mu)_{ba}}{\omega_{ab} - \omega - i\Gamma_{ab}} \right]. \tag{2.89}$$

Introducing $\pm\, i\Gamma_{ab}$ changes the frequency ω from the real frequency $\pm\omega_{ab}$ to a pair of complex frequencies $\pm\omega_{ab} \pm i\Gamma_{ab}$. Thus, the frequency singularity of $\chi_{\mu\alpha}{}^{(1)}(-\omega;\omega)$ is moved from the real frequency axis into the lower half of the complex plane, as shown in Fig. 2.2. In this case, if time is inversed, i.e., $-\omega$ replaces $+\omega$, (2.89) becomes (2.90)

$$\chi_{\mu\alpha}^{(1)}(-\omega) = -\frac{N}{\varepsilon_0 \hbar} \sum_{a,b} \rho_{aa}^{(0)} \left[\frac{(R^\mu)_{ab}(R^\alpha)_{ba}}{\omega_{ab} - \omega + i\Gamma_{ab}} + \frac{(R^\alpha)_{ab}(R^\mu)_{ba}}{\omega_{ab} + \omega - i\Gamma_{ab}} \right]. \tag{2.90}$$

Comparing (2.90) and (2.89), we see that

$$\chi_{\mu\alpha}{}^{(1)}(-\omega;\omega) \neq \chi_{\mu\alpha}{}^{(1)}(\omega;-\omega). \tag{2.91}$$

It means the time-inversion symmetry of susceptibility tensors will be invalid, if the interaction between molecules is taken into account.

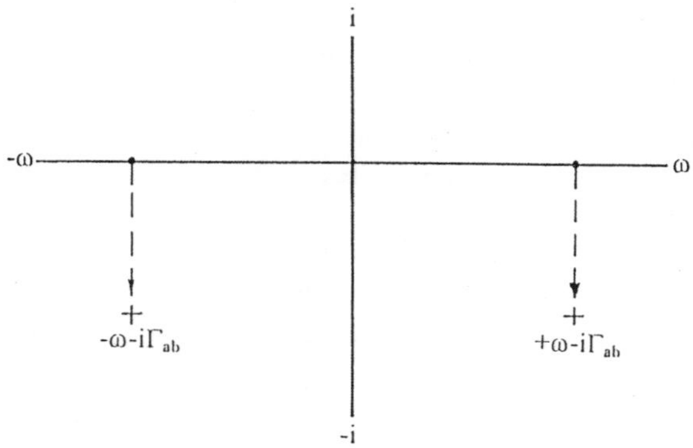

Fig. 2.2 Vanishing the singularity on the real-frequency axis by moving it into the lower complex-frequency plane.

2.5.2.2 Second and Third Orders of Susceptibility Considered Together with Intermolecular Interactions

In like manner, the modifications of second-order, third-order and even nth-order susceptibilities, allowing for intermolecular interaction, can be found. In (2.89), the transition frequency ω_{ab} is replaced by $\omega_{ab} \pm i\Gamma_{ab}$ to move the frequency singularity into the lower half of the complex frequency plane as

$$
\begin{aligned}
&\chi^{(2)}_{\mu\alpha\beta}(-\omega_1-\omega_2;\omega_1,\omega_2) \\
&= \frac{\widetilde{S}}{2!}\frac{N}{\varepsilon_0 \hbar^2}\sum_{abc}\rho^{(0)}_{aa}\cdot\left\{\frac{R^{\mu}_{ab}R^{\alpha}_{bc}R^{\beta}_{ca}}{(\omega_{ab}+i\Gamma_{ab}+\omega_1+\omega_2)(\omega_{ac}+i\Gamma_{ac}+\omega_2)}\right. \\
&\quad + \frac{R^{\alpha}_{ab}R^{\mu}_{bc}R^{\beta}_{ca}}{(\omega_{ab}-i\Gamma_{ab}-\omega_1)(\omega_{ac}+i\Gamma_{ac}+\omega_2)} \\
&\quad \left. + \frac{R^{\alpha}_{ab}R^{\beta}_{bc}R^{\mu}_{ca}}{(\omega_{ab}-i\Gamma_{ab}-\omega_1)(\omega_{ac}-i\Gamma_{ac}-\omega_1-\omega_2)}\right\}.
\end{aligned}
\tag{2.92}
$$

2.5.3 Feynman Diagram Representations

When the system is in steady state, the different orders of electric susceptibilities are described by (2.76), (2.77) and (2.78). But to write a complete representation, especially when $n \geq 3$, is a complicated problem. In this case, a Feynman diagram is used first, and then the relevant formula is written. Before starting the diagram, some rules must be defined:

(1) Propagation lines are described by horizontal lines;
(2) An arrow down toward a propagation line represents the positive frequency $+\omega$ and corresponds to a photon annihilation of the field;
(3) An arrow up from a propagation line represents the negative frequency $-\omega$ and corresponds to the photon creation of the field;
(4) The nodes a, b, ..., m represent the energy levels which join in the interaction processes.

By these rules, a term $(\omega_{ba} - \omega_1 - \omega_2 - \omega_3)$ in the denominator of (2.77) means that in a transition from levels b to a, a photon with frequency $\omega_\sigma = -(\omega_1 + \omega_2 + \omega_3)$ radiates from the medium to the field and is denoted by an arrow up from the horizontal line. In this way, the four terms of (2.77) can be expressed by the following four diagrams in Fig. 2.3. In practical applications, the process is reversed. First draw these diagrams line by line regualarly, and then write down the formula elements referring to each diagram. With this diagrammatic process, we can accurately write even higher orders of electric susceptibility.

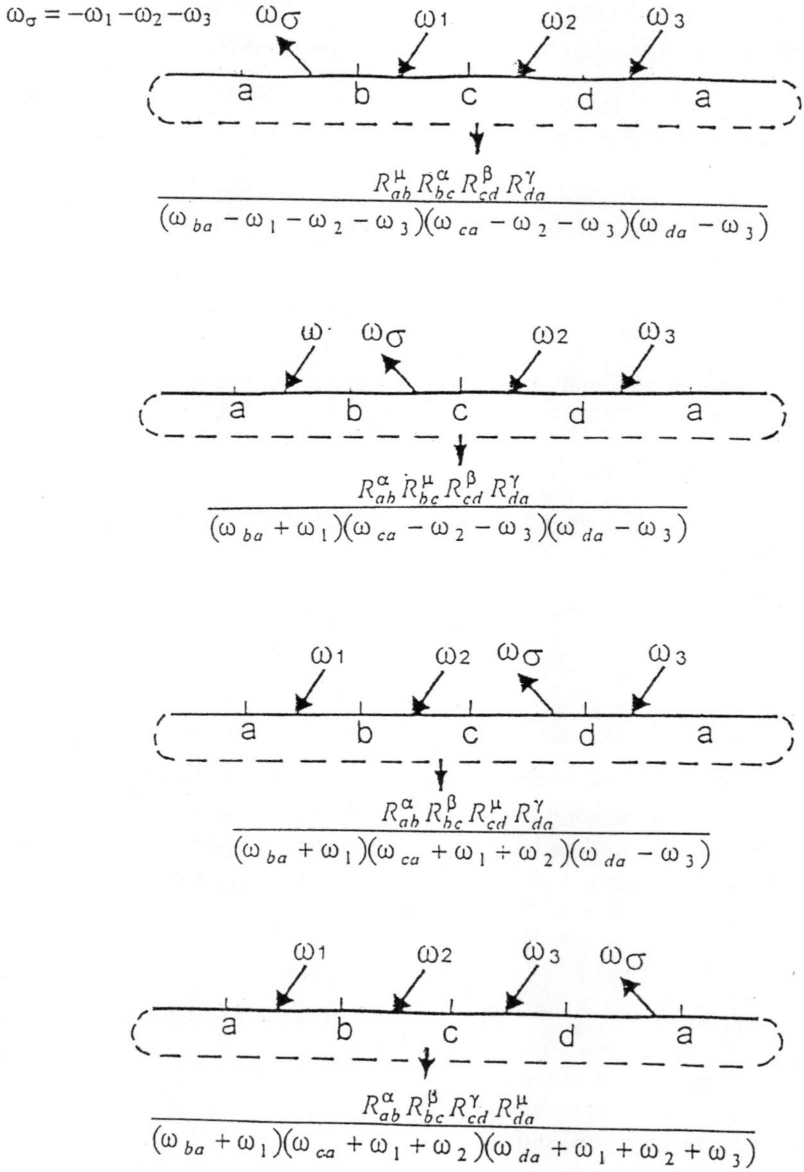

Fig. 2.3 The Feynman diagrammatic representation of $\chi^{(3)}$.

2.6 Nonlinear Susceptibility in Optical Fibers and Semiconductors, and the Kramers-Kronig Relations

2.6.1 Nonlinear Susceptibility in Silica Fibers

Since the silica molecule is centrosymmetric, then according to (2.85) the second and all higher even orders of electrical susceptibilities are zero as

$$\begin{aligned} &\boldsymbol{P}^{(2)}(\omega)=\boldsymbol{P}^{(4)}(\omega)=\cdots=\boldsymbol{P}^{(2r)}(\omega)=0 \\ &\chi^{(2)}(\omega)=\chi^{(4)}(\omega)=\cdots=\chi^{(2r)}(\omega)=0. \end{aligned} \tag{2.93}$$

Thus $\chi(\omega)$ in undoped silica fiber is, with even-order nonlinear effects, zero,

$$\chi_{fiber}(\omega)=\chi^{(1)}(\omega)+\chi^{(3)}(\omega)+\chi^{(5)}(\omega)+\cdots. \tag{2.94}$$

The lowest nonlinear optical effects, which are important, are related to the third order of susceptibility as (2.95)

$$\begin{aligned} \chi^{(3)}_{\text{fiber}}(\omega)=\chi^{(3)}_{\text{SPM}}(\omega)+\chi^{(3)}_{\text{XPM}}(\omega)+\chi^{(3)}_{\text{FWM}}(\omega)+\chi^{(3)}_{\text{SBS}}(\omega) \\ +\chi^{(3)}_{\text{SRS}}(\omega)+\cdots \end{aligned} \tag{2.95}$$

where $\chi_{\text{fiber}}{}^{(3)}=\chi_{\text{PARAMETRIC}}{}^{(3)}+\chi_{\text{NONPARAMETRIC}}{}^{(3)}$ (2.96)

Although $\chi_{\text{fiber}}{}^{(3)}$ is extremely weak in the transparent region of undoped silica fiber, the total nonlinear effects in the fiber practical transmission is still strong. The interaction length between the optic field and fiber medium is very long compared with other optical crystals. It is always at least a kilometer and so the total nonlinear effects are still important. Some of the nonlinear effects in the transmission fiber are stemmed from $\chi_{\text{PARAMETRIC}}{}^{(3)}$ in (2.96) which is based on the intensity-induced refractive index, $\delta n(\omega)=f(I)=F(|E|^2)$. Among them, there are self-phase modulation (SPM) and cross-phase modulation (XPM) as well as other nonlinear techniques applied in fibers, such as fiber soliton transmission, phase conjugation for dispersion compensations, nonlinear fiber-loop for switches, demultiplexers and so on.

We assume here all the fields are polarized in the x direction and all the frequencies are degenerated in the fiber, so the four-rank tensor

$$\chi_{\mu\alpha\beta\gamma}{}^{(3)}(-\omega_4;\omega_1,\omega_2,\omega_3)=\chi_{xxxx}{}^{(3)}(\omega)\equiv\chi^{(3)}.$$

The corresponding expression for polarization $\boldsymbol{P}(\omega)$ is

$$\begin{aligned} \boldsymbol{P}(\omega) &= \boldsymbol{P}^{(1)}(\omega) + \boldsymbol{P}^{(3)}(\omega) \\ &= \varepsilon_0 \left[\chi^{(1)}(-\omega;\omega)\boldsymbol{E}(\omega) + \frac{3}{4}\chi^{(3)}(-\omega;\omega,-\omega,\omega)\boldsymbol{E}(\omega)\boldsymbol{E}^*(\omega)\boldsymbol{E}(\omega) \right]. \end{aligned} \tag{2.97}$$

P will be shown in the next chapter:

$$\boldsymbol{P}(\omega) = (\varepsilon - \varepsilon_0)\boldsymbol{E}(\omega) = \varepsilon_0(\varepsilon_r(\omega) - 1)\, \boldsymbol{E}(\omega) \tag{3.23}$$

Compare (3.23) with (2.97)

$$\varepsilon_r(\omega) = 1 + \chi^{(1)}(-\omega;\omega) + (3/4)\chi^{(3)}(-\omega;\omega,-\omega,\omega)|\boldsymbol{E}|^2 \tag{2.98}$$

$$\begin{aligned} [Re\varepsilon_r(\omega)]^{1/2} &= 1 + (1/2)Re\chi^{(1)}(-\omega;\omega) \\ &\quad + (1/2)(3/4)Re\chi^{(3)}(-\omega;\omega,-\omega,\omega)|E|^2 - \end{aligned} \tag{2.99}$$

$$[Re\varepsilon_L(\omega)]^{1/2} = 1 + (1/2)Re\chi^{(1)}(-\omega;\omega).$$

According to (3.24) in the next chapter, linear refractive index $n_0(\omega)$ is given by

$$\begin{aligned} n_0(\omega) &= [Re\varepsilon_L(\omega)]^{1/2}, \\ &= [\,1 + Re\chi^{(1)}(-\omega;\omega)]^{1/2} \\ &= 1 + (1/2)Re\chi^{(1)}(-\omega;\omega) - \cdots\,. \end{aligned} \tag{2.100}$$

The total refractive index is the sum of a linear term and a nonlinear term

$$n(\omega) \equiv [Re\varepsilon_r(\omega)]^{1/2} \equiv n_0(\omega) + \delta n, \tag{2.101}$$

then

$$\begin{aligned} \delta n(\omega) &\equiv n(\omega) - n_0(\omega) \\ &\cong \frac{1}{2}\cdot\frac{\frac{3}{4}Re\chi^{(3)}(-\omega;\omega,-\omega,\omega)|E|^2}{n_0(\omega)} \\ &\equiv n_2 \cdot |E|^2 \end{aligned} \tag{2.102}$$

where we take the binomial expansion of (2.100) and (2.101) with the assumption that $\delta n(\omega) << n_0(\omega)$ is satisfied, if the nonlinear effect is the only perturbation. Then, in a Kerr medium, the refractive indices are

$$n(\omega) = \sqrt{Re\varepsilon_r(\omega)} \approx n_0(\omega) + n_2(\omega)|\mathrm{E}|^2 \tag{2.103}$$

$$n_0(\omega) = \sqrt{1 + Re\chi^{(1)}(-\omega;\omega)} \approx 1 + \tfrac{1}{2}Re\chi^{(1)}(-\omega;\omega) - \cdots \tag{2.104}$$

$$n_2(\omega) = \frac{3}{8} \frac{\text{Re}\,\chi^{(3)}(-\omega;\omega,-\omega,\omega)}{n_0(\omega)} . \tag{2.105}$$

In the optical fiber communications, lightwave signals propagate in the transparent and extremely low loss region of the fiber, so the imaginary part of $\chi^{(3)}(\omega)$ is zero and $Re\chi^{(3)}(\omega) = \chi^{(3)}(\omega)$. (The n_2 of silica fiber is 3.2×10^{-16} cm^2/W.) Also, all the fields are polarized in the x direction, then the nonlinear refractive index coefficient $n_2(\omega)$ is

$$n_2 = \frac{3}{8n_0}\chi^{(3)}_{xxxx} = \frac{3}{8n_0}\chi^{(3)} \tag{2.106}$$

The fact that the intensity-induced refractive index is proportional to intensity I ($\delta n \propto n_2|E|^2 \propto n_2 I$) is extremely important, as it is linked with many nonlinear effects in the fibers, such as SPM and XPM effects.

Example

If the known terms are $\chi_{\text{fiber}}^{(3)} = \chi^{(3)}{}_{xxxx} = 1.28\times10^{-19}$m^2/W, and $n_0 = 1.5$, then, we find the nonlinear refractive-index coefficient $n_2 = 3.2\times10^{-20}$m^2/W.

SPM is the self-induced phase-shift φ_{SPM} stemming from the intensity-induced refractive index shift by the lightwave signal itself,

$$\varphi_{SPM}(\omega) = n(\omega)k_0(\omega)L = \left[\ n_0(\omega) + n_2(\omega)|E_s|^2\right] k_0(\omega)L . \tag{2.107}$$

Cross-phase-modulation (XPM) stems from the field intensity-dependent mutual shift of the refractive indices of the interacting waves of different wavelengths. If two waves of ω_1, ω_2 copropagate inside the same fiber, the nonlinear phase shifts φ_{XPM} of ω_1 and ω_2 are given by (2.114) and (2.115):

$$\varphi_{\text{XPM1}} = n_2(\omega_1)k_0(\omega_1)L(|E_1(\omega_1)|^2 + 2|E_2(\omega_2)|^2) \tag{2.108}$$

$$\varphi_{\text{XPM2}} = n_2(\omega_2)k_0(\omega_2)L(|E_2(\omega_2)|^2 + 2|E_1(\omega_1)|^2) . \tag{2.109}$$

Example

In silica fiber, $n_2 = 3.2\times10^{-20}$m^2/W. If $n_0(\omega) = 1.5$, then, with $\lambda = 1.55\mu$m, $\chi_{\text{fiber}}^{(3)} = 1.28\times10^{-19}$m^2/W.

2.6.2 The Kramers-Kronig Relations

From (2.25) and (2.26), both the real and imaginary parts are functions of frequency. For a given material, they are all dependent on the same material properties, such as n, ω_0, and ζ. Consequently they relate to each other. According to the theory of complex variable functions, if a function is satisfied on causality and is finite, the relation between the real and imaginary parts can be calculated. In this case, at a time t, the polarization P is determined not only by the interaction of $E(t)$ at t, but also by the resultant field interaction cumulative from τ to t, where τ is the time at which the field begins to interact (Fig. 2.4). Thus, from the responsibility of the medium, the polarization is

$$P(t) = \varepsilon_0 \int_0^{\infty} f(\tau) E(t-\tau) d\tau \tag{2.110}$$

where response $f(\tau)$ is a time function. For an arbitrary time τ, the response function is certainly a finite value, which depends on the medium properties. The polarization P and field E expressed in Fourier integrals are:

$$P(t) = \frac{1}{2\pi} \int_{-\infty}^{+\infty} P(\omega') e^{i\omega' t} d\omega' \tag{2.111}$$

$$E(t) = \frac{1}{2\pi} \int_{-\infty}^{+\infty} E(\omega') e^{i\omega' t} d\omega' . \tag{2.112}$$

Substitution into (2.110) yields

$$P(\omega') = \varepsilon_0 \chi(\omega') E(\omega') \tag{2.113}$$

where $\chi(\omega')$ is defined as the medium susceptibility,

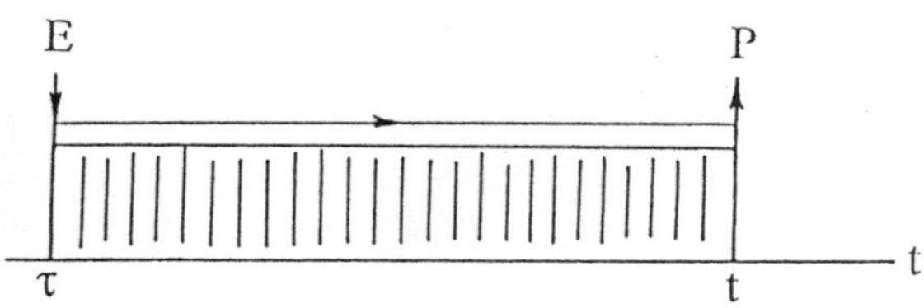

Fig. 2.4 The polarization $P(t)$ determined by the resultant interaction of the time accumulated from τ to t.

$$\chi(\omega') = \int_0^\infty f(\tau) e^{i\omega'\tau} d\tau \,. \tag{2.114}$$

$\chi(\omega')$ is a complex function and can be expressed as

$$\chi(\omega') = \chi'(\omega') + i\chi''(\omega'). \tag{2.115}$$

From (2.114),

$$\chi(-\omega') = \chi^*(\omega'). \tag{2.116}$$

Equation (2.116) implies that $\chi'(\omega')$ is an even function, and $\chi''(\omega')$ is an odd function. The relation between $\chi'(\omega')$ and $\chi''(\omega')$ can be deduced from the fundamentals of complex variable functions. Take ω' as a complex $\omega'=\omega_1' + i\omega_2'$. From (2.114), in the lower half of the complex plane, $\chi(\omega')$ is a single-valued and finite function with no singularities. $f(\tau)$ is also a finite function in the integration range, consequently the integral is convergent and finite.

We deduce the meaning of the relations of $\chi'(\omega')$ and $\chi''(\omega')$. We select a real value of $\omega' = \omega$ and find the contour integral of the function $(\chi/(\omega' - \omega))$ round a closed contour consisting of the real axis indented by a small semicircle which is centered on ω in the lower half-plane. Then the contour integral is closed in the lower half-plane by a large semicircle centered on the origin, as shown in Fig. 2.5. Since this contour encloses no singularities, its integral vanishes. When $|\omega| \to \infty$, $\chi \to 0$, and $(\chi/(\omega' - \omega))$ decreases to zero more quickly than does $1/\omega'$, hence the integral $\int_{c'}(\chi/(\omega' - \omega))d\omega'$ is convergent. The contribution from the clockwise large semicircle vanishes in the limit as the radius of the semicircle tends to infinity. According to the Cauchy formula, the contribution from the counter-clockwise integral around the infinitely small semicircle equals to $i\pi\chi(\omega)$. Thus, the integral taken to lie along the real axis is (2.117)

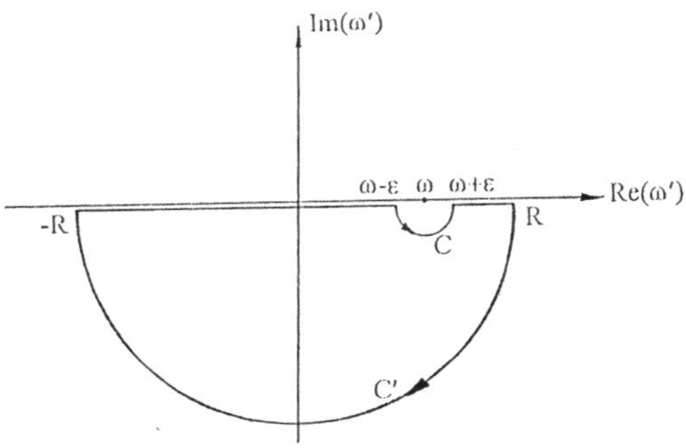

Fig. 2.5 The integral path of the Kramers-Kronig Relations.

$$\lim_{\varepsilon\to 0}\left[\int_{-\infty}^{\omega-\varepsilon}\frac{\chi}{\omega'-\omega}d\omega' + \int_{\omega+\varepsilon}^{\infty}\frac{\chi}{\omega'-\omega}d\omega'\right] + i\pi\chi(\omega) = 0\,. \tag{2.117}$$

The first part is the Cauchy principal value, written as

$$P.V.\int_{-\infty}^{+\infty}\frac{\chi}{\omega'-\omega}d\omega' \equiv \lim_{\varepsilon\to 0}\left[\int_{-\infty}^{\omega-\varepsilon}\frac{\chi}{\omega'-\omega}d\omega' + \int_{\omega+\varepsilon}^{\infty}\frac{\chi}{\omega'-\omega}d\omega'\right], \tag{2.118}$$

and so

$$\chi(\omega) = \frac{i}{\pi}P.V.\int_{-\infty}^{+\infty}\frac{\chi}{\omega'-\omega}d\omega'\,. \tag{2.119}$$

According to (2.115), express $\chi(\omega')$ and $\chi(\omega)$ in complex form, and express (2.119) as a real and an imaginary part, giving (2.120) and (2.121). These are the Kramers-Kronig Relations, where $P.V.$ is the Cauchy principal value. These relations are very important throughout all aspects of optical engineering. The real and imaginary parts of linear susceptibility are mutually dependent. If the absorption $\alpha(\omega)$ is known, the refractive index $n(\omega)$ can be found, and vice versa. Equations (2.120) and (2.121) are:

$$\chi'(\omega) = \frac{1}{\pi}P.V.\int_{-\infty}^{+\infty}\frac{\chi''(\omega')}{\omega'-\omega}d\omega' \tag{2.120}$$

$$\chi''(\omega) = -\frac{1}{\pi}P.V.\int_{-\infty}^{+\infty}\frac{\chi'(\omega')}{\omega'-\omega}d\omega'\,. \tag{2.121}$$

Using the properties of (2.116), $\chi'(\omega)$ is an even function and $\chi''(\omega)$ is an odd function. Then by multiplying the factor $(\omega' + \omega)/(\omega' + \omega)$ by (2.120) and (2.121), the Kramers-Kronig Relations become (2.1220 and (2.123):

$$\chi'(\omega) = \frac{2}{\pi}P.V.\int_{0}^{+\infty}\frac{\omega'\chi''(\omega')}{\omega'^2-\omega^2}d\omega' \tag{2.122}$$

$$\chi''(\omega) = -\frac{2\omega}{\pi}P.V.\int_{0}^{+\infty}\frac{\chi'(\omega')}{\omega'^2-\omega^2}d\omega'\,. \tag{2.123}$$

In more general cases, the representation of $\chi(\omega')$ is always replaced by the $F(\omega')$ which is an analytic function of a complex frequency ω'. Then (2.122) and (2.123) become (2.124) and (2.125). Equations (2.120) and (2.121), or (2.122) and (2.123) are the well known Kramers-Kronig Relations, simplified as K-K Relations. This is a fundamental relation not only in nonlinear photonics, but also in optical science and engineering.

$$\operatorname{Re} F(\omega) = \frac{2}{\pi} P.V. \int_0^{+\infty} \frac{\omega' \operatorname{Im} F(\omega')}{\omega'^2 - \omega^2} d\omega' \tag{2.124}$$

$$\operatorname{Im} F(\omega) = -\frac{2\omega}{\pi} P.V. \int_0^{+\infty} \frac{\operatorname{Re} F(\omega')}{\omega'^2 - \omega^2} d\omega' . \tag{2.125}$$

When $F(\omega') \to 0$ and $|\omega'| \to \infty$, all poles lie in the upper half-plane.

2.6.3 Nonlinear Susceptibility and Light Induced Nonlinear Refractive Index in Semiconductors

In nonlinear photonics, K-K Relations can be used in a regime in which the function $F(\omega')$ is itself a function of the optical intensity I. Thus, for example, an optical field of intensity I may cause the absorption coefficient $\alpha(\omega,I)$ to differ from its linear value $\alpha(\omega,0)$ by an induced amount $\Delta\alpha(\omega,I)$. An example of such a change is the bleaching of interband absorption in semiconductors. As a result of this induced change, the refractive index $n(\omega,I)$ will differ by the refractive index change $\Delta n(\omega,I)$ from its linear value. By substituting

$$F(\omega) = \alpha(\omega,0) + \Delta\alpha(\omega,I) - i\omega[n(\omega,0) + \Delta n(\omega,I)] . \tag{2.126}$$

In the linear cases, the absorption coefficient $\alpha(\omega,0)$ or the gain coefficient $g(\omega,0)$ is the counterpart of $\chi''(\omega)$. When the absorption $\alpha(\omega,0)$ or gain $g(\omega,0)$ of semiconductor materials is known, mostly by experiments, then the Kramers-Kronig Relations are always used to find the linear $n(\omega,0)$.

$$\begin{aligned} n(\omega,0) &= \frac{1}{\pi}\int_0^{\infty} \frac{\alpha(\omega',0)}{\omega'^2 - \omega^2} d\omega' \\ or \quad &= \frac{1}{\pi}\int_0^{\infty} \frac{g(\omega',0)}{\omega'^2 - \omega^2} d\omega'. \end{aligned} \tag{2.127}$$

In the nonlinear cases, $\alpha(\omega,I)$ or $g(\omega,I)$ is always known from experimental data. There is a perturbed change $\Delta\alpha_{NL}(\omega,I)$ or $\Delta g_{NL}(\omega,I)$ caused by photonic fields I, we have

$$\Delta\alpha_{NL}(\omega,I) = \alpha(\omega,I) - \alpha(\omega,0). \tag{2.128}$$

Then we will have related change of refractive index $\Delta n(\omega,I)$ from the measured values of the absorption change $\Delta\alpha(\omega,I)$ in semiconductors, in which the absorption of the carriers, such as electrons or electron-hole pairs, is a dominant process coupled with the parameters of some opto-electronic devices.

$$\Delta n_{\mathrm{NL}}(\omega, I) = \frac{1}{\pi} \int_0^\infty \frac{\Delta \alpha_{\mathrm{NL}}(\omega', I)}{\omega'^2 - \omega^2} d\omega' . \tag{2.129}$$

Thus in the nonlinear cases, the refractive index $n(\omega,I)$ of the semi-conductors is

$$n(\omega,I) = n(\omega,0) + \Delta n_{\mathrm{NL}}(\omega,I) \tag{2.130}$$

where I is the optical field of intensity and it is always omitted in the following chapters.

References

[2.1] N. Bloembergen: *Nonlinear Optics,* Benjamin, New York (1965)

[2.2] Y.R. Shen: *The Principles of Nonlinear Optics*, John Wiley, New York (1984)

[2.3] P.N. Butcher and D. Cotter: *The Elements of Nonlinear Optics*, Cambridge University Press, New York (1990)

[2.4] A.C. Newell and J.V. Moloney: *Nonlinear Optics*, Addision-Wesley, Redwood City, CA (1992)

[2.5] C. Flytzanis: Theory of nonlinear optical susceptibilities. In: H. Rabin and C.L. Tang (eds.): *Quantum Electronics: A Treatise*, Vol. 1, Academic Press, New York (1975)

[2.6] Qun Wu, Yili Guo and Bingkun Zhou: Investigation of the simplified diagrammatic representation of high-order nonlinear optical susceptibility, *Acta Optica Sinica* **6**, 339–347 (1986)

3. Wave Equations in Nonlinear Photonics

In nonlinear cases, when intense lightwaves propagate in and interact with the optical medium, the waves' properties are quite different from those in linear cases. These waves couple to each other, that is, a wave's nonlinear characteristics are determined not only by the medium and the wave itself, but also by the other waves propagating simultaneously in the medium. The physical quantity linked with all the waves is the susceptibility tensor $\chi_{\mu\alpha\beta\gamma\ldots\xi}{}^{(n)}(-\omega_{n+1};\omega_1,\omega_2,\omega_3,\ldots,\omega_n)$ discussed in Chap. 2. Through this tensor, other waves contribute to and influence a given wave.

In nonlinear photonics, nonlinear effects in optical fibers, semiconductors, and quantum wells are of great interest in optical communications and optoelectronics. The wave equations and the coupled-wave equations in these media derive from the same Maxwell's equations and will be developed below in Sects. 3.1 through 3.3.

The theory of polarization $\boldsymbol{P}^{(n)}(\omega)$, susceptibility $\chi^{(n)}(\omega)$, and K-K Relations in these media have been discussed in the unified medium equations developed in Chap. 2. The resulting wave equations reveal that the propagation characteristics are determined also by the phase relations between these waves. These are the phase-matching conditions in the parametric process, and will be discussed in Chap. 4.

3.1 Electromagnetic Wave Propagation Equations in Nonlinear Photonics

The unified theory of electric and magnetic fields was first developed by Maxwell in the 1860s. His equations were different forms of Gauss's, Faraday's, and Ampere's laws. These equations in vector form are:

$$\nabla \times \boldsymbol{E} = -\frac{\partial \boldsymbol{B}}{\partial t} \tag{3.1}$$

$$\nabla \times \boldsymbol{H} = \frac{\partial \boldsymbol{D}}{\partial t} + \boldsymbol{J} \tag{3.2}$$

$$\nabla \cdot \boldsymbol{D} = \rho \tag{3.3}$$

$$\nabla \cdot \boldsymbol{B} = 0 \tag{3.4}$$

where $\boldsymbol{J}$ is the conductive current density, ρ is the free charge density in medium in the weak field case. The electric and magnetic properties of medium can be written as:

$$\boldsymbol{J} = \sigma \boldsymbol{E} \tag{3.5}$$

$$\boldsymbol{D} = \varepsilon \boldsymbol{E} = \varepsilon_0 \boldsymbol{E} + \boldsymbol{P} \tag{3.6}$$

$$\boldsymbol{B} = \mu \boldsymbol{H} = \mu_0 \boldsymbol{H} + \boldsymbol{M} \tag{3.7}$$

where σ, ε, and μ are the conductivity, dielectric, and magneto-conductivity constants respectively. In the anisotropic medium they are all tensors, and in the isotropic medium they are scalars. $\boldsymbol{P}$ is the electric polarization intensity and $\boldsymbol{M}$ is the magneto-polarization intensity. In the linear case, it can be supposed that σ, ε, and μ are independent of $\boldsymbol{E}$ and $\boldsymbol{H}$. But if the fields are strong, σ, ε, and μ are dependent on $\boldsymbol{E}$. This is the domain of nonlinear photonics. Most cases in nonlinear photonics use non-magnetic materials, in which the conductive current $\boldsymbol{J}$ and the free charge density ρ are always zero. Hence, in a non-magnetic medium, without free current and free space charge, the above equations become (3.8–3.13):

$$\nabla \times \boldsymbol{E} = -\frac{\partial \boldsymbol{B}}{\partial t} \tag{3.8}$$

$$\nabla \times \boldsymbol{H} = -\frac{\partial \boldsymbol{D}}{\partial t} \tag{3.9}$$

$$\nabla \cdot \boldsymbol{D} = 0 \tag{3.10}$$

$$\nabla \cdot \boldsymbol{B} = 0 \tag{3.11}$$

$$\boldsymbol{B} = \mu_0 \boldsymbol{H} \tag{3.12}$$

$$\boldsymbol{D} = \varepsilon \boldsymbol{E} = \varepsilon_0 \boldsymbol{E} + \boldsymbol{P}\,. \tag{3.13}$$

3.1.1 Wave Equations in the Time-Space Domain

The electric field vector $\boldsymbol{E}(r,t)$ and polarization intensity vector $\boldsymbol{P}(r,t)$ are generally expressed in the complex form

$$\boldsymbol{E}(r,t) = \frac{1}{2}[\boldsymbol{E}(r)e^{-i\omega t} + \boldsymbol{E}^*(r)e^{+i\omega t}] \tag{3.14}$$

where

$$\boldsymbol{E}(\boldsymbol{r}) = \widetilde{\boldsymbol{E}}(\boldsymbol{r})e^{i[k\cdot r+\varphi(r)]} \tag{3.15}$$

and $\tilde{\boldsymbol{E}}$ is the real amplitude. Similarly, the expression for the polarization is

$$\boldsymbol{P}(r,t) = \tfrac{1}{2}[\boldsymbol{P}(r)e^{-i\omega t} + \boldsymbol{P}^*(r)e^{+i\omega t}] \tag{3.16}$$

or

$$\boldsymbol{P}(r,t) = \boldsymbol{P}^{(1)}(r,t) + \boldsymbol{P}^{(2)}(r,t) + \boldsymbol{P}^{(3)}(r,t) + \ldots = \boldsymbol{P}_{\mathrm{L}}(r,t) + \boldsymbol{P}_{\mathrm{NL}}(r,t) \tag{3.17}$$

as in Chap. 2.

From Maxwell's equations and the equations describing the medium's electromagnetic properties, it is easy to deduce the wave propagation equations. First, apply curl to both sides of (3.8) to obtain

$$\nabla\times\nabla\times\boldsymbol{E}(t) = -\frac{\partial}{\partial t}(\nabla\times\boldsymbol{B}) \tag{3.18}$$

where $\partial/\partial t$ and $\partial/\partial r$ are exchanged. Substituting (3.9) and (3.12) into (3.18) leads to the wave equation for a source-free and non-magnetic medium in real, time-dependent form

$$\nabla\times\nabla\times\boldsymbol{E}(t) = -\mu_0\frac{\partial^2\boldsymbol{D}(t)}{\partial t^2}. \tag{3.19}$$

Substituting (3.13) into (3.19), we have

$$\nabla\times\nabla\times\boldsymbol{E}(t) + \frac{1}{c^2}\frac{\partial^2\boldsymbol{E}(t)}{\partial t^2} = -\mu_0\frac{\partial^2\boldsymbol{P}(t)}{\partial t^2}. \tag{3.20}$$

Suppose $\nabla\cdot\boldsymbol{E} = 0$. Then we have the equation of electromagnetic wave propagation in nonlinear photonics in the time-spatial domain in vector form:

$$\nabla^2\boldsymbol{E} - \frac{1}{c^2}\frac{\partial^2\boldsymbol{E}}{\partial t^2} = \mu_0\left(\frac{\partial^2\boldsymbol{P}_{\mathrm{L}}}{\partial t^2} + \frac{\partial^2\boldsymbol{P}_{\mathrm{NL}}}{\partial t^2}\right). \tag{3.21}$$

3.1.2 Wave Equations in the Frequency Domain

In nonlinear photonics, numerous effects are discussed in the frequency domain, since most parameters have dispersion. By expressing $\boldsymbol{E}(t)$ and $\boldsymbol{P}(t)$ in (3.20) in terms of their Fourier transforms, we obtain a vector wave equation for the frequency ω in a non-magnetic medium without current and free charge

$$\nabla \times \nabla \times \boldsymbol{E}(\omega) = \frac{\omega^2}{c^2}\boldsymbol{E}(\omega) + \omega^2 \mu_0 \left[\boldsymbol{P}_{\mathrm{L}}(\omega) + \boldsymbol{P}_{\mathrm{NL}}(\omega)\right]. \tag{3.22}$$

From (3.13), we obtain the basic form of the polarization $\boldsymbol{P}$,

$$\boldsymbol{P}(\omega) = \boldsymbol{P}_{\mathrm{L}}(\omega) + \boldsymbol{P}_{\mathrm{NL}}(\omega) = (\varepsilon - \varepsilon)_0 \cdot \boldsymbol{E} = \varepsilon_0(\varepsilon_r(\omega) - 1) \cdot \boldsymbol{E} \tag{3.23}$$

and in the linear regime,

$$\boldsymbol{P}_{\mathrm{L}}(\omega) = (\varepsilon - \varepsilon)_0 \cdot \boldsymbol{E} = \varepsilon_0(\varepsilon_{\mathrm{L}}(\omega) - 1) \cdot \boldsymbol{E}. \tag{3.24}$$

Substituting $\boldsymbol{P}^{(1)}(\omega) = \varepsilon_0 \chi^{(1)}(-\omega;\omega)\boldsymbol{E}(\omega)$ into (3.22) yields

$$\nabla \times \nabla \times \boldsymbol{E}(\omega) = \frac{\omega^2}{c^2}\varepsilon(\omega) \cdot \boldsymbol{E}(\omega).$$

The possible solutions are the running waves in the quasi-plane wave approximation:

$$\boldsymbol{E}(\omega) = \widetilde{\boldsymbol{E}}(\omega) e^{iK \cdot r}. \tag{3.25}$$

We turn to the nonlinear regime, in which, from (3.17), $P(\omega)$ is always in two terms:

$$\boldsymbol{P}(\omega) = \boldsymbol{P}^{(1)}(\omega) + \boldsymbol{P}^{(2)}(\omega) + \boldsymbol{P}^{(3)}(\omega) + \ldots \tag{3.26}$$

$$= \boldsymbol{P}_{\mathrm{L}}(\omega) + \boldsymbol{P}_{\mathrm{NL}}(\omega). \tag{3.27}$$

Substituting $\boldsymbol{P}(\omega) = \boldsymbol{P}_{\mathrm{L}}(\omega) + \boldsymbol{P}_{\mathrm{NL}}(\omega)$ in (3.22) yields

$$\nabla \times \nabla \times \boldsymbol{E}(\omega) = \frac{\omega^2}{c^2}\varepsilon(\omega) \cdot \boldsymbol{E}(\omega) + \omega^2 \mu_0 \boldsymbol{P}_{\mathrm{NL}}(\omega). \tag{3.28}$$

Inserting the electric field (3.25) into (3.28), we have

$$\begin{aligned} \nabla \times \nabla \times \left[\widetilde{\boldsymbol{E}}(\omega) \exp(iK \cdot r) \right] = & \frac{\omega^2}{c^2}\varepsilon_{\mathrm{L}}(\omega) \cdot \widetilde{\boldsymbol{E}}(\omega) \exp(iK \cdot r) \\ & + \omega^2 \mu_0 \boldsymbol{P}_{\mathrm{NL}}(\omega) \end{aligned} \tag{3.29}$$

where $\boldsymbol{E}(\omega)$ is the electric field vector in the frequency domain. $\tilde{\boldsymbol{E}}(\omega)$ is the complex envelope function of the vector $\boldsymbol{E}(\omega)$ including both the amplitude and phase performance and generally is a function of coordinates. k is the wave vector $\{\omega[-(\omega)+i\kappa(\omega)]/c\}\cdot\boldsymbol{u}$, $\boldsymbol{u}$ is a real unit vector in the propagation direction, and $-(\omega)$ and $\kappa(\omega)$ are the real and imaginary refractive indices respectively. Equations (3.28) and (3.29) are fundamental in our further discussions. The equations of the wave properties and medium responsibilities discussed here are all defined as averages in a small volume, the dimension d of which must be much smaller than the optical wavelength λ.

If λ = 1000nm, d may be 100nm or so, and a large amount of atoms or particles may be in it.

3.2 Slowly Varying Amplitude and Nearly Planar Wave Approximations

Equation (3.29) is a nonlinear differential equation with a second derivative of $\tilde{\boldsymbol{E}}(\omega)$. To solve it analytically is extremely difficult, and it is in most cases solved numerically by computer. In this section, some assumptions and approximations are introduced to simplify (3.29) and to find the analytical solutions. Such approaches are convenient for those practical problems for which these approximations are satisfied. Even for conditions far different from these approximations, the analytical results deduced here still offer some useful physical concepts.

First, we assume a quasi-monochromatic planar wave $\boldsymbol{E}(\omega,z)$ propagating along z only. For writing convenience, we shall leave out henceforth the z~x on the top of $\boldsymbol{E}$ and write $\boldsymbol{K}$ as $\boldsymbol{k}$. Then $\boldsymbol{k}\cdot\boldsymbol{r} = \pm kz$ (z for a forward-traveling wave, and $-z$ for a backward-traveling wave).

Usually the waves discussed consist of a linear combination of planar waves, whose wave vectors and frequencies lie in a narrow band, in the order of $0 < \Delta k/k_0 << 1$ and $0 < \Delta\omega/\omega_0 << 1$ about the central value k_0 and ω_0. Continuous waves and pulse waves always satisfy these conditions. For pulse width τ = 100fs, the corresponding bandwidth is about $\Delta\omega \cong 10^{13}$/s, much smaller than the central frequency $\omega_0 = 10^{15}$/s. Thus $\Delta\omega/\omega_0 = 0.01$.

In isotropic, cubically symmetric and birefringent media, the field unit vibration $\boldsymbol{e}$ and wave propagation vector $\boldsymbol{u}$ are orthogonal ($\boldsymbol{e}\cdot\boldsymbol{u}$ = 0). This allows us to replace $\nabla\times\nabla\times\boldsymbol{E}$ in (3.21), (3.22) and (3.29) by $-(\partial^2/\partial z^2)\boldsymbol{E}$, where $\boldsymbol{E}$ denotes either $\boldsymbol{E}(t)$ in (3.21) or its Fourier transform $\boldsymbol{E}(\omega)$ in (3.22).

Second, a well-known approximation of the slowly-varying envelope is superimposed on these equations. We suppose that the variations of complex amplitude $\boldsymbol{E}(\omega)$, both in real amplitude and phase, and of $\partial\boldsymbol{E}(\omega)/\partial z$ and its derivative $(\partial^2/\partial z^2)\boldsymbol{E}(\omega)$, are sufficiently small with the distance z, being comparable with λ. This slowly-varying amplitude approximation can be represented by

$$\left|\frac{\partial^2}{\partial z^2}\boldsymbol{E}(\omega,z)\right| << \left|k\frac{\partial}{\partial z}\boldsymbol{E}(\omega,z)\right| << \left|k^2\boldsymbol{E}(\omega,z)\right| . \tag{3.30}$$

Another version of the explanation is that, since $k = 2\pi/\lambda$, then

$$\begin{aligned}\frac{\partial}{\partial z}\boldsymbol{E}(\omega) &= \frac{i\omega^2\mu_0}{2k}\boldsymbol{P}_{\mathrm{NL}}(\omega)\exp(-ikz)\\ &= \frac{i\omega}{2cn\varepsilon_0}\boldsymbol{P}_{\mathrm{NL}}(\omega)\exp(-ikz).\end{aligned} \tag{3.31}$$

The slowly-varying amplitude implies the propagating directon can not vary rapidly, thus the oppositely propagating field component generated by $\boldsymbol{P}_{\mathrm{NL}}$ can be neglected. (3.31) is equivalent to

$$|\,\lambda\cdot(\boldsymbol{E}''/\boldsymbol{E}')\,| \ll 1, \quad \text{and} \quad |\lambda\cdot(\boldsymbol{E}'/\boldsymbol{E})| \ll 1. \tag{3.32}$$

Thus within λ, the changes of electric field $|\partial \boldsymbol{E}/\partial z|$ and its derivative $|\partial^2\boldsymbol{E}/\partial z^2|$ are very small. This approximation implies that the second derivative $|\partial^2\boldsymbol{E}/\partial z^2|$ may be neglected with respect to the small distance z. Thus, for a forward-traveling wave, where $k = -\omega/c$, and with the quasi-planar wave and slowly-varying envelope approximations, we have

$$\begin{aligned}\frac{\partial}{\partial z}\boldsymbol{E}(\omega,z) &= \frac{i\omega^2\mu_0}{2k}\cdot\boldsymbol{P}_{\mathrm{NL}}(\omega)e^{(-ikz)}\\ &= \frac{i\omega}{2cn\varepsilon_0}\cdot\boldsymbol{P}_{\mathrm{NL}}(\omega)e^{(-ikz)}.\end{aligned} \tag{3.33}$$

Thus, the second-order differential equation (3.29) reduces to a simpler first-order differential equation.

3.3 Coupled-Wave Equations in Multiwave Interactions

In nonlinear photonics, many nonlinear effects are related to the monochromatic planar multiwaves which interact with each other in the medium. In physics, coupling between multiwaves is due to the susceptibility $\chi^{(n)}(\omega)$. Mathematically, these coupling processes are described by the coupled-wave equations.

3.3.1 Coupled-Wave Equations in Three-Wave Interactions

The coupled three-wave equations are one of the fundamental equation groups and are linked with the lowest order of nonlinear effects. Some important nonlinear techniques, such as optical second harmonic generation (SHG), optical sum frequency and difference frequency, optical parametric oscillators (OPO) and amplifiers (OPA), are all described by these coupled three-wave equations.

When $\boldsymbol{E}_1(\omega_1)$ and $\boldsymbol{E}_2(\omega_2)$ interact simultaneously with the medium, the nonlinear polarization $\boldsymbol{P}^{(2)}(\omega_3)$ is induced through the nonlinear susceptibility $\chi^{(2)}(\omega_3)$, which will radiate the third radiation field $\boldsymbol{E}_3(\omega_3)$. So, all three waves $\boldsymbol{E}_1(\omega_1)$, $\boldsymbol{E}_2(\omega_2)$ and $\boldsymbol{E}_3(\omega_3)$, including their amplitudes, frequencies, and phases, are mutually coupled through the susceptibility $\chi^{(2)}(-\omega_3; \omega_1, \omega_2)$. In the parametric processes, all the frequencies and their combinations are far from the intrinsic resonance ω_0 of the medium, and both the energy and momentum are conserved.

The resulting polarization intensity $\boldsymbol{P}_3(\omega_3)$ will be related to one of the following processes, such as $\omega_3 = \omega_1+\omega_2$, $\omega_3 = \omega_1-\omega_2$, $\omega_3 = 2\omega_1$, and $\omega_3 = 2\omega_2$, or others. The resultant field $\boldsymbol{E}_3(\omega_3)$ radiated by $\boldsymbol{P}_3(\omega_3)$ will also include one of these. If $\omega_3 = \omega_1+\omega_2$ is expected, then the phase-matching condition $k_1+k_2-k_3 = 0$ must occur. If the orientation of the crystal or semiconductor is carefully arranged to keep a certain spatial relation between the optical field vector and the crystal principal axis, the sum-frequency process can be enhanced, while other processes are weakened, since the phases are far from the matching conditions. From (2.60) and (2.80) which govern the polarization $\boldsymbol{P}$,

$$\boldsymbol{P}^{(2)}(\omega_3) = \varepsilon_0\chi^{(2)}(-\omega_3;\omega_1,\omega_2){:}\boldsymbol{E}(\omega_1)\boldsymbol{E}(\omega_2).$$

In the steady state, the field-interactive times are much longer than the nonlinear effect constructive time. We have three polarization waves for the nonlinear sum-frequency $\omega_3 = \omega_1+\omega_2$, $(\omega_1 = \omega_3-\omega_2,\ \omega_2 = \omega_3-\omega_1)$.

$$\boldsymbol{P}^{(2)}(\omega_1,z) = \varepsilon_0\chi^{(2)}(-\omega_1;\omega_3,-\omega_2){:}\boldsymbol{E}(\omega_3,z)\boldsymbol{E}^*(\omega_2,z) \tag{3.34}$$

$$\boldsymbol{P}^{(2)}(\omega_2,z) = \varepsilon_0\chi^{(2)}(-\omega_2;\omega_3,-\omega_1){:}\boldsymbol{E}(\omega_3,z)\boldsymbol{E}^*(\omega_2,z) \tag{3.35}$$

$$\boldsymbol{P}^{(2)}(\omega_3,z) = \varepsilon_0\chi^{(2)}(-\omega_3;\omega_1,\omega_2){:}\boldsymbol{E}(\omega_1,z)\boldsymbol{E}(\omega_2,z) \tag{3.36}$$

where $\boldsymbol{E}(\omega_1,z)$, $\boldsymbol{E}(\omega_2,z)$, and $\boldsymbol{E}(\omega_3,z)$ are all quasi-monochromatic planar waves propagating in z, with t neglected due to the steady-state condition. $\chi^{(2)}(-\omega_3;\omega_1,\omega_2)$ is defined in (2.82) and the permutalion symmetry is as shown in (2.80). Substituting (3.34–3.36) into (3.33) results in the scalar complex amplitude equations, which are the three wave coupled equations of polarization (3.37–3.39):

$$\frac{\partial}{\partial z}E_1(\omega_1,z) = \frac{i\omega_1}{2cn_1}e(\omega_1)\cdot \boldsymbol{P}_{\mathrm{NL}}(\omega_1,z)e^{-i\Delta k_1 z} \tag{3.37}$$

$$\frac{\partial}{\partial z}E_2(\omega_2,z) = \frac{i\omega_2}{2cn_2}e(\omega_2)\cdot \boldsymbol{P}_{\mathrm{NL}}(\omega_2,z)e^{-i\Delta k_2 z} \tag{3.38}$$

$$\frac{\partial}{\partial z} E_3(\omega_3, z) = \frac{i\omega_3}{2cn_3} \boldsymbol{e}(\omega_3) \cdot \boldsymbol{P}_{\mathrm{NL}}(\omega_3, z) e^{+i\Delta k_3 z}. \tag{3.39}$$

Here, $\boldsymbol{P}_{\mathrm{NL}}(\omega_1,z)$ is the polarization wave with ω_1 radiated. This polarization $\boldsymbol{P}_{\mathrm{NL}}(\omega_1,z)$ is caused by the interaction of $\boldsymbol{E}(\omega_2,z)$ and $\boldsymbol{E}(\omega_3,z)$ with the medium, thus

$$\boldsymbol{P}_{\mathrm{NL}}(\omega_1,z) = \varepsilon_0 \chi^{(2)}(-\omega_1;\omega_3,-\omega_2):\boldsymbol{e}(\omega_3)\boldsymbol{e}(\omega_2)E(\omega_3,z)E^*(\omega_2,z)\, e^{i(k_3-k_2)z} \tag{3.40}$$

$$\boldsymbol{P}_{\mathrm{NL}}(\omega_2,z) = \varepsilon_0 \chi^{(2)}(-\omega_2;\omega_3,-\omega_1):\boldsymbol{e}(\omega_3)\boldsymbol{e}(\omega_2)E(\omega_3,z)E^*(\omega_1,z)\, e^{i(k_3-k_1)z} \tag{3.41}$$

$$\boldsymbol{P}_{\mathrm{NL}}(\omega_3,z) = \varepsilon_0 \chi^{(2)}(-\omega_3;\omega_1,\omega_2):\boldsymbol{e}(\omega_3)\boldsymbol{e}(\omega_2)E(\omega_1,z)E(\omega_2,z)\, e^{i(k_1+k_2)z}. \tag{3.42}$$

Substitute (3.40), (3.41) and (3.42) into (3.37), (3.38) and (3.39), respectively, and define a phase-mismatching parameter Δk.

$$\Delta k \equiv k_1 + k_2 - k_3 \tag{3.43}$$

$$-\Delta k \equiv k_3 - k_1 - k_2. \tag{3.44}$$

Then the scalar-amplitude equations (3.37–3.39) become (3.45–3.47):

$$\frac{\partial}{\partial z} E_1(\omega_1, z) = \frac{i\omega_1}{2cn_1}[\chi^{(2)}(-\omega_1;\omega_3,-\omega_2)\vdots\boldsymbol{e}(\omega_1)\boldsymbol{e}(\omega_2)\boldsymbol{e}(\omega_3)] \cdot E_3(z)E_2^*(z)e^{-i\Delta kz} \tag{3.45}$$

$$\frac{\partial}{\partial z} E_2(\omega_2, z) = \frac{i\omega_2}{2cn_2}[\chi^{(2)}(-\omega_2;\omega_3,-\omega_1)\vdots\boldsymbol{e}(\omega_1)\boldsymbol{e}(\omega_2)\boldsymbol{e}(\omega_3)] \cdot E_3(z)E_1^*(z)e^{-i\Delta kz} \tag{3.46}$$

$$\frac{\partial}{\partial z} E_3(\omega_3, z) = \frac{i\omega_3}{2cn_3}[\chi^{(2)}(-\omega_3;\omega_1,\omega_2)\vdots\boldsymbol{e}(\omega_1)\boldsymbol{e}(\omega_2)\boldsymbol{e}(\omega_3)] \cdot E_1(\mathrm{z})E_2(z)e^{-i\Delta kz} \tag{3.47}$$

where $\chi^{(2)}(-\omega_l;\omega_m,\omega_n)\vdots\boldsymbol{e}(\omega_m)\boldsymbol{e}(\omega_n)\boldsymbol{e}(\omega_l)$ is a scalar. In most cases, the scalar susceptibility is used instead of the tensor forms and we use $\chi^{(n)}(\omega)$ as the scalar susceptibility for convenience. Also we use the wave-vector $k_i = \omega_i n_i/c$ in these equations since it is quite important. Then, with scalar susceptibilities, we have the three wave coupled equations (3.48–3.50):

$$\frac{\partial}{\partial \mathrm{z}} E_1(\omega_1, z) = \frac{ik_1}{2n_1^2}\chi^{(2)}(-\omega_1;\omega_3,\omega_2)E_3(z)E_2^*(z)e^{-i\Delta kz} \tag{3.48}$$

$$\frac{\partial}{\partial z}E_2(\omega_2,z)=\frac{ik_2}{2n_2^2}\chi^{(2)}(-\omega_2;\omega_3,-\omega_1)E_3(z)E_1^*(z)e^{-i\Delta kz} \tag{3.49}$$

$$\frac{\partial}{\partial z}E_3(\omega_3,z)=\frac{ik_3}{2n_3^2}\chi^{(2)}(-\omega_3;\omega_1,\omega_2)E_1(z)E_2(z)e^{+i\Delta kz} \tag{3.50}$$

where $E_m(z)$ is the simplified form of $E(\omega_m,z)$. From the three wave coupled equations, the susceptibility $\chi^{(2)}$ reflects the coupling intensity between these waves and depends substantially on the intrinsic characteristics of medium.

These susceptibilities are equal in the lossless region of the medium, according to the overall permutation symmetry (2.80). Then for a given lossless medium, we have

$$\chi^{(2)}(-\omega_1;\omega_3,-\omega_2)=\chi^{(2)}(-\omega_2;\omega_3,-\omega_1)=\chi^{(2)}(-\omega_3;\omega_1,\omega_2)=\chi^{(2)}. \tag{3.51}$$

Thus (3.48), (3.49) and (3.50) can be simplified as the three wave coupled equations, where $k_i=\omega_i n_i/c$ is the wave-vector of $\boldsymbol{E}_i$

$$\frac{\partial}{\partial z}E_1(z)=\frac{ik_1}{2n_1^2}\chi^{(2)}E_3(z)E_2^*(z)e^{-i\Delta kz} \tag{3.52}$$

$$\frac{\partial}{\partial z}E_2(z)=\frac{ik_2}{2n_2^2}\chi^{(2)}E_3(z)E_1^*(z)e^{-i\Delta kz} \tag{3.53}$$

$$\frac{\partial}{\partial z}E_3(z)=\frac{ik_3}{2n_3^2}\chi^{(2)}E_1(z)E_2(z)e^{+i\Delta kz}. \tag{3.54}$$

The waves vary with z and are coupled through the identical susceptibility $\chi^{(2)}$. We shall solve these equations analytically for some nonlinear effects in the following chapters.

3.3.2 Coupled-Wave Equations in Four-Wave Interactions

Coupled four-wave equations are another fundamental group in nonlinear photonics. They are the mathematical and physical model corresponding to the third-order of nonlinear effects. Important nonlinear techniques, such as optical third-harmonic generation (THG), four-wave mixing (FWM), optical Kerr effects, SRS, SBS, optical self-focusing, and two-photon absorption (TPA), are all described by coupled four-wave equations.

Generally, the coupled four-wave equations describe the processes in which three injected fields $E_1(z)$, $E_2(z)$, and $E_3(z)$ interact simultaneously with the medium. Through the susceptibility $\chi^{(3)}(\omega)$, the optical polarization $P^{(3)}(\omega)$ is induced in the medium, in which the amplitude, frequency, and phase are all related to $\chi^{(3)}(\omega)$. The medium then radiates the fourth field $E_4(z)$.

Corresponding to (2.60), for the process $\omega_1+\omega_3 = \omega_2+\omega_4$, we have the third-order polarization $P^{(3)}(\omega_4)$:

$$\begin{aligned} \boldsymbol{P}^{(3)}(\omega_4) = \tfrac{3}{2}\varepsilon_0 \Big[& \chi^{(3)}(-\omega_4;\omega_1,-\omega_2,\omega_3)\vdots \boldsymbol{E}(\omega_1)\boldsymbol{E}^*(\omega_2)\boldsymbol{E}(\omega_3) \\ & + \chi^{(3)}(-\omega_4;\omega_1,-\omega_1,\omega_4)\vdots \boldsymbol{E}(\omega_1)\boldsymbol{E}^*(\omega_1)\boldsymbol{E}(\omega_4) \\ & + \chi^{(3)}(-\omega_4;\omega_2,-\omega_2,\omega_4)\vdots \boldsymbol{E}(\omega_2)\boldsymbol{E}^*(\omega_2)\boldsymbol{E}(\omega_4) \\ & + \chi^{(3)}(-\omega_4;\omega_3,-\omega_3,\omega_4)\vdots \boldsymbol{E}(\omega_3)\boldsymbol{E}^*(\omega_3)\boldsymbol{E}(\omega_4) \Big] \\ & + \tfrac{3}{4}\varepsilon_0 \chi^{(3)}(-\omega_4;\omega_4,-\omega_4,\omega_4)\vdots \boldsymbol{E}(\omega_4)\boldsymbol{E}^*(\omega_4)\boldsymbol{E}(\omega_4) \end{aligned} \tag{3.55}$$

where we choose the relation $\omega_1+\omega_3 = \omega_2+\omega_4$ as an example of non-degenerate four-wave mixing, and thus the phase-mismatching factor $\Delta k = k_1+k_3-k_2-k_4$. The polarizations of other frequencies $P^{(3)}(\omega_2)$, $P^{(3)}(\omega_3)$ and $P^{(3)}(\omega_1)$ can also be written in like manner.

Introducing the real unit vector $\boldsymbol{e}$ of the polarized field, susceptibility $\chi^{(3)}(\omega_1)$ can be written in the scalar form:

$$\boldsymbol{e}_4\chi^{(3)}(-\omega_4;\ \omega_1,-\omega_2,\omega_3)\vdots\ \boldsymbol{e}_1\boldsymbol{e}_2\boldsymbol{e}_3 = \chi^{(3)}$$

$$\boldsymbol{e}_4\chi^{(3)}(-\omega_4;\omega_1,-\omega_1,\omega_4)\vdots\ \boldsymbol{e}_1\boldsymbol{e}_1\boldsymbol{e}_4 = \chi_{41}{}^{(3)}$$

$$\boldsymbol{e}_4\chi^{(3)}(-\omega_4;\omega_2,-\omega_2,\omega_4)\vdots\ \boldsymbol{e}_2\boldsymbol{e}_2\boldsymbol{e}_4 = \chi_{42}{}^{(3)}$$

$$\boldsymbol{e}_4\chi^{(3)}(-\omega_4;\omega_3,-\omega_3,\omega_4)\vdots\ \boldsymbol{e}_3\boldsymbol{e}_3\boldsymbol{e}_4 = \chi_{43}{}^{(3)}$$

$$\boldsymbol{e}_4\chi^{(3)}(-\omega_4;\omega_4,-\omega_4,\omega_4)\vdots\ \boldsymbol{e}_4\boldsymbol{e}_4\boldsymbol{e}_4 = \chi_{44}{}^{(3)}.$$

For the other three frequencies, ω_1, ω_2, and ω_3, the scalar form of $\chi^{(3)}$ can also be written. Then, with $\Delta k = k_1+k_3-k_2-k_4$ ($k_i = \omega_i n_i/c$ is the wave-vector of $\boldsymbol{E}_i$), the scalar form of the four-wave coupling-wave equations, e.g., $\omega_1+\omega_3 = \omega_2+\omega_4$, can be written from (3.33) as the scalar-form four-wave coupled-wave equations:

$$\begin{aligned} \frac{\partial E_1(\omega_1,z)}{\partial z} = \frac{i3k_1}{4n_1^2}\Big[& \chi^{(3)} E_2 E_3^* E_4 e^{-i\Delta kz} + \tfrac{1}{2}\chi_{11}^{(3)}|E_1|^2 E_1 \\ & + \chi_{12}^{(3)}|E_2|^2 E_1 + \chi_{13}^{(3)}|E_3|^2 E_1 + \chi_{14}^{(3)}|E_4|^2 E_1 \Big] \end{aligned} \tag{3.56}$$

$$\frac{\partial E_2(\omega_2,z)}{\partial z}=\frac{i3k_2}{4n_2^2}\big[\chi^{(3)}E_1E_3E_4^*e^{i\Delta kz}+\tfrac{1}{2}\chi_{22}^{(3)}|E_2|^2E_2 \\ +\chi_{21}^{(3)}|E_1|^2E_2+\chi_{23}^{(3)}|E_3|^2E_2+\chi_{24}^{(3)}|E_4|^2E_2\big] \tag{3.57}$$

$$\frac{\partial E_3(\omega_3,z)}{\partial z}=\frac{i3k_3}{4n_3^2}\big[\chi^{(3)}E_1^*E_2E_4e^{-i\Delta kz}+\tfrac{1}{2}\chi_{33}^{(3)}|E_3|^2E_3 \\ +\chi_{31}^{(3)}|E_1|^2E_3+\chi_{32}^{(3)}|E_2|^2E_3+\chi_{34}^{(3)}|E_4|^2E_4\big] \tag{3.58}$$

$$\frac{\partial E_4(\omega_4,z)}{\partial z}=\frac{i3k_4}{4n_4^2}\big[\chi^{(3)}E_1E_2^*E_3e^{i\Delta kz}+\tfrac{1}{2}\chi_{44}^{(3)}|E_4|^2E_4 \\ +\chi_{41}^{(3)}|E_1|^2E_4+\chi_{42}^{(3)}|E_2|^2E_4+\chi_{43}^{(3)}|E_3|^2E_4\big] \tag{3.59}$$

These coupled-wave equations will be solved in Chap. 5 for third-harmonic generation, four-wave mixing (FWM), and stimulated light-scattering, and also in Chaps. 6, 7 and 8 for self-phase modulations (SPM), cross-phase modulations (XPM) and optical-phase conjugations (OPC).

3.4 Nonlinear Wave-Equations in Optical Fibers

Nonlinear phenomena in optical fibers, including SPM, XPM, four-wave mixing, and light scattering, are physically like those in other media. However, optical fiber is both extremely thin and long, and has very low susceptibility in undoped fibers. Hence in fibers the forms of wave equations and their solutions, such as the nonlinear Schrödinger equation and its soliton solution, may differ from those in bulk crystals, which is the traditional nonlinear medium. In these cases, the effective area A_{eff} and the effective length L_{eff} are important transmission parameters. In addition, even if the polarization mode dispersion (PMD) caused by the stress-induced birefringence in the fiber affects rather seriously the OFC system with high bit-rate, the fiber is still treated generally as a isotropy material in basic nonlinear equations.

3.4.1 Basic Nonlinear Equations in Fibers

Starting from the basic wave equation in the non-magnetic medium (3.21),

$$\nabla^2 \boldsymbol{E} - \frac{1}{c^2}\frac{\partial^2 \boldsymbol{E}}{\partial t^2} = \mu_0\left(\frac{\partial^2 \boldsymbol{P}_{\mathrm{L}}}{\partial t^2} + \frac{\partial^2 \boldsymbol{P}_{\mathrm{NL}}}{\partial t^2}\right), \tag{3.21}$$

we use the monochromatic planar wave condition and the slowly-varying envelope approximation (Sect. 3.2) in the fibers. In most fiber transmissions, $\lambda \cong 1\mu\text{m}$, and the central frequency $\omega_0 \cong 10^{15}/\text{s}$. For pulses with pulsewidth $\geq$ 100fs, the corresponding spectrum width $\Delta\omega$ is $\leq 10^{13}/\text{s}$. Consequently, $\Delta\omega/\omega_0 = 0.01 << 1$. Therefore, the monochromatic condition is almost valid. By using the slowly-varying envelope approximation, we separate the rapidly varying part of the field from the slowly varying part. $E(r,\omega)$ is the slowly-varying envelope in the space and frequency domain,

$$\nabla^2 E(r,\omega) + \varepsilon(\omega) k_0^{\,2} E(r,\omega) = 0 \tag{3.60}$$

$$k_0 = \omega/c \text{ and } \varepsilon(\omega) = 1 + \chi^{(1)}{}_{xx}(\omega) + \varepsilon_{\mathrm{NL}} = (n_0 + \Delta n_{\mathrm{NL}})^2 \tag{3.61}$$

where

$$\varepsilon_{\mathrm{NL}} = (3/4)\chi^{(3)}{}_{xxxx}(\omega)|E(r,t)|^2. \tag{3.62}$$

Conventionally, (3.60) can be solved by separation of variables. The general propagation equations of optical pulses in fibers were found by many authors in the past decades. An analysis and some of the following equations (3.63–3.68) are contributed by Agrawal [3.5]. If the solution is taken to be

$$E(r,\omega-\omega_0) = \xi(x,y)\, E_{\mathrm{SV}}(z,\omega-\omega_0)\exp(i\beta_0 z) \tag{3.63}$$

where $\xi(x,y)$ corresponds to the transverse distribution of the fundamental fiber mode, $E_{\mathrm{SV}}(z,\omega-\omega_0)$ is the slowly-varying envelop, β_0, the dispersion-free part of mode-propagation coefficient.

Use a retarded frame of reference moving with the pulse at the group velocity v_g, by making the transformation

$$T = (t - z/v_g) = t - \beta_1 z.$$

This puts the first-order GVD parameter β_1, which is in the propagation equation, into T, so (3.60) becomes

$$\frac{\partial E_{SV}}{\partial z} = -\frac{i}{2}\beta_2\frac{\partial^2 E_{SV}}{\partial T^2} - \frac{1}{2}\alpha E_{SV} + i\gamma\left|E_{SV}\right|^2 \cdot E_{SV}$$

$$T = t - \beta_1 z \tag{3.64}$$

in the special case of $\alpha = 0$, this equation is referred to as the nonlinear Schrödinger equation.

$$2i\beta_0 \partial E_{SV} / \partial_Z + (\bar{\beta}^2 - \beta_0^2) E_{SV} = 0 \tag{3.65}$$

where γ and effective core area A_{eff} are defined as:

$$\gamma = \frac{\omega_0}{c} \cdot \frac{n_2}{A_{eff}} = k_0 \cdot \delta_{fiber}$$

$$A_{eff} = \frac{\left[\int_{-\infty}^{+\infty} \int_{-\infty}^{+\infty} |\xi(x,y)|^2 \, dxdy \right]^2}{\int_{-\infty}^{+\infty} \int_{-\infty}^{+\infty} |\xi(x,y)|^4 \, dxdy}$$

where nonliner fiber factor δ_{fiber} is intrinsically propotional to the nonlinear refractive index n_2 and inversely propotional to the effective core area A_{eff}, k_0 is the light wave vector in the vaccum.

$$\begin{aligned} \bar{\beta}(\omega) &= \beta(\omega) + \Delta\beta \\ &= \beta(\omega) + \frac{k_0 \int_{-\infty}^{+\infty} \int_{-\infty}^{+\infty} \Delta n \, |\xi(x,y)|^2 \, dxdy}{\int_{-\infty}^{+\infty} \int_{-\infty}^{+\infty} |\xi(x,y)|^2 \, dxdy} . \end{aligned} \tag{3.66}$$

The Fourier transform of the slowly varying amplitude $E_{SV}(z,\omega-\omega_0)$ satisfies (3.64) which can be written as:

$$\partial E_{SV}(z,\omega-\omega_0)/\partial z = i[\beta(\omega) + \Delta\beta - \beta_0] \, E_{SV}(z,\omega-\omega_0) \tag{3.67}$$

where we used (3.65) and approximated ($\bar{\beta}^2 - \beta_0^2$) by $2\beta_0(\bar{\beta} - \beta_0)$. For this purpose, it is useful to expand the mode-propagation dispersion coefficient $\beta(\omega)$ with the frequency ω, where ω_0 is conventionally chosen so that $\beta_2(\omega_0) = 0$. β_1 and β_2 are the mode-propagation dispersion coefficient of the series. In (3.65), for $\Delta\beta$ we have Δn related to α, the nonlinear refractive index $n_2|E|^2$ and the effective core area A_{eff}. The latter two are combined with γ. Then (3.66) becomes the propagation equation of optical pulses in fibers with pulse width > 0.1ps

$$\frac{\partial E_{SV}}{\partial z} + \beta_1 \frac{\partial E_{SV}}{\partial t} + \frac{i}{2}\beta_2 \frac{\partial^2 E_{SV}}{\partial t^2} + \frac{\alpha}{2} E_{SV} = i\gamma |E_{SV}|^2 \cdot E_{SV} \tag{3.68}$$

(3.64) is valid for pulse width T_0 > 100fs. In the special case of $\alpha = 0$, this equation is the nonlinear Schrödinger equation, and is extensively studied in relation to solitons which will be discussed in Sect. 7.3. Equation (3.68) is the basic nonlinear equation for fibers. Mathematically, the general form of nonlinear polarization given by (2.11) must be used for the propagation equation extended as generalized nonlinear Schrödinger equations. The solutions of these equations and their nonlinear effects in fibers will be discussed in Chap. 7.

Here, a nonlinear fiber factor δ_{fiber} is defined as

$$\delta_{\text{fiber}} = \frac{n_2}{A_{\text{eff}}}$$

and is dependent on the fiber's intrinsic characteristics of the nonlinear refractive index n_2 and the effective core area A_{eff} of the fibers.

In order to decrease the nonlinearities in high speed and long distance fiber communications, this nonlinear fiber factor δ_{fiber} have to be lowered (see the chirp $\delta\omega_{\text{SPM}}$ in Sect. 7.2.2, and large effective area fiber in Sect. 8.4.3).

It is expected to design and fabricate the fiber with larger effective core area A_{eff} (such as from 50 to 85μm^2) or with smaller nonlinear refractive index n_2.

3.4.2 Lengths in Fibers, Nonlinear Fiber Coefficient δ_{fiber} and Nonlinear Photon Coefficient δ_{photon}

In a long distance fiber system, the high bit-rate signals are to propagate along the fiber as long and as far as possible, but limitations on such a system stem from various factors. Generally, the transmission length of pulses in the fiber is limited by the fiber's loss, dispersion, and nonlinearity, especially when the dispersion is strong and nonlinearity is important. Hence, we must find which factor is dominant. We first consider the intrinsic attenuation length L_{IA} and the effective attenuation length L_{eff} (which belong to the linear domain), the dispersion length L_{D} and the nonlinearity length L_{NL} (which are related to the nonlinear domain) and then evaluate their relative magnitudes. If L_{eff} is the least, the transmission should be limited by the nonlinearity-induced attenuation. If L_{D} is the least, transmission is limited by fiber dispersion, which is the broadening of short pulses caused by velocity dispersion. If L_{NL} is the least, the transmission is limited by nonlinearity-induced phase changes, which are the signals' deterioration caused by intensity-induced refractive index n_2.

3.4.2.1 Intrinsic Attenuation Length L_{IA} of the Fiber

The intrinsic attenuation of the fiber, caused by spontaneous Rayleigh scattering, waveguide imperfections, and absorption in silica glass, is represented by the coefficient α, in a standard length of 1km, and $\alpha = \ln(P/P_0)\ \text{km}^{-1}$. Also,

$$P = P_0 e^{-\alpha L}$$

and

$$\ln(P/P_0) = -\alpha L. \tag{3.69}$$

In communication, α is conventionally expressed as

$$-10 \log_{10} e \cdot \ln(P/P_0) = \alpha_{\text{dB}} L$$

where

$$\alpha = \frac{\alpha_{\text{dB}}}{4.34}\left(\frac{1}{km}\right). \tag{3.70}$$

From (3.69) and (3.70), we have the intrinsic attenuation length L_{IA}. At that length, the received power is $1/e$ of the transmitted power

$$L_{\text{IA}}(1/e) = 1/\alpha = 4.34/\alpha_{\text{dB}}. \tag{3.71}$$

At λ = 1.55μm, the minimum fiber loss α_{dB} is about 0.2dB/km. From (3.71) L_{IA} is 21.7km. If we define the attenuation length as that length at which the received power is one-tenth of the transmitted power, then

$$\ln(P/P_0) = -\alpha L = -\alpha_{\text{dB}} L/4.34. \tag{3.72}$$

When α_{dB} is 0.2dB/km, $-\alpha L = -\alpha_{\text{dB}} L/4.34 = -0.046L$. If $P/P_0 = 0.1$, then $\ln(0.1) = -2.3$. With α_{dB} = 0.2dB/km, $L_{\text{IA}}(0.1) = 2.3\times 4.34/0.2 = 49.9$km, which is much longer than that in copper cable, and the later is correspondingly only about several hundreds of meters.

3.4.2.2 The L_{eff} Considering the Nonlinearity-Induced Attenuation

In high-capacity and long-distance fiber communications, nonlinearities in fibers will be the final limitations. The effects of nonlinearity will be discussed in Chap. 7. Nonlinearities affect not only the spectrum broadening or shape deformation, but also induce a power penalty which also influences the effective interactive length and thus the transmission length.

The effective length L_{eff} of nonlinearities in fibers is defined of two types. One is related to the intensity-induced refractive index and to parametric processes. The other is related to stimulated light scattering and to non-parametric processes.

In the third-order of nonlinear effects, the signal penalty $\alpha_{\text{Int-Ref}}$ due to the intensity-induced refractive index (the real part of $\chi^{(3)}$) is always caused by the parametric processes. In this case (refer to the general expression (3.56–3.59)), the signal wave equation can be formed as

$$\partial E_{\text{sig}-1} / \partial L = \xi_{\text{Int-Ref}} \cdot E_{\text{sig}-1} \left| E_{\text{sig}-2} \right|^2 \cdot \exp(-\alpha_{\text{Int-Ref}} L) \tag{3.73}$$

where $E_{\text{sig-1}}$ and $E_{\text{sig-2}}$ are degenerated if in the SPM case, but non-degenerated if in the XPM and FWM cases. The solution in the degenerated case is composed by the following two parts of amplitude and phase,

$$\begin{aligned} E_{\text{sig}-1}(L,T) &= E_{\text{sig}-1}(0,T) \cdot \exp(-\varphi_{\text{Int-Ref}}) \\ &= E_{\text{sig}-1}(0,T) \cdot \exp(-\zeta_{\text{Int-Ref}} \cdot L_{\text{eff}}). \end{aligned} \tag{3.74}$$

In order to match (3.74) with the form of $\exp(-\alpha z)$ in (3.73), the effective length form of the nonlinearity induced attenuation in (3.74) can be presented as:

$$L_{\text{eff}} = \frac{1}{\alpha}[1 - \exp(-\alpha L)]\,. \tag{3.75}$$

In the same way, the intensity equation of the nonlinear light scattering has the form

$$\partial I_{\text{Stokes}} / \partial Z = g_{\text{Raman}} I_{\text{P}} I_{\text{Stokes}} - \alpha_{\text{Stokes}} I_{\text{Stokes}} \tag{3.76}$$

where I_{Stokes} is Stokes intensity of Raman scattering, I_{P} is the pump intensity, g_{Raman} is the gain coefficient, and α is the absorption coefficient of the medium. The solution is

$$I_{\text{S}}(L_{\text{Raman}}) = I_{\text{S}}(0) \exp(g_{\text{Raman}} I_{\text{P}} L_{\text{eff}} - \alpha_{\text{Stokes}} L_{\text{Raman}}) \tag{3.77}$$

where L_{eff} is also conveniently defined as

$$L_{\text{eff}} = (1/\alpha_{\text{Stokes}})[1 - \exp(-\alpha_{\text{Stokes}} L_{\text{Raman}})]\,. \tag{3.78}$$

Thus, from (3.77) and (3.78), the nonlinearity-induced attenuation will decrease the effective fiber length. The nonlinearity results in a real fiber-length L transmission in which the pump decays exponentially, and which can be replaced by transmission in an effective fiber-length L_{eff} in which the pump is constant.

Cascaded EDFA (erbium-doped fiber amplifier) has been used for long-distance transmission. Each EDFA compensates for the attenuation in each previous segment, and so the total effective length $L_{\text{eff},T}$ of transmission line is the summation of n effective-length segments.

$$L_{\text{eff},T} = \Sigma L_{\text{eff}} = -L_{\text{eff}}\,. \tag{3.79}$$

Physically, the resulting multiplication of nonlinear gain is just the exponential summation.

3.4.2.3 Dispersion Length L_D of the Fiber

In high-capacity communication, signal pulses and their corresponding spacing are quite narrow. The pulse-broadening and spatial deviation stemming from fiber dispersion will be discussed in Chaps. 7 and 8.

Commonly, the initial Gauss pulse width T_0 is considered as the $(1/e)$-high half-width of $E(0,T) = \exp[-T^2/2T_0^2]$. Thus, the dispersion length L_D is defined as a length, after which the transmitted Gauss pulse width T_0 is broadened by the factor $\sqrt{2}$, and hence L_D indicates the pulse-broadening rate stemmed from the fiber-dispersion β_2 [3.5].

$$L_D = \frac{T_0^2}{|\beta_2|}. \tag{3.80}$$

3.4.2.4 Nonlinearity Length L_{NL} of the Fiber

The nonlinearity length L_{NL} of the fiber as defined here connects to the nonlinear phase-shift-affected pulse distortion related to the intensity-induced refractive index of the parametric processes. The nonlinear contribution from the fiber material is associated with the nonlinear refractive-index coefficient n_2 from (2.106),

$$n_2 \equiv (3/8n_0)\chi^{(3)}xxxx. \tag{3.81}$$

Thus, the nonlinearity length L_{NL} is defined as

$$L_{NL} = \frac{1}{\gamma P_0} = \frac{c}{\omega_0} \cdot \frac{1}{P_0} \cdot \frac{A_{eff}}{n_2}$$

or

$$L_{NL} = \frac{c}{\omega_{photon} P_0} \cdot \frac{1}{\delta_{fiber}} = \frac{1}{\delta_{photon}} \cdot \frac{1}{\delta_{fiber}} \tag{3.82}$$

where

$$\delta_{photon} = \frac{\omega_{photon} P_0}{c} = k_{photon} P_0 \tag{3.83}$$

$$\delta_{fiber} = n_2/A_{eff.} \tag{3.84}$$

Thus the nonlinearity length L_{NL} connected to the nonlinear phase-shift-affected pulse distortion is inversely proportional to the nonlinear factors δ_{photon} and δ_{fiber}. δ_{photon} is propotional to the transmitted light frequency ω_{photon} and the light initial peak power P_0, and the nonlinear fiber factor δ_{fiber} is dependent on the fiber's intrinsic characteristics of the nonlinear refractive index n_2 and the effective core area A_{eff} of the fibers. Equation (3.88) shows that the length L_{NL} is limited by the parameters of δ_{fiber} and δ_{photon}. The larger effective core area A_{eff} or the smaller nonlinear refractive index n_2 will extend the length L_{NL} and hence lessen the fiber nonlinearity. For self-phase modulation, it means that at length L_{NL}, from (3.80), the maximum phase-shift $\Phi_{\text{NL,max}} = (L_{\text{eff}}/L_{\text{NL}}) = 1$.

Table 3.1 Effective lenghts of fibers.

Designation	L_{eff}	L_{NL}	L_D
Formula	$L_{\text{eff}} = \frac{1}{\alpha}[1 - \exp(-\alpha L)]$	$L_{\text{NL}} = \frac{1}{\gamma P_0} = \frac{1}{\delta_{\text{photon}}} \cdot \frac{1}{\delta_{\text{fiber}}}$	$L_D = \frac{T_0^2}{\lvert\beta_2\rvert}$
Courses in Physics	A real L with $\chi^{(3)}$ induced power attenuation decayed exponentially can be replaced by L_{eff} in which the power is constant.	The effect distance defined by phase-shift $\Phi_{\text{NL,max}} = 1$, which is caused by light-induced nonlinear refractive index change $\delta n = n_2\lvert E\rvert^2$	It is defined by β_2-induced pulse broadening to the extent of $\sqrt{2}T_0$.
Parameter in Formula	Loss coefficient $\alpha = \frac{\alpha_{dB}}{4.34}\left(\frac{1}{km}\right)$ $L_{\text{IA}}(1/e) = 1/\alpha$ $= 4.34/\alpha_{\text{dB}}$	$\gamma = (n_2\omega_0)/(cA_{\text{eff}})$ P_0 is the initial peak power $\delta_{\text{photon}} = \omega_{\text{photon}}P_0/c$ $\delta_{\text{fiber}} = n_2/A_{\text{eff}}$.	T_0 is initial Gauss pulse width of $(1/e)$-high, half-width.
Calculated length if $T_0 = 20$ps $P_0 = 1$mW $\alpha_{\text{dB}} = 0.2$	$L_{\text{eff}} = 21.7\text{km} = L_{\text{IA}}$ (if $L = \infty$ km).	$L_{\text{NL}} = 50$km.	$L_D = 20$km.

Example

Using $L = 100$km, $\lambda = 1.55\mu$m, $\alpha = 0.2$dB/km, $\lvert\beta_2\rvert = 20\text{ps}^2/\text{km}$, $\gamma = 20/\text{W}\cdot\text{km}$, $T_0 = 20$ps, $P_0 = 1$mW, calculate the L_{D}, L_{NL} and L_{eff}.

$L_{\text{D}} = T_0^2/\lvert\beta_2\rvert = 400/20 = 20$km,

$L_{\text{NL}} = (1/\gamma P_0) = 1/20\times10^{-3} = 50$km, and

$L_{\text{eff}} = (1/\alpha)[1 - \exp(-\alpha L)] = 21.7\times[1 - \exp(-1.38)] = 21.48$km.

Thus, $L_D < L_{eff} < L_{NL}$. Dispersion is the important limit in this case. Changing P_0 to 5mW with all else unchanged, we have $L_{NL} = (1/\gamma P_0) = 1/20\times5\times10^{-3} = 10$km. Thus, $L_{NL} < L_D < L_{eff}$. Nonlinearity becomes dominant since the power is larger.

3.5 Nonlinear Wave Equations in Semiconductors

A more extensive description of the optically linear and nonlinear response, where semiconductor nonlinearity plays key roles in the bulk semiconductors and quantum wells, will be treated in Chap. 6. This topic on optical nonlinearities of semiconductors has been discussed already in some books, for example, by Butcher and Cotter [2.3].

For many years, bulk multilayer semiconductors and quantum wells have served as major light sources in fiber communications, compact disks, laser printers, and laser diode pumping devices. In recent years, nonlinear techniques in semiconductors have become more significant, not only because of the small controlling power due to the gain of semiconductor lasers and amplifiers, but also the high and quick nonlinear response due to the population pulsation in the standing-wave cavity devices and traveling wave amplifiers. All optical networks need different kinds of novel semiconductor devices, such as modulators, switches, gates, phase conjugators, wavelength converters and WDM demultiplexers, which have exceptionally high conversion efficiency and output power far superior to the conventionally passive components. In recent years, nearly-degenerate FWM and gain cross-modulation experiments in semiconductor optical amplifiers (SOA) show that SOA is very attractive for use in wavelength converters, phase conjugators, as well as the SPM or XPM devices in nonlinear fiber loop (NFL) used for DEMUX and clock recovery.

The multiwave propagation equation has characteristics both of wave propagation and wave coupling, through the nonlinear susceptibility. The wave equations in semiconductors need parameters belonging to semiconductor materials, to semiconductor waveguide properties, and to quantum well structures. As mentioned in Sect. 3.4, there are the mode-propagation coefficient β, nonlinearity coefficient γ and attenuation coefficient α in the fiber wave-equations. The parameters in semiconductors are the optical confinement factor in the active layer Γ, the linewidth enhancement factor α_{enh}, and the gain coefficient g.

Some steps are traditionally used to determine the output signal by solving the coupled wave equations. First, find the carrier density deviation δn from the carrier rate equation. Second, find the nonlinear polarization P_{NL} from the relation between P_{NL} and δn, and third, substitute the P_{NL} as a driving source into Maxwell's equations to obtain the wave-equations in the semiconductors.

There are two kinds of processes in physics, in which the needed population pulsation are driven by the pump-probe field to produce the nonlinear effects. One is the carrier-density induced refractive index change, the response speed of which is decided by the spontaneous carrier lifetime τ_s of the interband relaxation. Accordingly, the detuning δf of the pump-probe frequency is about several GHz. In this case, phase mismatch is ignored since the wavelength difference is small. The other process is the spectral hole burning induced nonlinear gain effect, and its response speed is decided by the intraband relaxation time τ_c of the electron in the conduction bands, about 0.3 to 1.0ps. Hence the pump-probe frequency detuning δf related with τ_c can be as high as several hundred GHz. In this case, the detuning δf is limited by the phase mismatch rather than by the medium response time.

In semiconductors, nearly-degenerate FWM is always used as the base for some devices. The output of FWM is strongly dependent on both amplitude and sign of the multiwave frequency detuning. This reflects the fact that dispersion of nonlinear susceptibility in semiconductors is extremely strong. If we need larger frequency detuning δf, we must choose the more convenient physical process.

For carrier-saturation induced refractive index change from interband relaxation, theoretical analysis is carried out as follows. First, the optical injection induced carrier density deviation is found, using the carrier rate equation for the total electric field composed of the three angular frequency components: ω_0, $\omega_0+\Omega$, and $\omega_0-\Omega$. Then the nonlinear polarization in FWM is

$$P_{\mathrm{NL}} = \varepsilon_0 \frac{\partial \chi}{\partial v} \delta n E \tag{3.85}$$

where ε_0 is dielectric constant of vacuum, χ is susceptibility, n is carrier number, δn is carrier density deviation and E is total electric light fields.

The optical multiwave propagation of pulses in the semiconductors has been discussed already by many authors in the past decades. An analysis and the equations were contributed by Agrawal [3.5]. In the FWM, three coupling equations (since two pump wave equations are degenerated, and the three angular frequency components: ω_0, $\omega_0+\Omega$, and $\omega_0-\Omega$.) are obtained by substituting the polarization P_{NL} as a nonlinear driving source into Maxwell's equations.

Another approach is NDFWM. The injection probe frequency detuning caused is several hundred GHz, and is suitable for use in wavelength converters and phase conjugators. In this approach, the physical process that leads to population pulsation is the modulation of the nonlinear gain arising from spectral hole burning.

References

[3.1] Y.R. Shen: *The Principles of Nonlinear Optics*, John Wiley, New York (1984)

[3.2] H.A. Haus: *Waves and Fields in Optoelectronics*, Prentice-Hall, Englewood Cliffs, N.J. (1984)

[3.3] D. Marcuse: *Principles of Quantum Electronics,* Academic Press, New York (1980)

[3.4] D.C. Hanna, et al.: *Nonlinear Optics of Free Atoms and Molecules*, Springer-Verlag, Berlin (1979)

[3.5] G.P. Agrawal: *Nonlinear Fiber Optics*, Academic Press, New York (1989)

[3.6] K. Inoue, et. al.: Nearly degenerate four-wave mixing in a traveling semiconductor laser amplifier, *Appl. Phys. Lett.* **51**, 1051–1053 (1987)

[3.7] Zhiyuan Li, Ben-Yuan Gu and Guo-Zhen Yang: Slowly varying amplitude approximation appraised by transfer-matrix approach, *Physical Review* **B60**, 10644-10647 (1999)

[3.8] J.M. Tang and K.A. Shore: Analysis of nearly degenerate four-wave mixing in multisection laser diodes, *IEEE J. Selected Topics in Quantum Electronics* **5**, 866-872 (1999)

[3.9] A. Berzanskis, W. Chinaglia, et al.: Spatial structures in optical parametric amplification, *Physical Review* **A60**, 1626–1635 (1999)

4. Energy and Phase Relations in Nonlinear Photonics

Energy transfers and phase-matching relations between multiwaves are fundamental to nonlinear photonics. With respect to these, nonlinear processes are of two types. In nonparametric processes, the energy of the whole adiabatic system, which includes the medium, is conservative. In parametric processes, in which the medium's energy does not take part, only the multiwaves' energy is conservative.

Another topic in parametric processes is momentum conservation or phase-matching conditions. These conditions are naturally satisfied in ideally lossless and dispersion-free media. In reality, considerable dispersion exists in most optical media, and the group velocities of multiwaves are quite different. Hence, the waves disperse and lose spatial interaction during propagation processes, and the resulting nonlinear effects cannot accumulate and enhance efficiently. It is necessary to compensate for the dispersion and thus satisfy the phase-matching conditions.

There are two approaches to dispersion compensation. One is perfect matching throughout propagation, and the other is the phase mismatching states, whether leading or lagging, be periodically offset during propagation.

Conventionally, the phase-matching condition for most parametric processes is implemented by using anisotropic crystals which have enough birefringence. At a certain angle or temperature, the phase-matching is satisfied between the "o" light and the "e" light (a certain tolerance is permitted) on the whole interactive distance. Because of this perfect phase-matching (PPM), the parametric transfer efficiency is remarkably high.

In recent years, quasi-phase-matching (QPM) is prominent in waveguides. The multiwaves interaction is not strictly phase-matched in the near range. The phase mismatching is offset periodically and with reversing sign in propagation to obtain the average QPM in the long range. QPM is based on sub-micron technologies in the size range 0.1 to 1.0μm, and that is why it is noticeable from late 1980s. QPM adapts to planar waveguides, grating-like fibers, and to distributed-feedback (DFB) semiconductor devices in which the micro structures are similar in size to light wavelengths. Generally, QPM's efficiency is less than that of PPM's, but is significant in compact optical systems.

4.1 Parametric and Nonparametric Processes in Nonlinear Photonics

When the power density of one light wave or more is great enough, the energy transfers between these waves exhibit nonlinear properties. In standard nonparametric nonlinear processes, nonlinear features of energy transfer between waves and with medium exist simultaneously. In such processes, part of the wave energy transfers to other energy types in the medium, such as transfer to acoustic waves in the medium as SBS (Fig. 4.1(1)), to molecular vibration in the medium as SRS (Fig. 4.1(2)), and to the electron level by two-photon absorption (TPA) (Fig. 4.1(3)). All these processes will change some features of the lightwaves themselves, such as their frequencies, phases, and temporal patterns. These energy transfers will degrade the signal by such effects as channel power penalties or signal crosstalk between different channels.

In parametric nonlinear effects between the multiwaves, no energy transfer occurs between lightwaves and the medium, hence the total energy of multiwaves is conservative. Notice that the energy transfer can occur between waves if certain conditions are satisfied. When these effects are highly efficient, energy conservation and momentum conservation (the phase-matching condition) must be satisfied:

$$\sum_{i=1}^{m} \hbar\omega_i = \sum_{j=1}^{n} \hbar\omega_j \tag{4.1}$$

$$\Delta k = \sum_{r=1}^{m+n} k_r = 0 \,. \tag{4.2}$$

Equation (4.2) can be satisfied perfectly with proper choice of multiwave interactive parameters.

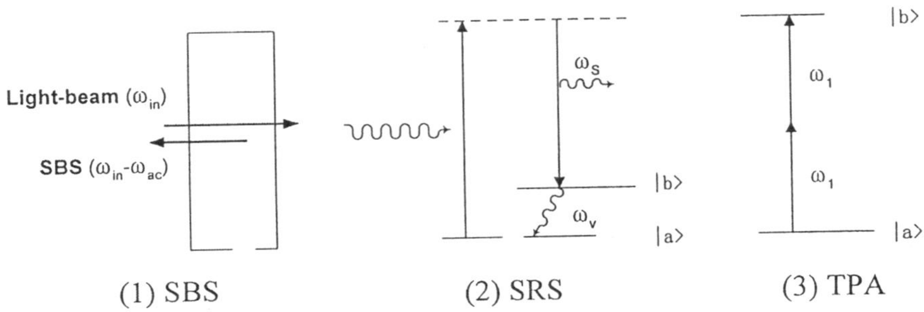

Fig. 4.1 Some types of energy transfers in the nonparametric processes.

In quantum mechanics, when these coherent photons act with different levels in the medium, some are annihilated and others are created. If the transitional particle population, such as the electrons, on the energy levels return to the initial condition, as when |a⟩ → |b⟩ → ... → |n⟩ → |a⟩, it is a parametric process. If the population on the energy levels are changed, as |a⟩ → |b⟩ → ... → |n⟩, it is a nonparametric process. Correspondingly, for interactions of light waves with energy levels in the medium, as well as the spatial patterns of wave vectors, eleven kinds of parametric and non-parametric processes occur (Table 4.1).

Table 4.1 Some parametric and nonparametric processes in nonlinear photonics.

Designation	Schematic Diagram of Nonlinear Processes	Energy Conservation	Momentum Conservation
SHG	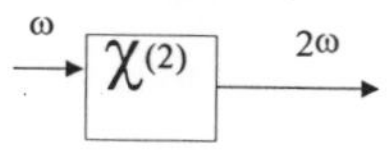	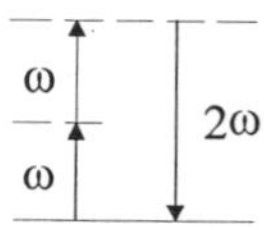	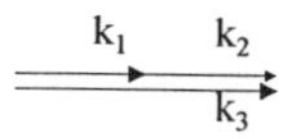
Sum Frequency	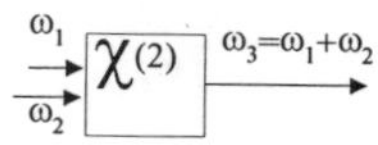	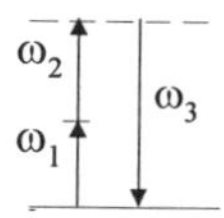	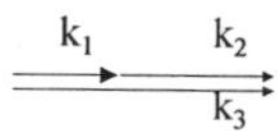
Difference Frequency	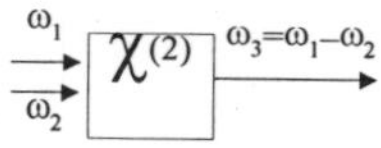	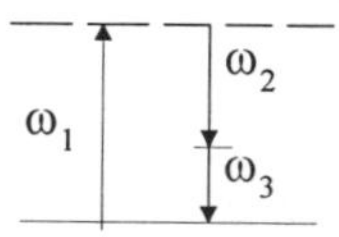	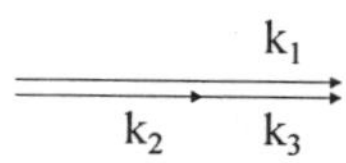
OPO	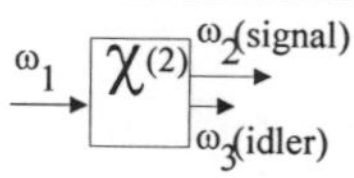	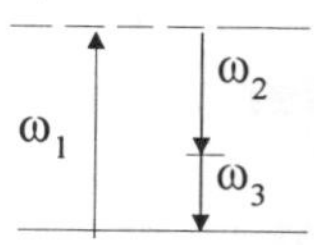	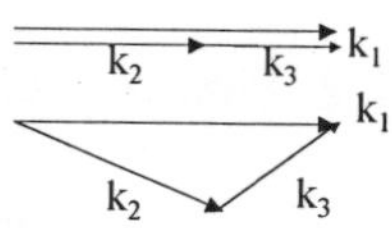
THG	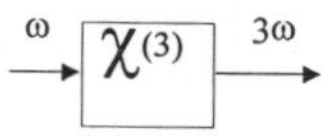	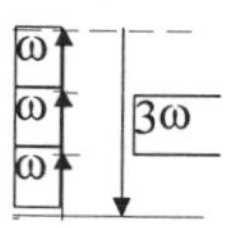	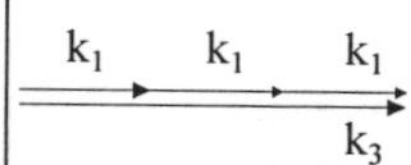
FWM	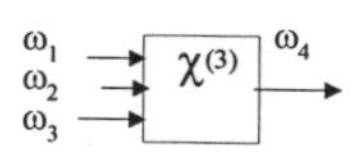	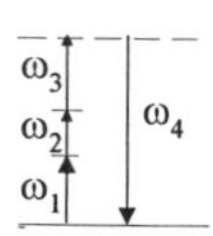	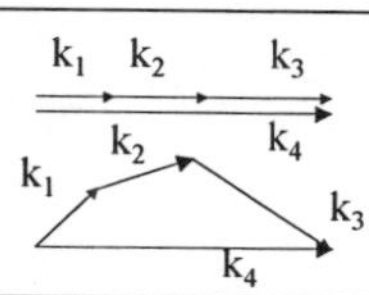

Table 4.1 (continued)

Designation	Schematic Diagram of Nonlinear Processes	Energy Conservation	Momentum Conservation
FWM	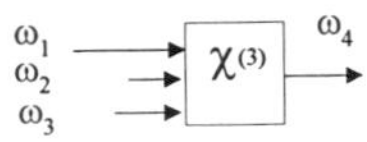	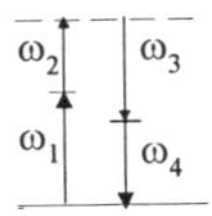	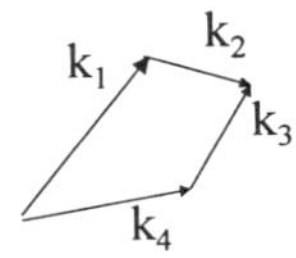
TPA	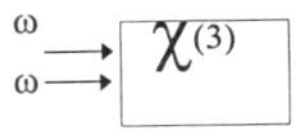	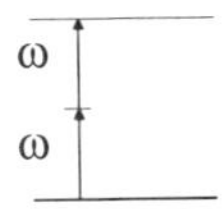	
SRS	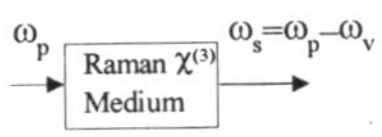	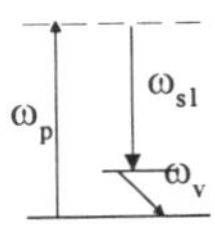	
SARS	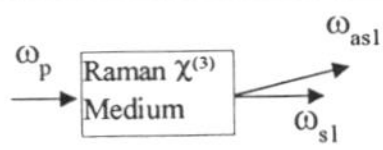	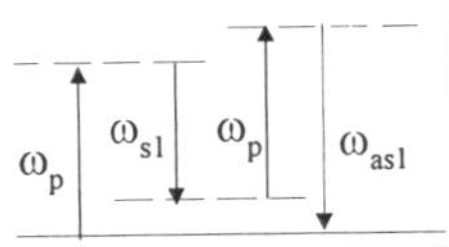	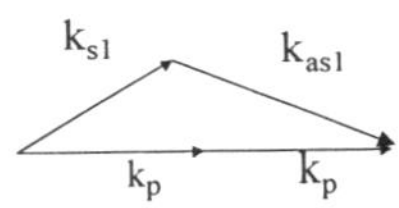
SBS		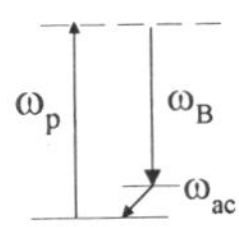	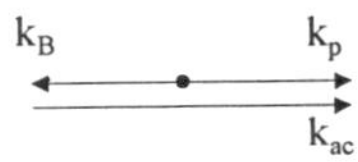

4.2 Energy Conservation and Manley-Rowe Relations

Traditionally, energy-conservation and Manley-Rowe Relations are discussed together with the three-wave sum-frequency coupled equations (3.52), (3.53) and (3.54) since they are typical and always used. The three wave coupled equations, varying with z and coupled through an identical susceptibility, are (3.52–3.54):

$$\frac{\partial}{\partial z}E_1(z) = \frac{ik_1}{2n_1^2}\chi^{(2)}E_3(z)E_2^*(z)e^{-i\Delta kz} \tag{3.52}$$

$$\frac{\partial}{\partial z}E_2(z) = \frac{ik_2}{2n_2^2}\chi^{(2)}E_3(z)E_1^*(z)e^{-i\Delta kz} \tag{3.53}$$

$$\frac{\partial}{\partial z} E_3(z) = \frac{ik_3}{2n_3^2} \chi^{(2)} E_1(z) E_2(z) e^{+i\Delta kz} \tag{3.54}$$

where $k_i = \omega_i n_i / c$ is the wave-vector of E_i

We have (3.52) multiplied by $(n_1/c)E_1^*(z)$, and (3.53) multiplied by $(n_2/c)E_2^*(z)$. These equations become

$$(n_1/c)E_1^*(dE_1/dz) = (i\omega_1/2c^2)\chi^{(2)} \cdot E_1^* E_2^* E_3 e^{-i\Delta kz} \tag{4.3}$$

$$(n_2/c)E_2^*(dE_2/dz) = (i\omega_2/2c^2)\chi^{(2)} \cdot E_1^* E_2^* E_3 e^{-i\Delta kz}. \tag{4.4}$$

Since in sum-frequency $\omega_1 + \omega_2 = \omega_3$, adding these gives

$$(n_1/c)E_1^*(dE_1/dz) + (n_2/c)E_2^*(dE_2/dz) = (i\omega_3/2c^2)\chi^{(2)} \cdot E_1^* E_2^* E_3 e^{-i\Delta kz}. \tag{4.5}$$

Multiplying the complex conjugation of (3.54) by $(n_3/c)E_3(z)$ gives

$$(n_3/c)E_3^*(dE_2/dz) = -(i\omega_3/2c^2)\chi^{(2)} \cdot E_1^* E_2^* E_3 e^{-i\Delta kz}. \tag{4.6}$$

The sum of (4.5) and (4.6) is

$$(n_1/c)E_1^*(dE_1/dz) + (n_2/c)E_2^*(dE_2/dz) + (n_3/c)E_3^*\,(dE_3/dz) = 0. \tag{4.7}$$

The complex conjugate of (4.7) is

$$(n_1/c)E_1\,(dE_1^*/dz) + (n_2/c)E_2\,(dE_2^*/dz) + (n_3/c)E_3(dE_3^*/dz) = 0. \tag{4.8}$$

Adding (4.7) and (4.8), we have

$$\begin{aligned} &(n_1/c)[E_1(dE_1^*/dz) + E_1^*(dE_1/dz)] + (n_2/c)[E_2(dE_2^*/dz) + E_2^*(dE_2/dz)] \\ &\quad + (n_3/c)[E_3(dE_3^*/dz) + E_3^*(dE_3/dz)] \\ &= (n_1/c)(d/dz)|E_1|^2 + (n_2/c)(d/dz)|E_2|^2 + (n_3/c)(d/dz)|E_3|^2 \\ &= 0. \end{aligned}$$

Integrating this with respect to time, we have

$$(1/c)\cdot(n_1|E_1|^2 + n_2|E_2|^2 + n_3|E_3|^2) = \text{constant}. \tag{4.9}$$

Since the energy flux density passing through the unit normal area in unit time is

$$U_{\text{flow-density}} = (1/2)\varepsilon_1|E_1|^2 \cdot v = (1/2)\varepsilon_0 n_1 c|E_1|^2, \tag{4.10}$$

then

$$U_{\text{flow-density}} = \frac{1}{2}\varepsilon_0 c\left(n_1 |E_1|^2 + n_2 |E_2|^2 + n_3 |E_3|^2\right) = \text{constant}. \tag{4.11}$$

In the lossless region, (4.9) implies that the total energy which the waves carry in parametric processes is constant at any cross-section in the medium. Multiply (4.3) by $1/\omega_1$, and multiply (4.4) by $1/\omega_2$ and subtract (4.4) from (4.3) to obtain

$$(1/\omega_1)(n_1/c)E_1^*(dE_1/dz) - (1/\omega_2)(n_2/c)E_2^*(dE_2/dz) = 0. \tag{4.12}$$

The complex conjugate of (4.12) added to (4.12) yields

$$(1/\omega_1)E_{\text{flow-density}}(\omega_1) - (1/\omega_2)E_{\text{flow-density}}(\omega_2) = \text{constant}, \tag{4.13}$$

or

$$N(\omega_1) - N(\omega_2) = \text{constant}. \tag{4.14}$$

Equation (4.14) shows that in sum-frequency processes, if a photon ω_1 vanishes, a photon ω_2 vanishes also. Consequently, the difference in the number of photons of ω_1 and of ω_2 is constant. Thus, we have the Manley-Rowe Relations:

$$N(\omega_1) - N(\omega_2) = \text{constant} \tag{4.14}$$

$$N(\omega_1) + N(\omega_3) = \text{constant} \tag{4.15}$$

$$N(\omega_2) + N(\omega_3) = \text{constant}. \tag{4.16}$$

Equation (4.15) shows that creation of a photon ω_3 will annihilate a photon ω_1, and the total number of photons ω_1 and photons ω_3 is constant. The same result holds for photons ω_2 and photons ω_3. These relations are significant for nonlinear parametric processes, such as sum-frequency and difference-frequency, as well as OPA and OPO, which will be discussed in Chap. 5.

4.3 Momentum Conservation and Perfect Phase-Matching

4.3.1 Momentum Conservation and the Necessity of Phase-Matching

Momentum conservation in dispersion-free media for three-wave parametric processes is given by

$$\hbar k_1 + \hbar k_2 = \hbar k_3 \tag{4.17}$$

or

$$\Delta k = k_1 + k_2 - k_3 \tag{4.18}$$

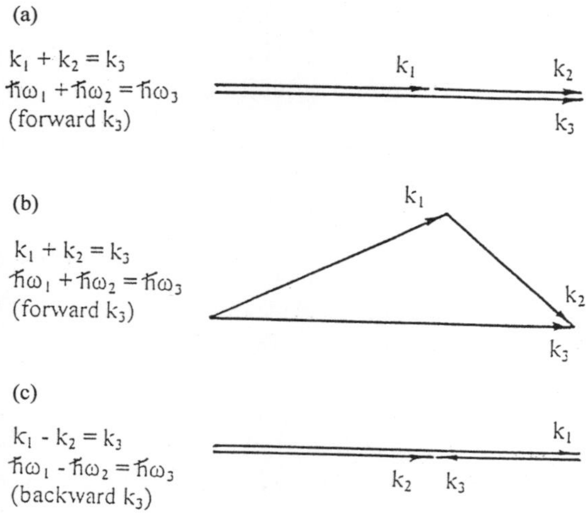

Fig. 4.2 a–c The momentum conservation (phase-matching condition) in the dispersion-free media for the three-wave processes.

Figure 4.2 shows the spatial patterns of wave vectors which satisfy (4.18). In Fig. 4.2(a) and (c), the collinear k_1 and k_2 are in the same direction for sum-frequency and in the opposite direction for difference-frequency. Figure 4.2(b) shows the usual case in three-wave mixing, where the wave vectors k_1+k_2–k_3 constitute an arbitrary triangle. Second harmonic generation (SHG) is used to discuss this topic, since it is the most common technique and is simple in principle.

Energy conservation in SHG is $\hbar\omega+\hbar\omega = 2\hbar\omega$, and the photon momentum conservation in dispersion-free regions is

$$\hbar k_1 + \hbar k_1 = \hbar k_2, \tag{4.19}$$

where k is the wave vector,

$$k = \omega/v = \omega n/c = 2\pi/\lambda. \tag{4.20}$$

Hence

$$2k_1 = k_2. \tag{4.21}$$

This condition can be satisfied if the wave vectors and crystal orientation are arranged carefully. Then SHG will be the dominant effect. Similarly, in conventional dispersion cases, the phase velocities of different frequency waves always disperse without special arrangements. Synchronized accumulation and amplification are impossible. When two $E(\omega)$ waves with fundamental frequency ω interact, they form a polarization wave $P(2\omega)$ through the susceptibility $\chi^{(2)}(2\omega)$. Consequently, wherever $E(\omega)$ arrives, the $P(2\omega)$ also simultaneously arrives. The phase velocity is

$$v_{p(2\omega)} = v_{E(\omega)} = \omega/k(\omega). \tag{4.22}$$

The phase velocity of the SHG wave $E(2\omega)$ radiating from $P(2\omega)$ is

$$v_{E(2\omega)} = 2\omega/k(2\omega). \tag{4.23}$$

After the proper choice of crystal orientation and wave vectors, the SHG wave $E(2\omega)$ propagates with the same velocity as the fundamental wave $E(\omega)$, or

$$v_{E(2\omega)} = v_{E(\omega)} = v_{p(2\omega)}\,. \tag{4.24}$$

In this case, throughout the entire length of interaction, the SHG wave $E(2\omega)$ is enhanced in phase. The energy transfer from $E(\omega)$ to $E(2\omega)$ is continuous and efficient, for which perfect phase matching (PPM) is necessary. From (4.22) and (4.23),

$$2\omega/k(2\omega) = \omega/k(\omega),$$

or

$$k(2\omega) - 2k(\omega) = \Delta k = 0. \tag{4.25}$$

From (4.20),

$$2\omega \cdot n_2/c = 2\cdot \omega n_1/c. \tag{4.26}$$

Thus,

$$n_2(2\omega) = n_1(\omega). \tag{4.27}$$

But because of dispersion, $n_2(2\omega)$ is always greater than $n_1(\omega)$ in a normal-dispersion region, and so

$$n_2(2\omega) > n_1(\omega),$$

then

$$2\omega \cdot n_2/c > 2\cdot \omega n_1/c, \tag{4.28}$$

or

$$k(2\omega) - 2k(\omega) = \Delta k > 0.$$

This is phase-mismatching. In the normal-dispersion region,

$$v_{E(2\omega)} < v_{p(2\omega)}. \tag{4.29}$$

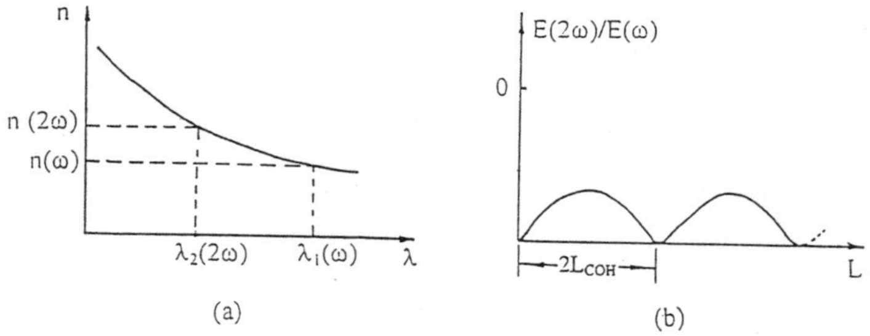

Fig. 4.3 a, b The phase mismatching in the normal dispersion region. $E(2\omega)/E(\omega)$ is a periodical function with very low amplitude and a period of $2L_{\mathrm{COH}}$ (L_{COH} is the coherent length).

When the wave $E(2\omega)$ arrives at a new position, it will cancel, or partly cancel, the newly generated wave $E_{\mathrm{new}}(2\omega)$ there. In fact, the output $E(2\omega)$ and the ratio of $E(2\omega)/E(\omega)$ are very low-amplitude functions of period $2L_{\mathrm{COH}}$ (L_{COH} is the coherent length), as shown in Fig. 4.3(b). As a result, little if any SHG output occurs, and only the original field is observed.

4.3.2 Perfect Phase-Matching Techniques in Nonlinear Crystals

In optical transmission, compensation for refractive index dispersion is a major problem, for which there are different methods of solution. Optical birefringence in anisotropic crystals is used to compensate for crystal dispersion. Buffer gas is always added to compensate for the dispersion of original gas. Waveguide dispersion of refractive index is used to compensate for the dispersion in fibers, and the changes in the absorption coefficient distribution is used to compensate for the dispersion of refractive index dispersion in III-V compound semiconductor materials. Some of these will be considered in Chaps. 6, 7 and 8. Here we deal only with SHG in anisotropic crystals, and introduce some standard procedures of phase-matching. The small-signal and large-signal solutions will be developed in Sect. 5.1.

4.3.2.1 Angle (or Critical) Phase-Matching

At a particular angle θ between the wave propagation direction and the optical axis in the anisotropic crystals (Fig. 4.4), two orthogonal light waves, "o" and "e", are used for dispersion compensation by optical birefringence. Since this kind of phase-matching depends on θ, even a small deviation from the matching angle θ_m will cause considerable mismatching of wave-vector Δk.

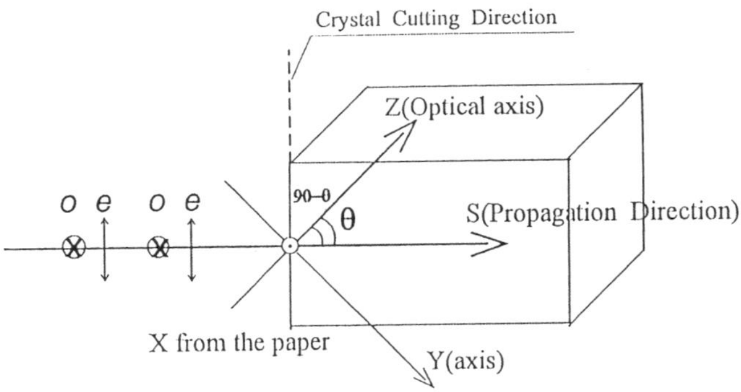

Fig. 4.4 Relation between the directions of light-beam propagtion, crystal cutting and crystal axes.

4.3.2.2 Type-I (or Parallel) Phase-Matching

Here, two fundamental wave fields, "o" and "e", are parallel. The equations of refractive- index dispersion in this case are:

$$n^{(o)}(\omega) + n^{(o)}(\omega) = n^{(e)}(2\omega) \text{ [negative-uniaxial crystal oo}\rightarrow\text{e]} \qquad (4.30)$$

or

$$n^{(e)}(\omega) + n^{(e)}(\omega) = n^{(o)}(2\omega) \text{ [positive-uniaxial crystal ee}\rightarrow\text{o].} \qquad (4.31)$$

From (4.30), the type-I phase-matching condition in negative-uniaxial crystals is

$$\Delta k = k^{(e)}(2\omega) - 2k^{(o)}(\omega) = (2\omega/c)[n_e^{2\omega}(\theta) - n_o^{\omega}] = 0 \text{ [oo}\rightarrow\text{e],} \qquad (4.32)$$

and from the refractive index ellipsoid, we have

$$n_o^{\omega} = n_e^{2\omega}(\theta^I{}_m) = \{[\cos^2(\theta^I{}_m)/(n_o^{2\omega})^2] + [\sin^2(\theta^I{}_m)/(n_e^{2\omega})^2]\}^{-1/2}. \quad (4.33)$$

Hence,

$$\theta^{\text{type-I}}_{\text{m,negative-u.c}} = \sin^{-1}\left[\frac{\left(n_e^{2\omega}\right)^2}{\left(n_o^{\omega}\right)^2}\cdot\frac{\left(n_o^{2\omega}\right)^2-\left(n_o^{\omega}\right)^2}{\left(n_o^{2\omega}\right)^2-\left(n_e^{2\omega}\right)^2}\right]^{1/2}. \qquad (4.34)$$

This condition in negative-uniaxial crystals is described here by the refractive-index ellipsoid, the refractive-index dispersion curve, and the phase-matching equations, which are all shown in Fig. 4.5.

Similarly, from (4.31), this condition in positive-uniaxial crystals is

$$\Delta k = k^{(o)}(2\omega) - 2k^{(e)}(\omega) = (2\omega/c)[n_o^{2\omega} - n_e^{\omega}(\theta)] = 0 \text{ [ee}\rightarrow\text{o].} \qquad (4.35)$$

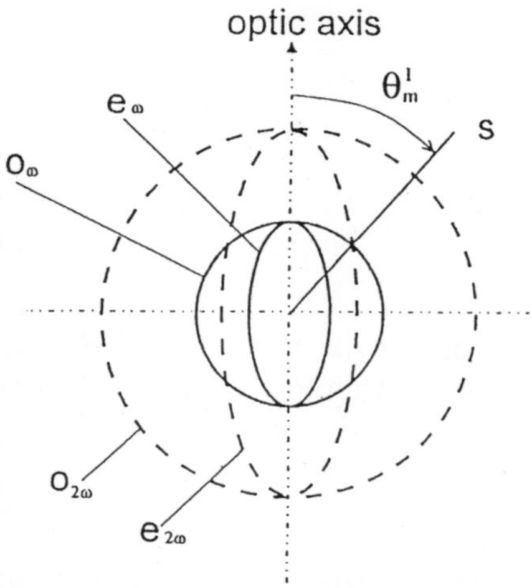

(a) Index Elliptical-Sphere

$$o_\omega + o_\omega \rightarrow e_{2\omega}$$

$$\Delta k = k_2 - k_1 = \frac{2\omega}{c}\left[n_e^{2\omega}(\theta) - n_o^{\omega}\right] = 0$$

$$n_o^{\omega} = n_e^{2\omega}(\theta_m^I)$$

$$n_o^{\omega} = \left[\frac{\cos^2\theta^m}{\left(n_o^{2\omega}\right)^2} + \frac{\sin^2\theta^m}{\left(n_e^{2\omega}\right)^2}\right]^{-1/2}$$

(b) Equations of phase-matching

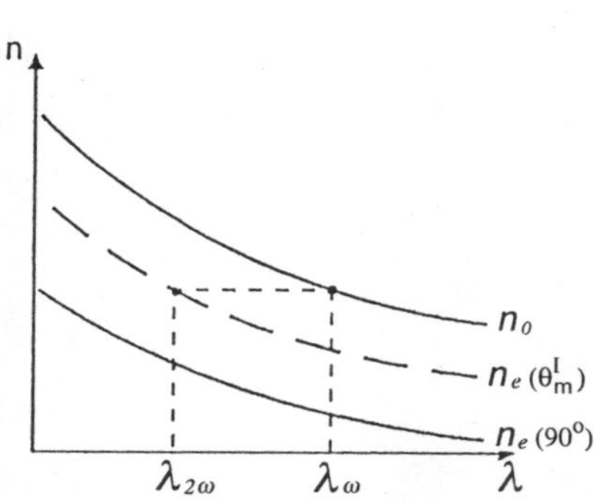

(c) Dispersion curve of Index

Fig. 4.5 a–c The SHG parallelly critical phase-matching condition on negative uniaxial optical crystals.

From the refractive index ellipsoid, we have

$$n_O^{2\omega} = n_e^{\omega}(\theta^I{}_m) = \{[\cos^2(\theta^I{}_m)/(n_O{}^{\omega})^2] + [\sin^2(\theta^I{}_m)/(n_e{}^{\omega})^2]\}^{-1/2}. \quad (4.36)$$

Hence,

$$\theta^{\text{type-I}}_{\text{m,positive-u.c}} = \sin^{-1}\left[\frac{\left(n_e^{\omega}\right)^2}{\left(n_o^{2\omega}\right)^2}\cdot\frac{\left(n_o^{2\omega}\right)^2-\left(n_o^{\omega}\right)^2}{\left(n_e^{\omega}\right)^2-\left(n_o^{\omega}\right)^2}\right]^{1/2}. \quad (4.37)$$

Two fundamental wave fields, in this case both "o" or both "e", are again parallel. The refractive index ellipsoid, the refractive index dispersion curve, and the phase-matching equations, are all shown in Fig. 4.6.

4.3.2.3 Type-II (or Orthogonal) Phase-Matching

In this type, two waves, "o" and "e", are orthogonal. The equations of refractive index dispersion are:

$$n^{(O)}(\omega) + n^{(e)}(\omega) = n^{(e)}(2\omega) \text{ [negative-uniaxial crystal oe}\rightarrow\text{e]}, \quad (4.38)$$

or

$$n^{(e)}(\omega) + n^{(O)}(\omega) = n^{(O)}(2\omega) \text{ [positive-uniaxial crystal eo}\rightarrow\text{o]}. \quad (4.39)$$

From (4.38), the type-II phase-matching condition in negative-uniaxial crystals is (4.40),

$$\begin{aligned}\Delta k &= [k^{(O)}(\omega) + k^{(e)}(\omega)] - k^{(e)}(2\omega)\\ &= (2\omega/c)\{[n_O{}^{\omega} + n_e{}^{\omega}(\theta)]/2 - [n_e{}^{2\omega}(\theta)]\} \text{ [oe}\rightarrow\text{e]}. \end{aligned} \quad (4.40)$$

From the refractive index ellipsoid, we have

$$n_e{}^{2\omega}(\theta^{II}{}_{m.nag.}) = (1/2)[n_e{}^{\omega}(\theta^{II}{}_{m.nag}) + n_O{}^{\omega}], \quad (4.41)$$

so, as shown in Fig. 4.7, the type-II critical phase-matching condition in negative-uniaxial crystals is:

$$\left[\frac{\cos^2\theta^{II}_{m-nag.}}{(n_o^{2\omega})^2}+\frac{\sin^2\theta^{II}_{m-nag.}}{(n_e^{2\omega})^2}\right]=\frac{1}{2}\left[n_o^{\omega}+\left(\frac{\cos^2\theta^{II}_{m-nag.}}{(n_o^{\omega})^2}+\frac{\sin^2\theta^{II}_{m-nag.}}{(n_e^{\omega})^2}\right)^{-1/2}\right]. \quad (4.42)$$

Similarly from (4.39), the type-II phase-matching condition in positive-uniaxial crystals is

$$\begin{aligned}\Delta k &= [k^{(e)}(\omega) + k^{(O)}(\omega)] - k^{(O)}(2\omega)\\ &= (2\omega/c)\{[n_e{}^{\omega}(\theta) + n_O{}^{\omega}]/2 - [n_O{}^{2\omega}]\} = 0. \end{aligned} \quad (4.43)$$

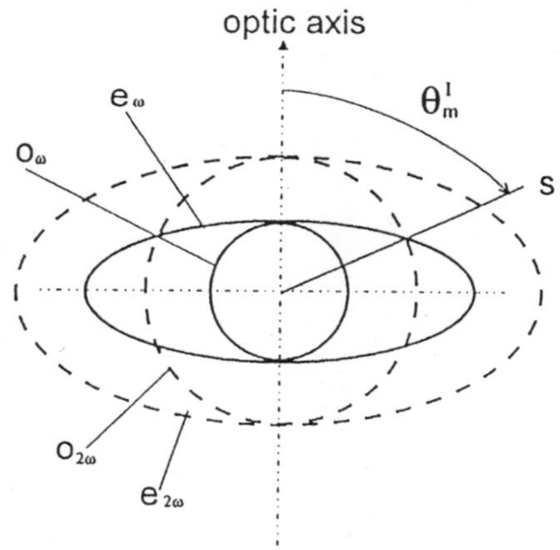

(a) Index Elliptical-Sphere

$$e_\omega + e_\omega \to o_{2\omega}$$

$$\Delta k = k_2 - 2k_1 = \frac{2\omega}{c}\left[n_o^{2\omega} - n_e^{\omega}(\theta)\right] = 0$$

$$n_o^{2\omega} = n_e^{\omega}(\theta_m^I)$$

$$n_o^{2\omega} = \left[\frac{\cos^2\theta^m}{\left(n_o^{\omega}\right)^2} + \frac{\sin^2\theta^m}{\left(n_e^{\omega}\right)^2}\right]^{-1/2}$$

(b) Equations of phase-matching

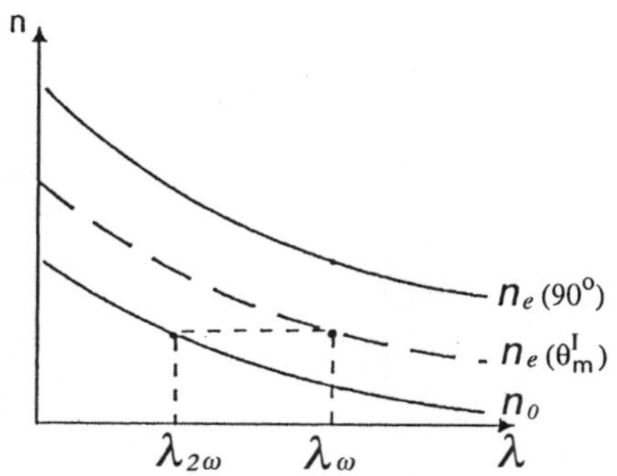

(c) Dispersion curve of Index

Fig. 4.6 a–c The SHG parallelly critical phase-matching condition on positive uniaxial optical crystals.

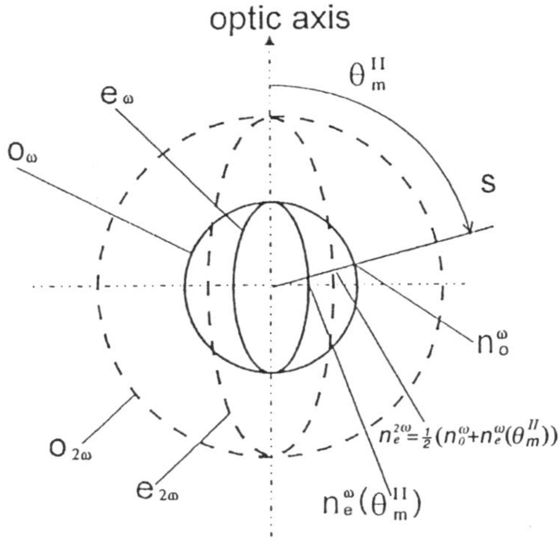

(a) Index Elliptical-Sphere

$$o_\omega + e_\omega \to e_{2\omega}$$

$$\Delta k = \frac{2\omega}{c}\left[n_e^{2\omega}(\theta) - \frac{1}{2}n_e^{\omega}(\theta) - \frac{1}{2}n_o^{\omega}\right] = 0$$

$$n_e^{2\omega}(\theta_m^{II}) = \frac{1}{2}(n_e^{\omega}(\theta_m^{II}) + n_o^{\omega})$$

$$\left[\frac{\cos^2\theta^m}{\left(n_o^{2\omega}\right)^2} + \frac{\sin^2\theta^m}{\left(n_e^{2\omega}\right)^2}\right]^{-1/2} = \frac{1}{2}\left\{n_o^{\omega} + \left[\frac{\cos^2\theta^m}{\left(n_o^{\omega}\right)^2} + \frac{\sin^2\theta^m}{\left(n_e^{\omega}\right)^2}\right]^{-1/2}\right\}$$

(b) Equations of phase-matching

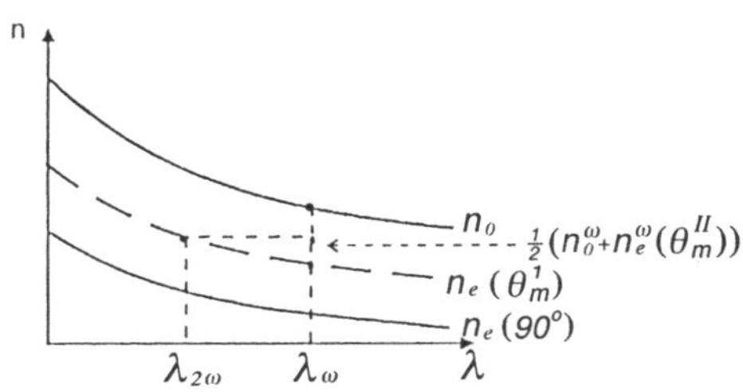

(c) Dispersion curve of Index

Fig. 4.7 a–c The SHG orthogonally critical phase-matching condition on negative uniaxial optical crystals.

Hence,

$$\theta_{\text{m,positive-u.c.}}^{\text{type-II}} = \sin^{-1}\left\{\frac{\left[\dfrac{n_o^{\omega}}{2n_o^{2\omega} - n_o^{\omega}}\right]^2 - 1}{\left[\dfrac{n_o^{\omega}}{n_e^{\omega}}\right]^2 - 1}\right\}^{1/2} . \tag{4.44}$$

The elliptical sphere of refractive index, the dispersion curve of refractive index, as well as the phase-matching equations of type-II critical phase-matching condition in the positive-uniaxial crystal are shown in the Fig. 4.8.

4.3.2.4 Temperature (or Noncritical) Phase-Matching

When two spheres of n_o^{ω} and $n_e^{2\omega}$ are tangent at $\theta_m = 90°$ in negative-uniaxial crystals, or two spheres of n_e^{ω} and $n_o^{2\omega}$ are tangent at $\theta_m = 90°$ in positive-uniaxial crystals, these are also phase-matching conditions, as they all satisfy the condition $n(2\omega) = n(\omega)$ in (4.27). Generally, when the temperature varies, the changes of "*o*" light's refractive index n_o and "*e*" light's refractive index n_e are unequal. Thus the two spheres can be tangent at $\theta_m = 90°$ by controlling the crystal temperature as shown in Fig. 4.9(a) and (b).

In temperature phase-matching, if the cross-angle between the light beam and the optical axis of crystal departs slightly from 90°, the phase-matching is still maintained. Thus, there is greater tolerance of the beam divergence angle, also called "noncritical" phase-matching, and this is why it is always used in engineering.

Example

For a 1.06μm Nd:YAG laser, when we use the nonlinear crystal CD^*A for SHG, the 90° phase-matching can occur if the temperature is 100 to 112°C.

4.3.2.5 The Impossible Situation for Phase-Matching

Only some crystals can be used for phase-matching because optical birefringence is too weak in some crystals. For these, the pair of spheres n_o^{ω} and $n_e^{2\omega}$, or n_e^{ω} and $n_o^{2\omega}$, have neither cross points nor tangential points (Fig. 4.9(c) and (d)). Such crystals, e.g., quartz (SiO_2), cannot be used for SHG.

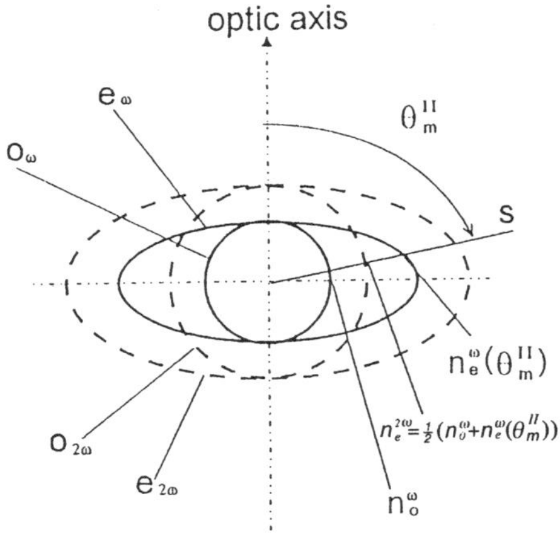

(a) Index Elliptical-Sphere

$$o_\omega + e_\omega \rightarrow e_{2\omega}$$

$$\Delta k = \frac{2\omega}{c}\left[n_e^{2\omega} - \frac{1}{2} n_e^{\omega}(\theta) - \frac{1}{2} n_o^{\omega} \right] = 0$$

$$n_o^{2\omega} = \frac{1}{2}(n_e^{\omega}(\theta_m^{II}) + n_o^{\omega})$$

$$n_o^{2\omega} = \frac{1}{2}\left\{ n_o^{\omega} + \left[\frac{\cos^2\theta^m}{\left(n_o^{\omega}\right)^2} + \frac{\sin^2\theta^m}{\left(n_e^{\omega}\right)^2} \right]^{-1/2} \right\}$$

(b) Equations of phase-matching

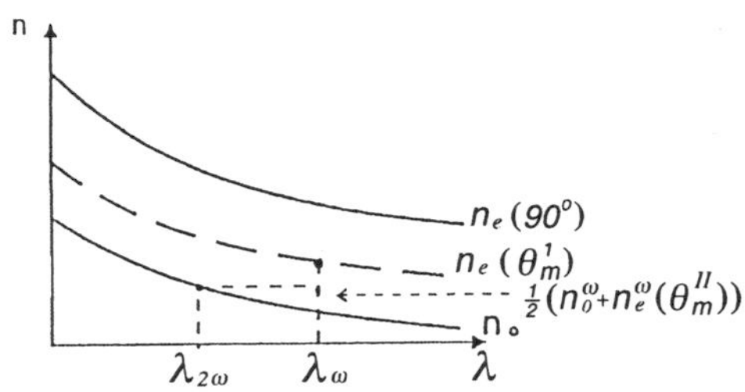

(c) Dispersion curve of Index

Fig. 4.8 a–c The SHG orthogonally cratical phase-matching condition on positive uniaxial optical crystals.

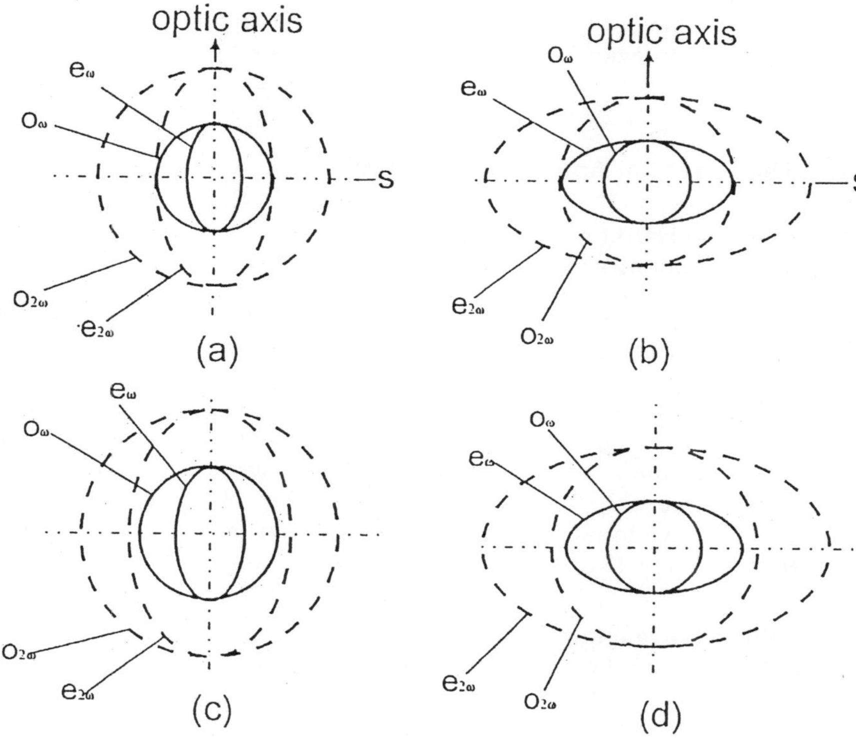

Fig. 4.9 a–d **(a)** The SHG noncritical phase-matching condition on negative uniaxial optical crystals. **(b)** The SHG noncritical phase-matching condition on positive uniaxial optical crystals. **(c)** The impossibility of the SHG phase-matching condition on negative uniaxial optical crystals. **(d)** The impossibility of the SHG phase-matching condition on positive uniaxial optical crystals.

4.4 The Concept and Use of Quasi-Phase Matching (QPM)

In many situations, to satisfy the uniformed perfect-phase matching SHG is really unreliable. Many media have the same problems as the isotropic materials do, with too large or too small a birefringence at the expected λ, or with waves in the same polarization in the medium. Traditional PPM-SHG cannot be used for them, but QPM-SHG can, and this is why this technique is superior. Also, the required wavelength λ, the working temperature T, and the angle θ can have greater tolerances in QPM. In this QPM case, it is essential to make some concessions with alternately over-

and under-matching, and alternate offset of phase-mismatched states, whether leading or lagging; this is the phase-matching managed QPM-SHG, which has an averaged efficiency and makes the system stiffness enhanced.

4.4.1 The Potential Applications of QPM

4.4.1.1 Second Harmonic Generation (SHG) and Optical Parametric Amplifier (OPA) by QPM

The development of QPM technology is rather rapid. Some results of the multiple-channel wavelength conversion by the use of engineered quasi-phase-matching structures in $LiNbO_3$ waveguides [4.11] and of the quasi-phase-matched harmonic generation through coupled parametric processes in a quasi-periodic optical superlattice [4.12] are reported recently. In addition, by combining the erbium-doped fiber sources and periodically-poled lithium niobate, high conversion efficiency of as much as 83% (energy efficiency) for second harmonic generation into the near IR (768 nm) was obtained [4.13].

4.4.1.2 WDM Wavelength Converter of QPM for the Multi-Channel Communication Applications

The wavelength conversion based on the QPM differential frequency generation (DFG) was reported. The phase mismatching

$$\begin{aligned}\Delta k_{\mathrm{QPM}} &= k_{\mathrm{pump}} - k_{\mathrm{signal}} - k_{\mathrm{signal}} \\ &= 2\pi n_{\mathrm{pump}}/\lambda_{\mathrm{pump}} - 2\pi n_{\mathrm{signal}}/\lambda_{\mathrm{signal}} - 2\pi n_{\mathrm{conversion}}/\lambda_{\mathrm{conversion}}\end{aligned}$$

is compensated by using the QPM grating with grating period of $\Lambda = 2\pi/\Delta k_{\mathrm{QPM}}$.

Theory estimates that the conversion efficiencies of input multi-channels into output multi-channels by the $LiNbO_3$ waveguide QPM-DFG are rather uniform in a wide wavelength range.

4.4.1.3 The Compression of Short Pulse-Width by QPM

The compression technique of short pulse-width by QPM-grating frequency conversion is quite different with the conventional methods by the dispersion linear compensations or interactions with nonlinear effects.

The technique is based on the group velocity mismatching of the fundamental and SHG waves, together with the different spatial positions of SHG conversion in the non-periodical QPM grating (chirped QPM grating). When the spatial position of each different frequency component is selected in the chirped QPM grating, these positions determine the different time-delays between the fundamental and SHG waves of this frequency. If the chirp of the QPM grating is matching with the chirp of the light pulse, then the SHG pulse will be compressed.

The concepts and uses of quasi-phase matching (QPM) have been discussed for many years, and the QPM techniques and devices are also well developed in recently years, but its engineering application will face the challenges of many other devices. Only when the technologies, efficiencies and marketing benefit are convenient and satisfactory, the QPM techniques and devices will be used in the integrated systems.

4.4.2 The Periodic and Nonperiodic QPM

4.4.2.1 The Efficiency of Periodic QPM

Many papers have expounded on the theory and experiments of QPM-SHG. Figure 4.10 and its explaination are published in a significant paper by Fejer, Byer, et al. [4.6].

The fundamental wave $E_1(\omega_1)$ and the induced SHG polarization wave $P(2\omega_1)$ have the same velocity v_1, determined by the fundamental wave's refractive index $n_1(\omega_1)$. The SHG wave $E_2(2\omega_1)$ radiated by $P(2\omega_1)$ has the velocity v_2, determined by the SHG wave's refractive index $n_2(\omega_2)$. The sign + or – of the power-flow transferred from one wave to another is determined by the relative phase difference between these two waves. In phase mismatching, their different phase velocities cause the continuously varying phase shift between these two waves. The direction changes of power flow caused by this phase shift produces are shown as curve C in Fig. 4.10.

The curve C shows that along the propagation, the SHG power-flow grows and reduces periodically due to the alternation of sign. The distance in which the phase difference changes by π is the coherent length, given by $L_{\mathrm{COH}} \equiv (\lambda/4)(n_2 - n_1)$

The curve A in Fig. 4.10 shows that in traditional PPM theory, the SHG power-flow increases as the square of the device length.

Another approach to cause SHG power-flow to increase with the device length is QPM-SHG. In QPM, the relative phase, or phase difference of the polarization wave, and the SHG wave is inverted after each odd multiple of L_{COH} (L_{COH}, $3L_{\mathrm{COH}}$, $5L_{\mathrm{COH}}$, etc.). If in one segment, the phase of $P(2\omega_1)$ leads that of $E_2(2\omega_1)$ and in the next segment the phase of $E_2(2\omega_1)$ leads that of $P(2\omega_1)$, then the relative phase is periodically zero, and the average relative phase is zero. This special condition is needed for SHG power-flow to increase with the device length.

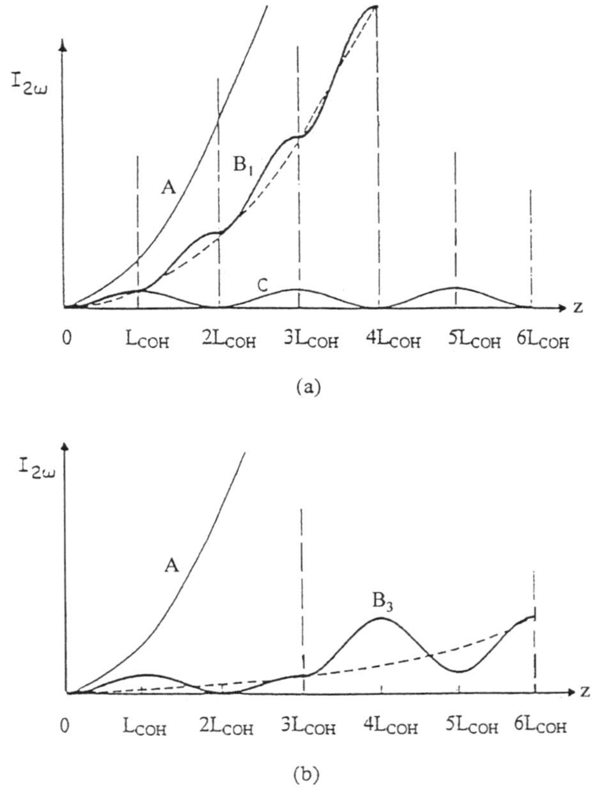

Fig. 4.10 a, b Phase-matching and the coherent length L_{COH} in QPM-SHG.

In recent years, the periodic domain inversion of the nonlinear coefficient or polarization, with period Λ and generated in $LiNbO_3$ waveguides by material growth and periodic-doping, has become promising. The spatial domain of nonlinear coefficient or of polarization domain periodically inverts since the sign alternates. This is an efficient but unnecessary approach. Another technique is the domain grating, in which the condition needed for QPM is that any periodic grating structure, with a given nonlinear coefficient, and which interacts with a given travelling-wave vector, will induce the vector's spatial harmonic wave. Hence, The amplitude-modulated grating of the nonlinear coefficient is sufficient.

The interactive processes of the fundamental and SHG waves, the output wave of SHG, SHG efficiency, tolerance for Δk, the bandwidth, and the tolerance of the periodic structure are significant parameters in QPM-SHG. In the lossless region, under the conditions of the slowly-varying amplitude approximation and the steady-state small-signal, the analysis is from the three-wave scalar-amplitude coupled equations of (3.52) and (3.54):

$$(d/dz)E_1(z) = (ik_1/2n_1^2)\cdot\chi^{(2)}\cdot E_3(z)E_1^*(z)\cdot e^{-i\Delta kz} \tag{3.52}$$

$$(d/dz)E_{SHG}(z) = (ik_3/2n_3^2)\cdot\chi^{(2)}\, E_1^2(z)\cdot e^{+i\Delta kz} \tag{3.54}$$

where $E_1(z)$ = constant and $(d/dz)E_1(z) \equiv 0$, due to the small-signal condition. The $\chi^{(2)}$ is the nonlinear susceptibility. Δk is the wave-vector mismatching permitted in the case of QPM-SHG as $\Delta k_{QPM} = 2k_\omega - k_{2\omega}$, where $L_{COH} = \lambda/4(n_\omega - n_{2\omega})$.

In the conventional PPM case, since $\Delta k_{PPM} = 0$ and $\chi^{(2)} = \chi^{(2)}{}_{PPM}$ = constant, then the output wave amplitude of SHG is

$$E_{PPM-SHG}(L) = \frac{ik_{2\omega}}{2n_{2\omega}{}^2}\cdot\chi^{(2)}_{PPM}\cdot E_1^2(0)\cdot L\,. \tag{4.45}$$

In QPM-SHG, $\chi^{(2)}(z)$ is a spatially varying function and $\Delta k \neq 0$. Hence we take the integral of (3.54), then

$$E_{QPM-SHG}(z) = (ik_3/2n_3^2)\cdot E_1^2(0)\cdot\int \chi^{(2)}{}_{QPM}(z)\cdot e^{+i\Delta kz}dz. \tag{4.46}$$

For this integral, $\chi^{(2)}QPM(z)$ can be expressed by either a spatially real-function or a Fourier transform function.

From the spatial real-function expression of $\chi^{(2)}QPM(z)$, it is easy to find the SHG output wave field $E_{SHG-QPM}(L)$ by integrating (4.46),

$$E_{QPM-SHG}(L_{out}) = \frac{1}{m\pi}\cdot\frac{ik_{2\omega}}{n_{2\omega}{}^2}\cdot\chi^{(2)}\cdot E_1^2(0)\cdot L_{out}\,, \tag{4.47}$$

where $m = \Lambda/2L_{COH}$, the number of L_{COH} in a half domain period $\Lambda/2$.

Comparing (4.47) with (4.45), the SHG output $E_{QPM-SHG}(L)$ is $(2/m\pi)E_{PPM-SHG}(L)$. If m is 1 (first-order QPM-SFG, curve B_1 in Fig. 4.10), the decreasing factor $(E_{QPM-SHG}/E_{PPM-SHG})$ is 2/3, and 4/9 for power or efficiency $(E^2{}_{QPM-SHG}/E^2{}_{PPM-SHG})$. When m = 2, the factor is 1/3, and 1/9 for efficiency. In the periodical QPM-SHG, the phase-matching band-width of the fundamental frequency $\Delta\lambda_{QPM}$ is close to the phase-matching band-width of the fundamental frequency $\Delta\lambda_{PPM}$ in the PPM-SHG case [4.6].

4.4.2.2 The Efficiency of Nonperiodic QPM

The non-periodical QPM has more variations and patterns. One purpose is to enlarge the phase-matching bandwidth of the fundamental frequency $\Delta\lambda_{QPM}$ in the QPM-SHG. For this purpose, the multi-periodical QPM is reported.

The whole QPM grating is divided into several segments, a certain grating period Λ_i was settled in the i-th segment. In this case, the total SHG intensity is composed of all the intensity contributions from Λ_i segments themselves and by all intensity contributions from the mutual-interaction of different period segments. By this way, the phase-matching bandwidth of fundamental frequency $\Delta\lambda_{QPM}$ is enlarged obviously and keep almost the same SHG efficiency.

4.4.3 QPM-SHG in Waveguides

Generally, the higher the SHG efficiency, the stronger the optical intensity needed. Hence a planar-strip waveguide has a very small cross-section, and consequently a very small power can cause enough field intensity to produce the nonlinear effects. As the pump wave and the resultant nonlinear wave, such as an SHG wave, are guided waves, they are independent of each other. The converted efficiency will seriously decrease if there is even a small phase-mismatch.

However, in QPM-SHG, as long as the QPM grating periodically has an additional compensation $k_m=2\pi m/\Lambda$, the total mismatch $\Delta k_{QPM} \equiv 2k_1-k_2-k_m$ remains 0, even for some originally phase-mismatched ($2k_1-k_2 \neq 0$) conditions. Hence, QPM-SHG is important in optical waveguides.

Since the period of nonlinear coefficient domain inversion or modulation is about 3 to 6μm and the width and depth of the periodical grating structures are in the range of 0.5 to 1.0μm, micron and sub-micron technologies of electronics or optoelectronics are needed. There are various means to cause periodical domain inversion in waveguides. One is electron beam (EB) scanning, another is proton exchanging and reactive ion etching. These methods are stable and repeatable, and the results show the applied perspectives.

4.4.3.1 Domain Inversion Caused by Electron Beam (EB) Scanning

The implementation of ferroelectric domain inversion grating by EB scanning is a typical example for understanding the technologies. In this case, the $+z$ surface of the $LiNbO_3$ crystal is metal-electrode grounded. The domain inversion gratings are obtained by EB scanning on the $-z$ free surface.

The inverted zones penetrate the depth of crystal plate, while the gratingss boundary is normal to the crystal surface. Here, EB uses several tens of kV with 0.3μm focal region diameter. When the grating period is 6.459μm and the output green light wave is SHG from a 1.064μm Nd:YAG laser, experiment shows that the normalized QPM-SHG tangent efficiency is 50%/W. The other parameters are the output near-field FWHM pattern, 1.0μm wide and 0.9μm blue light SHG from Ti:Al_2O_3 laser, experiment shows that the normalized QPM-SHG tangent efficiency is 70%/W [4.7]. The other parameters are the output near-field FWHM pattern, 1.2μm wide and 0.8μm deep.

4.4.3.2 Domain Inversion Caused by Proton Exchange and Reactive Ion Etching (PERIE)

The implementation of ferroelectric domain inversion grating by PERIE is an another typical example. The $LiNbO_3$ device includes a ferroelectric domain inversion grating and a channel waveguide array. The grating periods of the waveguide channels vary slightly, and a group of channels with the minimum phase mismatching is selected from the multi-period gratings. As a result, the radius of mismatching is much less than that of single-period grating.

When the grating period is 6.4 μm and the output green light wave is SHG from a 1.064μm Nd:YAG laser, an experiment with the domain grating sample by PERIE [4.8] shows that the normalized QPM-SHG tangent efficiency is 17%/W. The other parameters are the 3μm width and 10μm spacing of the ridge-channel waveguides with 300 channels.

Example

The substrate of ferroelectric domain inversion grating waveguide is the $LiNbO_3$ crystal. The refractive index of the fundamental e light n_e^{ω} is 2.156, and the refractive index of SHG e wave n_e^{ω} is 2.234. Thus the output of the refractive index of e, by QPM-SHG, is estimated as $N_e^{\omega} \cong n_e^{\omega} + (\Delta n_e^{\omega})/2 = 2.161$, while the $N_e^{2\omega} \cong n_e^{2\omega} + (\Delta n_e^{2\omega})/2 = 2.244$. The difference $N_e^{2\omega} - N_e^{\omega}$ is about 0.08. When annealing proton exchanging is used, the standard data are $(\Delta n_e^{\omega}) = 0.01$ and $(\Delta n_e^{2\omega}) = 0.02$. Thus the difference $N_e^{2\omega} - N_e^{\omega}$, estimated as $\leq (\Delta n_e^{2\omega})/2$, is only about 0.01.

4.4.3.3 Domain Inversion Caused by Periodical Poling in Silica Fibers and Ferroelectrics Waveguides

When the periodically poled technologies by electrical or thermal fields is used, the periodical domain inversion of nonlinear susceptibility is caused in some optical media. Then by using the QPM or the SHG, the difference-frequency and other nonlinear optical effects can be obtained.

One of the topics with bright potential is the periodical poling in ferroelectrics crystal $LiNbO_3$ waveguides. The periodical domain inversion of $\chi^{(2)}$ is caused. The parametric difference-frequency conversion by QPM then obtained has been discussed in [4.11], [4.16], [4.17]. The potential application of this is an all-optical wavelength-conversion device in WDM optical fiber communication systems.

Another topic of great interest is the periodical poling in fused silica fiber. Since the silica is a kind of central-symmetric media, the lowest of the even order of susceptibility $\chi^{(2)}$ is vanished, then no SHG or difference-frequency (which has been discussed already in (2.85), Chap. 2) can be produced. But after periodical poling, a certain domain inversion of $\chi^{(2)}$ will be caused. A very attractive example of an efficient SHG in poling fiber was obtained by means of QPM-SHG [4.18]. Large SHG signals had been observed in a silica waveguide with thermal poling and silver ions injected simultaneously [4.19].

References

[4.1] L.M. Magid: *Electromagnetic Fields, Energy and Waves*, John Wiley, New York (1972)

[4.2] J.M. Manley and H.E. Rowe: Some general properties of nonlinear elements, Part I, General energy relations, *Proc. IRE* **44**, 904–913 (1956)

[4.3] E.P. Wigner: *Group Theory*, Academic Press, New York (1959)

[4.4] F. Zernike and J.E. Midwinter: *Applied Nonlinear Optics*, John Wiley, New York (1973)

[4.5] R.H. Pantell: *Fundamentals of Quantum Electronics*, John Wiley, New York (1969)

[4.6] M.M. Fejer, R.L. Byer, et al.: Quasi-phase-matched second harmonic generation: tuning and tolerance, *IEEE J. Quantum Electron* **28**, 2631–2654 (1992)

[4.7] M. Fujimura, H. Nishihara, et. al.: $LiNbO_3$ waveguide quasi-phase matching second harmonic generation devices with ferroelectric domain-inverted gratings formed by electron-beam scanning, *J. Lightwave Technol.* **11**, 1360–1368 (1993)

[4.8] Y. Ishigame, H. Nishihara, et al.: $LiNbO_3$ Waveguide second-harmonic-generation device phase-matched with a fan-out domain-inverted grating. *Opt. Lett.* **16**, 375–377 (1991)

[4.9] Yili Guo, et al.: Parametric and nonparametric processes in high pressure hydrogen SRS, *Acta Physica Sinica* **34**, 25–33 (1985)

[4.10] Yili Guo, et al.: The transient energy transfer effects in four wave mixing, *Kexue Tongbao* (J. of Academia Sinica, English edition) **29**, 731–734 (1984)

[4.11] M.H. Chou, M.M. Fejer, et al.: Multiple-channel wavelength conversion by use of engineered quasi-phase matching structure in $LiNbO_3$ waveguides, *Opt. Lett.* **24**, 1157–1159 (1999)

[4.12] Yiqiang Qin, Naiben Ming, et al.: Quasi-phase matched harmonic generation through coupled parametric processes in a quasiperiodic optical super-lattice, *J. of Applied Physics* **84**, 6911–6916 (1998)

[4.13] D. Tavemer, D.C. Hanna, et al.: Highly efficient second-harmonic and sum-frequency generation of nanosecond pulses in a cascaded erbium-doped filter: periodically-poled lithum niobate source, *Opt. Lett.* **23**, 162–164 (1998)

[4.14] M. Mlejnek, E.M. Wright, et al.: Second harmonic generation of femto-second pulses at the boundary of a nonlinear dielectric, *Physical Review Letter* **83**, 2934–2937 (1999)

[4.15] G. Imeshev, M.M. Fejer, et al.: Generation of dual-wavelength pulses by frequency doubling with quasi-phase-matching gratings, *Opt. Lett.* **26**, 268–270 (2001)

[4.16] Xueming Liu, Hanyi Zhang and Yili Guo: Theoretical analyses and optimizations for wavelength conversion by quasi-phase-matching difference frequency generation, *J. Lightwave Technology* **19**, 1785–1792 (2001)

[4.17] D. Hafmann, G. Schreiber, et al.: Quasi-phase-matched difference-frequency generation in periodically poled Ti:$LiNbO_3$ channel waveguides, *Opt. Lett.* **24**, 896-898 (1999)

[4.18] V. Pruneri, G. Bonfrate and P.G. Kazansky: Greater than 20%-efficient frequency doubling of 1532-nm nanosecond pulses in quasi-matched germanosilicate optical fibers, *Opt. Lett.* **24**, 208-210 (1999)

[4.19] J. Arentoft, K. Pedersen, et al.: Large second harmonic generation in thermally poled silica waveguide, *OFC'2001,* paper MC6

5. Some Standard Topics in Nonlinear Optics

Nonlinear optics has progressed since the first ruby laser appeared in 1960, and always develops rapidly once a new device or new medium appears. Although nonlinear absorption and nonlinear scattering were known before the laser was invented, most applied nonlinear devices and techniques are based on powerful lasers.

Second harmonic generation by ruby laser was invented in 1961. When the ruby laser with λ_{FUND} = 694.3nm injects into the quartz crystal, its SHG radiation λ_{SHG} = 347.15nm is detected at the output. At that time, the SHG efficiency was only 10^{-8} of input, then stayed at about 10% for a relatively long time, which is why it was usually discussed as a perturbation or a small signal process. In the 1980s, the SHG efficiency became 60 to 70% with new nonlinear crystals and it is now a large-signal effect. High-efficiency solid-state SHG devices are used in research and industry, especially diode-pumped compact SHG devices. Other important second-order nonlinear effects are optical parametric conversion and the optical parametric oscillator (OPO) or amplifier (OPA). By these approaches the optical signals can be shifted from one frequency to another. The pump laser is used to produce tunable radiation by OPO, or to amplify a weak coherent optical signal by OPA. The products of OPO are in the wavelength range of 0.2 to 10μm and some of the high power OPOs can offer 10W. Rough tuning is by angle and temperature, and fine tuning is by electric field or by pressure. However, because of costs and efficiency, these devices face commercial challenge from other devices.

Third harmonic generation (THG) and four-wave mixing (FWM) are two topics in third-order nonlinear effects. THG in nonlinear optics crystals and in metal vapors is well understood and has been used commercially for many years. THG combined with the fundamental wave and the SHG can generate coherent radiation of new frequencies by sum- or difference-frequency mixing. The FWM has the best efficiency and adaptability, and so is used most often. It can be used in frequency replacement and in spectroscopy, but sometimes is treated as a crosstalk or a signal penalty in optical transmission.

Optical phase conjugation (OPC) is very attractive due to its unique features. On the one hand, OPC simulates holography, so it has the properties of the conjugate image. On the other hand, OPC has the time-reversal feature, and so can be considered as a wave-front or wave-distortion compensator. It succeeds in improving beam quality in laser systems and in disturbed-air transmissions. One must find an efficient and fitting medium for OPC mirrors. At least 20% efficiency is needed for commercial use and the OPC device must be compatible with the whole system, as well as compact, simple, and stable.

Stimulated light scattering were observed long before the laser. They were used to measure medium structure and properties, and became important in laser techniques in the 1960s to the 1980s. By frequency mixing of the fundamental wave with SHG and THG, and then stimulating dye lasers and Raman shifters, we have the commercial wide-band tunable lasers. Recently, Brillouin and Raman scatterings are of much concerned in high capacity OFC, which will be discussed in Chap. 8.

5.1 Optical Second Harmonic Generation

The Efficiency of SHG

From (2.73), we have the polarization element on the axis μ

$$P^{(2)}{}_{\mu}(\omega_3) \equiv \varepsilon_0 \chi^{(2)}{}_{\mu\alpha\beta}(-\omega_3;\omega_1,\omega_2)E_\alpha(\omega_1)E_\beta(\omega_2)$$

where $(-\omega_3\,;\,\omega_1,\omega_2)$ means that in this process, two photons with frequencies ω_1 and ω_2 are absorbed by the medium and one photon with frequency ω_3 is emitted from the polarization in the medium. In the SHG process, two injected photons with same frequency ω are absorbed and one photon with frequency 2ω is emitted. The polarization of 2ω is

$$P^{(2)}{}_{\mu}(2\omega) \equiv (1/2)\varepsilon_0 \chi^{(2)}{}_{\mu\alpha\beta}(-2\omega;\omega,\omega)E_\alpha(\omega)E_\beta(\omega). \tag{5.1}$$

In the quasi-monochromatic planar-wave approximation in Sect. 3.2, the polarizations can be expressed. According to the coupled wave equations in three-wave interactions, we have the scalar complex amplitude equations:

$$(\partial/\partial z)E_1(z) = (i\omega_1/2cn_1)\chi^{(2)}\, E_2(z)E^*{}_1(z)e^{-i\Delta kz} \tag{5.2}$$

$$(\partial/\partial z)E_2(z) = (i\omega_2/2cn_2)\chi^{(2)}\, E_1{}^2(z)e^{+i\Delta kz}. \tag{5.3}$$

where $\Delta k = k_2 - 2k_1$ is the phase mismatching parameter in Sect. 4.3.

In the small-signal case, the generated SHG wave is much weaker than the fundamental wave and so $E_1(z)$ = constant. Integrating (5.3) with the initial $E_2(0) = 0$, we have the SHG field at the nonlinear medium exit (L),

$$E_2(L) = -(\omega_2/2cn_2\Delta k)\chi^{(2)}\, E_1{}^2(0)(e^{+i\Delta kz} - 1). \tag{5.4}$$

Using the relation between the light intensity I with field E,

$$I(L) = (1/2)\varepsilon cn|E(L)|^2. \tag{5.5}$$

Then,

$$
\begin{aligned}
I_2(L) &= \frac{\varepsilon c n_2}{2} |E_2(L)|^2 \\
&= \frac{\omega_1^2}{2c^3 n_1^2 n_2 \varepsilon_0} \chi^2 L^2 I_1^2 \cdot \frac{\sin^2(\Delta kL/2)}{(\Delta kL/2)^2} \\
&= \xi \cdot \chi^2 L^2 I_1^2 \cdot \sin c^2(\Delta kL/2)
\end{aligned} \tag{5.6}
$$

where χ^2 is the square of the second-order of susceptibility $\chi^{(2)}$, L is the length of the SHG crystal, I_1 is the injected intensity of fundamental wave and the factor ξ is linked with the linear parameters of material and the beam,

$$
\xi = \frac{\omega_1^2}{2c^3 n_1^2 n_2 \varepsilon_0}. \tag{5.7}
$$

Hence the SHG efficiency at the nonlinear crystal exit with $\Delta k = k_2 - 2k_1$,

$$
\begin{aligned}
\eta_{\text{SHG-small-signal}} &= \frac{I_2(L)}{I_1} \\
&= \frac{\omega_1^2}{2c^3 n_1^2 n_2 \varepsilon_0} \chi^2 L^2 I_1 \cdot \frac{\sin^2(\Delta kL/2)}{(\Delta kL/2)^2} \\
&= \xi \cdot \chi^2 L^2 I_1 \cdot \sin c^2(\Delta kL/2).
\end{aligned} \tag{5.8}
$$

From (5.8), when $\Delta k \neq 0$, $I_2(z)$ and η_{SHG} vary periodically with increasing Δkz (Fig. 5.1).

Since the wave vector $k = \omega/v = \omega n(\omega)/c$, then for the SHG and the fundamental, we have

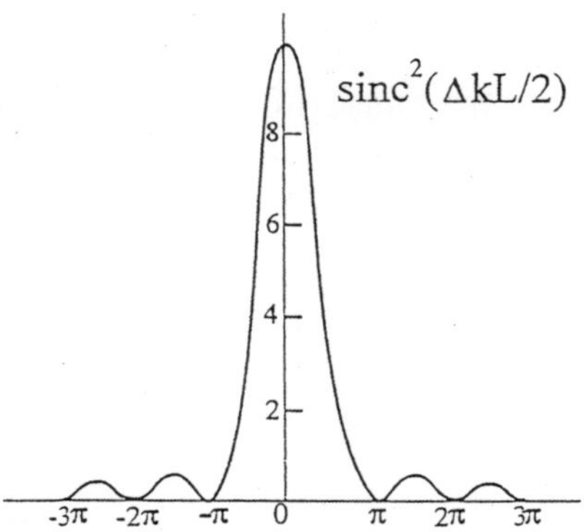

Fig. 5.1 Graph of the Sin c^2 ($\Delta kL/2$) function.

$$k_2 = 2\omega\, n_2(2\omega)/c \tag{5.9}$$

$$k_1 = \omega \cdot n_1(\omega)/c \tag{5.10}$$

$$\Delta k = k_2 - 2k_1 = 2\omega \cdot [n_2(2\omega) - n_1(\omega)]/c. \tag{5.11}$$

In most normal-dispersion media, the condition $[n_2(2\omega) - n_1(\omega)] = 0$ cannot be satisfied. For anisotropic crystals, the phase-matching commonly used is dispersion compensation by birefringence, as in angular (critical) phase-matching and temperature (noncritical) phase-matching (see Sect. 4.3 of Chap. 4).

Due to the crystal size tolerance, assembly errors and misalignment of lasers, as well as the divergence angle of the laser itself, there is some tolerance for phase-matching.

In angular matching (type-I and type-II) for negative uniaxial crystals, the relation of Δk and $\Delta\theta$ was mentioned in Sect. 4.3 of Chap. 4. In SHG, the angular tolerance $\Delta\theta_{\text{angular-matching}}$, which is also the tolerance of the injection laser beam divergence angle $\Delta\theta_{\text{beam-divergence}}$, can also be obtained. In the same way, in temperature phase-matching (90° matching), the differential $[\partial n_2^e(\theta)/\partial\theta] = 0$ (see Chap. 4), and we can also have the angular tolerance for 90° matching in SHG. Hence the 90° phase-matching has superiority in the angle tolerance for the divergence angle of injection laser beam.

High Efficiency SHG

With advanced techniques and novel crystals, such as BBO or LBO and so on, the applied SHG is already 60 to 70% efficient. In these cases, the small-signal analysis is invalid, as is the approximation that $E_1(z)$ = constant in (5.2) and (5.3), and we must directly integrate (5.3). The only way is to solve these coupled wave equations on three-wave interactions simultaneously. Accordingly we have the scalar complex amplitude equations

$$(\partial/\partial z)E_1(z) = (i\omega_1/2cn_1)\chi^2\, E_2(z)E^*{}_1(z) \tag{5.12}$$

$$(\partial/\partial z)E_2(z) = (i\omega_2/2cn_2)\chi^2\, E_1^{\,2}(z) \tag{5.13}$$

where $\Delta k = k_2 - 2k_1 = 0$ is always satisfied for high-efficiency SHG. The exact solutions of (5.12) and (5.13) can be obtained in rather simple form. The final SHG output intensity at the nonlinear medium exit is

$$I_2(L) = I_1(0)\cdot \tan h^2\left(\frac{L}{L_{\text{SH}}}\right). \tag{5.14}$$

Hence, the large-signal SHG efficiency is

$$\eta_{\text{SHG-LS}} = \frac{I_2(L)}{I_1(0)} = \tan h^2\left(\frac{L}{L_{\text{SH}}}\right) \tag{5.15}$$

where $L_{\text{SH}} = [(\omega_1\chi/2cn_1)E(0)]^{-1}$ is the characteristic length of the SHG.

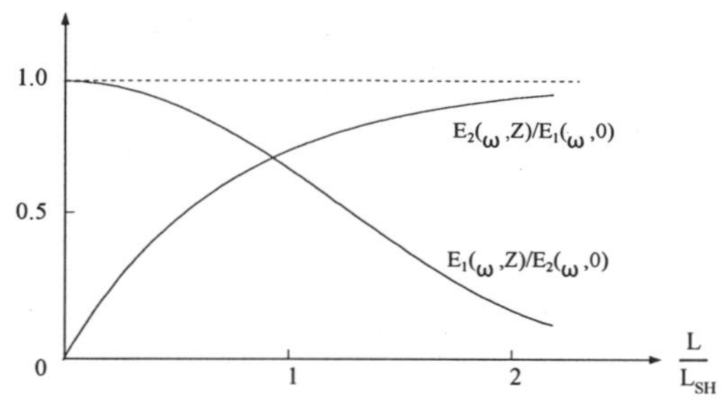

Fig. 5.2 The SHG and fundamental wave varied with the distance (L/L_{SH}).

The SHG and fundamental wave intensities vary with the distance (L/L_{SH}) (Fig. 5.2). From (5.15), when (L/L_{SH}) > 3, the large-signal SHG efficiency approaches 100%. In practice, there are some deviations from the exact phase-matching conditions and the monochromatic planar-wave approximations. Usually, Gauss focusing beams are used. Accordingly, good SHG efficiency is 60 to 70%.

5.2 Three-Wave Mixing (TWM) Optical Parametric Processes

5.2.1 Optical Parametric Amplifiers (OPA)

Modern optical parametric amplifiers are extensions of microwave parametric amplifiers in optics and use the same principles. A strong laser with higher frequency ω_{pump} is used as the pumping source, while a weak input optical signal with relatively low frequency ω_{signal} is amplified. At the same time, another so-called optically idle signal with lower frequency $\omega_{idle} = \omega_{pump} - \omega_{signal}$ is yielded.

According to theories of nonlinear susceptibilities, the three-wave coupled equations and the phase-matching conditions (see Chaps. 2, 3 and 4), the optical parametric amplification can be treated as a three-wave mixed amplifying process. When the pump and signal are injected into the nonlinear medium, they induce the polarization with frequency $\omega_{idle} = \omega_{pump} - \omega_{signal}$ through the second-order electric susceptibility. At the same time, the idle signal coupled with the pump, induces through susceptibility the polarization with frequency $\omega_{signal} = \omega_{pump} - \omega_{idle}$. The latter radiates and enhances the signal light. Because the pump is much stronger than the signal and idle signal, if the phase-matching condition is well satisfied, this nonlinear frequency mixing will act continuously along the space. Then the optical parametric amplifying will occur. The condition of energy conservation is

$$\omega_{\text{pump}} = \omega_{\text{signal}} + \omega_{\text{idle}}. \tag{5.16}$$

The condition of momentum conservation is

$$k_{\text{pump}} = k_{\text{signal}} + k_{\text{idle}}. \tag{5.17}$$

If the three wave vectors are all in the same direction, this becomes

$$n_p \omega_p = n_s \omega_s + n_i \omega_i. \tag{5.18}$$

This analysis is from the three-wave coupled equations in Sect. 3.3.1, (3.52–3.54). The waves vary with z coupled through an identical susceptibility.

$$\frac{\partial}{\partial z} E_1(z) = \frac{i\omega_1}{2cn_1} \chi^{(2)} E_3(z) E_2^*(z) e^{-i\Delta kz} \tag{3.52}$$

$$\frac{\partial}{\partial z} E_2(z) = \frac{i\omega_2}{2cn_2} \chi^{(2)} E_3(z) E_1^*(z) e^{-i\Delta kz} \tag{3.53}$$

$$\frac{\partial}{\partial z} E_3(z) = \frac{i\omega_3}{2cn_3} \chi^{(2)} E_1(z) E_2(z) e^{+i\Delta kz} \tag{3.54}$$

where $\omega_i/c = k_i/n_i$ *and* $\omega_3 = \omega_1 + \omega_2$. In OPA, $\omega_3 = \omega_{\text{pump}}$, $\omega_1 = \omega_{\text{signal}}$, and $\omega_2 = \omega_{\text{idle}}$. The small-signal solution equals the solution limited in small-distance z. The small-signal criterion is the pump field decreased by a small amount, e.g., about 10%, but considered as a constant. Hence $dE(\omega_{\text{pump}}, z)/dz$ is neglected in the three-wave coupled equations and only (3.52) and (3.53) needed to be solved.

5.2.1.1 The Gain Coefficient g_{OPA}

From the Manley-Rowe Relations, each amplified signal-photon created will have an accompanying idle-photon, thus

$$N(\omega_{\text{signal}}, z) - N(\omega_{\text{idle}}, z) = N(\omega_{\text{signal}}, 0) - N(\omega_{\text{idle}}, 0) = \text{constant}. \tag{5.19}$$

We suppose at the initial $N(\omega_{\text{idle}}, 0) = 0$, then

$$\begin{aligned} N(\omega_{\text{signal}}, z) &= N(\omega_{\text{signal}}, 0) + N(\omega_{\text{idle}}, z) \\ &= N(\omega_{\text{signal}}, 0)\{1 + \sin h^2[1 - (\Delta k/2\sigma)^2]^{1/2}(\sigma z)\}/[1 - (\Delta k/2\sigma)^2]) \end{aligned} \tag{5.20}$$

$$\text{where} \quad \sigma = [(1/2c^2)\cdot(\omega_{\text{idle}}{}^2 \omega_{\text{signal}}{}^2/k_2 k_1)^{1/2} |\chi^{(2)}| \cdot E(\omega_{\text{pump}})]. \tag{5.21}$$

This obviously amplifies the signal wave. If z is large but the pump is not greatly decreased, we can use the relation $\sinh(\sigma z) \cong (1/2)\exp(\sigma z)$. Therefore we have the field solutions of amplified signal photons in OPA under small signal and phase-matching conditions

$$N(\omega_{\text{signal}}, z) = N(\omega_{\text{signal}}, 0)\left[1 + \tfrac{1}{2} \cdot e^{g_{\text{OPA}}\, z/2}\right], \tag{5.22}$$

and the gain coefficient $g_{\text{OPA}} = 2\,\sigma$ or

$$g_{\text{OPA}} = \left[\left(\frac{1}{c^2}\right) \cdot \left(\frac{\omega_{\text{idle}}^2 \cdot \omega_{\text{signal}}^2}{k_2 k_1}\right)^{1/2} \cdot \left|\chi^{(2)}\right| \cdot E\left(\omega_{\text{pump}}\right)\right] \tag{5.23}$$

and is propotional to nonlinear susceptibility $\chi^{(2)}$ and pump field strength $E(\omega_{\text{pump}})$.

5.2.2 The Optical Parametric Oscillators (OPO)

OPO is a standard second-order nonlinear effect. If some conditions are satisfied, then when a strong pump injects into an optical cavity including a nonlinear crystal, two (and at least one) new optical waves will oscillate in and exit from the cavity. OPO creates coherent optical waves, and its tunable range is commonly wider than that of the conventional lasers. Figure 5.3 is a schematic diagram of a double resonant parametric oscillator (DRPO). If the pump with higher frequency ω_{pump} is strong enough, the cavity gain will ensure that both ω_{signal} and ω_{idle} oscillate beyond their thresholds.

It is convenient to find the oscillation condition by using the three-wave coupled equations linked with the energy-flow density which passes through the unit area normal to the direction of wave propagation in unit time.

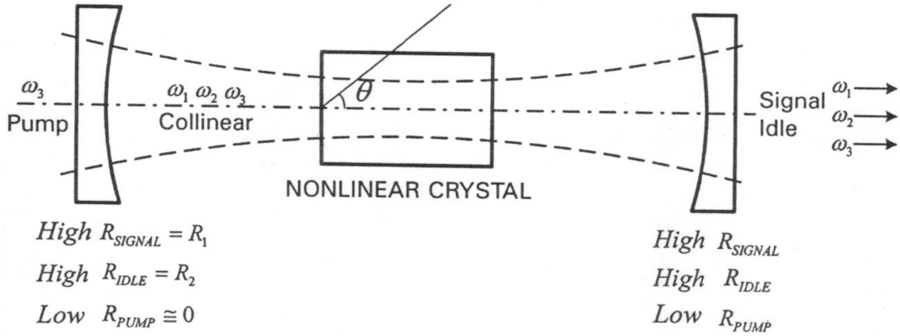

Fig. 5.3 Schematic diagram of a double resonance optical parametric oscillator.

The method to find the self-adapted condition is very similar to the standard method for lasers, in which:

(1) The nonlinear crystal is an optical cavity, the intensity reflectance of the two surfaces are $R_1 = |r_1|^2$ for signal and $R_2 = |r_2|^2$ for idle, where r_1 and r_2 are the field reflectance.

(2) A reference plane at z is set in the cavity. The light starts rightward from the z, then is reflected by the right mirror, transmitted leftward, and then reflected by the left mirror. The light returns to z after a round-trip. If the complex field does not change after this round-trip, the signal and idle oscillations are stable.

(3) The satisfied resultant amplitude condition and the satisfied phase conditions of oscillation can be found in this case.

The essential map of the double resonant parametric oscillator (DRPO) is shown in Fig. 5.3. The waves with frequencies ω_{signal} and ω_{idle} simultaneously oscillate beyond their threshold. If in one reflection G is to be maximum, the momentum conservation of multiwaves (phase-matching $\Delta k = 0$) must be satisfied. In DRPO, the $k_{\text{signal}} = (n_1\omega_{\text{signal}}/c)e_{\text{signal}}$, $k_{\text{idle}} = (n_2\omega_{\text{idle}}/c)e_{\text{idle}}$, and $k_{\text{pump}} = (n_3\omega_{\text{pump}}/c)e_{pump}$ are all collinear in the cavity, thus in scalar form, $n_3\omega_{\text{pump}} = n_1\omega_{\text{signal}} + n_2\omega_{\text{idle}}$, as in (5.18). If the medium is isotropic and of normal dispersion, in the case of phase-mismatching, the quantity on the left side in the above equation will be larger than that on the right side. Hence, anisotropic crystals and angle or temperature phase-matching have to be used. Table 5.1 shows the phase-matching forms for OPO in uniaxial optical crystals.

The primary sketch of single resonant parametric oscillators (SRPO) is shown in Fig. 5.4. Only one wave with frequency ω_{signal} oscillates, due to multiple round trips in the optical cavity and beyond the threshold, but another wave with frequency ω_{idle} escapes from the cavity after one trip.

The standard noncollinear phase-matching condition is:

$$k_3 = k_1 + k_2,$$

where $k_3 = k_{\text{pump}}$ (oblique), $k_1 = k_{\text{signal}}$ (output collinear with cavity axis) and $k_2 = k_{\text{idle}}$ (oblique escape).

Table 5.1 OPO phase-matching in uniaxial crystals.

Crystal category	Phase-matching form	$\omega_p \to \omega_s + \omega_i$
Negative uniaxial crystal	Type-I	$e \to o + o$
Negative uniaxial crystal	Type-II	$e \to o + e$
Positive uniaxial crystal	Type-I	$o \to e + e$
Positive uniaxial crystal	Type-II	$o \to o + e$

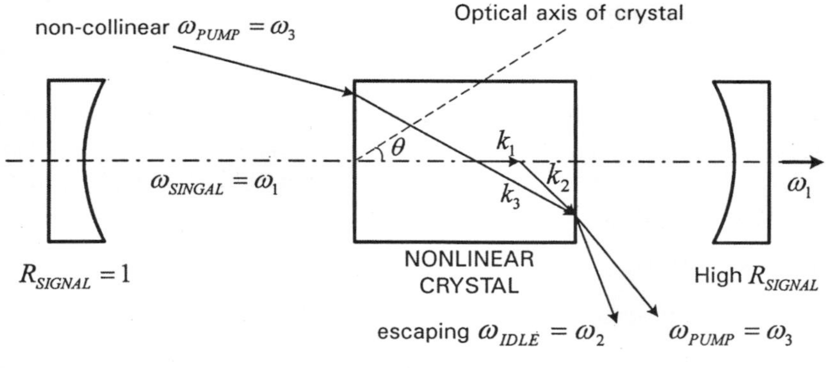

Fig. 5.4 A non-collinear phase-matched signal resonance parametric oscillator.

When $r_2(\omega_{\text{idle}}) = 0$, the oscillation condition of SRPO can be divided into the amplitude condition and the phase condition. The phase φ, wave-vector k_1 and cavity length L have to be satisfied in this condition. The signal intensity reflectance at both mirrors is $R = 1$, hence the SRPO threshold condition is obtained. The needed threshold pump of SRPO is stronger than that of DRPO, and so the output signal is also stronger. Also, only one phase condition must be satisfied, hence the oscillating frequency of SRPO is more stable, but simple structure and easy alignment are DRPO's advantages.

5.2.2.1 Frequency Tuning in OPO

OPO is quite different from conventional lasers. It is not stimulated emission caused by resonant transition, but is caused by the nonlinear three-wave parametric process. Hence the tuning range is notably wide. Frequency tuning is discussed in this section in the case of collinear phase-matching (DRPO). The phase-matching condition is

$$\omega_{\text{pump}} n_{3(e)} = \omega_{\text{signal}} n_{1(o)} + \omega_{\text{idle}} n_{2(o)}, \tag{5.24}$$

and the energy conservation is

$$\omega_{\text{pump}} = \omega_{\text{signal}} + \omega_{\text{idle}}. \tag{5.25}$$

If angular tuning is used and if the oscillation occurs with frequencies $\omega_{\text{signal}(o)}$ and $\omega_{\text{idle}(o)}$ when the angle between the crystal optical axis and the cavity axis is θ_0, the indices are $n_{1(o)}$ and $n_{2(o)}$ respectively. From (5.24), the momentum conservation is

$$\omega_{\text{pump}} n_{3(e)}(\theta_0) = \omega_{\text{signal}(o)} n_{1(o)} + \omega_{\text{idle}(o)} n_{2(o)}. \tag{5.26}$$

Keeping ω_{pump} fixed and turning the crystal by $\Delta\theta$ (i.e. $\theta_0 \rightarrow \theta_0 + \Delta\theta$) will lead to $n_{3(e)} \rightarrow n_{3(e)} + \Delta n_3$. Then from (5.26)

$\omega_{\text{signal(o)}} \rightarrow \omega_{\text{signal(o)}} + \Delta\omega_{\text{signal}}$ and

$\omega_{\text{idle(o)}} \rightarrow \omega_{\text{idle(o)}} + \Delta\omega_{\text{idle}}$.

Frequency variations from medium dispersion cause refraction index changes: $n_{1(o)} \rightarrow n_{1(o)} + \Delta n_1$ and $n_{2(o)} \rightarrow n_{2(o)} + \Delta n_2$. When ω_{pump} is fixed, then from (5.25), $-\Delta\omega_{\text{idle}} = +\Delta\omega_{\text{signal}}$. Hence (5.24) must be satisfied by a new group of frequencies:

$$\omega_{\text{pump}}(n_{3(e)} + \Delta n_3)$$
$$= (\omega_{\text{signal(o)}} + \Delta\omega_{\text{signal}})(n_{1(o)} + \Delta n_1) + (\omega_{\text{idle}(o)} + \Delta\omega_{\text{idle}})(n_{2(o)} + \Delta n_2).$$

Using (5.24),

$$\omega_{\text{pump}} n_{3(e)} - \omega_{\text{signal}} n_{1(o)} - \omega_{\text{idle}} n_{2(o)} = 0 \quad \text{and}$$

$$n_{2(o)}\Delta\omega_{\text{idle}} = -n_{2(o)}\Delta\omega_{\text{signal}}.$$

Neglecting the second-order quantity $\Delta n \Delta\omega$, the signal frequency tuning is, in collinear phase-matching (DRPO),

$$\Delta\omega_{\text{signal}} = \frac{(\omega_{\text{pump}}\Delta n_3 - \omega_{\text{signal}(o)}\Delta n_1 - \omega_{\text{idle}(o)}\Delta n_2)}{n_{1(o)} - n_{2(o)}}. \tag{5.27}$$

Since the pump is the e light and ω_{pump} is fixed, $n_{3(e)}$ is a function only of θ, $n_{3(e)} = n_{3(e)}(\theta)$. The $n_{1(o)}$ and $n_{2(o)}$ are the refractive indices independent of θ, and only depend on ω. Thus

$$\Delta n_1 = (\partial n_{1(o)}/\partial\omega|_{\omega\text{signal}(o)})\cdot\Delta\omega_{\text{signal}}, \tag{5.28}$$

$$\Delta n_2 = (\partial n_{2(o)}/\partial\omega|_{\omega\text{idle}(o)})\cdot\Delta\omega_{\text{idle}}, \tag{5.29}$$

$$\Delta n_3 = (\partial n_{3(e)}/\partial\theta|_{\theta=\theta_0})\cdot\Delta\theta. \tag{5.30}$$

5.3 Optical Four-Wave Mixing (FWM) and Optical Phase Conjugation (OPC)

OPC is a magic and fantastic technique in the optics, but the condition for high efficiency is always a little critical. So, it acts almost as a hopeful candidate.

5.3.1 Conditions Required in Parametric FWM

Four-wave mixing exposes the processes in which the three fields $E_1(z)$,

$E_2(z)$ and $E_3(z)$ inject into and interact simultaneously with the medium. Through the susceptibility $\chi^{(3)}(\omega)$, the optical polarization $P^{(3)}(\omega)$ is induced in the medium. $P^{(3)}(\omega)$'s amplitude, frequency, and phase are all related to $\chi^{(3)}(\omega)$, and $P^{(3)}(\omega)$ radiates the fourth field $E_4(z)$. Actually, FWM shows all the four waves interact with each other through the optical medium. Of course, in the lossless regions of the medium, the processes are inverted. Any three of them could generate the fourth.

If we choose the relation $\omega_1 + \omega_3 = \omega_2 + \omega_4$ as an example of non-degenerated four-wave mixing, the phase-mismatching factor $\Delta k = k_1 + k_3 - k_2 - k_4$. The polarization of frequencies $P^{(3)}(\omega_1)$, $P^{(3)}(\omega_2)$, $P^{(3)}(\omega_3)$ and $P^{(3)}(\omega_4)$ can be written as well. The scalar-form equations of four-wave mixing (3.56–3.59) are listed in Sect. 3.3.

Each of the four equations has five terms on the right side. The first one is a standard FWM term, its effeciency is determined by Δk. The second term is a self-action with the same frequency, such as SPM. The third, fourth and fifth terms are mutual actions with different frequencies, as XPM and SRS.

The solutions of these coupled-wave equations reveal third-order nonlinearities, such as third-harmonic generation, four-wave mixing (FWM), and stimulated light-scattering. Since there is one more wave in FWM than in TWM, more nonlinear processes occur in FWM. The standard example of nonparametric FWM is stimulated Raman scattering (SRS) which will be discussed in the next section. Many FWM is optical parametric four-wave mixing.

These are the four-wave mixing equations (3.56–3.59) with the relation $\omega_1 + \omega_3 = \omega_2 + \omega_4$ in nondegenerate four-wave mixing $\Delta k = k_1 + k_3 - k_2 - k_4$. The relation $\omega_i/c = k_i/n_i$ is used in these equations.

$$\frac{\partial E_1(\omega_1,z)}{\partial z} = \frac{i3\omega_1}{4cn_1}\Big[\chi^{(3)} E_2 E_3^* E_4 e^{-i\Delta kz} + \tfrac{1}{2}\chi_{11}^{(3)}|E_1|^2 E_1 + \chi_{12}^{(3)}|E_2|^2 E_1 + \chi_{13}^{(3)}|E_3|^2 E_1 + \chi_{14}^{(3)}|E_4|^2 E_1\Big]$$

$$\frac{\partial E_2(\omega_2,z)}{\partial z} = \frac{i3\omega_2}{4cn_2}\Big[\chi^{(3)} E_1 E_3 E_4^* e^{i\Delta kz} + \tfrac{1}{2}\chi_{22}^{(3)}|E_2|^2 E_2 + \chi_{21}^{(3)}|E_1|^2 E_2 + \chi_{23}^{(3)}|E_3|^2 E_2 + \chi_{24}^{(3)}|E_4|^2 E_2\Big]$$

$$\frac{\partial E_3(\omega_3,z)}{\partial z} = \frac{i3\omega_3}{4cn_3}\Big[\chi^{(3)} E_1^* E_2 E_4 e^{-i\Delta kz} + \tfrac{1}{2}\chi_{33}^{(3)}|E_3|^2 E_3 + \chi_{31}^{(3)}|E_1|^2 E_3 + \chi_{32}^{(3)}|E_2|^2 E_3 + \chi_{34}^{(3)}|E_4|^2 E_4\Big]$$

$$\frac{\partial E_4(\omega_4,z)}{\partial z}=\frac{i3\omega_4}{4cn_4}\left[\chi^{(3)}E_1E_2^*E_3e^{i\Delta kz}+\tfrac{1}{2}\chi_{44}^{(3)}|E_4|^2E_4\right.$$
$$\left.+\chi_{41}^{(3)}|E_1|^2E_4+\chi_{42}^{(3)}|E_2|^2E_4+\chi_{43}^{(3)}|E_3|^2E_4\right].$$

Parametric FWM has degenerate four-wave mixing (DFWM) with all the four waves at the same frequency, nearly degenerate four-wave mixing (NDFWM) with very small frequency differences, dual degenerate four-wave mixing (DDFWM), with two frequencies ($\omega_1=\omega_3$, $\omega_2=\omega_4$) and non-degenerate four-wave mixing ($\omega_1\neq\omega_3\neq\omega_2\neq\omega_4$).

In parametric processes, both energy conservation and momentum (wave-vector k) conservation must be satisfied simultaneously. Since the energy $\hbar\omega$ and momentum $\hbar k=\hbar\omega/v$ are related through by ω, the magnitudes of the wave-vectors k_1, k_2, k_3, k_4 are all restricted by the photon energy. To ensure that the vector-polygon is closed ($\Delta k=0$), the directions of k_s are restricted.

5.3.2 Optical Phase Conjugation (OPC)

5.3.2.1 Phase Distortion and Phase-Conjugated Waves

Most optical media are not absolutely homogeneous. Avoidance of phase distortion is very important for an intense pulse beam in the air and for a series of very short and compact light pulses in glass fiber. In microwave transmission, the wavefront is distorted by inhomogeneity of the air density. A distributed adjustment antenna transmits a wavefront which is phase-conjugate to the distorted wavefront. This is an adaptive microwave system. Similarly, phase-conjugation optical reflective mirrors, or adaptive optics, have already been used in engineering. However, these adaptive systems are extremely complicated, as they need a phase-distortion detection system, servo-control mechanisms, distributed adjustable antennas or optical reflective mirrors, and so on.

The techniques of nonlinear OPC are very interesting. The OPC waves generated in a nonlinear medium have the time-reversal feature, and so may be considered as a wavefront (or wave-distortion) compensator in long-distance transmission. Only a compact nonlinear OPC mirror must be installed in the optical path. The original signal wave $E_s(r,t)$ in the transmitted optical path is

$$E_s(r,t)=\mathrm{Re}\{E_s(r)\exp[-i(\omega_s t-k_s z)]\}\tag{5.31}$$

where $E_s(r)$ is the complex amplitude of $E_s(r,t)$. The phase conjugation wave $E_{\mathrm{OPC}}(r,t)$ is

$$E_{\mathrm{OPC}}(r,t)=\mathrm{Re}\{E_s^*(r)\exp[-i(\omega_s t\pm k_s z)]\}\tag{5.32}$$

where $E_s^*(r)$ is the complex amplitude of $E_{OPC}(r,t)$. The + sign corresponds to the backward phase conjugation wave propagated in the $-z$ direction, and the – sign corresponds to the forward phase conjugation wave propagated in the $+z$ direction, the direction of the original signal wave.

If the signal wave is an OPC wave, its complex amplitude is also the phase conjugation of $E_s(r)$. Suppose the complex amplitude of $E_s(r,t)$ is

$$E_s(r) = A(r) \exp[i\varphi(r)].$$

Then the complex amplitude of $E_{OPC}(r,t)$ is

$$E_{OPC}(r) = A(r) \exp[-i\varphi(r)], \tag{5.33}$$

where the real quantities $A(r)$ and $\varphi(r)$ are the amplitude and phase of $E_s(r,t)$. Hence, the phase-conjugated wave is not the phase conjugation of the total equation of $E_s(r,t)$, but only the phase conjugation of the complex amplitude $E_s(r)$. It is not concerned with the temporal phase factor $\exp[-i(\omega_s t - k_s z)]$.

Since the optical field is a real field, $E_s(r,t)$ can be written as

$$E_s(r,t) = (1/2)E_s(r) \exp[-i(\omega_s t - k_s z)] + (1/2)E_s^*(r) \exp[i(\omega_s t - k_s z)] \tag{5.34}$$

and its backward phase conjugation wave is

$$E_{OPC}(r,t) = (1/2)E_s^*(r) \exp[-i(\omega_s t + k_s z)] + (1/2)E_s(r) \exp[i(\omega_s t + k_s z)]. \tag{5.35}$$

On comparing these two equations, we see that $E_{opc}(r,t) = E_s(r,-t)$. Hence, $E_{opc}(r,t)$ is normally called the time-inversion wave of $E_s(r,t)$.

The scalar wave equation of the linear polarized signal field is

$$\nabla^2 E_s(r) + [\omega_s^2 \mu\varepsilon(r) - k_s^2]\, E_s(r) + 2ik_s[\partial E_s(r)/\partial z] = 0. \tag{5.36}$$

The complex conjugate of (5.36) is

$$\nabla^2 E_s^*(r) + [\omega_s^2 \mu\varepsilon(r) - k_s^2]\, E_s^*(r) + 2ik_s[\partial E_s^*(r)/\partial z] = 0. \tag{5.37}$$

It is obvious that (5.37) is the wave equation satisfied by $E_{opc}(r,t) = \mathrm{Re}\{E_s^*(r) \exp[-i(\omega_s t + k_s z)]\}$. Its direction is opposite to that of $E_s(r,t)$. Hence (5.37) is the wave equation of the backward OPC.

5.3.2.2 Wavefront Distortion Compensation by OPC Mirror

We treat the nonlinear optical medium, in which the OPC wave was generated when the original signal wave interacted with it, as an OPC mirror. To compare the OPC mirror with conventional mirrors helps in understanding the compensation of the wavefront distortion by OPC mirrors.

1. A point light source

When a light beam from a point source travels to a conventional mirror, it is reflected through an angle of 2θ, where θ is the angle between the ray and the normal to the mirror, and continues to travel away from the point source (Fig. 5.5(a)); but should the beam strike an OPC mirror, it will return to its source on exactly the same path which the ray first took to the OPC mirror. If one were to look into an OPC mirror, one would always see oneself, no matter where one is standing and no matter how large θ is. Any distortions imposed on the beam by inhomogeneities between the source and the OPC mirror will then be exactly compensated for by the returning beam, which will encounter all these inhomogeneities in reverse order and from the other side of each. Thus, there will be antidistortions in the returning beam, which will exactly cancel the distortions in the original beam. Thus the distortions are always completely eliminated (Fig. 5.5(b)). Also, the beam will diverge as it travels to the mirror, but on reflection will converge again back on the point from which it came!

Figure 5.5(c) shows the forward OPC wave generated from the transparent OPC mirror of which the wave-vectors are well arranged, hence the phase-conjugated wave can transmit forward through the optically transparent medium. It takes the + sign in $E_{\mathrm{OPC}}(r,t)$ and propagates in the same direction as does $E_s(r,t)$. Finally it converges to the mirror-image of the source. This wave is expected to be a compensator in OPC transmission.

2. A perfect planar wave

Here a perfectly planar wave is transmitted in an inhomogeneous air medium of which the inhomogeneity is lumped as a plate of glass. When the wave is reflected by a conventional mirror and again passes through the glass plate, it has double distortion, as shown in Fig. 5.6(a). Fig. 5.6(b) shows that after a round trip, the wavefront distórtion is compensated by the PC mirror. Figure 5.6(c) shows that the wavefront of the PC wave is compensated by the phase conjugation transparent (PCT) mirror.

5.3.2.3 Degenerate Four-Wave Mixing OPC

OPC can be generated in both parametric three-wave and parametric four-wave processes. In Chap. 2, parametric three-wave processes are based on second-order electric susceptibility $\chi^{(2)}(-\omega_3;\omega_1,\omega_2)$. In Chap. 4, the OPC wave in three-wave processes is a forward wave, considering the phase-matching condition. After solving the three-wave coupled equations in Chap. 3 with the small-signal or large-signal conditions, we could find solutions of forward OPC waves at exit surfaces of nonlinear crystals, with $E_{\mathrm{OPC}}(L) \propto E_s^*(0)$. But if $\Delta k \neq 0$, the intensity of the OPC wave will be considerably weakened.

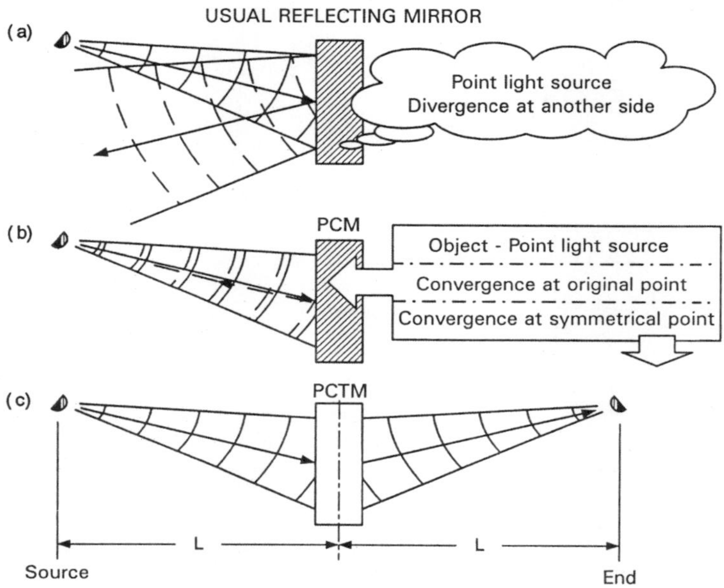

Fig. 5.5 a–c A comparison of the spatial properties of the PCM, the PCTM and a usual reflecting mirror.

Four-wave mixing, especially degenerate four-wave mixing, are the convenient approaches to use OPC, as they are quite simple. Consequently in some literature, FWM and OPC are closely linked. According to the different combinations of wave vectors (k_1, k_2, k_3, k_4), the OPC wave is either a forward wave or a backward wave. Figure 5.7(a) shows the nearly-degenerate FWM-OPC wave is a forward wave in cases such as dispersion compensation in long-distance optical fiber. Figure 5.7(b) shows the degenerate FWM-OPC wave is a backward OPC wave in some experiments for transmitted image restoration in nonlinear crystals.

A schematic sketch of the conventionally degenerate FWM-OPC is shown in Fig. 5.8. E_1 and E_2 are two pumps propagating opposite to each other. Usually there is a small angle between the pumps and the signal. Their wave vectors are

$$k_1 + k_2 = 0,\ k_3 + k_4 = 0,\ \text{or}\ k_1 + k_2 + k_3 + k_4 = 0 \tag{5.38}$$

if all the waves satisfy the quasi-monochromatic planar-wave and slowly-varying conditions. Then, with the scalar condition $\Delta k = k_1 + k_3 - k_2 - k_4 = 0$, $\omega/c = k/n$, the scalar forms of the four-wave coupling-wave equations are:

$$\frac{\partial E_{1pump}(\omega, z)}{\partial z} = \frac{i3\omega}{4cn}\left[\chi^{(3)}_{OPC} E^*_{2pump} E_{3OPC} E_{4signal}\right] \tag{3.56}$$

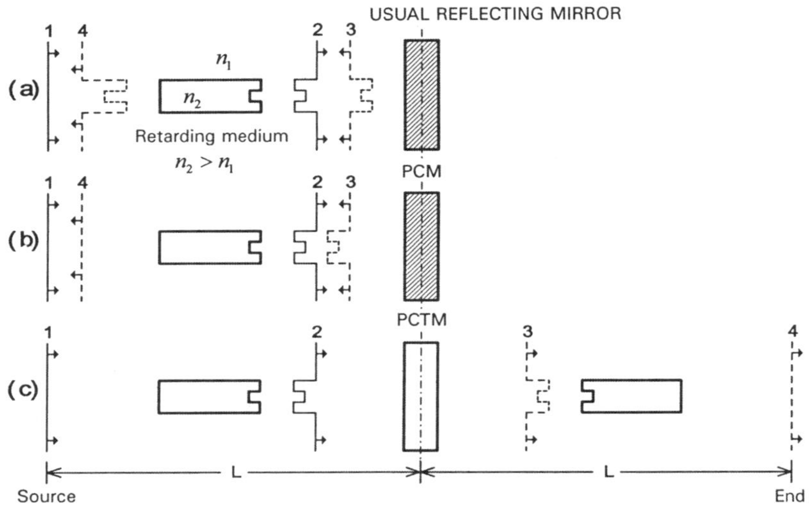

Fig. 5.6 a–c Wave front distortions caused and compensated by the usual reflecting mirror, the PC mirror and the PCT mirror.

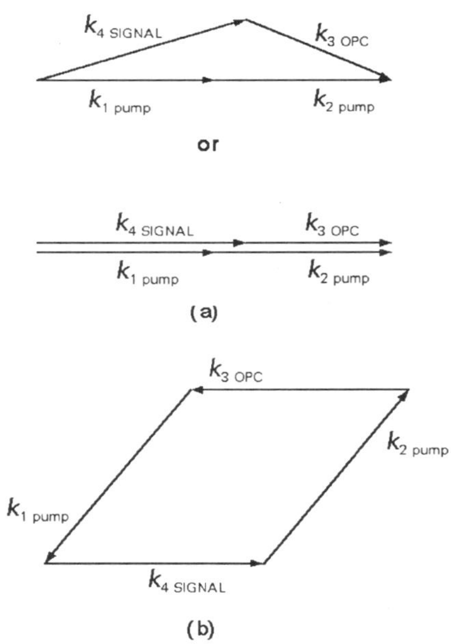

Fig. 5.7 a, b The wave-vector patterns in the FWM-OPC. **(a)** The forward nearly degenerated FWM-OPC. **(b)** The backward degenerated FWM-OPC.

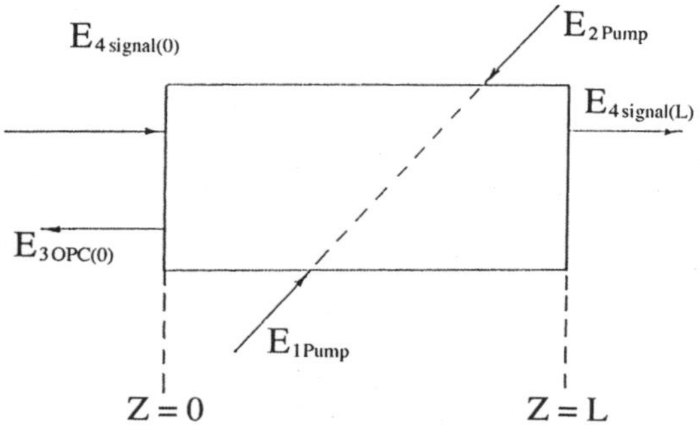

Fig. 5.8 A schematic sketch of the conventionally degenerated FWM-OPC.

$$\frac{\partial E_{2pump}(\omega, z)}{\partial z} = \frac{i3\omega}{4cn}\left[\chi^{(3)}_{OPC} E^{*}{}_{1pump} E_{3OPC}\, E_{4signal}\right] \tag{3.57}$$

$$\frac{\partial E_{3OPC}(\omega_3, z)}{\partial z} = \frac{i3\omega}{4cn}\left[\chi^{(3)}_{OPC} E_{1pump} E_{2pump}\, E^{*}_{4signal}\right] \tag{3.58}$$

$$\frac{\partial E_{4signal}(\omega, z)}{\partial z} = \frac{i3\omega}{4cn}\left[\chi^{(3)}_{OPC} E_{1pump} E_{2pump}\, E^{*}_{3OPC}\right. \tag{3.59}$$

When the pump depletion is neglected (small-signal condition), if all the waves are polarized in the x direction, then the equations (3.58) and (3.59) become

$$(\partial/\partial z)E_{3\ \mathrm{OPC}}(z) = i(\zeta\chi^{(3)}{}_{\mathrm{opc}})\cdot E^{*}{}_{4\ \mathrm{signal}}\ (z) \tag{5.39}$$

$$(\partial/\partial z)E^{*}{}_{4\ \mathrm{signal}}\ (z) = i(\zeta\chi^{(3)}{}_{\mathrm{opc}})^{*}\cdot E_3(z) \tag{5.40}$$

where

$$\chi^{(3)}{}_{\mathrm{opc}} = \chi^{(3)}{}_{xxxx}(-\omega_4;\omega_1,-\omega_2,\omega_3)$$

and

$$(\zeta\chi^{(3)}{}_{\mathrm{opc}})^{*} \equiv -(3\,\omega/4cn)\cdot E_{1\ \mathrm{pump}}(L)\cdot E_{2\ \mathrm{pump}}(0)\cdot\chi^{(3)}{}_{\mathrm{opc}}\ .$$

The magnitude of ζ indicates the boundary condition of the pumps, the light beam parameter (ω) and the material parameter (n). The higher the nonlinear susceptibility $\chi^{(3)}{}_{\mathrm{opc}}$, the larger the growth rate of OPC wave, $(\partial/\partial z)E_{3\ \mathrm{OPC}}(z)$, in (5.39).

If we define the $E_{4\ \mathrm{signal}}(0)$ ($+z$ direction) as the input signal and $E_{3\ \mathrm{opc}}(z)$ as the backward OPC wave, then the initial condition is $E_{3\ \mathrm{opc}}(L) = 0$. Thus the efficiency of the FWM-OPC process is defined as

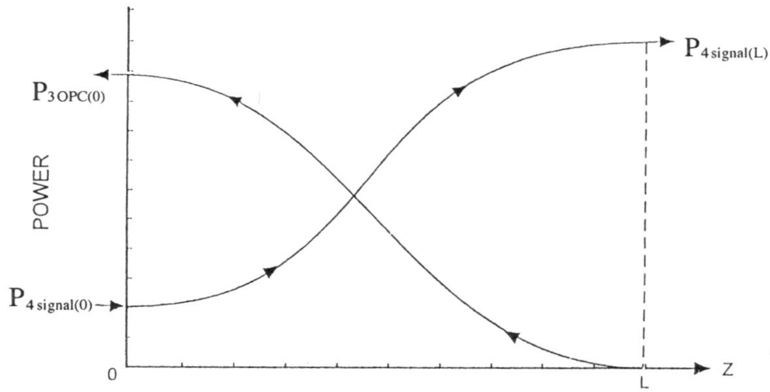

Fig. 5.9 The amplifying property of the degenerate four-wave mixing.

$$\eta_{\text{opc}} \equiv |E_{3\ \text{opc}}(0)|^2/|\ E_{4\ \text{signal}}(0)|^2, \tag{5.41}$$

and we have the FWM-OPC efficiency

$$\eta_{\text{opc}} = \tan^2 (|\zeta\cdot\chi^{(3)}{}_{\text{opc}}|L). \tag{5.42}$$

When $|\zeta\cdot\chi^{(3)}{}_{\text{opc}}|L$ is comparatively small, the OPC efficiency is relatively low. If we use a segment of the conventional fiber as an OPC device, the η_{opc} becomes larger only with increasing fiber length L and increasing pump power. At the 1.31μm or 1.55μm band of the conventional fiber, the fiber OPC efficiency of the signal wave to OPC wave is only a few percent and sometimes about 10% because of the low $\chi^{(3)}$. The $\chi^{(3)}$ is high in semiconductors, hence it is always considered as a candidate of OPC device. From (5.42), when $3\pi/4 > |\zeta|L > \pi/4$, $\eta_{\text{opc}} > 1$. That means the OPC light reflected from the mirror could be stronger than the injection signal. Figure 5.9 shows this amplifying property. Notice that the energy amplified is obtained from pumps through four-wave mixing.

5.4 Nonlinear Light Scattering

5.4.1 Stimulated Brillouin Scattering (SBS)

Brillouin scattering is the light wave's inelastic scattering by acoustic waves in the medium. In this process, a very small part of photon energy transfers to the phonon, consequently a tiny frequency shift of light wave is induced. When an intense laser beam or pulses propagates in the medium, the coherent acoustic wave will be enhanced by the electrostriction and the scattered light wave is amplified coherently. As a result, the scattering process becomes stimulated Brillouin scattering (SBS). This was predicted and calculated by Brillouin in 1922, and observed experimentally in liquid by Gross. The wave number of frequency shift is only about 0.1 cm^{-1}.

5.4.1.1 Acoustic Wave and Light Wave Coupled Equations in Brillouin Scattering

Suppose $q(z,t)$ is the departure from balance of a particle at position z. Three kinds of force act on the particle while the optical field exists. These are the elastic restoring force $(1/\alpha)(\partial^2 q/\partial z^2)$, the damping force $-D(\partial q/\partial t)$, and the electrostrictive force $(S/2)(\partial E^2/\partial z)$. The electrostriction coefficient or optoelastic coefficient is $S \equiv \rho(d\varepsilon/d\rho)_T$, where ρ is the medium's density, T is the temperature, and α is the medium's elastic coefficient. Then the following equations (5.43) and (5.44) of one-dimensional acoustic wave are derived [5.1],

$$(1/\alpha)(\partial^2 q/\partial z^2) - D(\partial q/\partial t) + (S/2)(\partial E^2/\partial z) = \rho(\partial^2 q/\partial t^2). \tag{5.43}$$

Thus the electromagnetic field E is linked with the acoustic wave q through the electrostriction coefficient S. With assumptions and simplification, the final acoustic wave equation is

$$\frac{dq_s}{dz_s} = -\frac{D}{2\rho v_s}\cdot q_s - \frac{S}{8\rho v_s^2}\cdot E_2 E_1^*, \tag{5.44}$$

where v_s is the acoustic speed in the medium, q_s is the acoustic-wave field amplitude, z_s is the propagation distance of the acoustic wave along its wave-vector k_s, E_2 is the injection light wave (pump) field, and E_1 is the Brillouin scattering light wave field.

When the pump laser interacts with the medium, the induced strongly acoustic field changes the dielectric constant by $d\varepsilon$. The additional nonlinear polarization term caused is

$$P_{NL} = (d\varepsilon)E(z,t). \tag{5.45}$$

Because of the electrostriction, or optoelastic, coefficient, $d\varepsilon = S(d\rho/\rho)$, the relative change of mass density $(d\rho/\rho) \propto$ the strain of the medium, but the strain $\propto (\partial q/\partial z_s)$. Consequently

$$P_{NL} = -SE(z,t)\cdot[\partial q(z,t)/\partial z_s]. \tag{5.46}$$

Substituting (5.46) into Maxwell's equations in Sect. 3.1 yields

$$\nabla^2 E(z,t) = \mu\varepsilon[\partial^2 E(z,t)/\partial t^2] - \mu(\partial^2 P_{NL}/\partial t^2) \tag{5.47}$$

where μ is the megnetic permeability. We have the equations for the

Brillouin scattering light wave field E_1, the injection light wave (pump) field E_2, and the acoustic field amplitude q_s (both the light wave field amplitude E and acoustic field amplitude q_s coupled)

$$\frac{dE_{1Brillouin}}{dz_1} = -\frac{c\xi\omega_1\omega_s}{2n_1v_s}\cdot E_{2\,pump}q_s^* - \frac{\delta\cdot E_{1Brillouin}}{2} \tag{5.48}$$

$$\frac{dE_{2\,pump}}{dz_2} = -\frac{c\xi\omega_2\omega_s}{2n_2v_s}\cdot E_{1Brillouin}q_s^* - \frac{\delta\cdot E_{2\,pump}}{2} \tag{5.49}$$

In these, ω_1, ω_2, and ω_s are the frequencies of Brillouin light wave, pump wave and acoustic wave, respectively, v_s is the velocity of acoustic wave, $S \equiv \rho(d\varepsilon/d\rho)_T$ is optoelastic coefficient, $\xi = S\mu$ is a material parameter and δ is the medium's attenuation coefficient for the light waves.

5.4.1.2 SBS is Always Opposite in Direction to the Pump

Figure 5.10 exhibits the phase-matching condition $k_{\text{s-acoustic}} = k_{2\text{ pump}} - k_{1\text{ SBS}}$ between these wave-vectors. Since energy conservation $\omega_{2\text{ pump}} = \omega_{1\text{ SBS}} + \omega_{\text{s-acoustic}}$ in SBS, and $\omega_{\text{a-coustic}}$ is much less than the other two, consequently $\omega_{\text{pump}} \cong \omega_{\text{SBS}}$, and $k_{\text{pump}} \cong k_{\text{SBS}}$. Accordingly we have $k_{\text{s-acoustic}} = 2\cdot(k_{1\text{ SBS}}\sin\theta)$. When $\theta = \pi/2$, $k_{\text{s-acoustic}} = k_{\text{s-acoustic-max}}$ the directions of SBS and the pump are opposite in space, and the wave-vector $k_{\text{s-acoustic}} = k_{2\text{ pump}} - k_{1\text{ SBS}}$ is maximum. The relation between these amplitudes is

$$k_{\text{s-acoustic}} = k_{\text{s-acoustic-max}} = 2k_{1\text{ SBS}}. \tag{5.50}$$

From (5.50), the threshold will be at a minimum and consequently the most useful for attaining the $I_{\text{SBS-threshold}}$. Hence SBS is always opposite in direction to the pump.

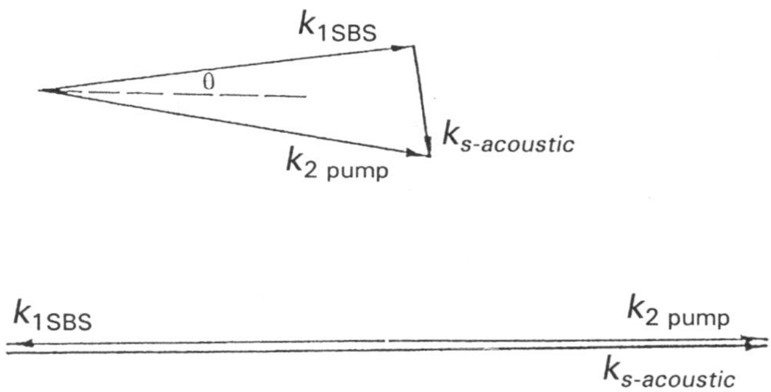

Fig. 5.10 The phase-matched SBS excited backwards.

Example

Some data for (5.50) in silica: $(1/\alpha) = 5\times10^{10}$ Newton/m^3, $\delta_s = 20\text{cm}^{-1}$, $\delta = 0.02\text{cm}^{-1}$, $\varepsilon_{1r} = \varepsilon_{2r} = 2.2$, $S \cong \varepsilon_0 = 10^{-11}$(MKS), $k_{1\ \text{SBS}} = (2\pi/\lambda_{1\ \text{SBS}})$. Since $k_{\text{s-acoustic}} = 2k_{1\ \text{SBS}}$, so $k_{1\ \text{SBS}}\ k_{\text{s-acoustic}} = 2k_{1\ \text{SBS}}^2$.

5.4.1.3 Stimulated Brillouin Scattering

When a strong pump laser beam or intense signal pulses interact with a Brillouin scattering medium, both the coherent acoustic wave and the coherent Brillouin scattering are amplified. If the pump depletion, due to the energy transfered from the pump to acoustic wave and Brillouin scattering, can be neglected, then solutions are needed only for (5.44) and (5.48). After some axis transforms and supposing that two waves q_s and $E_{1\ \text{Brillouin}}$ are increased exponentially, the gain coefficient of Brillouin scattering $G_{\text{Brillouin}}$ can be found by the setting the coefficient of the determinant of the equations equal to zero.

If $G_{\text{Brillouin}} \geq 0$, then the acoustic wave q_s and the Brillouin scattering light wave are both amplified. Since the power density $I_2 = 2\varepsilon_2 v_2 |E_2|^2$ (W/cm^2), the SBS threshold pump power density is

$$\begin{aligned} I_{\text{SBS-threshold}} &\geq \frac{c\delta_s \delta \varepsilon_1 \varepsilon_2}{2\pi^2 \alpha S^2} \cdot \frac{\lambda^2_{1\text{Brillouin}}}{n^2_{1\text{Brillouin}} \cdot n_{2\text{pump}}} \\ &= \varsigma \cdot \frac{\lambda^2_{1\text{Brillouin}}}{n^2_{1\text{Brillouin}} \cdot n_{2\text{pump}}} \end{aligned} \tag{5.51}$$

where

$$\varsigma = \frac{c\delta_s \delta \varepsilon_1 \varepsilon_2}{2\pi^2 \alpha S^2} \tag{5.52}$$

is the material coefficient for SBS, ε_1 and ε_2 are the medium's dielectric constants for the SBS and pump wavelengths respectively, v_2 is the pump-wave speed in the medium, α is the elastic coefficient of the medium, S is the electrostriction coefficient of the medium, and $n_{1\ \text{Brillouin}}$ and $n_{2\ \text{pump}}$ are the refractive indices of Brillouin scattering and acoustic wave-vectors in the medium respectively.

5.4.2 Stimulated Raman Scattering (SRS)

5.4.2.1 Physics of SRS

When a strong pump laser with frequency ω_{pump} injects into some medium, such as hydrogen gas, liquid CS_2, or solid silica, a scattering radiation ω_{Raman} with considerable frequency red-shift is generated. The Raman scattering is the light wave inelastically scattered by the vibration energy levels in the medium. In this process, a part of photon energy, which equals the vibration level energy $h\omega_{\text{vibration}}$, is transferred to the optic phonon. In this case a light wave, with a shift about 100 to 1000 cm^{-1}, called Stokes scattering, is induced. Figure 5.11 shows the spontaneous Raman scattering spectrum.

In Fig. 5.11(a), $v = 1$ is the real vibration level, and the left vertical line in Fig. 5.11(b) is the red-shifted Stokes line $\omega_s = \omega_{s1}$ corresponding to $v = 1$ level scattering. The anti-Stokes process in Fig. 5.11(a) is really a four-wave mixing, and the right vertical line in Fig. 5.11(b) is the blue-shifted anti-Stokes line ω_{as} for which no real level exists. Quantitatively, the 1st Stokes (stimulated) and anti-Stokes (FWM) are

$$\omega_{\text{Stokes}} = \omega_{\text{pump}} - \omega_{\text{vibration}} \tag{5.53}$$

$$\omega_{\text{anti-Stokes}} = 2\,\omega_{\text{pump}} - \omega_{\text{Stokes}} = \omega_{\text{pump}} + \omega_{\text{vibration}}. \tag{5.54}$$

If the injected laser beam or pulses are very intense, the scattered light wave will be amplified with significant gain G_{Raman} and enhanced coherently. As a result, the scattered process becomes SRS. At the same time, FWM reveals comparatively efficient and a series of spectrum lines corresponding to ω_{s2}, ω_{as2}, ω_{s3}, ... , are shown (Fig. 5.12). These are all generated by FWM, and the wave-vector diagrams of phase-matching are shown in Fig. 5.13.

Example

How large is the Raman shift? If it is in the magnitude of $(1/\lambda_{\text{pump}}) - (1/\lambda_{\text{Stokes}}) = 1000\text{cm}^{-1}$ (1000 wavelengths in 1cm), when $\lambda_{\text{pump}} = 694.3\text{nm}$ (ruby laser), the $1/\lambda_{\text{pump}} = 14400\text{cm}^{-1}$, so $(1/\lambda_{\text{Stokes}}) = (1/\lambda_{\text{pump}}) - 1000\text{cm}^{-1} = 13400\text{cm}^{-1}$. Hence $\lambda_{\text{Stokes}} = 746.3\text{nm}$.

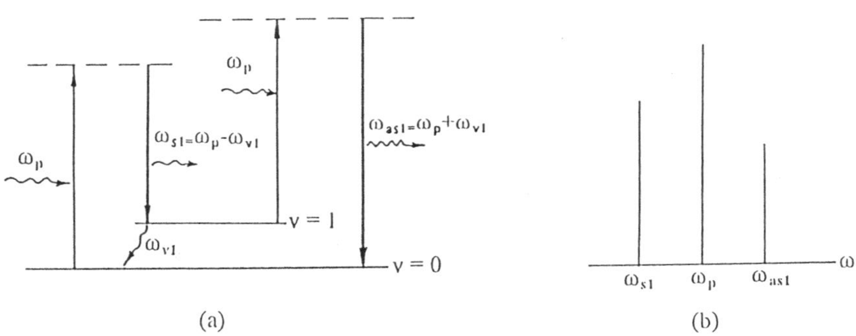

Fig. 5.11 a, b The energy levels and spectrum of the Raman scattering.

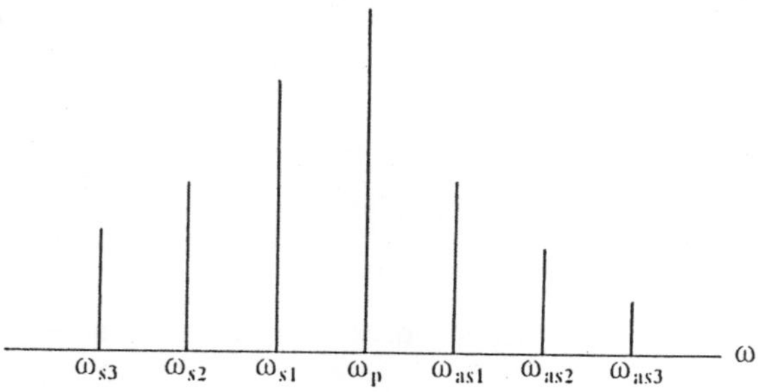

Fig. 5.12 The spectrum of stimulated Raman scattering (only ω_{s1} is from real level excited, the others of ω_{sn} and ω_{asn} are all from FWM).

5.4.2.2 Nonlinear Susceptibilities in SRS

Call the amplitude of vibration y, and the molecular polarization of the scattering medium, F. The molecular scattering can be described by a system of two energy levels: the upper energy level $|b\rangle$, and the lower energy level $|a\rangle$, with the interval $\Delta E = h\omega_v$. In this case, the molecules' Hamiltonian can be expressed by the molecular Hamiltonian H_v and the Hamiltonian of interaction between the molecular vibrations and the optical field H_{int},

$$H = H_v + H_{\text{int}}.$$

From the electric-dipole approximation $H_{\text{int}} = -\int FEdE$, where the scattering molecular polarization F, defined by H_{int} but not the susceptibility $\chi^{(3)}$, is a function of y, the Taylor series of which can be expanded as

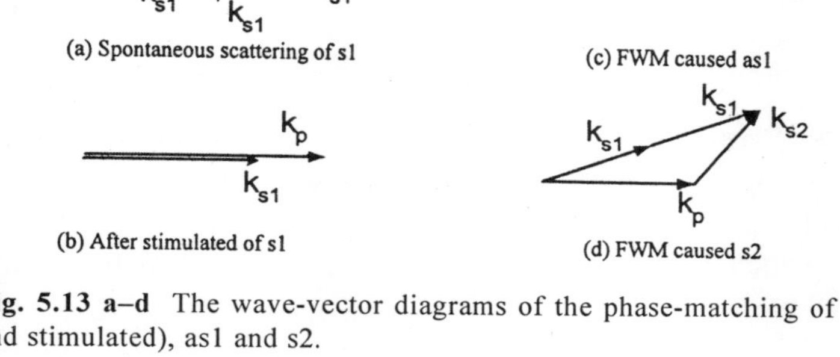

Fig. 5.13 a–d The wave-vector diagrams of the phase-matching of s1(spontaneous and stimulated), as1 and s2.

$$F = F_0 + (\partial F/\partial y)\cdot y + (1/2)(\partial^2 F/\partial y^2)\cdot y^2 + \ldots \tag{5.55}$$

and

$$H_{\text{int}} = -(1/2)\Sigma(\partial F/\partial y)\, y(r_i)\, E(r_i)\, E(r_i)\,. \tag{5.56}$$

The equation of the number of excited state populations, n, is

$$\partial n/\partial t + (1/T_1)(N - n_1) = (1/2h\omega_0)(\partial F/\partial y)\, E^2(\partial y/\partial t). \tag{5.57}$$

The vibration amplitude y satisfies a second-order differential equation with respect to t,

$$\partial^2 y/\partial t^2 + (2/T_2)(\partial y/\partial t) + \omega_0^{\,2} y = (\omega_0 y_{ab}^{\,2}/h)(\partial F/\partial y)\, E^2 \tag{5.58}$$

where $n_1 \equiv n_{\text{excited}} \equiv n_{\text{upper}}$, $n_0 \equiv n_{\text{ground}} \equiv n_{\text{lower}}$, $N \equiv n_{\text{total}} \equiv n_1 + n_0$.

The left sides of (5.58) and (5.57) have two damping terms with time constants T_1 and T_2, introduced to refer to the optical Bloch equations in spin systems. The intramolecular relaxations and intermolecular interactions are considered in these terms which are not included in (5.56).

T_1 is the energy relaxation time for the dipole oscillator, or the longitudinal relaxation time. It is linked with the diagonal elements of ρ_{aa} and ρ_{bb} in the density matrix ρ. T_1 reveals the population lifetimes of the excited states and is linked with the stored energy or population relaxation in (5.57).

T_2 is the dephasing time for the dipole oscillator or the transverse relaxation time. It is linked with the off-diagonal elements of ρ_{ab} and ρ_{ba} in the density matrix ρ. T_2 is sensitive to the phase relations between molecules. It reveals the storage time of the phase coherence and is linked with the amplitude relaxation in (5.58).

The phase coherence is more easily changed than is the population of the excited state. Hence, T_2 is much less than T_1.

Finally, the solution of the nonlinear polarization $P_{\text{NL}}(\omega)$ in the frequency domain is

$$P_{NL}(\omega) = \int_{-\infty}^{+\infty} d\omega_1 d\omega_2 \left[\frac{\left(\dfrac{\partial F}{\partial y}\right)^2 \cdot N \dfrac{(1-2\xi)}{4m}}{\omega_0^2 - (\omega_1 - \omega)^2 + \dfrac{2i(\omega_1 - \omega)}{T_2}} + \chi_{NL}^{(3)} \right] \tag{5.59}$$
$$\cdot E_{pump} E^*_{pump} E_{Stokes}(\omega - \omega_1 + \omega_2).$$

When $\xi \equiv (n_{\text{upper}}/N)$, and when $\xi = 1/2$, no more molecules can be excited to the upper level.

From (2.20), which defines polarization $P_{\text{NL}}(\omega)$ with respect to $\chi^{(3)}(\omega)$, and gives $P_{\text{NL}}(\omega)$ as

$$P_{NL}(\omega) = \int_{-\infty}^{+\infty} d\omega_1 d\omega_2 \chi^{(3)}(-\omega;\omega_1,-\omega_2,\omega-\omega_1+\omega_2) \cdot E_{pump}E^*_{pump}E_{Stokes}(\omega-\omega_1+\omega_2). \tag{5.60}$$

Comparing (5.59) and (5.60), where ξ is the population rate of vibration upper level, $\omega = \omega_s$, $\omega_1 = \omega_p$, $-\omega_2 = -\omega_p$, we have the nonlinear SRS susceptibility $\chi^{(3)}(\omega)$ described by the medium properties

$$\chi^{(3)}_{Raman}(-\omega;\omega_1,-\omega_2,\omega-\omega_1+\omega_2) = \frac{1}{4m}\left(\frac{\partial F}{\partial y}\right)^2 \cdot \frac{N(1-2\xi)}{\omega_0^2-(\omega_1-\omega)^2+2i\dfrac{(\omega_1-\omega)}{T_2}} + \chi^{(3)}_{NR} \tag{5.61}$$

The real part, linked to the refractive index, and the imaginary part, linked to the Raman gain, of $\chi^{(3)}(\omega) = \chi' - i\chi''$, for pump scattering and Stokes scattering in SRS, are respectively

$$\chi^{(3)'}_{Raman}(-\omega_s;\omega_p,-\omega_p,\omega_s) = \frac{1}{4m}N(1-2\xi)\left(\frac{\partial F}{\partial y}\right)^2 \cdot \frac{\omega_0^2-(\omega_p-\omega_s)^2}{\left[\omega_0^2-(\omega_p-\omega_s)^2\right]^2+4\dfrac{(\omega_p-\omega_s)^2}{T_2^2}} + \chi^{(3)}_{NR} = \chi'_{Stokes} + \chi^{(3)}_{NR} \tag{5.62}$$

and

$$\chi^{(3)''}_{Raman}(-\omega_s;\omega_p,-\omega_p,\omega_s) = \frac{1}{2m}N(1-2\xi)\left(\frac{\partial F}{\partial y}\right)^2 \cdot \frac{(\omega_p-\omega_s)/T_2}{\left[\omega_0^2-(\omega_p-\omega_s)^2\right]^2+4\dfrac{(\omega_p-\omega_s)^2}{T_2^2}} + \chi^{(3)}_{NR} = \chi''_{Stokes}\,. \tag{5.63}$$

Figure 5.14 demonstrates the profiles of the nonlinear susceptibility of SRS. The $\chi^{(3)}(\omega)$ exhibits the dispersion pattern and $\chi^{(3)''}(\omega)$ shows the Lorentz line shape. In regions departing from the resonant frequency ω_0, the dashed line is the nonresonant susceptibility $\chi^{(3)}{}_{NR}$, a real constant independent of ω. At the resonance $\omega_{pump} - \omega_{Stokes} = \omega_0$, $\chi^{(3)''}{}_{Stokes} = \chi^{(3)''}{}_{max}$, $\chi^{(3)'}{}_{Stokes} = 0$, and $\chi^{(3)'}(\omega) = \chi^{(3)}{}_{NR}$.

5.4.2.3 SRS Generators

SRS is a significant nonlinear effect, which can be efficiently caused in many different media. There are Raman generators, amplifiers, and oscillators. Due to space limitations, we discuss here only SRS generators. It is not difficult to extend the principles to the other two processes. In SRS generators, in which there are no input signals or cavities, there exists only quantum noise, which is amplified coherently by the interaction of the pump laser with vibration levels and becomes an intense SRS output.

In this process, the Stokes field E_{Stokes} and the pump field E_{pump} are described as

$$\partial E_s(z,t)/\partial z + (\delta_s/2)E_s = (2\pi\omega_s/\mu_s c)\chi^{(3)}(-\omega_s;\omega_p,-\omega_p,\omega_s)\cdot|E_p|^2\cdot E_s \tag{5.64}$$

$$\partial E_p(z,t)/\partial z + (\delta_p/2)E_p = (2\pi\omega_p/\mu_p c)\chi^{(3)*}(-\omega_s;\omega_p,-\omega_p,\omega_s)\cdot|E_s|^2\cdot E_p \tag{5.65}$$

where $\delta_s(\omega_s)$ and $\delta_p(\omega_p)$ are the medium absorption coefficients; and $\mu_s(\omega_s)$ and $\mu_p(\omega_p)$ are the medium refractive indices of Stokes scattering and pump scattering respectively.

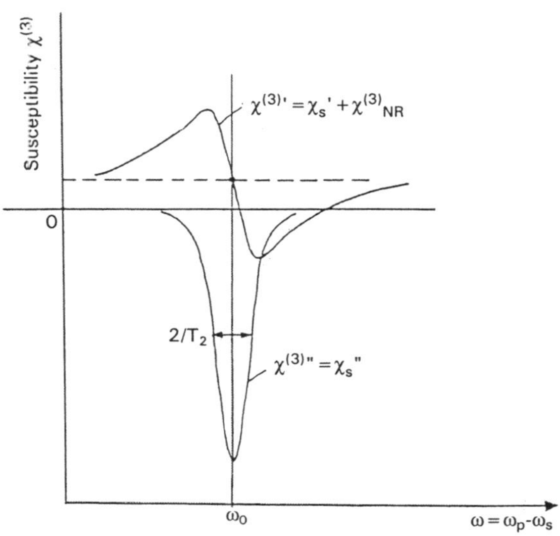

Fig. 5.14 The profiles of the nonlinear susceptibility of SRS.

Suppose the pump intensity depletion and the Stokes absorption are both neglected. Then the solution of (5.64) can be written as

$$E_s(z,t) = E_s(0,t)\,\exp(i2\pi\omega_s\chi^{(3)}{}_s\cdot|E_p|^2\cdot z\,/\mu_s c)\ . \tag{5.66}$$

The initial Raman-Stokes noise signal at injection cannot be determined by classical theory, but only by quantum mechanics. The number of Stokes photons in the population rate equation of single mode (ω_s, k_s) is

$$\partial n_s(k_s)/\partial z = g(\omega_s)I_p[1 + n(k_s)] \tag{5.67}$$

where the gain coefficient $g(\omega_{\text{Stokes1}})$ of the first-order Stokes scattering component is found from quantum mechanics. Using (5.63) on $\chi^{(3)\prime\prime}{}_{\text{Stokes}}$, the gain is

$$\begin{aligned} g(\omega_{Stokes}) &= -\frac{32\pi^2\omega_s}{\mu_s\mu_p c^2}\cdot\chi^{(3)\prime\prime}_{Stokes} \\ &= \frac{16\pi^2\omega_s N}{\mu_s\mu_p c^2 m T_2}\left(\frac{\partial F}{\partial y}\right)^2\cdot(1-2\xi)\frac{\omega_v}{\left(\omega_0^2-\omega_v^2\right)^2+4\left(\dfrac{\omega_v}{T_2}\right)^2} \end{aligned} \tag{5.68}$$

or

$$g(\omega_{Stokes}) = -\frac{32\pi^2 c^2\omega_0}{\mu_s\mu_p\hbar\omega_s^3}N\left(\frac{\partial\sigma}{\partial\Omega}\right)(1-2\xi)\frac{\omega_v/T_2}{\left(\omega_0^2-\omega_v^2\right)^2+4\left(\dfrac{\omega_v}{T_2}\right)^2}\ . \tag{5.69}$$

In thermal balance, the population of upper energy-level $\xi_{\text{upper}} << 1/2$, so $g(\omega_{\text{Stokes1}}) > 0$, and the Stokes spectrum is a gain spectrum and the Stokes scattering is collinearly stimulated and amplified. As long as the anti-Stokes scattering is proportional to the factor ($2\xi_{\text{upper}} - 1$), then in thermal balance $g(\omega_{\text{antiStokes1}}) < 0$. In this case the Stokes spectrum is a loss spectrum, and the anti-Stokes scattering observed cannot be stimulated output, but is the four-wave mixing of $\omega_{\text{antiStokes}} = 2\omega_{\text{pump}} - \omega_{\text{Stokes}}$.

The factor ($\partial\sigma/\partial\Omega$) is the Raman scattering differential cross-section. The larger the cross-section, the higher the Raman gain $g(\omega_{\text{Stokes1}})$ for a given pump intensity. The input effective noise is

$$I_{\text{noise(Stokes)}}(\omega_S)\cdot\,\omega_S = [h\omega_S^{\,3}\mu_s^{\,3}/(2\pi)^3\cdot c^2]\cdot\Delta\Omega\ . \tag{5.70}$$

At the output of the Raman medium of length L, the resulting Stokes intensity is

$$\begin{aligned} I_{\text{Stokes1}}(L) &= I_{\text{noise(Stokes)}}\cdot\left[e^{g_{\text{Stokes}}\cdot I_{\text{pump}}\cdot L}-1\right] \\ &= I_{\text{noise(Stokes)}}\cdot\left[e^{G_{\text{Stokes}}\cdot L}-1\right] \end{aligned}$$

and

$$I_{\text{noise(Stokes)}} \cong I_{\text{noise(Stokes)}}(\omega_S)\cdot(\Delta\omega_{\text{Stokes}})$$

where $I_{\text{noise(Stokes)}}$ is the effective noise in the Raman linewidth, g_{Stokes1} is the gain coefficient at the spectrum center, and $\Delta\omega_{Stokes}$ is the stimulated Raman linewidth, which is narrower than the spontaneous Raman linewidth.

Example

Calculate the input effective noise $I_{\text{noise(Stokes)}}(\omega_S)$. If $(\omega_S/2\pi c) = 15{,}000\text{cm}^{-1}$, and $(\Delta\omega_S/2\pi c) = 1000\text{cm}^{-1}$, $\Delta\Omega = 2\times10^{-4}$ Sr, $I_{\text{noise(Stokes)}}(\omega_S) \cong 10^{-3}\text{W/cm}^3$.

For the SRS strong transition, the standard differential Raman cross-section $(\partial\sigma/\partial\Omega) \cong 5\times10^{-30}\text{cm}^2$. If $N = 10^{22}/\text{cm}^3$, and $(\omega_S/2\pi c) = 15{,}000\text{cm}^{-1}$, $(\Delta\omega_S/2\pi c) = 1000\text{cm}^{-1}$, and $T_2 \cong 10^{-12}$s, then $g(\omega_{\text{Stokes1}}) = 10^{-9}\text{cm/W}$.

If $I_{\text{Stokes}}/I_{\text{noise(Stokes)}} = e^{25}$ and the conversion ratio $I_{\text{Stokes}}(L)/I_{\text{pump}} \cong 1$ %, the laser peak power density $I_{\text{pump}} \cong 10^9\text{W/cm}^2$, and the Raman medium length is 25cm.

References

[5.1] A. Yariv: *Quantum Electronics,* 3rd ed., John Wiely, New York, Chaps. 4–16 (1989)

[5.2] R.L. Byer: Optical parametric oscillators. In: H. Rabin and C.L. Tang (eds.): *Quantum Electronics: A Treatise,* Vol. 1, Academic Press, New York (1975)

[5.3] D.M. Pepper and A. Yariv. In: R.A. Fisher (ed.): *Optical Phase Conjugation,* Academic Press, New York, Chap. 2 (1983)

[5.4] R.K. Jain and M.B. Klein. In: R.A. Fisher (ed.): *Optical Phase Conjugation,* Academic Press, New York, Chap. 10 (1983)

[5.5] J. Feinberg. In: R.A. Fisher (ed.): *Optical Phase Conjugation,* Academic Press, New York, Chap. 11 (1983)

[5.6] W. Kaiser and M. Maier: Stimulated Rayleigh, Brillouin and Raman spectroscopy. In: F.T. Arecchi (ed.): *Laser Handbook*, Vol. 2, North-Holland, Amsterdam, New York (1972)

[5.7] K. Kato and M. Masutani: Widely tunable 90° phase-matched KTP parametric oscillator, *Opt. Lett.* **17**, 178–179 (1992)

[5.8] S.K. Wong and P.J. Mathieu: High-energy hybrid Raman optical parametric amplifier eye safe laser source, *Appl. Optics* **33**, 1686–1690 (1994)

[5.9] A. D'Ottavi, F. Martelli, et al.: Very high efficiency FWM in a single semiconductor traveling-wave amplifier, *Appl. Phys. Lett.* **68**, 2185–2188 (1996)

[5.10] Yili Guo, et al.: High efficiency up-conversion and real-time storage by four wave mixing in BDN dye, paper presented at *CLEO'83,* Baltimore, USA, 1983, *IEEE J. Quantum Electronics* **20**, 328–331 (1984)

[5.11] Lichen Zhong and Yili Guo: PAM-to-PDM conversion: a novel function of a bistable optical system, *J. Optics Communications* (North-Holland) **61**, 410–414 (1987)

[5.12] Yili Guo, Caiyun Lou, et al.: Threshold and energy transfer efficiency in high pressure hydrogen SRS, *Acta Physica Sinica* **34**, 24–31 (1985)

[5.13] Minyan Yao and Yili Guo: Analysis of parameters of Raman laser frequency of SRS in H_2, *J. of Tsinghua University* **26(1)**, 52–61 (1986)

[5.14] P. Lodahl and M. Saffman: Pattern formation in singly resonant second harmonic generation with competing parametric oscillation, *Physical Review A* **60**, 3251–3261 (1999)

[5.15] M. Betz, G. Goger, et al.: Ultrafast electron-phonon scattering in semiconductors studied by nondegenerate four-wave mixing, *Physical Review B* **60**, R11265–11268 (1999)

[5.16] S. Bischoff, A. Buxens, et al.: Bidirectional four-wave mixing in semiconductor optical amplifiers: theory and experiment, *Journal of Lightwave Technology* **17**, 1617–1625 (1999)

[5.17] E. Peral and A. Yariv: Degradation of modulation and noise characteristics of semiconductor lasers after propagation in optical fiber due to a phase shift induced by stimulated Brillouin scattering, *IEEE J. Quantum Electronics* **35**, 1185–1195 (1999)

[5.18] E. Rotermund, V. Petrov, et al.: Laser-diode-seeded operation of a femtosecond optical parametric amplifier with $MgO{:}LiNbO_3$ and generation of 5-cycle pulses near 3 μm, *J. Optical Society of America B* **16**, 1539–1545 (1999)

6. Semiconductor and Quantum Well Nonlinear Photonics

After many advances in about forty years, semiconductor laser diodes, quantum well devices, and photodetectors have become major devices in optoelectronics, and are still developing. Because of fiber communications and optical information processing, more new devices, especially active or passive functional, are needed. Polarization-independent optoelectronic or all-optical modulators, switches, amplifiers, logic components, memories, filters, wavelength converters, polarization converters, with speeds of about 10ps, with low driving power of about $10^2\mu$W, or low driving energy of femto-joule, or low driving voltage of about 1–5V are expected to be commercial products for the coming generation of large capacity optical fiber communication (OFC).

The nonlinear relations between the absorption coefficient α (or refractive index n) and the extra-field E are always the bases of these devices. In quantum well and strained-layer super-lattice $(S\text{-}L)^2$ structures, these nonlinear relations appear in new and unique ways.

The concept of super-lattice was first proposed in 1969 and the first super-lattice structure $Ga_{1-x}Al_xAs/GaAs/Ga_{1-x}Al_xAs$ was made by molecular beam epitaxy (MBE) technology in 1970. These ushered in a new epoch of quantum-sized material science and promoted electronics. A series of quantum-well devices based on quantum-sized effects, or quantum-confined effects, became the core devices of optoelectronics. Recently, because of the introduction of strained layers, the formation of the band-gap, the modification of the band structure, and the adjustment of the band-deviation, are done by electric and mechanical means. They will offer a more solid foundation for nonlinear photonics.

In this chapter, we shall first discuss the absorption of transitions of interbands and intrabands, then the changes of absorption coefficient $\Delta\alpha$ and the changes of refractive index Δn caused by the photon excitation, or by the extra-field acting. These effects are really the bases of nonlinear devices. Also we shall discuss the comparison of bulk semiconductors and quantum well structures.

A very interesting topic in nonlinear photonics is the optical properties and photonic nonlinearities in some novel quantum well devices. One kind of them is the interdiffused quantum well device which will be discussed in the Chaps. 10 and 11.

6.1 Photon-Induced Nonlinearities and Some Fast Photonic Processes in Semiconductors

The nonlinear absorption coefficient $\Delta\alpha$ and refractive index Δn excited by optical fields(or extra-electric fields) are always the bases of most nonlinear semiconductor devices.

In nonlinear photonics, the nonlinear susceptibility of semiconductors is always much greater than that of optical crystals and metal vapors. In semiconductors $\chi^{(3)}\cong 10^{-10}$esu [$\cong 1.4\times 10^{-18}$(MKS)], while in optical crystals $\chi^{(3)} \cong 10^{-20}$esu [$\cong$ 1.4×10^{-28}(MKS)], and in metal vapors $\chi^{(3)} \cong 10^{-30}$esu [$\cong 1.4\times 10^{-38}$(MKS)].

Most photonic nonlinearities stem from free-carrier nonlinear motions in the semiconductors, bound-electron enharmonic motions in the semiconductors, carrier transitions from valence to conductive bands, nonlinearities of exciton absorption in quantum wells, the Franz-Keldysh effect, and other sources.

6.1.1 The General Cases of Photon-Induced Absorption and Refractive Index Changes

6.1.1.1 Nonlinearities Resulting from Free-Carrier Motions

Figure 6.1 shows primarily the nonlinearities stemming from the optical excitation. The rather large third-order susceptibility $\chi^{(3)}$ can be caused by free-carrier motion and energy relaxation in mid-dopant level semiconductor excited by optical fields. Its conduction band characteristics can be expressed by the electron energy E_C in the conduction band, the electron's effective mass m_e^*, and the electron's velocity $v(p)$. These are associated with the momentum p.

The moderately-doped common material GaAs and InAs have narrower forbidden bands and smaller effective electron masses, so the values of $\chi^{(3)}{}_{NL}$ (in order of magnitude) for free electrons in a conduction band are larger. It is 10^{-19}MKS for GaAs or 10^{-18}MKS for InAs, that is about approximately 100 times that from bound electrons.

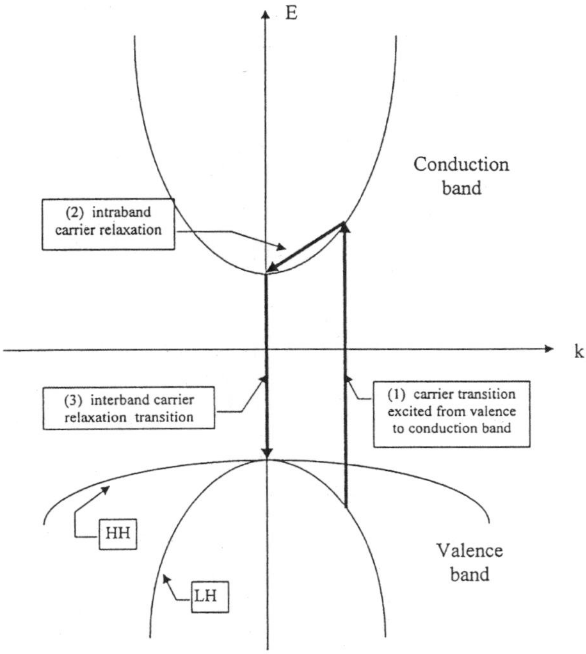

Fig. 6.1 The semiconductor nonlinearities stemmed from light excitation.

6.1.1.2 Nonlinearities of Carrier Transition from Valence to Conduction Bands

There are various energy states in semiconductors which can be revealed by energy bands. The characteristics of intrinsic semiconductor bulk can be expressed by a group of broadened valence bands, conduction bands, and numerous acoustic modes. Sketched diagrams of these effects are shown in Fig. 6.2, which are presented by the relations between energy E and wave-vector k.

Different transitions will occur between these states. These transitions include transitions of electrons from valence to conduction bands, then produced the free carriers and photoelectric conductance; transitions of free carriers between intervalence bands, interconduction bands, and in a band; and transitions between exciton levels. If transitions in Fig. 6.2 occur close to $k = 0$, they will be similar to those in III-V compound semiconductors. Note that the photon-induced plasmas in semi-conductors make the dielectric constant $\varepsilon(\omega)$, which is a complex function including a real part $\varepsilon'(\omega)$ and an imaginary part $\varepsilon''(\omega)$, change considerably.

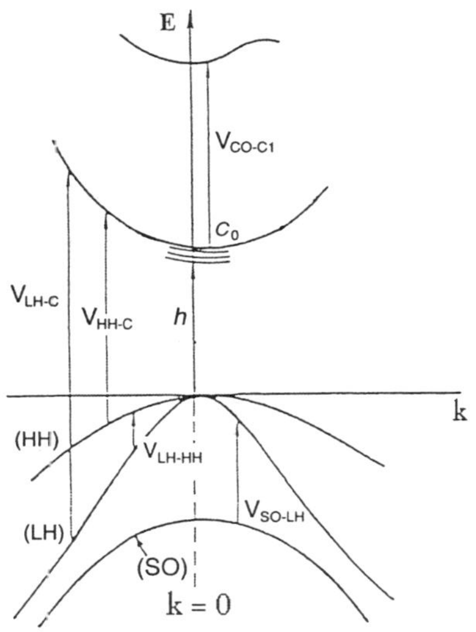

Fig. 6.2 The different transitions occurred between the miscellaneous states in bulk semiconductors (SO: separate-orbit band, LH: light-hole band, HH: heavy-hole band).

6.1.1.3 Absorption Coefficient α(ω) of Transitions from Valence to Conduction Band

The plasma density is supposed to be low, so that the multibody interaction effects can be neglected. Then consider only a single electron transition. Such a severe condition limits this approach in practical use.

The absorption coefficient $\alpha(\omega)$ of transition from the valence to the conduction band can be expressed by $\varepsilon''(\omega)$, which is the imaginary part of the dielectric function. If E_g is the direct-energy gap of a permitted-transition and $\hbar\omega > E_g$, we have,

$$\begin{aligned}\alpha(\omega) &= \frac{\varepsilon''(\omega)}{n(\omega)}\cdot\frac{\omega}{\mathrm{c}} = \xi\cdot\left|R_{vc}\right|^2\cdot(\hbar\omega - E_g)^{1/2} \\ &= (C/\omega)\cdot(\omega-\omega_g)^{1/2}\end{aligned} \tag{6.1}$$

where ξ is a parameter linked with the masses of electrons and holes, refractive index $n(\omega)$, etc. R_{vc} is the transition momentum matrix element independent of k and ω. The constant C is a factor involving material parameters, electric field E, reduced effective mass of the electron and heavy holes, and reduced effective mass of the electron and light holes. From (6.1) physically, $\alpha(\omega) = 0$ for $\hbar\omega < E_g$.

6.1.1.4 Photon-Induced Refractive Index Change Δn

From the $\Delta\varepsilon'(\omega)$, the change of the real part of the dielectric function, caused by the absorption induced carrier-density near the band gap, a two-level system is supposed. Ordinarily, the perturbation model is used. In this case, $\Delta\varepsilon'_{\mathrm{NL}}(\omega)$ is a perturbation term, and

$$\varepsilon'(\omega) + \Delta\varepsilon'(\omega) = [n(\omega) + \Delta n(\omega)]^2$$

Then the carriers-induced refractive index change Δn caused by photon absorption $\alpha(\omega)$ (from (6.1)) near the band edge is

$$\Delta n(\omega) = -\zeta \cdot \frac{\omega_g^2}{\omega_g^2 - \omega^2}, \tag{6.2}$$

where ζ is a parameter linked with the masses of electrons and holes, refractive index n, etc. ω_g is the ω linked to the photon energy in the direct band gap E_g.

6.1.2 The Virtual-Transient State Nonlinearities in Semiconductors

Entering the 21st century, people benefit from the huge amount of information provided through the networks. The ultrafast photonics, devices and techniques with 100fs to 1ps running time have become important fields. Among these, some ultrafast gain and refractive index processes (in 100fs to 1ps) in the semiconductor (quantum well) materials call for our attention.

6.1.2.1 Ultrafast Nonlinear Gain and Refractive Index Dynamics

The speed of nonlinear semiconductor bulk or quantum well devices is determined by how fast the nonlinearity dynamics can be obtained. In nonlinear photonics, the nonlinearity dynamics of the functional devices, such as switching, modulating, converting, and logic operating devices, are always linked closely with $\chi^{(3)}(\omega)$. Here its real part $\chi^{(3)\prime}(\omega)$ is related with the refractive index $n_{\mathrm{NL}}(\omega)$ and the imaginary part $\chi^{(3)\prime\prime}(\omega)$ is related with the absorption coefficient $\alpha_{\mathrm{NL}}(\omega)$ or gain coefficient $g_{\mathrm{NL}}(\omega)$.

We can use a simple representation to show the third order of nonlinearity

$$n = n_{\text{linear}} + (n_2 + i\alpha_2)\cdot I_{\text{in}} \tag{6.3}$$

where n_{linear} is the linear refractive index, n is the total refractive index including linear and nonlinear index, n_2 is the intensity induced nonlinear refractive index, and I_{in} is the injected intensity.

Some non-resonant nonlinear gain change $\Delta g_{NL}(\omega)$, or nonlinear absorption change $\Delta\alpha_{NL}(\omega)$, mainly stems from the carrier heating and two photon absorption (TPA), which is a standard non-resonant nonlinear process always linked with the ultra-fast processes (150–250fs), and secondarily stems from the spectral hole burning (SHB, 120–600fs). The refractive index change $\Delta n_{NL}(\omega)$ dominantly stems from the carrier heating and instantaneous optical Stark Effects (OSE, 0.1–1ps). Thus the speed of nonlinearly active devices is always determined by the raising times of carrier heating, two photon absorption, spectral hole burning, optical Stark Effects, and the recovering time of carrier-carrier scattering (C-CS, 160fs) or carrier-phonon scattering (C-PS, 250fs). In this field, very meaningful results have been contributed by many authors, for example, K.L. Hall, E.P. Ippen, H.A. Haus and their colleagues at MIT [6.24] and K.K. Anderson [6.25].

When $\Delta g_{NL}(\omega)$ or $\Delta\alpha_{NL}(\omega)$ is caused by the carrier heating, the carrier distribution is heated by the carrier absorption and two photon absorption at high inject light power. These hot carriers can give up their excess energy and make a low energetic distribution in the timescale of 0.25ps. This distribution will cool back to the lattice temperature via the typical carrier-photon scattering, and this recovering time is also about 0.25ps. In the strained-layer multi-quantum well devices, the transient absorption bleaching is due to carrier heating, and its recovering time is 150fs or so. In the case of $\Delta g_{NL}(\omega)$ or $\Delta\alpha_{NL}(\omega)$ caused by the spectral hole burning, the quick recovery is achieved via carrier-carrier scattering. At high carrier densities, the carrier-carrier scattering time constant is believed to be 100fs or less.

From Sect. 2.6 in Chap. 2, when the absorption $\alpha(\omega)$ or gain $g(\omega)$ in semiconductor materials is known (mostly by experiments), the refractive index $n(\omega)$ can be obtained by the Kramers-Kronig (K-K) Relation,

$$n(\omega) = \frac{1}{\pi}\int_0^\infty \frac{g(\omega')}{\omega'^2 - \omega^2} d\omega' . \tag{2.127}$$

In nonlinear processes, there is a perturbed change $\Delta g_{NL}(\omega)$ or $\Delta\alpha_{NL}(\omega)$ caused by photonic fields. Then the related change of refractive index is

$$\Delta n_{NL}(w) = \frac{1}{\pi}\int_0^\infty \frac{\Delta g_{NL}(\omega')}{\omega'^2 - \omega^2} d\omega' . \tag{6.4}$$

The long-lived change or step change in the refractive index are related to the pump-induced carrier density changes. A decrease in the carrier density $-\Delta N(\omega)$ increases the refractive index $+\Delta n_{NL}(\omega)$ (gain regime), and an increase in the carrier density $+\Delta N(\omega)$ decreases the refractive index $-\Delta n_{NL}(\omega)$ (absorption regime). In addition to the above long-lived refractive index change, there is a transient increase in refractive index $+\Delta n_{NL\text{-tran.}}(\omega)$ recovering with a time constant of about 0.6ps due to heating of the carrier distribution, but a transient decrease in refractive index $-\Delta n_{NL\text{-tran.}}(\omega)$ is produced by the rapid electronic or virtual process.

In strained-layer multi-quantum well devices, a transient increase in refractive index $+\Delta n_{NL\text{-tran.-SLMQW}}(\omega)$ recovers with a 1ps time constant and is related to the heated distribution cooling back to the lattice temperature.

6.1.2.2 Real Ambience for Ultrafast Nonlinear Photonics

Table 6.1 shows the real ambience for ultrafast photonics and Table 6.2 shows the gain and refractive index dynamics of the above- and below-band-gap pump respectively.

Table 6.1 Real ambience for ultrafast photonics.

Self-action state	Carrier heating	Spectral-hole burning(SHB)	Two-photon absorption	Optical Stark effect
Mutual-action state	Carrier-carrier scattering(C-CS)		Exciton-exciton interaction (E-EI)	

Table 6.2 Gain and index dynamics of the above- and below-band-gap pump.

Above-band gain	• Negative transient (100 fs, such as a dynamic TPA). • Carrier heating and then carrier cooling (1ps). • SHB recovery by C-CS (100fs), C-PS (250fs).
Above-band index	• Positive refractive index by carrier heating (0.1~1ps). • Negative refractive index by virtual processes (such as OSE, 0.1~1ps).
Below-band gain	• The only gain dynamics is TPA negative transient. • No stimulated transition and spectral-hole burning.
Below-band index	• Positive index by the cooling of heated carriers(1.3 ps). • Negative refractive index by carrier heating of FCA.

6.1.2.3 Virtual-Transient Nonlinearity

The virtual-transient nonlinearity has the following features:

- The negative transient gain is caused by two-photon absorption (TPA).

- The negative transient index is caused by the optical Stark effect (OSE).
- Virtual photon tunneling (VPT): photon-assisted tunneling and n-photon transition between virtual states.
- Virtual carrier population (VCP): femtosecond nonlinear dynamics by virtual carrier excitation, the OSE induced resonantly-excited carriers and the TPA contributed to nonlinear refractive index.

6.2 Franz-Keldysh Effects in Bulk Semiconductors

In the 1990s, the associated quantum-confined Stark Effect (QCSE) has been emphasized in quantum well (QW) and strained-layer QW structures. Experiments have proven these effects are highly valuable in high-speed, low-driving, low-chirping electric-absorption (EA) intensity and phase modulators integrated with LD sources.

The story can be traced back to the Franz-Keldysh Effect in which the extra electric-field induced change of optical absorption in bulk semiconductors. In 1958, Franz and Keldysh predicted that optical absorption α in semiconductors should occur for photon energies below the band gap in the presence of an electric field [6.3 and 6.4]. Physically, this results from the fact that the electric field bends the energy band profile and forms a triangular potential barrier between the conduction band and the valence band. Hence the transition exists due to electron tunnelling. This is supported by the room-temperature experimental absorption data near the direct gap edge, both in the absence and presence of applied electric fields. By applying the Kramers-Kronig integrals to the calculated electro-absorption (EA) in Ge and GaAs, the electrorefraction (ER) was estimated numerically [6.5].

In nonlinear photonics, the electric-field induced refractive index change Δn will be the property expected to form the phase-dominant devices based on Franz-Keldysh Effects. As in Chap. 2, Δn and $\Delta\alpha$ are linked by the Kramers-Kronig Relations. By these, Δn in semiconductor devices is always achieved by numerically integrating $\Delta\alpha$ which can be easily found by experiment. Data from experiment and theory on $\Delta\alpha$, for fixed λ, is $4\pi\Delta\kappa/\lambda$, and the refractive index n shows strongly nonlinear relations. This effect is expected to give improved devices.

6.2.1 The Absorption Change $\Delta\alpha$ and Refractive Index Change Δn

There is a change of the complex dielectric constant ε due to the electric field induced energy-band bending and electron tunnelling

$$\Delta\varepsilon(\omega,E) = \Delta\varepsilon'(\omega,E) + i\Delta\varepsilon''(\omega,E). \qquad (6.5)$$

The relationship between the complex dielectric constant ε and the refractive index n is

$$\varepsilon = (n + i\kappa)^2. \tag{6.6}$$

Then

$$\varepsilon' = n^2 - \kappa^2, \tag{6.7}$$

$$\varepsilon'' = 2n\kappa. \tag{6.8}$$

Taking differentials,

$$\Delta\varepsilon' \cong 2n\Delta n - 2\kappa\Delta\kappa, \tag{6.9}$$

$$\Delta\varepsilon'' \cong 2\kappa\Delta n + 2n\Delta\kappa. \tag{6.10}$$

Near the direct gap of semiconductors, the refractive indices n are much larger than the extinction coefficients κ, whereas their perturbations are of the same order of magnitude. Thus

$$2n\Delta\kappa >> 2\kappa\Delta n, \tag{6.11}$$

and

$$\Delta\varepsilon'' \cong 2n\Delta\kappa. \tag{6.12}$$

Similarly,

$$2n\Delta n >> 2\kappa\Delta\kappa, \tag{6.13}$$

and so

$$\Delta\varepsilon' \cong 2n\Delta n. \tag{6.14}$$

Equations (6.12) and (6.14) allow us to estimate the field-induced changes in the extinction coefficient and refractive index. In semiconductor nonlinear devices, the most significant parameter is the absorption coefficient α, defined as

$$\alpha \equiv 4\pi\kappa/\lambda, \qquad \text{or} \qquad \kappa \equiv \alpha\lambda/4\pi. \tag{6.15}$$

At a fixed wavelength, the absorption change

$$\Delta\alpha = 4\pi\Delta\kappa/\lambda, \qquad \text{or} \qquad \Delta\kappa = \lambda\Delta\alpha/4\pi. \tag{6.16}$$

6.2.2 The Analysis on Franz-Keldysh Effect by the Effective-Mass Theory in the Direct-Gap III-V Semiconductors

The effective-mass theory is used to calculate the Franz-Keldysh Effect in direct-gap III-V semiconductor [6.06]. It will be profitable to review some typical studies on the semiconductor nonlinear photonics, such as electric field effects in bulk semiconductors and QCSE effects in 2D-QW materials. From [6.7] and [6.6], the dielectric coefficient perturbations of $\varepsilon'(\omega,E)$ and $\varepsilon''(\omega,E)$ by the electric field E are obtained. Then from the equations of n, κ, and α in terms of the electro-optic functions, the results on optical refractive index $\Delta n(\omega,0)$, extinction coefficient $\Delta\kappa(\omega,0)$, and absorption coefficient $\Delta\alpha(\omega,0)$ at zero-field are obtained. From [6.7],

$$\alpha(\omega,0) = (C/\omega)(\omega - \omega_g)^{1/2}, \qquad \text{for } \hbar\omega > E_g, \tag{6.17}$$

and

$$\alpha(\omega,0) = 0, \qquad \text{for } \hbar\omega < E_g, \tag{6.18}$$

where the constant C is a factor involving material parameters, electric field E, reduced effective mass of the electron and heavy holes, and reduced effective mass of the electron and light holes.

This is the well-known square root dependence for the absorption edge near a direct gap. C can be estimated by fitting experimental zero-absorption data [6.8] with (6.17). Figure 6.3 shows primarily the brief drawing of the zero-field optical absorption in bulk semiconductors. This brief drawing, reproduced with the data from [6.6], shows that even below the band gap the absorption is existed in some extent.

Figure 6.4 shows basically the sketch map, reproduced with the data from [6.6], that Δn and $\Delta\kappa$ change with photon energies $(\hbar\omega - E_g)$ below the band gap in the case of some different values of the extra-field.

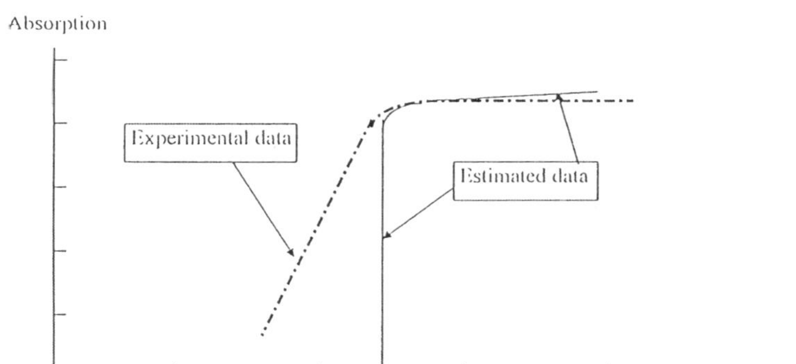

Fig. 6.3 Experimental data curve of the zero-field optical absorption for the bulk InP semiconductor (dash-dotted line), and the estimated absorption (solid line) from equations (6.17) and (6.18) [6.6] (by permission of IEEE).

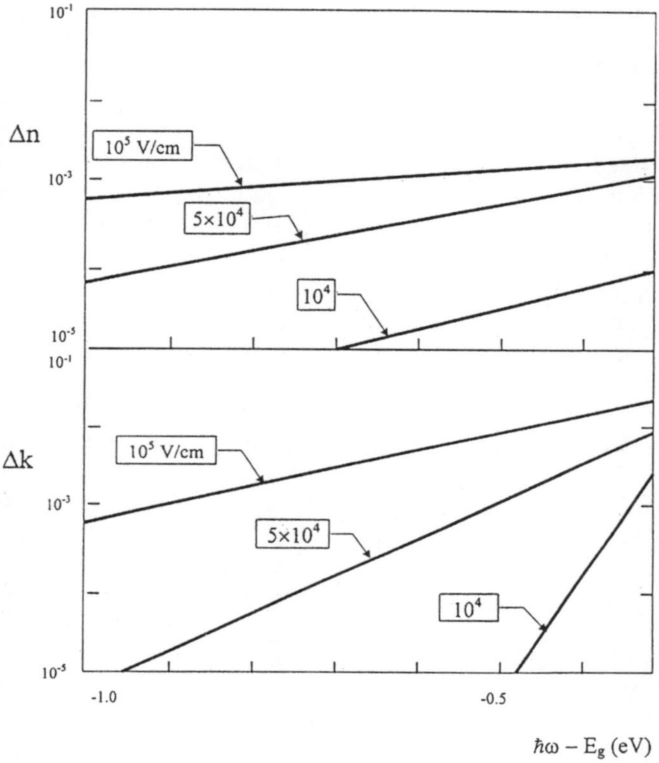

Fig. 6.4 Electric field induced changes Δn and Δk via photon energies below the band gap for fields of 10^4V/cm and 10^5V/cm [6.6] (by permission of IEEE).

The most practical way to apply the electric field to a guided-wave device is with a reverse-biased p-n junction or a reverse-biased Schottky barrier contact located directly on top of a high-purity channel waveguide. This allows high field strength to be reached in the depletion layer beneath the contact, a layer that fills most of the waveguide. The maximum allowable field strength, E_{max}, increases with increasing band gap and higher doping.

6.3 Particle State-Density Distributions in Quantum Wells

The conduction and valence bands split into overlapping subbands that become successively narrower as the electron motion is restricted in more dimensions. The particle state-density in two-dimensional quantum-well is lower than that in three-dimensional crystals.

The concept of super-lattice was first proposed in 1969 [6.10] and the first super-lattice structure $Ga_{1-x}Al_xAs/GaAs/Ga_{1-x}Al_xAs$ was made with molecular beam epitaxy (MBE) in 1970 [6.11]. These contributions usher in a new epoch in quantum-sized material science and promote the opto-electronics technologies to a new height. A series of quantum-well devices based on quantum-sized effects became central to optoelectronics. Strained layers allow formation of the band-gap, modification of the band-structure, and adjustment of the band deviation to be done by electrical and mechanical means. This allowed superior nonlinear lasing and nonlinear photonic operation, and offered a solid foundation for the integrated optoelectronics and integrated lightwave technologies.

Both multiple quantum well (MQW) and super-lattice (S-L) are ultra-thin layers periodically arranged with heterostructure junctions. When the barriers are thick enough, > 200Å, and the wells are deep enough, > 0.5eV, no overlap of the electron wave-functions occurs in adjacent wells. The total electron activity in such structures is simply the summation of the electron activity in separate wells, hence these structures are called multiple quantum wells, specially adapted to devices with low threshold and narrow linewidth. If the barriers are comparatively thin and the wells are fairly shallow, then from resonant tunnelling effects, there are large tunnelling probabilities in the wells, and the separate widthless levels of electrons in the well become the subbands with a certain width. This material is called a super-lattice (S-L) and is suitable for high power devices.

6.3.1 The Quantum-Size Effect in a Two-Dimensional Quantum Well

When the confined size (e.g. the thickness of material layer) of an electron is shorter than de Broglie wavelength λ_{deB} of the electron, the electron will show the quantum-sized effects. The wavelength $\lambda_{deB} = h/p$ (where h is the Planck constant and p its momentum) and is of the matter wave associated with the motion of a material particle. In semiconductor cases, λ_{deB} is about 200–300Å, so the thickness of the crystal layer is about 50–100Å which is only few times of an atom layer. The natural lattice period of atom layers (or one layer thickness) is about 10–20Å. In the 1960s, by liquid-phase epitaxy, the thickness precision of layers was about 0.1 to 0.2μm, making it impossible to precisely make crystals with such fine periodic structures. Only the MBE, MOCVD and other advanced technologies can control the thickness to 100Å in material growth. Energy band engineering is based on the control of the energy-level structures by control of material growth.

Example

The lattice period of an atom-layer in natural crystals $a \cong 10\text{–}20$ Å, the lattice period grown artificially by the "energy band engineering" $d \cong 50\text{–}100$Å, and de Broiglie wavelength of an electron free-distance $\lambda_{\text{deB}} \cong 200\text{–}300$ Å, consequently, $a < d < \lambda_{\text{deB}}$.

Today there are many methods to grow ultrathin multilayer materials, such as MBE, MOCVD, CBE, and ALE. By computer control, periodic structures of atomic thickness with extremely high quality have been grown, and their periods can be repeated precisely. Sometimes arbitrary change or periodic modulation of lattice structures can be realized by aid of computers. By these methods, two heterostructures are opposite and close together. Generally, two larger band gaps E_{g1}, such as GaAs, are arranged on both sides of the smaller band gap E_{g2}, such as GaAlAs. As a result, the potential configuration of $Ga_{1-x}Al_xAs/GaAs/Ga_{1-x}Al_xAs$ looks like a potential well (Fig. 6.5).

When the well width L_{wz} (z is the growth direction) is smaller than the de Broglie wavelength, the electron motion in z is confined by the well wall, but on the xy plane it still moves freely. Such a potential well is a two-dimensional quantum well or two-dimensional electron gas. Holes in the potential wells can also be considered as a gas. If only one well exists, it is called a single quantum well. If such wells are grown repeatedly and periodically, they become multiple quantum wells (MQW) (Fig. 6.6), where layer a is the well layer and layer b is the barrier layer.

MQW materials have many advantages over bulk semiconductors. MQW is used to make high quality QW lasers and QW modulators. QW structures have already been formed in the most active layers of traditional DFB-LDs.

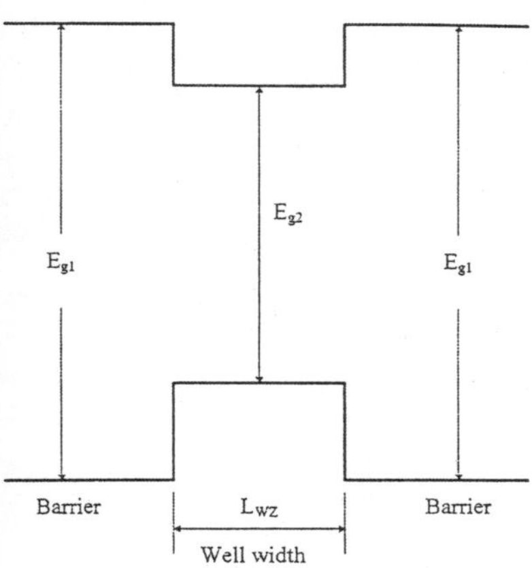

Fig. 6.5 The potential-well configuration of $Ga_{1-x}Al_xAs/GaAs/Ga_{1-x}Al_xAs$.

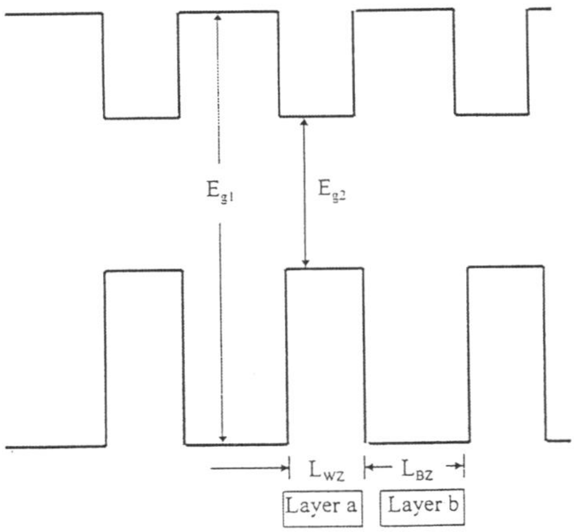

Fig. 6.6 The potential well configuration of multiple quantum well (MQW).

When the barrier width L_{bz} is thick enough, the coupling between adjacent wells can be neglected, and any one well can be treated as a single well. For an MQW structure, if the barrier is not too thick, the coupling between wells must be considered. The total potential contour becomes continuous over all the wells (Fig. 6.7). The MQW period, $L_{wz} + L_{bz}$, is about 20 to

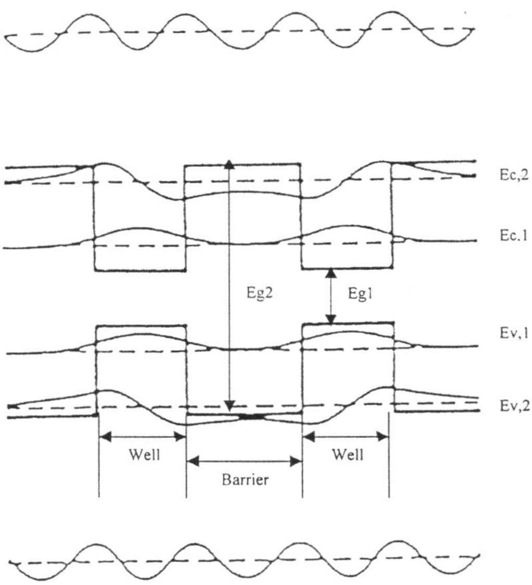

Fig. 6.7 Continuously over-well configuration and the totally potential contour of a superlattice.

30nm, much larger than the natural lattice period of about 1 to 3nm. Hence this structure is called a super-lattice (SL or S-L). Some practical MQWs are also super-lattice structures and there is considerable overlap between SL and MQW. If the discussions put no emphasis on the distinction between QW and SL, some people name it also as quantum well.

6.3.1.1 Some Types of the Composed Superlattices

Sometimes the composed super-lattices are divided into the following types:

Type-I of composed superlattices

The basical sketch of the potential contour is shown in Fig. 6.8. If both the bottom of the conduction-band and the top of the valence-band are of the same layer *a*, then the electrons and holes are confined to the same layer. Such a structure is a type-I super-lattice structure, or type-I quantum well structure. $Ga_{1-x}Al_xAs/GaAs/Ga_{1-x}Al_xAs$ is a standard example. Its band gaps are $E_g^I(GaAs)$ = 1.42eV, $E_g^{II}(Ga_{1-x}Al_xAs)$ = 1.92eV. Layer *a* is the well layer and layer *b* is the barrier layer.

Example

$Ga_{1-x}Al_xAs/GaAs/Ga_{1-x}Al_xAs$ is a standard example of type-I composed superlattice structure (or type-I quantum well structure), of which both the bottom of conduction-band and the top of valence-band lie in the identical layer *a*, shown in Fig. 6.8.

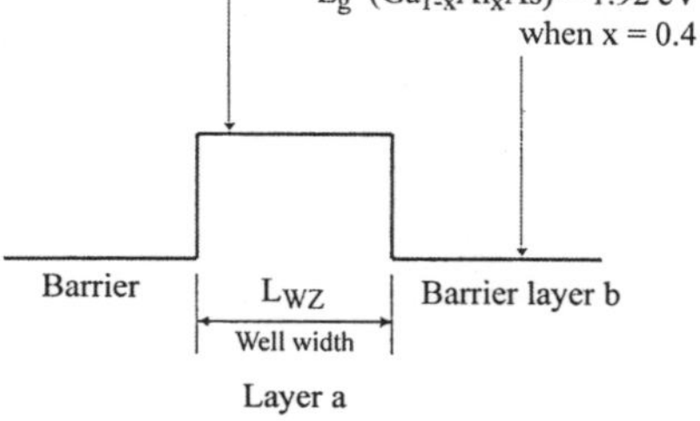

Fig. 6.8 A standard example of type-I composed superlattice structure (or type-I quantum well structure).

For it, we have

Band gap E_g^{I}(GaAs) = 1.42eV,

Band gap $E_g^{II}(Ga_{1-x}Al_xAs) = 1.42 + \Delta E_g = 1.42 + 1.247x$ eV (where $x < 0.45$).

When x = 0.4, ΔEg = 0.5eV, and consequently, Eg^{II} $(Ga_{1-x}Al_xAs)$ = 1.92eV.

Type-II of composed superlattices ($GaSb_{1ny}As_y/In_{1nx}Ga_xAs/GaSb_{1ny}As_y$ as an example)

If the bottom of the conduction-band and the top of the valence-band lie in different material layers *a* and *b*, the layer *b* becomes the well of the holes. Hence the electrons and holes are spatially separate in different layers (Fig. 6.9). This is a type-II super-lattice structure or a type-II quantum well structure. $GaSb_{1ny}As_y/In_{1nx}Ga_xAs/GaSb_{1ny}As_y$ is a standard example. Layer $GaSb_{1ny}As_y$ (layer *a*) is the well of holes, while the layer $In_{1nx}Ga_xAs$ (layer *b*) is the well of electrons. The electron well and the hole well are spatially separated.

In particular, in the strained layer case, GaAs/InGaAs/GaAs is a result of the adjustment of the band-deviation (Fig. 6.10). This structure is type-I for heavy holes, and is type-II for light holes.

6.3.1.2 Exciton Efficacy Enhanced by Quantum-Sized Effect

(1) In bulks, there are small effective masses m_{eh}^* and large dielectric constants ε_{ex}, consequently the excitons are weakly binding.

Excited by optical fields, the valence-band electrons absorb the photon energy $h\omega$ and transit to the conduction-band. Therefore, between the

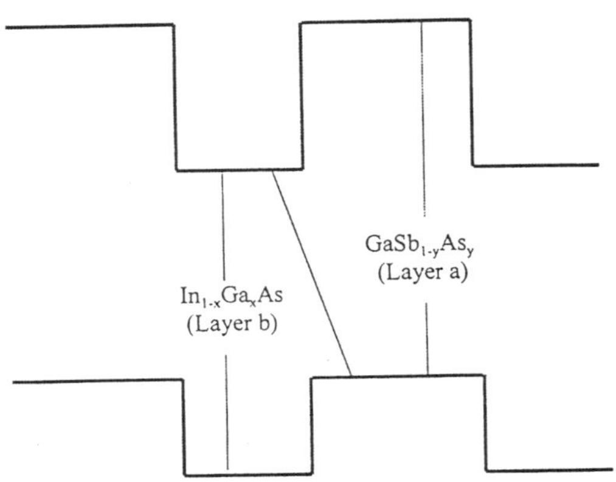

Fig. 6.9 $GaSb_{1-y}As_y/In_{1-x}Ga_xAs/GaSb_{1-y}As_y$ as a standard example of type-II structure.

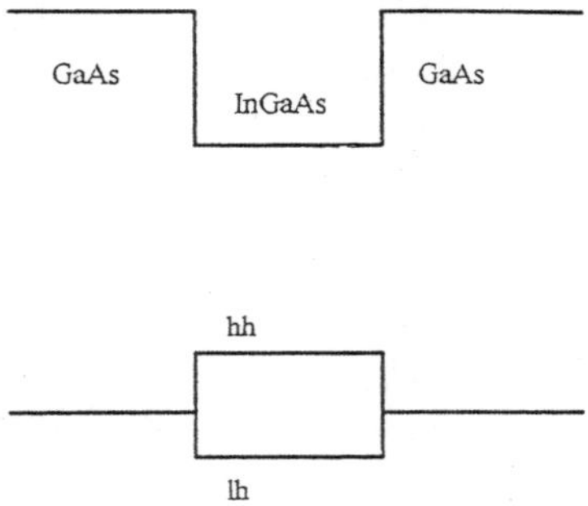

Fig. 6.10 The artificial adjustment of the band-deviation in the strained layer case.

positive hole in the valence-band and the passive electron in the conduction- band, exciton states exist, due to Coulomb's attraction. There are two kinds of excitons: free and bound. A electron-hole pair formed by a free electron and a free hole is a free exciton, or Motto-Wannier exciton, and an electron-hole pair bound on the impurities is a bound exciton.

In three-dimensional bulk (3D-bulk) materials, the Bohr's diameter (about 30nm for GaAs) of free excitons (the excited free electron-hole pairs) is spherically symmetrical. Naturally, such free excitons can move freely in these materials. In bulks, there are small effective masses m_{eh}^* and large dielectric constants ε_{ex}, consequently the excitons are weakly binding. The wave-function of free excitons overlap, and these excitons easily ionize at room temperature T_{room}. For example, the ionized energy is only about 4.2meV for GaAs bulk materials, but kT_{room} = 26meV. Because the wave-function overlaps over several lattice-spaces, two excitons can form a bi-exciton, which has very strong two-photon absorption (TPA), another origin of photonic nonlinearities.

For some materials, such as CdS in the II-VI group, in which the forbidden band has a larger band gap because of the actions of defects and impurities, the excitons are easily bound by them and become bound excitons.

(2) In two-dimensional QW materials, the mass of carriers is highly anisotropic, with extremely large mass along z. Consequently, ionization energy is 2 to 3 times higher than in 3D-bulks.

In two-dimensional QW materials, the exciton motion, limited by the well wall along the growth direction z, becomes a planar motion. In this case, the excitonic orbit-diameter will be modified significantly. Orbits of 3D excitons in bulk are spherical and isotropic, but those of 2D excitons in QW are nearly planar and anisotropic. Consequently, the mass of the carriers is highly anisotropic, with extremely large mass along z. Since the distance between electron and hole decreases in this case, the Coulomb attraction of excitons in the 2D-QW is much stronger. Accordingly its ionization energy is 2 to 3 times higher than in the 3D-bulks. In addition, the wave-functions are elliptically asymmetric in the 2D-QW. The narrower the well, the more drastic the asymmetry. When the barrier is large enough for GaAs wells, the corresponding ionization energy can exceed 10meV, and it is easier to observe the exciton actions at room temperature.

Photonic nonlinearities have many advantages, such as the low-intensity needed for saturation, absorption and high-speed nonlinear responsibility. When the energy of the injection photons equals the excitonic energy, a high absorption coefficient at room temperature occurs, due to the resonant absorption. Since this absorption takes only picoseconds, the response is extremely high for devices in which the two-dimensional exciton effect is used.

There is a very strong relation between the well width L_{WZ} and the exciton absorption coefficient α_{ex}. As the well width decreases, the free-excitonic absorption and formation become more efficient. The free-exciton absorption becomes more observable. The reason is not only the Coulomb interaction is increased, but also the excitonic two-dimensional features due to the well limitation on the electron-hole motion. When the well width is reduced, the exciton density is increased but the impurity density is decreased. As a result, the absorption supported from impurity is weakened but the stem from two-dimensional free-excitons is more enhanced.

6.3.2 The Particle State-Density Distributions $\rho_n(E)$ in Two-, One- and Zero-Dimensional Wells of Infinite Depth

6.3.2.1 Energy-Band Configuration of GaAs

Before introducing the energy-band configuration and the particle state-density distribution in $Ga_{1nx}Al_xAs/GaAs/Ga_{1nx}Al_xAs$ QWs, GaAs energy-bands are introduced, since they are very similar in their band contours, as shown in Fig. 6.11.

The conduction-band contour exhibits the S-state characteristics and the band-bottom is at $k = 0$, the Brillouin center. When the spin is not considered, the valence-band contour has P-state characteristics and has a triply degenerate band-top. When the spin is considered, the interaction between the spin and the orbit makes the state of $J = 1/2$ separated off and departured from the above valence band-top. The other two energy-bands are still dually degenerate. One is $J = 3/2$, $M_J = \pm 3/2$ ($E(k)$ curve of heavy-hole band with smaller curvature), the other is $J = 3/2$, $M_J = \pm 1/2$ ($E(k)$ curve of light-hole band with larger curvature). These two are degenerate in

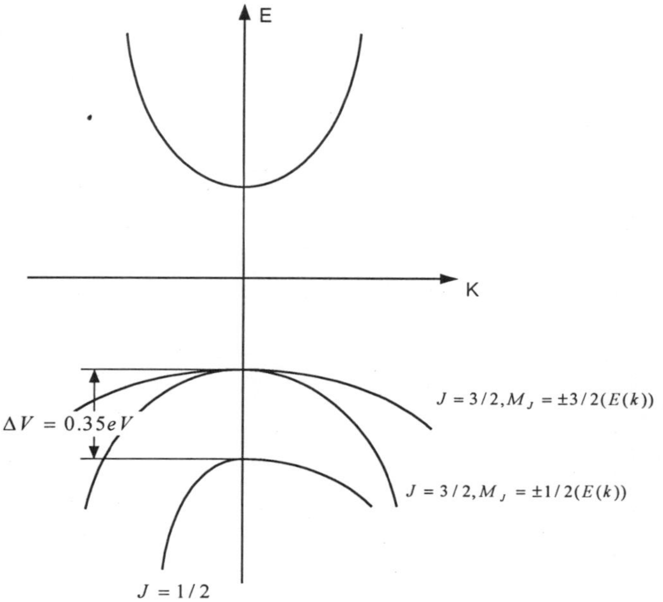

Fig. 6.11 The energy-band configuration of GaAs.

Fig. 6.11 at $k = 0$. Between $J = 1/2$ and $J = 3/2$ states, the band top difference is ΔV = 0.35eV.

Example

The valence-band $E(k)$ curvature is inversely proportional to the effective masses of electrons or holes. In GaAs, the effective mass of heavy holes $m_{hh}^* = 0.45$m, the effective mass of light holes $m_{lh}^* = 0.082$m, and the effective mass of electron $m_e^* = 0.067$m, where m is the free-electron mass. The effective mass of heavy holes is five times that of light holes, and heavy holes' curvature is about 0.2 that of light holes.

6.3.2.2 Particle State-Density Distribution in 3D Crystal and 2D QW/SL

In two-dimensional quantum-well/super-lattice materials (2D-QW/SL), the carrier motions in conduction-bands and filled-bands are quantized along the well growth direction z, but the motions in the planes, which are parallel to the well walls, are approximately continuous. Hence, the

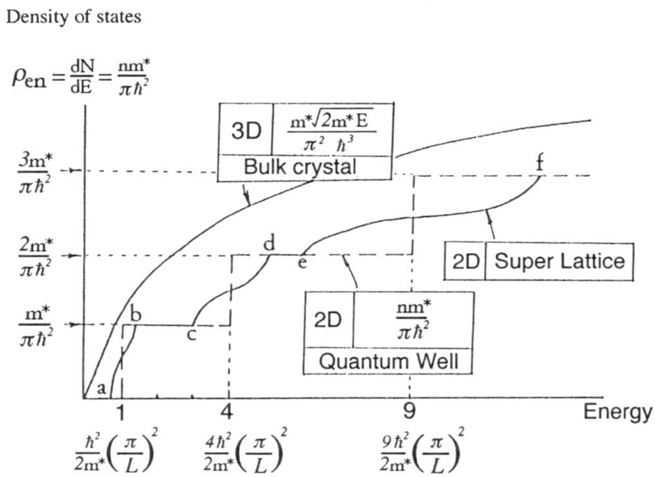

Fig. 6.12 Distinctions of particle state-density distribution in a 3D crystal and a 2D QW/SL.

distributions of the electron or hole energy in QW conduction-bands are governed by the sub-bands started from the quantized energy. In Fig. 6.12, because the motion-states in the sub-bands only include the two-dimensional motion-states in the well-plane, QW state-density is lower than in 3D-bulk crystals.

From the simplified calculation, if the 2D-well is deep enough and the barrier L_{bz} is wide enough, then the electron state-density ρ_{en} will be a constant,

$$\rho_{en} = dN/dE = nm^*/\pi\hbar^2 \tag{6.19}$$

where $m^* \equiv m_e^*$ is the electron's effective mass. The corresponding electron energy levels are

$$E_{n2D} = n^2(\hbar^2/2m^*)\cdot(\pi/L_{wz})^2, \qquad (n = 1,2,3, \dots). \tag{6.20}$$

The electron energy level in a 3D-bulk crystal is

$$E_{n3D} = m^*\cdot (2m^*E)^{1/2}/(\pi^2\eta^3)\ . \tag{6.21}$$

The profiles of electron state-density (dN/dE) versus E of 2D-QW/SL and 3D-bulk are shown in Fig. 6.12.

6.3.2.3 Comparison of Particle State Density Distributions $\rho_N(E)$ Between Two-, One- and Zero-Dimensional Wells with Infinite Depth

Electron state-density function $\rho_{e2D}(E)$ in 2D-QW with infinite-depth potential sink

Since the potential sink is infinitely deep, the wave function is zero at the well wall. The Schrödinger equation in 2D-QW with infinite potential depth is

$$\left[-\frac{\hbar^2}{2m^*}\cdot\frac{\partial^2}{\partial z^2}+V(z)\right]\Psi = E_z\Psi\,. \tag{6.22}$$

The energy eigenvalues are separately as:

$$E_{nz} = (\hbar^2/2m^*)\cdot(\pi n_z/L_{wz})^2, \qquad (n_z = 1,2,3,...), \tag{6.23}$$

which are the same as (6.29). Thus the total energies of electrons are

$$E_{n2d} = (\hbar^2/2m^*)(k_x^2 + k_y^2) + E_{nz} \tag{6.24}$$

$$= (\hbar^2/2m^*)(k_x^2 + k_y^2) + (\hbar^2/2m^*)\cdot(\pi n_z/L_{wz})^2, \quad (n_z = 1,2,3,...) \tag{6.25}$$

The first term on the right shows that the energy distribution along x and y is in continuous free states, and the second term is in separately bound states. With k_x and k_y varying continuously, then, if the quantum number $n_z = 1$, the occupied area in the (k_x, k_y) plane of the electron quantum states is L_{wz}^2. Also, the sum of the states under the $(k_x^2 + k_y^2)^{1/2}$ value is

$$2\left(\frac{L}{2\pi}\right)^2\int_0^k 2\pi\, kdk = \frac{s}{2\pi}\cdot\frac{2m^*}{\hbar^2}(E-E_{nz})\,. \tag{6.26}$$

Thus in the unit area ($S = L_{wz}^2 = 1$), the states in the range of E to $E + dE$ are

$$\rho_1(E)\, dE = m^*/\pi\hbar^2 dE\,, \tag{6.27}$$

or

$$\rho_1(E) = m^*/\pi\hbar^2 = \text{constant.} \tag{6.28}$$

Allowing for higher quantum numbers, the electron energy state-density $\rho_{e2D}(E)$ of 2D-QW with infinite-depth sinks is

$$\rho_{e2D}(E) = \sum_{n_z}\frac{m^*}{\pi\hbar^2}\cdot H(E-E_{nz})\,. \tag{6.29}$$

Electron state-density function $\rho_{e1D}(E)$ in 1D-QW with infinite-depth sink

If the well's size in z and in y is less than λ_{deB}, about 300Å in semiconductors, the motions of carriers in the conduction-band and filled-band are quantized along y and z. Thus, the corresponding electron and hole energy distributions are both separately quantized as bound states. Thus only along x is the energy distribution still continuous or nearly so, as in a bulk crystal. This structure is a one-dimensional quantum wire/super-lattice material (1D-QW). The electron total energy in 1D-QW is, where (n_y, n_z = 1,2,3, ...),

$$E_{n1d} = (\hbar^2/2m^*)k_x^2 + E_{ny} + E_{nz} \tag{6.30}$$

$$= (\hbar^2/2m^*)k_x^2 + (\hbar^2/2m^*)\cdot(\pi n_y/L_{wy})^2 + (\hbar^2/2m^*)\cdot(\pi n_z/L_{wz})^2, \quad (n_y, n_z = 1,2,3, \ldots)\,, \tag{6.31}$$

where

$$E_{ny} = (\hbar^2/2m^*)\cdot(\pi n_y/L_{wy})^2, \quad (n_y = 1,2,3, \ldots) \tag{6.32}$$

$$E_{nz} = (\hbar^2/2m^*)\cdot(\pi n_z/L_{wz})^2, \quad (n_z = 1,2,3, \ldots) \tag{6.33}$$

are the energy distributions of separately bound states. The total electron energy state-density of 1D-QW, or quantum wire, is

$$\rho_{e1D}(E) = \sum_{n_y n_z} \frac{\sqrt{2m^*}}{\pi\hbar} \cdot \frac{1}{\sqrt{E - E_{n_y} - E_{n_z}}}\,. \tag{6.34}$$

Electron state-density function $\rho_{e0D}(E)$ in 0D-QD with infinite-depth sink

If the well's dimensions in x, y, and z are all less than λ_{deB}, the motions of carriers in the conduction band and filled band are quantized in all the three dimensions. Hence, the corresponding electron and hole energy distributions are all separately quantized bound states. These structures are zero-dimensional quantum-dot materials (0D-QD). The electron total energy in 0D-QD is given by (6.35), where (n_x, n_y, n_z = 1,2,3, ...),

$$E_{n0d} = E_{nx} + E_{ny} + E_{nz} = (\hbar^2/2m^*)\cdot(\pi n_x/L_{wx})^2 + (\hbar^2/2m^*)\cdot(\pi n_y/L_{wy})^2 + (\hbar^2/2m^*)\cdot(\pi n_z/L_{wz})^2, \quad (n_x, n_y, n_z = 1,2,3, \ldots)\,, \tag{6.35}$$

where

$$E_{nx} = (\hbar^2/2m^*)\cdot(\pi n_x/L_{wy})^2, \quad (n_x = 1,2,3, \ldots) \tag{6.36}$$

$$E_{ny} = (\hbar^2/2m^*)\cdot(\pi n_y/L_{wy})^2, \qquad (n_y = 1,2,3, \ldots) \tag{6.37}$$

$$E_{nz} = (\hbar^2/2m^*)\cdot(\pi n_z/L_{wz})^2, \qquad (n_z = 1,2,3, \ldots) \tag{6.38}$$

are the energy distributions of separately bound states. The total electron energy state-density $\rho_{e0D}(E)$ in 0D-QD, or quantum dot, is

$$\rho_{e0D}(E) = 2\sum_{n_x n_y n_z} \delta(E - E_{n_x} - E_{n_y} - E_{n_z}) \,. \tag{6.39}$$

The distributions of the electron state-density function $\rho_{e0D}(E)$ in 0D-QD with infinite potential-depth sink are all separate in x, y and z. Fig. 6.13 shows basically the sketch map of the electron energy state-density distributions of 3D-bulk, 2D-QW, 1D-QW and 0D-WD.

6.4 Unique Features of QW Lasers

In this field, very meaningful results have been contributed by many authors, for example: the theory of gain, modulation response, and spectral linewidth in quantum well lasers by Arakawa and Yariv [6.26], the gain

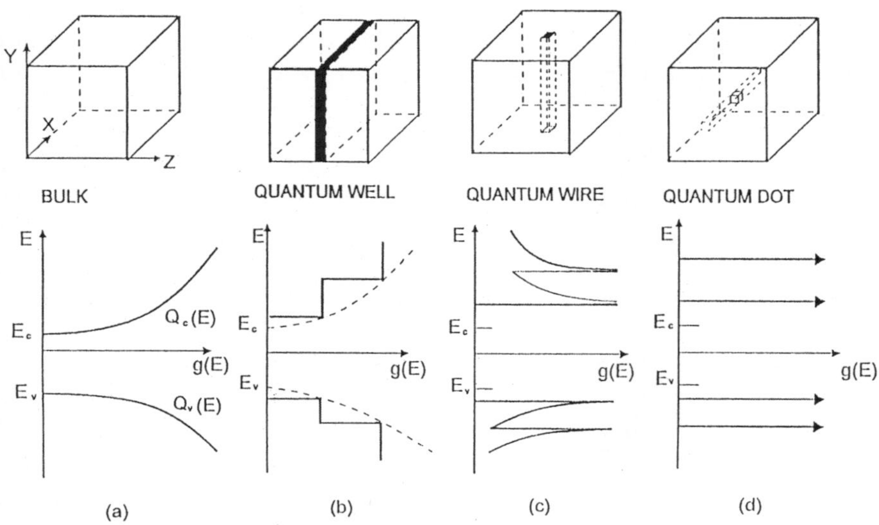

Fig. 6.13 a–d The density of states in different confinement configurations: **(a)** bulk; **(b)** quantum well; **(c)** quantum wire; **(d)** quantum dot. The conduction and valence bands split into over-lapping sub-bands that become successively narrower as the electron motion is restricted in more dimensions.

and carrier lifetime measurements in single quantum well lasers by W.T. Tsang, et al.[6.27], and the room temperature operation of MQW lasers by H. Okamoto, et al. [6.28].

6.4.1 The Effective Band Gap Shifts to a Shorter Wavelength

In 2D-QW, there is a strong limit on the electrons' and holes' motion along the direction perpendicular to the well wall, that is, the growth direction *z*. According to the wave equation, the motion will be the standing wave localized in the wells. Thus the energy state varies with the quantized feature. But in the *xy* plane parallel to the well wall, the motions of electrons and holes can still be treated as freely moving classical particles with quasi-continuous energy distribution. Since the motion along *z* is limited, the total state-density decreases for lack of one dimension's contribution. Hence its state-density distribution is quite different from that in bulk materials, and is altered from a parabolic to a step counter. The effective band gap shifts to a shorter wavelength (Fig. 6.14), an important feature of QW opto-electronics.

6.4.2 The Optical Gain-Spectrum and the Optical Gain-Peak

In QW material, the electron distribution in the conduction-band is quite different from that in the bulk material. In the 2D-QW, the electron

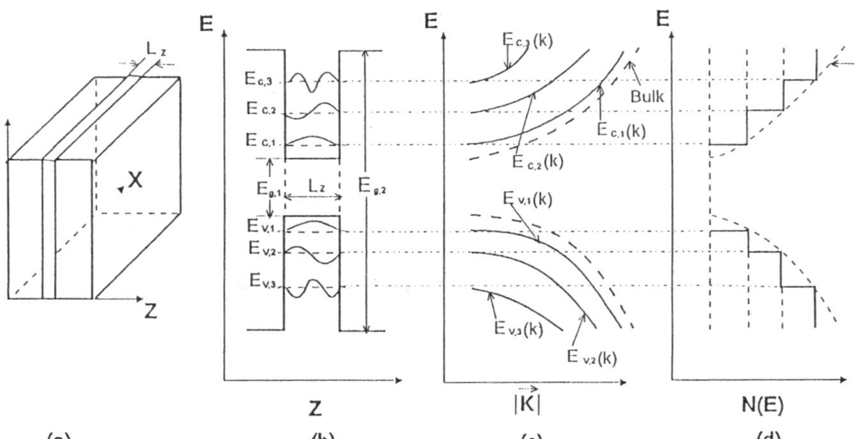

Fig. 6.14 a–d **(a)** Geometry of the quantum well structure. **(b)** Wave-functions in the quantum well. **(c)** Energy sub-bands. **(d)** Density of states in a quantum well.

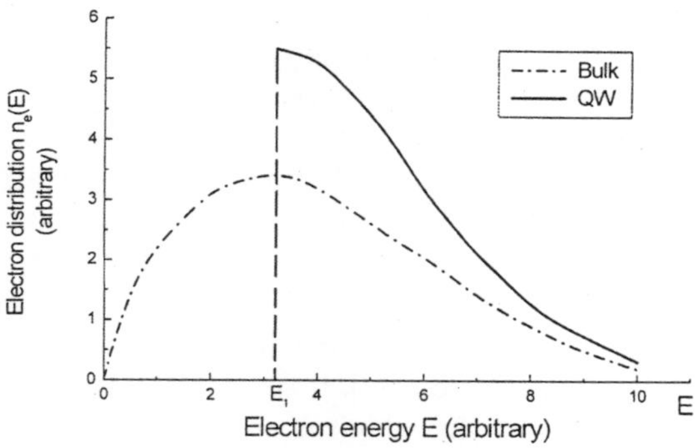

Fig. 6.15 The electron distribution $n_e(E)$ concentrated in the QW.

distribution $n_e(E)$ exists in the range only above the energy E_1, hence the electron distribution $n_e(E)$ is concentrated (Fig. 6.15), resulting in the gain-spectrum $g_e(E)$ of QW being narrowed and the gain-peak $g_{\text{peak}}(E)$ enhanced (Fig. 6.16).

6.4.3 Lower Threshold Current, Larger Quantum Efficiency, Higher Speed, Narrower Spectrum Linewidth, and Greater Output Power

The threshold current I_{th} of of 3D-bulk devices is usually about 20 mA, and that of 2D-QW devices is usually about 2 mA, one order of magnitude less. Amazingly, the I_{th} of strained-layer 2D-QW devices is only about 0.2mA. This is because the electron distribution $n_e(E)$ is concentrated, as

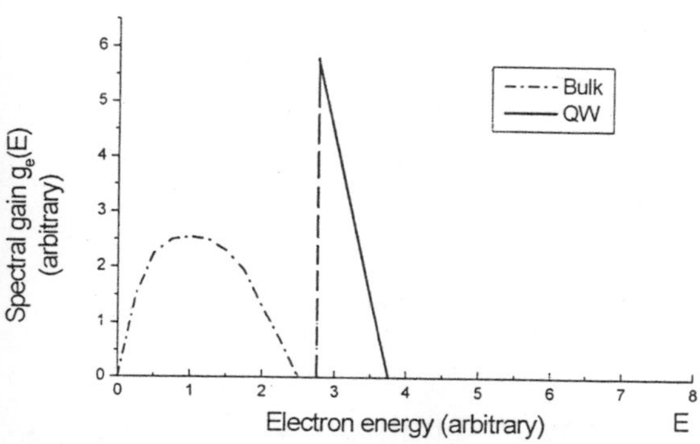

Fig. 6.16 The electron gain-spectrum $g_e(E)$ narrowed and the gain-peak $g_{\text{peak}}(E)$ enhanced in the QW.

mentioned above. Hence the use-efficiency of injection carriers is greatly enhanced, and for the same gain, the injection current density needed in 2D-QW devices is much less than that in 3D-bulk devices.

In laser diodes, because the electrons and photons have their own respective relaxation times and are coupling to each other, a self-relaxation oscillates at a frequency f_r less than 5GHz in the bulk laser diodes. In general, this limits the speed of laser diodes. By theory, $f_r = (1/2\pi)(g'P_0/\tau_p)^{1/2}$, and the differential quantum gain $g' = dg/dN$, which is the rate of change of optical gain dg induced by the injection carrier dN. The g' in QW is obviously higher than that of bulk material. The relaxation oscillating frequency f_r is as high as 30 GHz, which is appropriate for high-speed devices.

The narrower spectrum linewidth is the result of Fig. 6.16. As long as the gain spectrum in QW is much narrower and with sharper peaks, the higher-order cavity modes are mostly outside the gain spectrum. With the side-modes suppressed, single longitudinal mode characteristics with narrower linewidths can be obtained which are much superior to those of double heterostructure (DH) laser diodes.

6.4.4 High Stability of Threshold Current with Changing Temperature

In the QW potential well, the electrons and holes have step-like energy-state density distributions. With respect to threshold-gain fluctuations caused by temperature effects, the energy distribution change is comparatively small. Thus the threshold current I_{th} of QW-LD is highly stable with respect to temperature change. This effect is equivalent to that the QW-LD has a high characteristic temperature T_0.

Example

In a bulk double heterojunction DH-LD, $T_0 = 120–165^\circ C$.
In a $Ga_{1nx}Al_xAs/GaAs/Ga_{1nx}Al_xAs$ 2D-QW-LD, $T_0 = 170–250^\circ C$.
In a 1D-QW(quantum wire) LD, $T_0 = 480^\circ C$.

$$I_{th}(T) = I_{th}(0)\exp(T/T_0).$$

Thus the higher the T_0, the smaller the effect of changes in T.

6.5 Kramers-Kronig Relations in Quantum Well Materials

In 2D-quantum wells, the nonlinearly saturated absorption change $\Delta\alpha(\omega)$

induces the nonlinear refractive index change $\Delta n(\omega)$. Their relation is characterized by the Kramers-Kronig Relations.

6.5.1 Photonic Nonlinearities in Quantum Wells

Because of the quantum sized effects, the optical nonlinearities in the quantum well materials are different within bulk semiconductors. There are the 2D-excitonic nonlinear absorption effects, two-photon absorption effects and the shift of absorption band-edge by the electric field effect (quantum confined Stark effects).

One dominant origin of these nonlinearities is from the resonant absorption saturation of excitons. The ground-state wave functions overlap more in 2D-excitons than in 3D-excitons. Consequently QW has greater optical absorptions and stronger absorption peaks. The exciton absorption becomes saturated without a great photon density, such as $10^{17}/cm^3$. The dissociation time of 2D-excitons in the QW is only 0.3ps, much shorter than that of 3D-excitons in the bulk. The nonlinear saturated absorption of the 2D-QW compared with the 3D-bulk is shown as a schematic drawing in Fig. 6.17. When the injection optical power exceeds several 10^3 W/cm^2, the absorption in the bulk GaAs becomes saturated. In the GaAs 2D-QW, when this power exceeds several 10^2W/cm^2, the absorption becomes saturated. Thus, the injected power needed for saturated absorption is about one order lower than the power needed of 3D-excitons in the bulk.

Another major cause is the electric field effect. When extra inverted fields are applied to the QW, the exciton absorption band-edge shifts

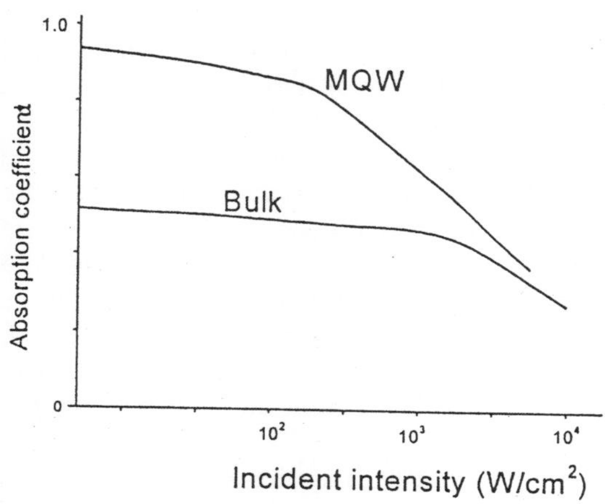

Fig. 6.17 The nonlinear saturated adsorption of a 2D-QW compared with a 3D-bulk at room temperature.

sensitively towards longer wavelengths (red-shift). This feature, used to form electroabsorption modulators, is called quantum confined Stark effect (QCSE). Although the intensities of exciton absorption peaks are somewhat weakened after shifting, the peaks remain even in strong fields. The 2D-exciton can bear very high intensity of inverted electric field. In general, this intensity is about 50 times higher than that the 3D-exciton can bear. The reason for such a strong Stark shift in QW is the limit of well walls. Both the electrons and holes cannot escape from the QW, hence the excitons are strongly polarized when the extra field is applied. The electric voltage and power needed to shift the exciton absorption band-edge are much less than in other methods of changing the absorption state. Thus QCSE is suited to the low controlling power and high-speed (< 1ps) devices.

6.5.2 Nonlinear Absorption α in the 2D Quantum Well

In the 2D-QW, the physical origins of the varying of absorption with input light intensity I_{in} (mW) include band-filling, exciton-formed plasma screening, Stark Effects, band-edge renormalization, and so on. In this section, we shall discuss only the excitonic plasma-screening induced nonlinear absorption and nonlinear refractive index in the 2D-QW, and the Stark Effects will be discussed in the next section.

B is the binding energy of the excitons. At room temperature, the condition is always $h\Omega_{L0} > B > \Gamma$, where $h\Omega_{L0}$ is the energy of the optical phonon and Γ is the linewidth. Because $h\Omega_{L0} > B$, the excitons ionize into electrons and holes due to collisions with optical phonons. Thus when the injection photons are absorbed and the excitons are formed, the excitons are ionized immediately by the optical phonons. The time the excitons exist can be estimated by the phonon bandwidth, which is of almost ps extent. Thus, the plasma of electrons and holes appears in the 2D-QW.

Physically, nonlinear absorption saturation can be considered as the follows. First, plasma screening of the Coulomb interaction between the charge carriers, especially between the attracting e-h pairs, causes the absorption saturation of excitons. The quantum multibody theory needed to process this action is somewhat complicated. Second, the band gap is reduced through exchange interaction and correlation effects. Third, decreasing the absorption at band-edge, which raises the plasma density sufficiently as a consequence of phase-space filling and Pauli blocking, causes the optical gain.

The exciton absorption saturation spectra for the various densities saturation due to the plasma screening, which is the major source of photonic nonlinearities in 2D-QWs, was contributed by G.R. Olbright, et al.

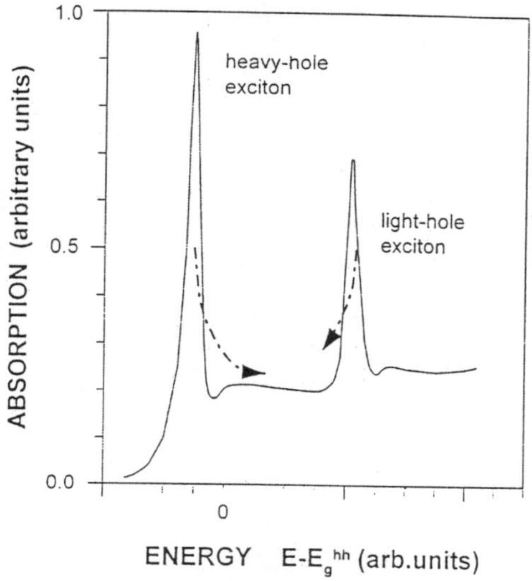

Fig. 6.18 The exciton absorption peaks move (in the arrow's directions) as the exciton densities increase.

[6.12]. The higher densities of heavy-hole plasmas cause the plasma-screening effect and induce the absorption saturation. Figure 6.18, reproduced here with the data in [6.12], shows the sketch map of the moving peak (the dash-dot line along the arrow's directions) as the exciton densities increase.

6.5.3 Kramers-Kronig Relations for 2D Quantum Wells

In 2D-quantum wells, the nonlinearly saturated absorption change $\Delta\alpha(\omega)$ induces the nonlinear refractive index change $\Delta n(\omega)$. Their relation is characterized by the Kramers-Kronig Relations:

$$\Delta n(\omega) = P.V. \int d\omega' \, [\Delta\alpha(\omega')]/(\omega'^2 - \omega^2) \tag{6.40}$$

where $P.V.$ is the Cauchy principal value. Figure 6.19, reproduced with the data presented by D.S. Chemla and D.A.B. Miller in [6.15], shows the sketch diagram of the absorption change $\Delta\alpha(\omega)_{\rm ch}$ and the induced refractive index change $\Delta n(\omega)_{\rm ch}$ of 2D-excitons in the 2D-QWs. This diagram, shown in arbitrary unit, is characterized by the Kramers-Kronig Relations. In the figure, maximum saturation absorption $\alpha_{\rm max}$ ($\Delta\alpha(\omega) = 0$) corresponds to the maximum refractive index change $|\Delta n|_{\rm max}$ and the maximum refractive index $n_{\rm max}(\Delta n(\omega) = 0)$ coincides with the maximum absorption change $|\Delta\alpha|_{\rm max}$.

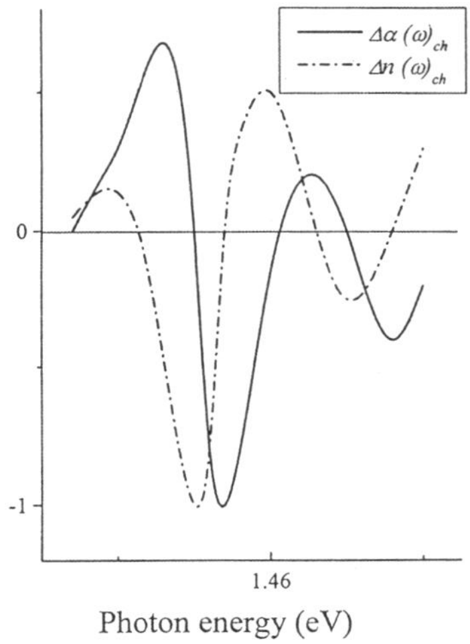

Fig. 6.19 The Kramers-Kronig Relations for a 2D-QW (the 2D-exciton absorption change $\Delta\alpha(\omega)_{ch}$ derived the refractive index change $\Delta n(\omega)_{ch}$) [6.15] (by permission of IEEE).

6.6 Quantum-Confined Stark Effects (QCSE) in Quantum Well Structures

The exciton absorption peak red-shifts in 2D-QW is called QCSE. As the rectangular well becomes inclined, there are variations of the transition selection rules making the forbidden transitions becoming possible.

6.6.1 QCSE with Extra Field

When an extra electric field biased inversely on the 2D-QW of which the absorption edge is shifted sharply, this makes the 2D-QW a superior electro-optic modulator. As such strong red-shift (Stark shift) is caused by the confinement of motion by well walls, this absorption edge red-shift in 2D-QW material is called Quantum Confined Stark Effects (QCSE). Since both the electrons and holes cannot escape from the QW, the excitons are forced to be strongly polarized even if the extra applied voltage is not great. The electric voltage needed to shift the band-edge is only 1 to 2V, and over 90% of the possible modulation is attainable. These electric-absorption (EA) modulators based on QCSE have high modulation linearity and 100GHz bandwidth. Because of the electric field effect and very low required-voltage, the power consumption of EA modulators is quite low, and they can be integrated with QW lasers.

6.6.2 Well-Bottom Inclined by the Extra Field and Electric Absorption

When the extra electric field is applied to a QW device and if the direction of the extra electric field is along the growth direction z of the QW, the electric potential counter of the QW is as exhibited in Fig. 6.20. Both the bottom of the conduction-band and the top of the valence-band are inclined.

The electric field, regarded as the perturbation, varies the electron function and the hole function. Both the ground-state energies are somewhat lowered, and the corresponding photon energy, resonant with the exciton absorption peak, is shifted to the lower energy. Figure 6.21 shows that this shift increases as the electric field becomes stronger, similar to the Stark Effect in hydrogen atoms. Due to the well-confining, the Stark Effects are stronger in a 2D-QW.

6.6.3 Exciton Linewidth Broadening and Lifetime Shortening Caused by the Extra Field

In the case of no extra field, background broadening of the exciton spectrum linewidth occurs. The resonant electron-absorbing photon becomes an exciton, which is scattered by an optical phonon and forms the free electron-hole pair. This is the uniform broadening process of the linewidth. This broadening is Γ_{SC} and the corresponding lifetime is τ_{ex}. In

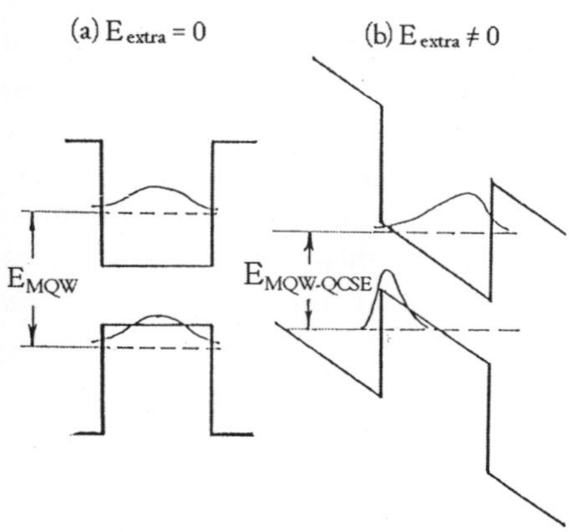

Fig. 6.20 a, b The inclined well bottoms caused by the extra-field.

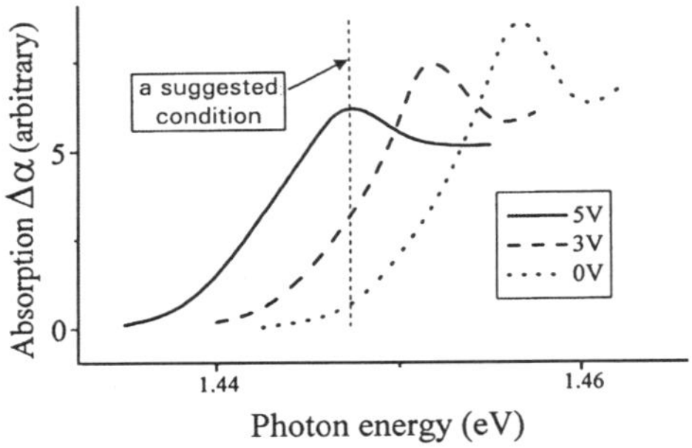

Fig. 6.21 The exciton absorption peak shifted to the lower energy (long wavelength shift) by the quantum confined Stark effect (QCSE).

addition, in a regular 2D-QW sample, even at extremely low temperature, there is still some residual broadening of the absorption linewidth. This residual broadening, induced by irregular fluctuations of QW thickness of atomic scale, is the non-uniform broadening Γ_{QW}, with the corresponding lifetime τ_{QW}. The total linewidth broadening is the sum of the uniform and non-uniform broadening.

If the spectrum line-shape is approximately of Lorentz shape, then $\Gamma = \Gamma_{SC} + \Gamma_{QW}$. If the spectrum line-shape is approximately of Gauss shape, then $\Gamma = (\Gamma_{SC}{}^2 + \Gamma_{QW}{}^2)^{1/2}$. The former is an algebraic sum and the latter is a vector sum.

The spectrum total linewidth broadening Γ is proportional to the spectrum energy width ΔE. From the uncertainty principle, $\tau\Delta E \geq h$ or $\tau\Gamma \geq h$, so $\tau = h/\Gamma$ and $\Gamma = h/\tau$.

When the extra field is applied, the absorption linewidth broadens. This is caused by the electric-field-induced spatial separation of the wave functions of electrons and holes, and the barrier-tunneling-induced dissociation of electrons and holes. Accordingly, the lifetime τ is shortened and absorption linewidth Γ is broadened. Accompanied by the voltage increase, the effective barrier-height is lowered and the effective barrier-width is narrowed until the 2D-exciton dissociates. In some cases, the non-uniformity of background impurities will induce the non-uniform distribution of the field which causes the additional broadening of $\Delta\Gamma$ and additional shortening of $\Delta\tau$.

Example

In the case of no extra field, the Gauss linewidth Γ is 4.587×10^{-22} Joule, and $h = 1.055\times 10^{-34}$ J·sec. Then the exciton lifetime $\tau = 230$fs.

6.6.4 QCSE Compared with Franz-Keldysh Effects

When an extra field is applied to the 2D-QW and the absorption spectrum is measured, the resultant QCSE phenomena are somewhat similar to those of of the Franz-Keldysh Effects in the 3D-bulk. However, there are some substantial differences induced by the excitons. At room temperature, the extra electric field needed in the 3D-bulk for observing the Franz-Keldysh Effects is ordinarily higher than 10^5V/cm. If the bulk active layer is 100nm, the ΔV is the same as the disturbed ground-state energy E_g of 1.43eV. But for 2D-QW, only 10^4V/cm is needed for QCSE. If the well width is 10nm, the ΔV is the same as the disturbed exciton bonding energy E_{bond} of 10meV. These distinctions are summarized in Table 6.3.

In semiconductor bulk, the 3D-exciton dissociates at about $E_{\text{extra}}=2\times10^3$V/cm, the bonding-energy of the 3D-exciton is one Ry (Rydberg energy). But in the 2D-QW, when E_{extra} is applied along the QW growth direction z, the 2D-exciton is dissociated at $E_{\text{extra}} > 10^5$V/cm. In this case, both the electron and hole in the QW are deviated by the extra field, but they are still confined as an exciton. If the extra field is parallel to the well plane, the 2D-exciton dissociated at $Eextra=2\times10^4$V/cm, the bonding-energy of 2D-exciton is 4 Ry. These field strengths are summarized in Table 6.4.

In QCSE, as the extra field becomes stronger, the photo-luminescence weakens, and the exciton absorption peak red-shifts and broadens. Also, as

Table 6.3 The Different Extra-field needed for Bulk-F-K Effect and 2D-QCSE Effect.

Materials	Effects	Thickness	Disturbed energy	Field strength E_{extra}	Absorption edge
3D-bulk	F-K	0.1μm	Eg=1.43eV	10^5V/cm	Gentle slope
2D-QW	QCSE	10nm	$Eexc$=10meV	10^4V/cm	Steep/red-shifted

Table 6.4 The Dissociated Field-strength for 3D-excitons and 2D-excitons.

Materials	Exciton diameter (nm)	Dissociated field-strength (V/cm)	Bonding energy
3D-bulk	~30	~2×10^3	1 Ry
2D-QW (E ‖ well-plane)	~10 (L_{WZ})	~2×10^4	4 Ry
(E ⊥ well-plane)	~10 (L_{WZ})	>10^5	4 Ry

the rectangular well becomes an inclined well, the local electrons settle on one side while the local holes on the other side (Fig. 6.20). Consequently, there are some variations on the selection rules of transition, with forbidden transitions corresponding to $\Delta n \neq 0$ becoming possible. For that reason, precise measurement of the level-spacing between the higher-order states and the ground-state in the valence-band is possible. We consider an infinite-depth well as a simple model. The effective mass Hamiltonian in an electric field is

$$H = H_0 + epz \tag{6.41}$$

where the H_0 is the Hamiltonian in the zero-field ($U = 0$). The energy eigenvalues are

$$E_n^{(0)} = (h^2/2m^*)\cdot(n\pi/L_{Wz})^2, \qquad (n = 1,2,3, \ldots). \tag{6.42}$$

If the field is only a perturbation,

$$eUL_{Wz} << E_n^{(0)} = (\hbar^2/2m^*)\cdot(n\pi/L_{Wz})^2, \qquad (n = 1,2,3, \ldots). \tag{6.43}$$

By the second-order perturbation, the field-induced ground-state energy change can be obtained as

$$\Delta E_{1\text{-SHIFT}} = E_1 - E_1^{(0)} = -S_{1\text{-SHIFT}} \cdot L_{Wz}^4 U^2 . \tag{6.44}$$

where the coefficient of energy shift $S_{1\text{-SHIFT}}$ is linked with the effective mass of electron m^* and Planck constant $\hbar^2$.

This shows that the larger the well width L_{Wz}, the stronger the QCSE and the larger the energy change $\Delta E_{1\text{-SHIFT}}$. But as the well width L_{Wz} increases, the excitons become more unstable. This is because the smaller are the spatial overlap coefficients of the electron wave function and hole wave function in 2D-excitons, the lower the exciton's binding energy becomes. Thus, between the needed 2D-QW effect and QCSE, we have to have a compromise choice of L_{Wz}, usually 5–15nm.

In more precise analysis, the finite-depth well must be considered. Using the calculus of variations, the energy levels and exciton binding energies are found, and the calculated results conform to the experimental data.

6.6.5 Semiconductor Optical Modulators of QCSE

In high-capacity fiber communication, the channels of 10Gbit/s bit-rate or over need high-speed optical modulators. Semiconductor optical modulator of QCSE is one hopeful candidate since QCSE modulator of III-V semiconductor can be integrated with 1.55μm lasers. A separately external modulator has its reflection induced frequency chirp of laser source and has larger insert loss. There are two dominant kinds of optical modulators: one is the electroabsorption (EA) optical modulator and the other is Mach-Zehnder (M-Z) optical modulator. High modulating rate, low driving voltage, high distinction as well as low frequency chirp are some key requisites for these modulators.

6.6.5.1 Quantum Well Electroabsorption (EA) Modulators of QCSE

The red shift of absorption band-edge due to the distinct change of transparent intensity induced directly vy QCSE is the principle of EA modulators of QCSE. Its physical patterns have been shown in Fig. 6.20 and Fig. 6.21. The III-V semiconductor EA optical modulator is based on the electroabsorption in the semiconductors.

Compared with the bulk material EA optical modulator, the quantum well (QW) EA optical modulator is much better because of its integrated technology, broad modulating bandwidth (such as 40GHz in 1.55μm) and high electroabsorption coefficient. The InGaAsP quaternary MQW structure shows the large Stark shifts in the 1–1.6μm range.

The key parameters of the QW-EA QCSE modulator are the on-off ratio, modulating bandwidth, inserting loss and frequency chirp. The on-off ratio of this modulator $R_{\text{on-off-EA}}$ is

$$R_{\text{on-off-EA}} = \exp(\Delta\alpha \cdot L), \tag{6.45}$$

where $\Delta\alpha$ is the absorption change between the on and off states, L is the absorption modulating arm. Hence the modulation efficiency is determined by the absorption change $\Delta\alpha$ per driving voltage applied. The chirp factor

$$\zeta_{\text{chirp-EA}} = \Delta n/\Delta\kappa, \tag{6.46}$$

where Δn is the refractive index change and $\Delta\kappa$ is the distinction coefficient change. In order to reduce the chirp factor $\zeta_{\text{chirp-EA}}$, some detuning between the resonant wavelength of exciton in the QW $\lambda_{\text{RES-EXCITON}}$ and the laser wavelength of the modulator $\lambda_{\text{MOD-LASER}}$ is settled that makes the smaller Δn at large $\Delta\kappa$. If the wavelength of the modulator $\lambda_{\text{MOD-LASER}}$ is 1.55μm, the resonant wavelength of exciton in the QW $\lambda_{\text{RES-EXCITON}}$ is always chosen at 1.51–1.52μm, and this means the wavelength detuning is about 30–40nm.

6.6.5.2 Quantum Well Mach-Zehnder (M-Z) Modulators of QCSE

The light is divided into two and injected into two waveguide arms of a M-Z interferometer. The arm's refractive index is adjusted by the extra electric field. When these two lights meet again at the output of M-Z interferometer, the result shows intensity modulation. Generally, this modulator is polarization-dependent but not the strained-layer MQW structures.

Since the QCSE in semiconductor MQW is a nonlinear effect, the refractive index change $\Delta n(\omega, E)$, varied with the extra electric field E, is also nonlinear

$$\Delta n_{\text{M-Z}}(\omega, E) = -n_0^3 (\gamma_1 E + \gamma_2 E^2)/2, \tag{6.47}$$

where n_0 is the intrinsic refractive index without any electric field, γ_1 and γ_2 are linear and second-order nonlinear optoelectric coefficient respectively. The field phase change of the modulating arm due to the drive-voltage V is

$$\Delta\varphi M\text{-}Z(\omega, V) = \xi \cdot k_0 L \Delta n_{\text{M-Z}}(\omega, E) = -\xi \cdot k_0 L_{\text{M-Z}} \cdot n_0 (\gamma_1 E + \gamma_2 E^2)/2 \tag{6.48}$$

where ξ is an overlap-factor of the electric and optic fields, k_0 is the optical wave vector in vacuum and $L_{\text{M-Z}}$ is the length of the modulating arm.

The key parameters of the QW-MZ QCSE modulator are also the on-off ratio, modulating bandwidth, inserting loss and frequency chirp. The on-off ratio of this modulator $R_{\text{on-off-M-Z}}$ is related with the device structure, operation style, arm ratio, etc. and is a complicated issue. When two routing meet at the output and if they are opposite in phases and equal in intensity, we can get the optimum $R_{\text{on-off-M-Z}}$.

The chirp factor $\zeta_{\text{chirp-M-Z}}$ is an important parameter of the QW-MZ QCSE modulator. It is always managed by the device structure, operation style and arm power ratio to give a negative chirp factor $\zeta_{\text{chirp-M-Z}} < 0$. It is meaningful for the large product of $B_{\text{bit-rate}} \cdot L_{\text{trans- distance}}$.

References

[6.1] C. Kittel: *Introduction to Solid State Physics,* John Wiley, New York (1986)

[6.2] R.K. Jain and M.B. Klein: Degenerate FWM in semiconductors. In: R.A. Fisher (ed.): *Optical Phase Conjugation,* Academic Press, Orlando, p. 333 (1983)

[6.3] Von Walter Franz: Einfluβ eines elektris en feldes auf eine optiks e absorptionskante, *Z. Naturforsch* Band **13A**, 484–491 (1958)

[6.4] W.L. Keldysh and Z.H. Eksperim: *Sov. Phys.— JETP (I Teor. Fiz.)* **34**, 788–790 (1958)

[6.5] B.O. Seraphin and N. Bottka: Franz-Keldysh Effect of the refractive index in semiconductors, *Phys. Rev.* **139**(2A), 560–565 (1965)

[6.6] B.R. Bennett and R.A. Soref: Electrorefraction and electroabsorption in InP, GaAs, GaSb, InAs, and InSb, *IEEE J. Quantum Electron* **23,** 2159–2166 (1987)

[6.7] D.E. Aspnes: Electric field effects in the dielectric constant of solid, *Phy. Rev.* **153**, 972–982 (1967)

[6.8] E.D. Palik (ed.): *Handbook of Optical Constants of Solids,* Academic Press, Orlando (1985)

[6.9] John D. Dow and David Redfield: Toward an unified theory of Urbach's Rule and exponential absorption edges, *Phys. Rev. B* **5**, 594–610 (1972)

[6.10] L. Esaki, et al.: *IBM Res. Note* RC-2418 (March 1969)

[6.11] L. Esaki and R. Tsu: Superlattice and negative differential conductivity in semiconductors, *IBM J. Res. Develop.* **14**, 61–65 (1970)

[6.12] G.R. Olbright, H.M. Gibbs, et. al.: Nonlinear optical properties of type-II quantum well, *Phy. Rev. B* **44**, 3043–3053 (1991)

[6.13] D.A.B. Miller, et. al.: Large room-temperature optical nonlinearity in GaAs/Ga1-xAlxAs multiple quantum well structures, *Appl. Phys. Lett.* **41**, 679–681 (1982)

[6.14] D.A.B. Miller, D.S. Chemla, et. al.: Electric field dependence of optical absorption near the band gap of quantum-well structures, *Phys. Rev. B* **32**, 1043–1060 (1985)

[6.15] D.S. Chemla, et al.: Room temperature excitonic nonlinear absorption and refraction in GaAs/AlGaAs multiple quantum well structures, *IEEE J. of Quantum Electron* **20**, 265–275 (1984)

[6.16] K. Kikuchi, Tien-Pei Lee, et al.: Observation of highly nondegenerate four-wave mixing in 1.5μm traveling-wave semiconductor optical amplifiers and estimation of nonlinear gain coefficient, *IEEE J. of Quantum Electron* **28**, 151–156 (1992)

[6.17] A.R. Adams, et al.: Band-structure engineering for Low-threshold high-efficiency semiconductor lasers, *Electron. Lett.* **22**, 249–250 (1986)

[6.18] C.E. Zah, Tien-Pei Lee, et al.: Low threshold 1.5μm tensile-strained single quantum well lasers, *Electron. Lett.* **27**, 1414–1415 (1991)

[6.19] G. Jones, E.P. O'Reilly, et al.: Improved performance of long-wavelength strained bulk-like semiconductor lasers, *IEEE J. of Quantum Electron* **29**, 1344–1354 (1993)

[6.20] B. Zhao, Chen, T.R., et al. Direct measurement of linewidth enhancement factors in quantum well lasers of different quantum well barrier heights, *Appl. Phys. Lett.* **62**, 1591–1593 (1993)

[6.21] Chaoyang Zhang and Yili Guo: A semiconductor nonlinear gain gate for photonic switching, *Acta Semiconductor Sinica* **16**, 358–364 (1995)

[6.22] Yili Guo, et al.: A kind of optoelectronic hybrid-integrated optically bistable device, *Acta Optica Sinica* **10**, 900–904 (1990)

[6.23] Lichen Zhong and Yili Guo: Analysis of steady-state characteristics of bistable laser diodes *J. Chinese Physics Lasers (AIP-OSA, USA)* **14**, 312–316 (1987)

[6.24] K.L. Hall, et al.: Subpicosecond gain and index nonlinearities in InGaAsP diode lasers, *Optics Communications* **111**, 589–612 (1994)

[6.25] K.K. Anderson, et al.: Femto-second dynamics of the nonlinear index near the band edge in AlGaAs waveguides, *Appl. Phys. Lett.* **56**, 1834 (1990)

[6.26] Y. Arakawa and A. Yariv: Theory of gain, modulation response, and spectral linewidth in AlGaAs quantum well lasers, *IEEE Journal of Quantum Electron* **21**, 1666–1674 (1985)

[6.27] N.K. Dutta, R.L. Hartman and W.T. Tsang: Gain and carrier lifetime measurements in AlGaAs single quantum well lasers, *IEEE Journal of Quantum Electron* **19**, 1243–1246 (1983)

[6.28] T. Saku, H. Okamoto, et al.: Room temperature operation of 650 nm AlGaAs multi-quantum-well laser diode grown by molecular beam epitaxy, *Japanese J. of Appl. Phys., Part 2 — Letters* **24**, 73–75 (1985)

7. Fiber Nonlinear Photonics

Photonic nonlinearties caused by the light intensity can be classified into two kinds: one is the nonlinear refractive index effect (SPM, XPM, FWM and others) which induced the phase effects, the other is the stimulated light scattering (SRS, SBS and others) which induced the gain or loss. Optical FWM and nonlinear light scattering in fibers will be discussed in Chap. 8. Another interesting subject, the nonlinearities in optical waveguide, will be explicated in Chap. 9.

The proposal that glass fiber can be used as a dielectric waveguide for communication was first published by Kao and Hockham in 1966 [7.1]. At that time, the attenuation of most known optical glass was as high as 1dB/m, but soon after this publication, great efforts were focused on reducing the attenuation in fused-silica fiber. Corning created firstly in 1970 the fiber with 20dB/km. For existing industrial fibers, the loss is only 0.2dB/km at 1.55μm and 0.4dB/km at 1.31μm. Other features associated with communications are all greatly enhanced. Up to the end of the 20th century, the underground fibers installed were about several hundreds of million kilometers, and more than tens of million kilometers of fibers are being installed yearly.

The demand for high capacity and ultra-broadband optical fiber transmission due to the dramatic growth of the internet data traffic has been put forward at the beginning of this century. Some of the most meaningful technologies in the coming years will be:

(1) Terabit/s of fiber transmission and IP/WDM smart optical networks.
(2) Long-haul and optically intelligent routing for internet data services.
(3) Ultra-broadband of 1430 to 1645nm and optical amplification devices.
(4) Narrow channel spacing of 25–50GHz for DWDM systems.
(5) 40Gbit/s/channel with D/D-slop and PMD compensation devices.
(6) Lower photonic nonlinearities in trunk fibers and in node devices.
(7) More higher fiber utilizing efficiency as 1 bit/s/Hz.
(8) Powerful routers formed by IP routers and optical wavelength routers.
(9) Less undesirable physical effects and accumulation of cascaded optical nodes.

It is a forever-interesting gambit how to increase the channel power to allow for a larger repeat-less span with postulated OSNR, thus lessening the cost. But the nonlinearities in fiber force a limit to the channel power by signal deformed due to some inter-channel effects such as XPM and FWM, as well as the intra-channel effects such SPM.

The fiber nonlinearity is always combined with the intrinsic dispersion of fibers, GVD and polarization-mode dispersion (PMD). Therefore, even if the dispersion is not in the nonlinear domain, we shall still discuss it here. The dominant effect of dispersion is the pulse-width broadening and it is linked with nonlinear SPM and XPM effects. This makes the phenomena more complicated. The XPM is intensely significant in the 50–25GHz spaced channels and the PMD is substantial in high bit-rate OFC.

Optical intensity-induced refractive index (IIRI) is one important part of fiber nonlinearity and will be discussed in Sect. 7.2. It is the counterpart of the real part of the electric susceptibility in the fiber. Self-phase modulation (SPM), cross-phase modulation (XPM), and other parametric processes in the fiber, such as FWM and phase conjugation, all belong to this domain. In long-haul systems, such as the trans-oceanic systems, the no regeneration range (with optical amplifiers only) is about 6000–10000km. Hence, the optical IIRI is dominant.

When the group velocity dispersion induced down-chirp $\delta\omega_{\mathrm{GVD}}$ and the SPM nonlinearity induced up-chirp $\delta\omega_{\mathrm{SPM}}$ compensate each other in the total uniform anomalous-dispersion transmission region, the chirp and deformation of the transmitted pulses can be completely offset in the first-order soliton. Based on this cognition, this perfect soliton is an ideal mode of the high bit-rate and long-distance transmission. The demonstrated transmissions have indicated very much astonishing results, such as 100Gbit/s $\times 10^4$ km in the straight-line and 100Gbit/s$\times 10^6$km in the loop. But the critical arrangements both for the uniform compensation of GVD and the elimination of time jitter cause the weak stiffness of the ideal first-order soliton and this means its stability is easy to be disturbed. Recently, some modified modes, such as dispersion-managed soliton transmission, are studied in experiments, in which the dispersion is alternately compensated. The stiffness of the system is therefore enhanced.

Some nonlinear techniques and components, such as nonlinear optical fiber-loops, are associated with the intensity-induced refractive index changes.

The fiber nonlinearities can be considered in three aspects:

(1) Nonlinearities, such as the FWM, XPM, SPM, SRS, and SBS, will deteriorate the transmission characteristics of fiber communications. These effects should be lowered.
(2) In some conditions, fiber nonlinearity is used to compensate other unfavorable effects to obtain a new solution, e.g. the soliton transmission is based on the SPM and GVD effects.

(3) Recently, some fiber nonlinearities are applied to realize the light leading light function, e.g. the XPM or FWM used for the demultiplexing in OTDM.

7.1 Group Velocity Dispersion (GVD) and Pulse Broadening

GVD and optical SPM are in different domains, GVD is an optically linear factor and the SPM is a nonlinear one, but they always affect each other in pulse transmission.

7.1.1 GVD-Caused Optical Pulse Broadening

Fiber telecommunication is already a mainstay of the world information infrastructure. WDM backbone transmissions at several hundred of Gbit/s and over a hundred wavelength channels are expected soon, because of the wide bandwidth of very low loss. In 1999, optical erbium-doped fiber amplifiers (EDFAs) had operated already in this region with 80 nm bandwidth by the combination of conventional band (C-band) and long-wavelength band (L-band) devices. Some new materials of optical amplifiers are being developed.

The Rayleigh scattering-induced attenuation is $\propto 1/\lambda^4$. The infrared absorption attenuation increases with λ and resonates at about 9 μm. Before 1990s, an OH absorption peaked at 1.39μm. There are two separate low loss bands at 1.31μm and 1.55μm. Figure 7.1 shows the dispersion and

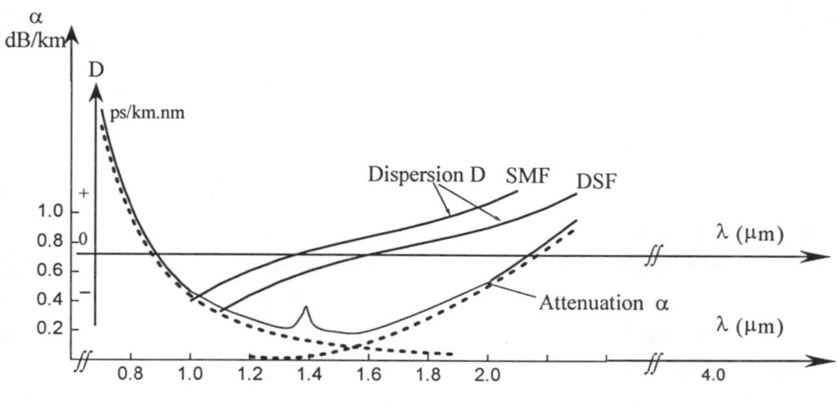

Fig. 7.1 Dispersion and attenuation of the conventional single-mode fibers (CSMF) and the conventional dispersion-shifted fibers (CDSF).

attenuation curves of the conventional silica single-mode fiber (CSMF). At the end of the 20th century, this OH peak almost vanished because the technologies had improved. Consequently, these two bands would join together. An extremely wide window of silica fiber with about 400nm or 40Terabit/s bandwidth will be offered for the 21st century communication. In this condition, more concern will be paid to the dispersion curve that seriously limits the product of the bit-rate B times by the distance L.

The problem involved in replacing the 1.3μm systems with those using 1.5μm is that conventional fiber (G652) has a GVD as high as D = +17ps/nm-km. The 10-Gbit/s transmission length is limited to around 60km. Over that distance, the pulse broadening due to GVD-SPM is such that the bit error rate (BER) obviously increases. Problems of fiber dispersion compensation are extremely significant and must be solved.

7.1.1.1 The Group Velocity Dispersion Curve *D(λ)*

Dispersion is an intrinsic and fundamental property of the fiber, and can be described by several parameters varying with ω, such as the refractive index n, the group index n_g, the group velocity v_g, and the mode-propagation coefficient β. The series expansion of the mode-propagation coefficient β with the frequency ω, conventionally ω_0 when $\beta_2(\omega_0) = 0$, is

$$\begin{aligned}\beta(\omega) &= \frac{n(\omega)\omega}{c} \\ &= \beta_0 + \beta_1(\omega-\omega_0) + \frac{1}{2}\beta_2(\omega-\omega_0)^2 + \frac{1}{6}\beta_3(\omega-\omega_0)^3 \\ &\quad + \ldots + \frac{1}{r!}\beta_r(\omega-\omega_0)^r\end{aligned}$$

where β_r is the r-th order of the mode-propagation coefficient dispersion or (r-1)-th order of GVD,

$$\beta_r = \left[\frac{d^r\beta}{d\omega^r}\right]_{\omega=\omega_0}$$

where β_0 is associated with the part of v_g independent of ω, β_1 is associated with the part of v_g dependent on ω, β_2 is the first-order dispersion of v_g, and β_3 is the nonlinear dispersion of v_g. The names, representations, units, and functions of the different orders of β are listed in Table 7.1.

In most cases, the linear dispersion coefficient curve, $D(\lambda)$, accompanies the attenuation curve $\alpha(\lambda)$.

The first-order of GVD (or the second-order of mode-propagation coefficient dispersion) is

Table 7.1 The features and functions of different orders of mode-propagation coefficient.

Item	*Name*	*Representation in Mathematics*	*Unit*	*Function*
β_0	Dispersion-free coefficient	$\beta_0 = \text{constant}$	1/km	A part of group velocity independent
β_1	First-order dispersion (of β)	$\beta_1 = \frac{d\beta}{d\omega} = \frac{1}{c}\left[n + \omega \frac{dn}{d\omega}\right] = \frac{n_g}{c}$	ps/km	Determing the group velocity v_g, hence the pulse broadening
	Group velocity parameter ($\sim 1/v_g$)	$= -\frac{1}{v_g}$		
β_2	Second-order dispersion (of β)	$\beta_2 = \frac{d^2\beta}{d\omega^2} = \frac{1}{c}\left[2\frac{dn}{d\omega} + \frac{d^2 n}{d\omega^2} \cong \frac{\omega}{c}\frac{d^2 n}{d\omega^2}\right]$	ps/km·nm ($d/d\lambda$)	Pulse broadening, deviating-off in spatial, $\beta_2 = 0$ is corresponding to the zero-dispersion wavelength λ_0 (1.27μm for bulks)
	First-order GVD $\left(of \frac{dv_g}{d\omega}\right)$	$= -\frac{1}{v_g}\left(\frac{dv_g}{d\omega}\right)$	ps^2/km ($d/d\omega$)	
β_3	Third-order dispersion (of β)	$\beta_3 = \frac{d^3\beta}{d\omega^3} = \cdots \approx \frac{d^3 n}{d\omega^3}$	ps/km·nm^2 ($d/d\lambda$)	Must be considered at λ_0 ($\beta_2 = 0$) or for ultra-short pulse ($\tau \le 100$fs)
	Second-order GVD $\left(of \frac{d^2 v_g}{d\omega^2}\right)$	$= +\frac{2}{v_g^{\ 3}}\left(\frac{dv_g}{d\omega}\right) - \frac{2}{v_g^{\ 2}}\left(\frac{d^2 v_g}{d\omega^2}\right)$	ps^3/km ($d/d\omega$)	

$$\beta_2(\omega) = (d^2\beta/d\omega^{2)} = (1/c)(dn_g/d\omega) = -(1/v_g)(dv_g/d\omega). \qquad (7.1a)$$

The dispersion coefficient is defined as

$$D(\lambda) = d\beta_1/d\lambda = -(2\pi c/\lambda^2)\beta_2 \cong -(2\pi c/\lambda^2)(dn_g/d\omega). \qquad (7.1b)$$

The region of $\lambda < 1.27\mu m$ for fused silica, in which n_g increases and v_g decreases with frequency ω ($dn_g/d\omega > 0$), is the normal-dispersion region of group index n_g (Sect. 2.2). In this region, $\beta_2 > 0$ and dispersion coefficient $D < 0$. The region of $\lambda > 1.27\mu m$ for fused silica, in which n_g decreases and v_g increases with frequency ω ($dn_g/d\omega < 0$), is the anomalous-dispersion region of group index n_g. In anomalous-dispersion region, $\beta_2 < 0$ and $D > 0$ (positive dispersion coefficient region). $\beta_2(\omega)$ and $D(\lambda)$ are always discussed as dominant factors in the fiber system.

The $D(\lambda)$ curve has a zero point at $\lambda = 1.27\mu m$, at which $\beta_2 = 0$, for fused silica bulk. But in the silica fiber, the zero point shifts toward $\lambda = 1.31\mu m$, due to the waveguide dispersion effect. From the electric susceptibility theory in Chap. 2, both $\chi'(\omega)$, related to the refractive-index dispersion, and $\chi''(\omega)$, related to the absorption, appear in (2.24–2.33) and in Figs. 2.1 and 2.2. From experiments on the fused silica fiber, $\chi''(\omega)$'s extremes are in the UV, at < 200nm, and in the IR, at about 9μm. A transparent region exists between these (Fig. 7.2(b)). The curve for $\chi'(\omega)$, related to $n^2(\omega) = \chi'(\omega) + 1$, is shown in Fig. 7.2(a). The curve $n(\omega)$ (not group index n_g) rises towards the UV, drops toward the IR, and has an

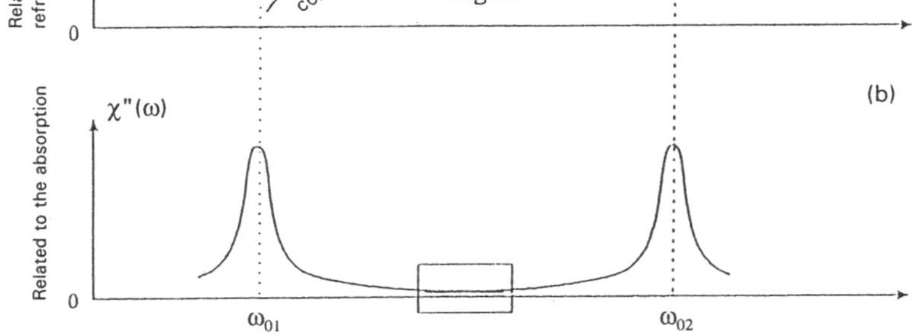

Fig. 7.2 a, b A schematic sketch of $\chi'(\omega)$, $\chi''(\omega)$ and the transparent region of the fussed silica fibers.

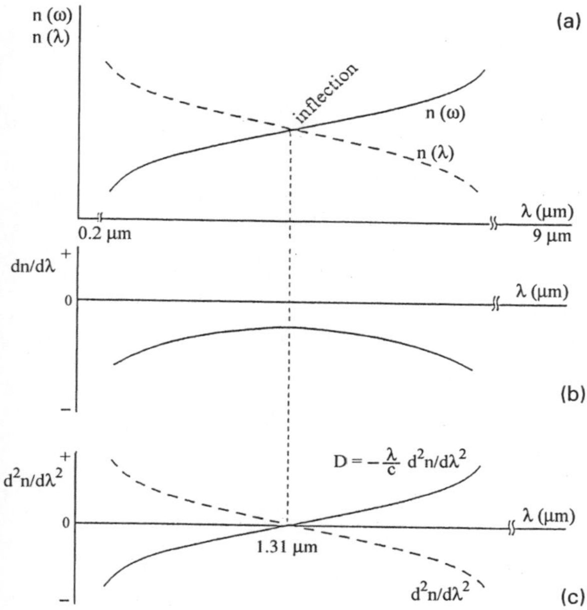

Fig. 7.3 a–c The curves of the refractive index $n(g)$ and its differentials for the fussed silica fiber near the zero dispersion (GVD) region.

inflection point. The values of the refractive index $n(\omega)$ in the transparent region are well approximated by the Sellmeier Equation. We transfer the curve $n(\omega)$ onto the curve $n(\lambda)$ in the region near the inflection point (Fig. 7.3(a)). We then take the first and second derivatives of this curve with respect to wavelength, as in Figs. 7.3(b) and (c). We use $D(\lambda) \cong -(\lambda/c)(d^2n/d\lambda^2)$ to get the expected curve $D(\lambda)$. This curve has $dD(\lambda)/d\lambda = 0$ at $\lambda = 1.27\mu\text{m}$, and this minimum shifts to $1.31\mu\text{m}$ due to the fiber waveguide effect. Finally, we plot the shape of curves $\alpha(\lambda)$ and $D(\lambda)$ in Fig. 7.1.

7.1.1.2 The Group Velocity Dispersion Equation

The following equations (7.2–7.5) and related explanations are presented in the book by G.P. Agrawal [7.2]. For a slow-varying amplitude E_{SV} transmitted along the propagating distance

$$\partial E_{SV}/\partial z = -(i/2)\beta_2\, \partial^2 E_{SV}/\partial T^2 - (\alpha/2)E_{SV} + i\gamma|E_{SV}|^2 E_{SV}. \tag{3.64}$$

By using a normalized amplitude U,

$E_{SV}(z,t) = (P_0)^{1/2} \cdot \exp(-\alpha z/2) \cdot U(z,\tau)$,

the transmission equation of the normalized amplitude $U(z,\tau)$ is

$$\partial U/\partial z = -[i \cdot sgn(\beta_2)/2L_D]\partial^2 U/\partial \tau^2 - [\exp(-\alpha z)/L_{NL}]|U|^2 U \tag{7.2}$$

where $sgn(\beta_2) = \pm 1$ depending on the sign of β_2. $\tau = T/T_0$ is the time normalized to the initial pulse width T_0. The length parameters L_D and L_{NL} were defined in Sect. 3.5.2. When the linear dispersion is dominant, that is, $L_D << L_{NL}$, and $L_D << L$, we neglect the nonlinear coefficient γ.

7.1.1.3 Broadening of a Gaussian Shape Pulse Without Initial Chirp

After transmission, the normalized-amplitude solution in the space-time domain for the Gaussian-shape pulse broadening without initial chirp is

$$U(z,T) = \frac{T_0}{\sqrt{T_0^2 - i\beta_2 z}} \cdot \exp\left[-\frac{T^2}{2(T_0^2 - i\beta_2 z)}\right] \tag{7.3}$$

by writing $U(z,T) = |U(z,T)|\exp[i\varphi_{GVD}(z,T)]$ where

$$\varphi_{GVD}(z,T) = -\frac{sgn(\beta_2)(z/L_P)}{1+(z/L_P)^2} \cdot \frac{T^2}{T_0^2} + \mathrm{arctg}\frac{z}{L_D}.$$

The instantaneous linear chirp $\delta\omega$ across the pulse edge from the central frequency ω_0 is induced from the time dependence of the phase $\varphi(z,T)$, and is given by

$$d\,\omega_{GVD} = -\frac{\partial \varphi_{GVD}}{\partial T} = -\frac{sgn(\beta_2)(z/L_D)}{1+(z/L_D)^2} \cdot \frac{T}{T_0^2}. \tag{7.4}$$

We see that even though the initial pulse is unchirped, the transmitted pulse becomes chirped. The sign of linear chirp $\delta\omega$ is dependent on the sign of β_2 and is associated with some features of GVD.

A. In the normal dispersion region of group index n_g ($dn_g/d\omega > 0$, $\beta_2 > 0$, $D < 0$)

Substitute $\beta_2 > 0$ in (7.4). Then on the pulse's leading edge, $T < 0$, it is the negative chirp $\delta\omega < 0$. On the pulse's trailing edge, $T > 0$, it is the positive chirp $\delta\omega > 0$. From the leading edge to the trailing edge, the dispersion-induced chirp $\delta\omega$ increases linearly and is called the up-chirp, as shown in Fig. 7.4(a). Thus, the leading edge is red-shifted and transmits fast, while the tailing edge is blue-shifted and transmits slow. Finally, the unchirped Gauss pulse is broadened by GVD, as shown in Fig. 7.4(b).

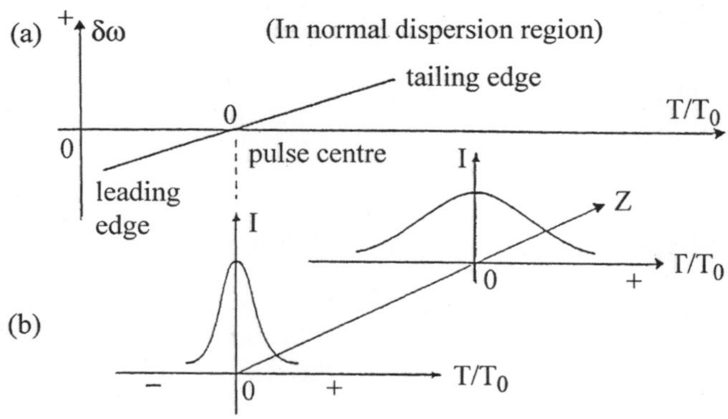

Fig. 7.4 a, b The up-linear chirp and pulse broadening stem from the GVD in the fiber normal dispersion region.

B. In the anomalous dispersion region of group index n_g ($dn_g/d\omega < 0$, $\beta_2 < 0$, $D > 0$)

Substitute $\beta_2 < 0$ in (7.4). Then on the leading edge, $T < 0$, it is the positive chirp $\delta\omega > 0$. On the trailing edge, $T > 0$, it is the negative chirp $\delta\omega < 0$. From the leading edge to the trailing edge, $\delta\omega$ decreases linearly and is called the down-chirp, as shown in Fig. 7.5(a). Thus, the leading edge has blue-shift, but still transmits fast, since the low refractive index is at high frequency and causes fast motion, while the trailing edge has red-shift and transmits slow. Finally, the unchirped Gauss pulse is also broadened by GVD (Fig. 7.5(b)).

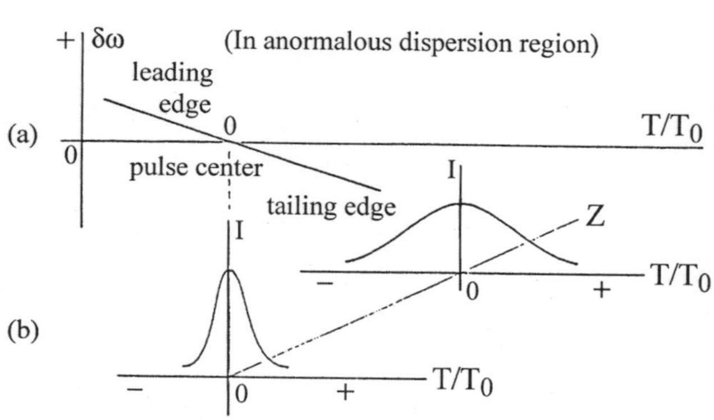

Fig. 7.5 a, b The down-linear chirp and pulse broadening stem from the GVD in the fiber anormalous dispersion region.

In conclusion, the GVD-induced pulse-broadening is independent of the sign of β_2. Also, for a given L_D, whether there is normal or anomalous dispersion, the amount of broadening is the same. Only if $\beta_2 = 0$, there is no linear chirp and no linear GVD broadening, but then the influence of the nonlinear or second-order of GVD $\beta_3 \neq 0$ must be considered.

7.1.1.4 Gaussian Shape Pulse Broadening with Initial Chirp

In practical systems, some chirps do not derive from the fiber dispersion of transmission but from the light source itself as the initial chirps. In a simple case, the injected current modulates directly the carrier density of the emitting layer (active layer) of the laser diodes. From this cause, the carrier density varies strongly at the leading and tailing edges of a near rectangular pulse (with several ns width). This causes the refractive index change in the emitting layer which induces the frequency chirp $\delta\omega$ (Fig. 7.6) at both the leading and tailing edges.

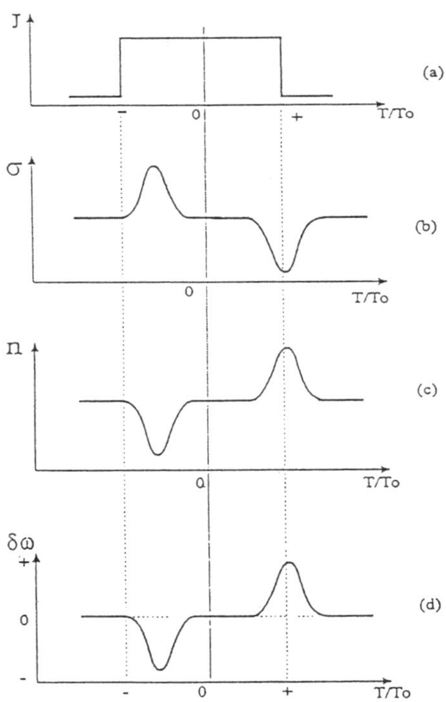

Fig. 7.6 a–d The frequency chirp caused by the free-carrier plasma effects.

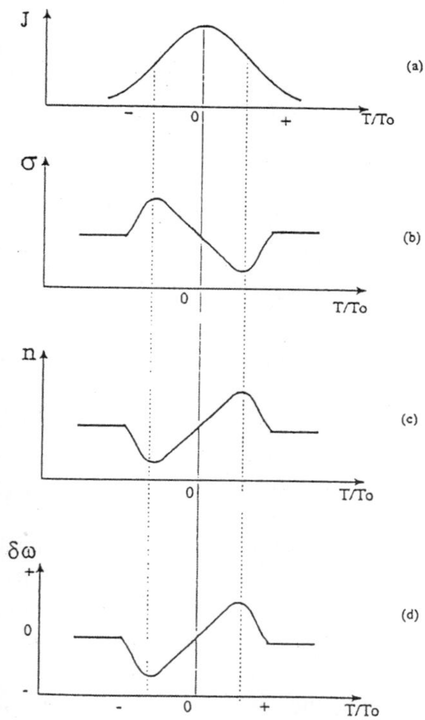

Fig. 7.7 a–d The frequency chirp distributed throughout the whole pulsewidth of a short Gauss shape pulse.

If the pulse is short enough, the frequency chirp $\delta\omega$ will distribute throughout the whole pulsewidth. From leading to tailing edge, the chirp increases as shown in Fig. 7.7. Especially for a direct modulating super Gaussian short pulse, it has an up-chirp for most of the energy amount (but with initial red-chirp and $C < 0$), hence it is called an up-chirp pulse.

Let the Gaussian pulse injected with initial chirp be

$$U(0,T) = \exp[-(1 + iC)T^2/(2\,T_0^2)], \tag{7.5}$$

where C is the initial chirp factor. The chirp of $C > 0$ is the initially positive chirp, and this means blue leading and red tailing and consequently a short pulse with initial down-chirp. Of course, $C < 0$ is the contrary condition.

If the pulse is short enough, the frequency chirp $\delta\omega$ will distribute throughout the whole pulsewidth. From leading to tailing edge, the chirp increases as shown in Fig. 7.7. Especially for a direct modulating super Gaussian short pulse, it has an up-chirp for most of the energy amount (but with initial red-chirp and $C < 0$), hence it is called an up-chirp pulse.

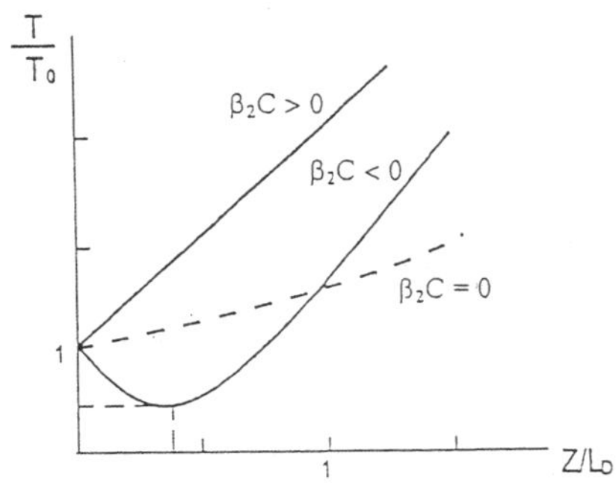

Fig. 7.8 Gaussian-shape pulse broadening with initial chirp.

Let the Gaussian pulse injected with initial chirp be

$$U(0,T) = \exp[-(1 + iC)T^2/(2\ T_0^2)], \tag{7.5}$$

where C is the initial chirp factor. The chirp of $C > 0$ is the initially positive chirp, and this means blue leading and red tailing and consequently a short pulse with initial down-chirp. Of course, $C < 0$ is the contrary condition.

The amplitude solution in the space-time domain shows that the broadening is determined by both the 1st-GVD β_2 and the initial chirp C. Since the fiber GVD chirp tends to broaden the pulse width more with distance, the convenient initial chirp of the pulse is selected in Fig. 7.8. At the beginning, these chirps compensate each other and induce pulsewidth narrowing, but once the GVD chirp exceeds the initial chirp, the pulse begins to broaden.

In some practical use, for example in the soliton-like cases, in order to match with the small anomalous dispersion of fiber (working wavelength in $D > 0$ region), the initial chirp selected is of $C > 0$ with blue leading. In this case of $\beta_2 C < 0$ (such as in the anomalous dispersion region $D > 0$ and $\beta_2 < 0$ at 1.560μm wavelength in the CSMF, with initial $C > 0$), the pulse-width is compressed to become narrower. In the case of $\beta_2 C > 0$ (such as $D < 0$ and $\beta_2 > 0$ at 1.545μm in the CSMF, with initial $C > 0$), the pulse broadening increases monotonically with normalized distance (Fig. 7.8).

Example

GVD caused pulse broadening is one of the dominant limits in OFC transmission capacity (BL or B^2L). If the CSMF (with the higher orders of β and initial chirp neglected) is used as a transmitted trunk at 1.55μm,

(1) Intrinsic linewidth $\delta\omega$ of LD is broad (suppose multi-mode LD, then whole spectrum linewidth W = 1THz), for a certain broadening, then the resultant B·L ≤ 5km·Gbit/s. If B = 100Mbit/s, then L = 50km.
(2) Iintrinsic linewidth $\delta\omega$ of LD is narrow (suppose single-mode LD, $\delta\omega$ = 80–100MHz, then the whole linewidth W ≅ B), then the resultant $B^2 \cdot L \leq 6000$km·(Gbit/s)^2. If B = 10Gbit/s, then L = 60km.

7.1.2 Dispersion Compensation and Dispersion-Slop Compensation

In a linear fiber system, the group velocity dispersion (GVD) and polarization mode dispersion (PMD) induced frequency chirp (shown in Figs. 7.4, 7.5 and 7.9) causes the transmitted pulse-shape distortion. For a long time, the dispersion compensation for the already installed trunk CSMF (D_{CSMF} = +17ps/nm·km at 1.55μm) by using a section of the dispersion compensating fiber (DCF_{CSMF}, with D_{CSMF} about –120 to –130 ps/nm·km) is a significant subject.

A high capacity OFC system is really a nonlinear fiber system or a nonlinear opto-electronics system. The light intensity induced nonlinear refractive index effects, SPM and XPM acted with dispersion, produce further pulse distortion and frequency spectrum broadening in the same fiber and limits the capacity of the OFC. Furthermore, in a DWDM system, the FWM has also considerable effect. For the Terabit/s DWDM transmission systems and networks, this technique is needed to lessen simultaneously the dispersion (GVD and PMD) and nonlinearities.

We can use, for example, the non-zero dispersion-shifted fiber (such as NZDSF-G655) as a low dispersion trunk in the dispersion management system to ease both the FWM and the GVD, and hence the SPM and XPM results. In a dispersion management system, the NZDSF (with D_{NZDSF} = 2–6ps/nm·km at 1.55μm) trunk needs also the dispersion compensating fiber DCF_{NZDSF} (with D_{CSMF} about –100 to –170ps/nm·km). The other design is alternately installing $D_{odd\text{-}sections}$ = –2 to –6ps/nm·km and $D_{even\text{-}sections}$ = +2 to +6 ps/nm·km, and it is the combination of equal lengths that makes the dispersions compensate each other.

There is another solution by inserting a chirped grating, but in this book, we shall discuss only those proposals based on the principles of arranging the fibers, as we just discussed above.

7.1.2.1 Dispersion Compensation and Management

The grounded CSMF, about several tens of million kilometers, has been used as a trunk fiber for a long time. For the high bit-rate transmission, such as 10Gbit/s or more, a segment of dispersion compensating fiber (DCF) is required to compensate the GVD of the trunk fibers.

Firstly, we show the process of the GVD-induced pulse in the spatial, time, and frequency domains in Fig. 7.9. At z_0, the optical wave with central frequency ω_0 and a certain spectrum width $I\sim\omega$, is enclosed in the envelope $I\sim T/T_0$. We suppose ω_0 is in the anomalous dispersion region that occurs in the communication ($\beta_2 < 0$, and $D = +17$ps/nm·km at 1.55 μm) (Fig. 7.9(a)). The GVD-induced phase shift $\varphi_{\mathrm{GVD}}\sim T/T_0$ from (7.8) is shown in Fig. 7.9(b), and the corresponding down-chirp $\delta\omega_{\mathrm{GVD}}\sim T/T_0$ from (7.9) is shown in Fig. 7.9(c). At a certain point z_1, the leading edge has high frequency, low refractive index and fast speed while the trailing edge has low frequency, high refractive index and low speed. As a result, the envelope broadens, as in Fig. 7.9(d).

We attach a second length of fiber at point z_1 in Fig.7.9, and its dispersion coefficient D has the opposite sign of D in the first length. In this case, the ω_0 is in the normal dispersion region. The blue leading edge of the initial chirp moves slowly and the red tailing edge moves fast. As a

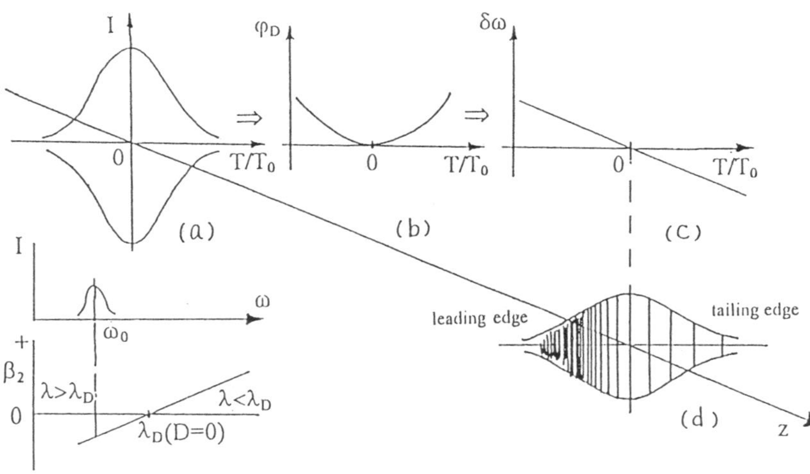

Z = 0 (initially)	Z = L (after a distance)
The pulse is chirp free but a certain linewidth in spectrum. The central ω_0 is located in the anormalous dispersion region.	Phase shift at leading and tailing edges of envelope and frequency chirp induced, accordingly the pulse-broadening is caused by chirp and GVD.

Fig. 7.9 a–d The GVD-induced pulse broadening related to frequency, space and time domain.

result, the envelop narrows, then reaches to the near-original pulsewidth and broadens again at another point z_2. This is similar to the curve $BC < 0$ in Fig.7.8. A third length of fiber with anomalous dispersion will be attached there. In recent years, a mode to arrange cautiously the DWDM fiber trunk, which manages both the low GVD and the photonic nonlinearities, is presented. For combating with the nonlinear FWM caused crosstalk, some new fibers (such as NZDSF) with low dispersion coefficient $D = \pm(2\text{–}6)$ps/nm·km are designed and installed as the trunk fibers. Of course, an accurate and perfect dispersion compensation is also needed for a long-haul NZDSF transmission system of about 2000km and $\geq$ 40Gbit/s. The dispersion compensation and management are shown in Tables 7.2 and 7.3. Recently, the dispersion-managed WDM system (DM-WDM) has become a hot topic in the OFC field and we shall discuss it later in Sect. 7.3.4.

7.1.2.2 Dispersion-Slop Compensation

In the 2000s, the wavelength channel number becomes several hundreds and the occupied bandwidth becomes 100nm or more, and we need to compensate the dispersion not only at a certain wavelength point but also at all the channel wavelengths. This means we have to compensate both the dispersion D and the dispersion slop DS = $dD(\lambda)/d\lambda$, or just as mentioned in equation (7.1b), $\beta_3(\lambda) = d\beta_2(\lambda)/d\lambda$, simultaneously in the whole broadband of the WDM systems. This is extremely significant to a 40Gbit/s DWDM transmission for practical distances. Some new kinds of compensating fiber, namely the $DCF_{CSMF\text{-}DDS}$ and $DCF_{NZDSF\text{-}DDS}$, are formed recently.

The added condition in these DCFs is that the relative dispersion-slop (RDS) of the $DCF_{NZDSF\text{-}DDS}$ (and the same for $DCF_{CSMF\text{-}DDS}$) equals the RDS

Table 7.2 The DCF and DMF systems.

	Dispersion compensation	*Dispersion management*
Views	The fiber trunk is treated as a linear system.	Fiber trunk is treated synthetically as a linear and nonlinear system.
Criterion	The sum dispersion (trunk + DC fiber) is compensated to zero.	Total dispersion $\neq 0$, but dispersion, nonlinearity and EDFA are arranged to make the pulse have minimum distortion.
Applied conditions	Keeps the existing hundreds of million km CSMF as the trunk fiber.	Adopts new trunk fibers and optimize the system.

Table 7.3 The dispersion compensation and management using fiber segment arrangement.

Method	Dispersion Compensation
Transmission Fibers	Conventional SMF, D = +17 ps/nm·km at 1.55 μm
Purpose	Reducing the GVD caused linearly pulse broadening

(1)

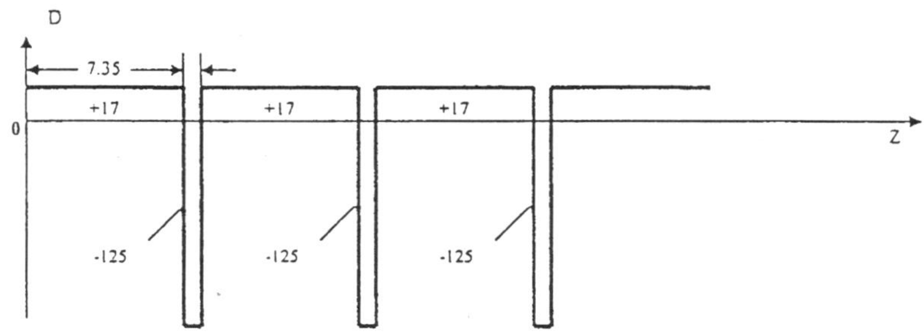

Method	Dispersion Management
Transmission Fibers	Low D Trunk Fiber with D ≅ -±2 ps/nm·km at 1.55 μm
Purpose	Reducing both the fiber nonlinearity and GVD effects

(2)

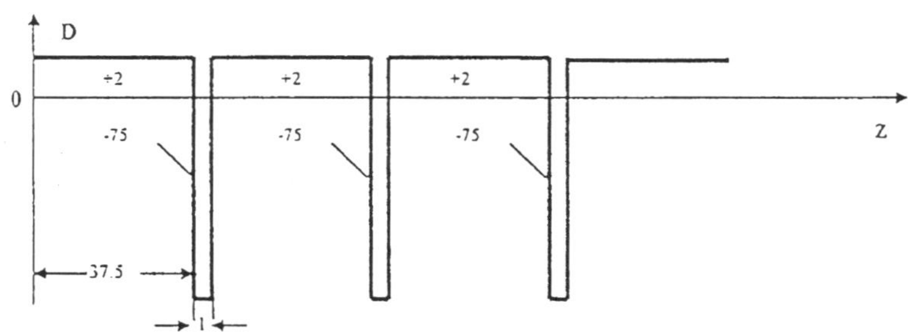

(3)

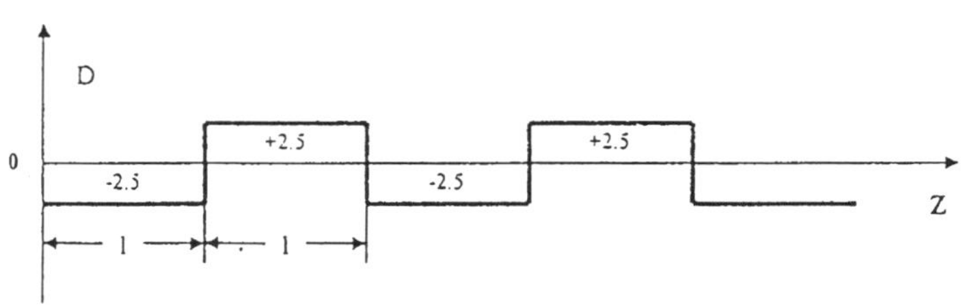

of the compensated NZDSF (or CSMF). Here, the RDS is defined as the dispersion-slop DS divided by the dispersion D (see [7.28]). The NZDSF is always with higher RDS since its dispersion is much lower than that of the CSMF.

We hope to use a single fiber to transmit signals within both the C-band (1530–1565nm) and the L-band (1565–1610nm). This means a DCF has to compensate simultaneously the D and DS both for the C- and L-bands [7.28]. A convenient NZDSF is with a small positive dispersion and a reduced DS at 1550nm (in the C-band region). The lower slop is much better for the DWDM, but it is difficult to reduce notably the slop of the NZDSF in the L-band. Table 7.4 compiles some imprecise sizes of D, DS and RDS demonstrated in recent years. Here, the DCF_{CSMF-2} (or $DCF_{NZDSF-DDS-L-2}$) has the equal length as its compensated fiber (so both are trunks), as shown in Table 7.3(3). The DCF_{CSMF-1} (or $DCF_{NZDSF-DDS-L-1}$) is a lump segment of the compensating fiber (one trunk and one DCF), as shown in Table.7.3(2).

Recently, a new type of reverse dispersion fiber (RDF-60#2) has been developed successfully for the SMF (SMF-E) trunk fiber in the broadband of 1500–1600nm [7.29]. The characteristics of the total fiber link are: $+0.03 ps/nm^2 \cdot km$ small slop, $82 \mu m^2$ enough A_{eff} for the FWM suppression and 0.205dB/km usual attenuation. In addition, a new module including an arrayed-waveguide grating (AWG) and a spatial filter was demonstrated lately as a dispersion slop compensator for all WDM channels in the 4THz bandwidth of C- plus L-bands [7.30]. This type of compensator has benefits of compact and compensating for the arbitrary figure of dispersion (2nd- and 3rd-order dispersion, positive and negative dispersion).

Table 7.4 The dispersion, dispersion-slop ($dD/d\lambda$), RDS ($[dD/d\lambda]\cdot[1/D]$) of the compensated and compensating fibers.

	Parameter dimension	Dispersion (ps/nm·km)	Dispersion-slop ($ps/nm^2 \cdot km$)	RDS (nm^{-1})
1	CSMF	+17	≅ +0.06	≅ +0.0035
2	DCF_{CSMF}	–(120–130)	–(0.42–0.45)	≅ +0.0035
3	$NZDSF_{at-C-band}$	+(4–5)	+(0.04–0.05)	≅ +0.01
4	$DCF_{NZDSF-DDS-C}$	–(160–180)	–(1.6–1.8)	≅ +0.01
5	$NZDSF_{at-L-band}$	≅ +8	≅ +0.08	≅ +0.01
6	$DCF_{NZDSF-DDS-L-2}$	≅ –8	≅ –0.07	≅ +0.008
7	$DCF_{NZDSF-DDS-L-1}$	–(80–90)	–(0.6–0.7)	≅ +0.07

7.1.3 PMD Impairment and PMD Mitigation

Polarization Mode Dispersion (PMD) impairment, a linear effect in fiber, is one of the final limitations for the 40Gbit/s or even higher bit-rate fiber transmission systems. Due to the degradation of the system performance, the PMD of the transmission, including the fiber trunk and the dispersion compensating fiber module (DCFM), has to be kept low.

In the fiber core, a small deviation from the ideal circular symmetry, caused probably by imperfect manufacturing and/or external stresses, results in a birefringent appearance, which is the physical origin of the PMD impairment. It breaks the degeneracy of the two orthogonal polarization modes of the fiber that makes their group velocities different. The difference in arrival time is called the differential group delay (DGD), which causes the inter-symbol interference. The PMD impairment fluctuates randomly from fiber to fiber, because of the randomness of the mentioned irregularities from the imperfect manufacturing and/or external stresses. These factors, which induce the PMD, vary with time and cause the PMD a random statistical effect. One can define mathematically a first-order PMD to which we neglect its frequency dependence and its value for a fiber in ps/√km is the mean value over time of the DGD.

The two main principles for the PMD mitigation are either inverting the linear transmission matrix of the system exhibiting the PMD in the optical domain, or equalizing the nonlinear transfer function of the system transmitter-fiber-receiver in the electrical domain [7.31]. Most of arrangements for the PMD compensation considered so far include a PMD counteractive element, a feedback signal and a tracking algorithm.

Some efficient techniques of the PMD mitigation are by using:

(1) the forward error correction coding and first-order compensator combined;
(2) the polarization scrambling and forward error correction coding combined;
(3) an electronic tunable equalizer with fast adaptive control;
(4) an optical equalizer with a variable lattice circuit;
(5) a fast eye monitor.

Recently, an automatic PMD compensator at 40 and 80Gbit/s by using polarization scrambling combined with a 3-dimensional analysis of the degree of polarization (DOP) was demonstrated in a field-trial experiment over 45km of installed fiber of the Deutsche Telekom [7.32].

7.2 Intensity-Induced Refractive Index and Phase Modulation

Light-intensity induces the SPM phase-shift, but the intensity-change induces SPM frequency-chirp.

7.2.1 Intensity-Induced Refractive Index in Optical Fibers

From Sect. 2.6, the total refractive index, including the intensity-induced term, is

$$n(\omega) \equiv [Re\varepsilon_r(\omega)]^{1/2} = n_0(\omega) + \delta n(\omega) \equiv n_0(\omega) + n_2(\omega)|E|^2. \quad (2.104)$$

Lightwave signals propagate in the transparent regions of the fiber. There the imaginary part of $\chi^{(3)}(\omega)$ is zero and $Re\chi^{(3)}(\omega) = \chi^{(3)}(\omega)$. The linear refractive index becomes

$$n_0(\omega) = [1 + Re\chi^{(1)}(-\omega;\omega)]^{1/2} \cong [1 + \chi^{(1)}(-\omega;\omega)]^{1/2}. \quad (2.100)$$

The intensity-induced nonlinear index change $\delta n = n_2(\omega)|E|^2$ in a Kerr medium is

$$\delta n = n_2(\omega)|E|^2 = \frac{3}{8}\frac{\mathrm{Re}\,\chi^{(3)}(-\omega;\omega,-\omega,\omega)}{n_0(\omega)}|E|^2. \quad (2.102)$$

In fiber communications, the field intensity E and hence $\chi^{(3)}$ are supposed in the direction x. The corresponding refractive index $n_2(\omega)$ of the refractive index by a unit field is

$$n_2 = \frac{3}{8n_0}\chi^{(3)}_{xxxx} = \frac{3}{8n_0}\chi^{(3)}. \quad (2.106)$$

Waves propagate in the extremely low loss region of the fiber, so the imaginary part of $\chi^{(3)}(\omega)$ is nearly zero. The n_2 of silica fibers is 3.2×10^{-16} cm^2/W. n_2 is related with the fiber structure and is decreased as A_{eff} is enlarged. Furthermore, n_2 is determined by the fiber material and is increased as the Ge concentration is increased. The total phase-shift, including the intensity induced phase-shift change, is

$$\varphi = nk_0L = (n_0 + n_2|E|^2)k_0L. \quad (2.107)$$

The intensity-induced nonlinear phase-shift change is

$$\varphi_{NL} = n_2|E|^2k_0L. \quad (7.6)$$

7.2.2 SPM-Induced and XPM-Induced Spectrum Broadening

7.2.2.1 SPM Phase Shift and Frequency Chirp

When the nonlinearities-induced self-phase-modulation is dominant (e.g. $L_{NL} << L_D$) and $L_{NL} << L$, we neglect the linear dispersion β_2 in (7.2). Then $U(z,T)$ satisfies a linear partial differential equation, which can be written as [7.2]

$$\partial U/\partial z = (i/L_{NL})[\exp(-\alpha z)]|U|^2 U, \tag{7.7}$$

where the field amplitude

$$U(z,T) = U(0,T)\exp[i\varphi_{SPM}(z,T)].$$

The intensity-induced SPM phase shift $\varphi_{SPM}(z,T)$ is

$$\varphi_{SPM}(z,T) = |U(0,T)|^2 \cdot L_{eff} \cdot \delta_{photon} \cdot \delta_{fiber}. \tag{7.8}$$

Two nonlinear factors δ_{photon} and δ_{fiber} depend on the photon's (beam's) and fiber's intrinsic characteristics respectively,

$$\begin{aligned} \delta_{photon} &= \frac{\omega_{photon} P_0}{c} \\ \delta_{fiber} &= \frac{n_2}{A_{eff}}. \end{aligned} \tag{7.9}$$

Many nonlinear effects are closely linked with and proportional to these two nonlinear factors.

The maximum self-phase shift φ_{SPMMAX} occurs at the pulse center $T = 0$, Thus $|U(0,0)| = 1$.

At different times T, from the leading edge to the trailing edge, there are different phase-modulations $\varphi_{SPM}(z,T)$. These temporal changes of phase cause the temporal, intensity-change-induced frequency chirp $\delta\omega_{SPM}(T)$ across the pulse from the central frequency ω_0,

$$\begin{aligned} \delta\omega_{SPM}(T) &= -\frac{\partial\varphi_{NL}}{\partial T} \\ &= -\frac{\partial |U(0,T)|^2}{\partial T} \cdot L_{eff} \cdot \delta_{photon} \cdot \delta_{fiber}. \end{aligned} \tag{7.10}$$

The chirp $\delta\omega_{SPM}$ is independent on the sign of β_2 in (7.4), and is different from the chirp $\delta\omega_{GVD}$ which is dependent on the sign of β_2. Equation (7.10) shows that the SPM chirp $\delta\omega_{SPM}$ increases with the transmission distance L_{eff}. Thus when the pulse moves along the fiber, new frequencies are generated continuously. This SPM-induced frequency spectrum exceeds the initial (at $z = 0$) spectrum.

7.2.2.2 The Physical Pattern on SPM-Induced Spectrum Broadening

Whether in the region with $\beta_2 > 0$ (normal dispersion) or $\beta_2 < 0$ (anomalous dispersion), the nonlinearity-caused SPM chirp is the up-chirp constantly at most parts of the pulse energy.

The spectrum broadening and spectrum multipeak processes of the Gauss pulse ($m = 1$), and of a super-Gauss pulse ($m = 3$) are shown in Fig. 7.10. Its upper right part is the SPM-induced $\varphi_{SPM}(z,T)$ described by (7.8). Its lower right part is the SPM-induced $\delta\omega_{SPM}(T)$ from (7.10). Its left part is the spectrum multipeaked structure, deduced from the former two parts. These have some important features. First, whether β_2 is positive or negative, the SPM chirp is always an up-chirp. Its leading frequency always red-shifts and its trailing frequency always blue-shifts. Second, the chirp $\delta\omega_{SPM}(T)$ of a super-Gaussian pulse is much larger than that of a Gaussian pulse, and always occurs in both the leading and trailing regions. Third, the spectrum multipeaked structures derive from the interference of different phases at a given value of $\delta\omega_{SPM}(T)$.

From (7.10) and Fig. 7.10, whether β_2 is positive (normal dispersion) or negative (anomalous dispersion) , the SPM chirp ($\delta\omega_{SPM}(T)$) is always an up-chirp ($\delta\omega/\delta T > 0$). But from (7.4) and Fig. 7.5, the GVD chirp ($\delta\omega_{GVD}(T)$) is a down-chirp ($\delta\omega/\delta T < 0$) in the anomalous dispersion ($\beta_2 < 0$, $D > 0$) region. Hence, the nonlinearly-induced up-chirp $\delta\omega_{SPM}$ and the dispersion-induced down-chirp $\delta\omega_{GVD}$ can be compensated each other in the uniform anomalous dispersion region. In this case it is possible to use a short pulse over long distances, and is called fiber soliton transmission.

7.2.2.3 XPM-Induced Refractive Index Modulation and Phase-shift Coupling

When two or more intense pulse waves propagate synchronously along the fiber, each one will induce a refractive index change Δn_{SPM}. Accordingly, the phase change of one wave $\Delta\varphi_1$ will be influenced by the term Δn_{SPM2} of the other. As a result, the XPM is caused by the modulation of the Δn by the total power.

As an example, we shall discuss two waves of different frequencies. The resulting polarizations are:

$$P(\omega_1) = \varepsilon_0\varepsilon_1 \cdot E_1 \quad \text{and} \quad P(\omega_2) = \varepsilon_0\varepsilon_2 \cdot E_2 , \tag{7.11}$$

where

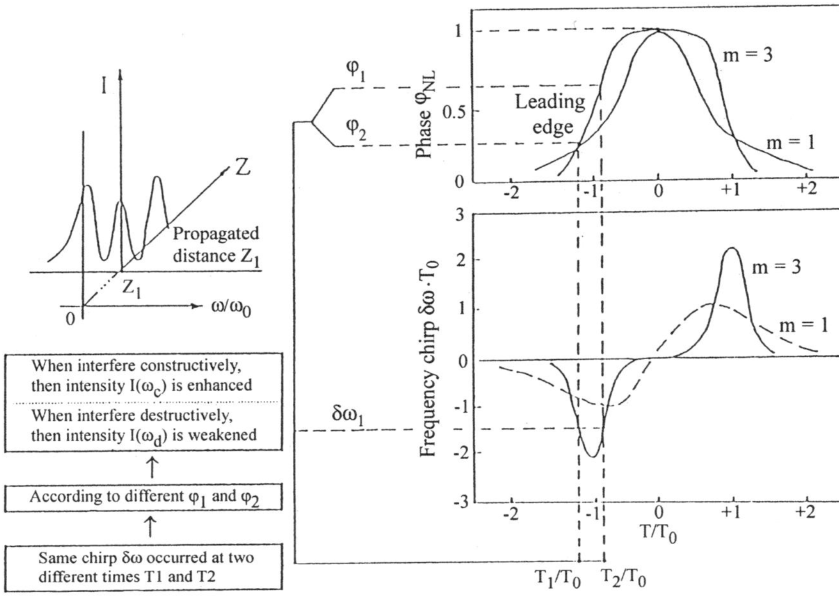

Fig. 7.10 The SPM-induced frequency chirp $\delta\omega_{\mathrm{SPM}}$ and the physical origin of the pulse spectum distribution.

$$\varepsilon_1 = \varepsilon_1 L + \varepsilon_{1\mathrm{NL}} = (n_1 + \Delta n_1)^2 \quad \text{and} \quad \varepsilon_2 = \varepsilon_2 L + \varepsilon_{2\mathrm{NL}} = (n_2 + \Delta n_2)^2 , \quad (7.12)$$

where n_1 and n_2 are the linear parts of the total refractive index from $\chi^{(1)}(\omega_1)$ and $\chi^{(1)}(\omega_2)$, and Δn_1 and Δn_2 are the nonlinear parts from $\chi^{(3)}(\omega_1)$ and $\chi^{(3)}(\omega_2)$. The nonlinear part is always the perturbation, i.e., $\Delta n_1 << n_1$ and $\Delta n_2 << n_2$. Hence the nonlinear parts of the total refractive index are:

$$\Delta n_1 \cong \varepsilon_{1\mathrm{NL}}/2n_1 \cong n_{\mathrm{NL}}(|E_1|^2 + 2|E_2|^2) \quad \text{and}$$

$$\Delta n_2 \cong \varepsilon_{2\mathrm{NL}}/2n_2 \cong n_{\mathrm{NL}}(|E_2|^2 + 2|E_1|^2) \quad (7.13)$$

where the nonlinear refractive index coefficient is

$$n_{\mathrm{NL}} = (3/8n)\chi^{(3)}{}_{xxxx} \quad (2.106)$$

and $n \cong n_1 \cong n_2$.

Thus, when two intense short pulses propagate together along the same fiber, each nonlinear phase shift φ_{NL} is associated with both intensities. We have the XPM-induced nonlinear phase shifts

$$\varphi_{\mathrm{NL}1} = (\omega_1 z/c)\Delta n_1 = (\omega_1 z n_{\mathrm{NL}}/c)(|E_1|^2 + 2|E_2|^2) = \varphi_{\mathrm{SPM}1} + \varphi_{\mathrm{XPM}1}$$

$$\varphi_{NL2} = (\omega_2 z/c)\Delta n_2 = (\omega_2 z n_{NL}/c)(|E_2|^2 + 2|E_1|^2) \tag{7.14}$$

$$= \varphi_{SPM2} + \varphi_{XPM2}. \tag{7.15}$$

In the absence of GVD, the XPM will only result in the modulation of the optical phase, but in dispersion fibers, the nonlinear phase shifts in (7.14–7.15) will translate into the intensity distortion or Q-factor reduction, especially in ≥ 10Gbit/s per channel transmission. In the CSMF, the XPM is inversely proportional to $\Delta\lambda$ and hence is heavier in the DWDM. In NZDSF, the XPM is even much severe than that in the CSMF because of its lower dispersion. However, the XPM is quite significant for the nonlinear fiber loop, in which one light wave is always guided by the other controlling light wave through the XPM effects.

7.2.2.4 The Broad-Band Super-Continuum (SC) Generation in Fiber and a Single Super-Continuum WDM Source

In the 21st century, people will need more convenient and low-cost communication means to get more information and to communicate with anybody in any place and at any time.

A light source component with high bit-rate (10–40Gbit/s or higher), high channel density (32–100 per fiber) and high quality (high stable, low-chirp) is a required technology of the optical layer for the coming fiber transmission or optical networking. For these purposes, super-continuum (SC) source is a kind of potential source which can offer few pico-second short pulses in a very wide spectrum range.

When a very strong short pulse beam is injected into an optical nonlinear medium, the output short pulse has a super-continuum (SC) broadening of spectrum which includes a lot of new frequency constituents. The width of the output spectrum is much broader than that of the input. In earlier years, SC was found in bulks, liquids and gases. The results indicate that the SC caused deals with many nonlinear effects, such as the SPM, XPM, FWM, and SRS, etc. Since the SC in optical fibers is a possible potential source, people have paid much attention to this topic in the recent years.

A real super-continuum (SC) short pulse source is consisted of several parts attached as the sequence shown in Fig.7.11:

(1) A short pulse pump laser, such as a mode-locking laser or a chirp compensated gain switched laser, offer the pump pulses with few ps.
(2) After the EDFA, the short pump pulses (with a few watts peak power) inject into a segment of the so-called SC fiber (a few meters long) of a dispersion-shifted fiber, or newly, a dispersion-flattened-decreasing fiber, in which the pulse wavelength is at the anomalous dispersion region with small dispersion (~1 ps/nm·km). Then the spectrum is broadened into a super-continuum (SC).

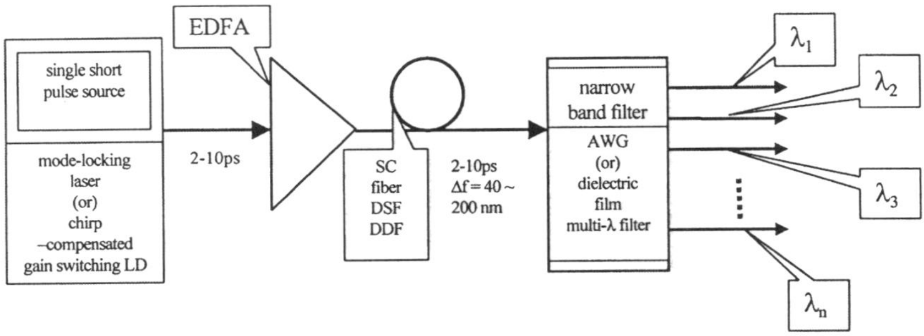

Fig. 7.11 A schematic diagram of the super continuum (SC) short pulse sources.

(3) A series of optical-band filters (OBF) or an arrayed-waveguide grating (AWG) can cut out simultaneously a series of WDM wavelengths in the wide SC spectrum range of 40–200nm. Since the figure of the time-bandwidth product is really the nearly transform-limited, the output SC short pulses are close to chirp-free.

The physical pattern of the SC in the fibers is considered as a compound consequence, caused by the nonlinear effects of the SPM, FWM and XPM, and the linear effect of the fiber dispersion. Principally, it can be numerically analyzed by the nonlinear Schrodinger equation with multi-wavelength complex amplitudes and multi-order dispersions in fibers.

Up to 1998, the experimental result of laboratory level is the 100Gbit/s× 4 channels×100km repeater-less OTDM/WDM transmission, using a single super-continuum source. Wider and more flattened characteristics of the fiber SC are needed to improve henceforth. In addition, the future of the SC pulse source depends on the challenge from other sources in the real communication engineering.

7.3 Periodically-Adjusted and Macro-Managed Transmission Systems

In a complicated linear-nonlinear-mixing system, some of the undesirable linear effects are always compensated or balanced by means of some suitable nonlinear processes. The perfect balancing or compensating conditions are frequently critical for a practical engineering system, especially if uniform and steady balancing are needed in the spatial domain (everywhere), temporal domain (anytime) and frequency domain (broadband). We can guess reasonably that this system is not only critical but also weak in stiffness. The smart way is to modify this uniform and steady criterion and replace it by a periodically-adjusted and macro-balancing linear-nonlinear-managed scheme.

The following descriptions on the perfect soliton systems in Sect. 7.3.1 and the dispersion-managed soliton systems in Sect. 7.3.4 are the demands of this modification to a periodically-adjusted and macro-managed transmission system. A similar example, which we have already discussed in Chap. 4, is about a perfect phase-matching multi-wave process (in Sect. 4.3), and a quasi-phase-matching multi-wave process (in Sect. 4.4) for some applied nonlinear optics use.

A periodically-adjusted and macro-managed transmission system is always easy in maintenance and management. Furthermore, the system should have a higher stiffness coefficient. In the design of such a system, we divide first the full range into numerous stretches, and then relax the constraints and allow a certain extent of disturbance, thus tolerating some deviation from the steady balancing or perfect compensation in the local (or micro) map of the individual stretches, but totally the whole characters are satisfied and are stable for a practical engineering system.

The fiber soliton results from the interactions of the GVD and SPM. From (7.18) and Fig. 7.10, whether β_2 is positive (normal dispersion) or negative (anomalous dispersion), the SPM chirp ($\delta\omega_{SPM}(T)$) is always an up-chirp ($\delta\omega/\delta T > 0$), but from (7.9) and Fig. 7.5, the GVD chirp ($\delta\omega_{GVD}(T)$) is a down-chirp ($\delta\omega/\delta T < 0$) in the anomalous-dispersion ($\beta_2 < 0$) region. Hence, the nonlinearly induced up-chirp $\delta\omega_{SPM}$ and the dispersion-induced down-chirp $\delta\omega_{GVD}$ can compensate each other in the uniform anomalous-dispersion region. In this case it is possible to use a short pulse over long distances, and is called the fiber soliton transmission. When the effect of uniform-dispersion is compensated by the fiber nonlinearlity SPM in the total soliton transmission, the chirp and deformation of the transmitted pulses can be completely offset in the first-order soliton (discussed in Sect. 7.3.1). The demonstrated transmissions have indicated very much astonishing achievements, such as the 100Gbit/s $\times 10^4$ km in the straight-line and 100Gbit/s$\times 10^6$km in the loop.

The critical arrangements both for the uniform compensation of the GVD and the elimination of time jitter cause the weak stiffness of the ideal first-order soliton. Recently, some modified modes, such as the dispersion-managed transmission, are in experiment in which the dispersion is periodically compensated. The stiffness of the system is therefore enhanced.

7.3.1 The Performances of a Perfect Fiber Soliton

7.3.1.1 The Mathematical Expressions of a Perfect Fiber Soliton

The equations describing perfect solitons can be deduced directly from the wave propagation equation (3.96). The following equations (7.16–7.19) are contributed by G.P. Agrawal [7.2]. In the special case of α = 0, (3.96) becomes the nonlinear Schrödinger equation [7.2]:

$$\partial E_{SV}/\partial z = -(i/2)\beta_2\,(\partial^2 E_{SV}/\partial T^2) + i\gamma|E_{SV}|^2 E_{SV}\,. \tag{3.64}$$

After some parameters are introduced for normalization, (3.96) becomes the standard nonlinear Schrödinger equation which characterizes the solitons [7.2]

$$\frac{\partial E}{\partial \zeta} - \frac{i}{2}\frac{\partial^2 E}{\partial \tau^2} - i|\,E\,|^2\,E = 0 \tag{7.16}$$

where $E \equiv E_{SV}[(\gamma\,T_0^{\,2})/|\beta_2|]^{1/2}$ is the normalized envelop function and $\zeta \equiv z/LD$, and $\tau \equiv T/T_0$.

The parameter of soliton order N, is defined as [7.2]:

$$N^2 \equiv \frac{L_D}{L_{\rm NL}} = \frac{\gamma P_0 T_0^2}{|\,\beta_2\,|} \tag{7.17}$$

where L_D and $L_{\rm NL}$ are defined by (3.80) and (3.82), P_0 is the pulse peak power and T_0 is the initial pulse-width at $1/e$ amplitude.

When an initial hyperbolic secant pulse $E(0,\tau) = N\cdot\text{sech}(\tau)$ is injected into the fiber, the first-order stable solution can be found from (7.16) as [7.2]:

$$E_{\text{soliton}-1\text{st}}(\varsigma,\tau) = \text{sech}(\tau)\cdot e^{(i\varsigma/2)}\,. \tag{7.18}$$

7.3.1.2 The Peak Power Needed for Maintaining a Perfect First-Order Soliton in the Dispersion-Uniform Compensation

If a perfect first-order soliton (7.18) is injected into a lossless and uniform dispersion fiber, and when the dispersion-nonlinearity compensation is satisfied, the solitons can be transmitted arbitrarily long, with no deformation of the pulse shape. The peak power $P_{\text{soliton-1st}}$ needed for maintaining such a perfect first-order solitons can be found from (7.17).

When N = 1 we have

$$P_{\text{soliton-1st}} = 0.776\lambda^3|D|A_{\text{eff}}/(\pi^2 c n_2 T^2_{\text{FWHM}}), \tag{7.19}$$

and $P_{\text{soliton- Nth}} = N^2\cdot P_{\text{soliton-1st}} = N^2\cdot 0.776\lambda^3|D|A_{\text{eff}}/(\pi^2 c n_2 T^2_{\text{FWHM}})$

where λ is the soliton wavelength, D is the fiber GVD coefficient at λ, and T_{FWHM} is the soliton pulse width.

When N = 2, 3, ..., we have the second-, third-, and higher-order solitons. The high peak power necessary to inject the Nth-order soliton is

N^2 times that required for the first-order soliton. Hence, the first-order soliton is the only candidate for the factual communication.

For a practical communication, it needs at least a certain transmitted power to keep the soliton shape and an enough signal-noise ratio (SNR) which is required from a reasonable bit-rate error (BER) of the system. An enough anomalous uniformed dispersion β_2 (matching with the power) known from (7.17) is indispensable for this pulse-shape keeping and SNR keeping. If the dispersion is too low, it is unfavorable for either the pulse-shape or SNR keeping.

However, this enough dispersion occupies the time-jitter and hence makes more BER in the soliton receiving. Nowadays, this time-jitter is a terrible obstacle for applying the soliton to a really commercial system. From the view of time-jitter eliminating, the lower the dispersion, the better the jitter eliminating. This is very much against the needed power. Thus an ideal and perfect dispersion-uniformed soliton system needs more critical conditions in practical use, because it is sensitive to the power, shape-change or the time-jitter disturbing. This means the stiffness of this ideal system is weak. More detail discussions on this will be found in later sections.

Example

If the GVD coefficient of the dispersion-shifted fiber (DSF) at 1.55μm window D = 3ps/nm.km, the effective area of the fiber core A_{eff} = 50μm^2, the nonlinear refractive index of fiber $n_2 = 3.2\times10^{-16}$ cm^2/W, and the pulse width of soliton T_{FWHM} = 20ps. Then the peak power needed for the first-order soliton $P_{\text{soliton-1st}}$ = 11.3mW.

7.3.1.3 CRZ, RZ and NRZ Coding in High-Speed Transmission

Though the soliton was not used in communication systems in the 20th century, it gives us an inspiration that, for very short pulse transmission, the chirped-return-to-zero (CRZ) and the return-to-zero (RZ) coding appears to be more beneficial than the conventional non-return-zero (NRZ) coding in some cases.

Many different viewpoints exist on using the RZ pattern in high-speed transmission. A certain theoretical calculation shows that the RZ coding is more superior in some high-speed systems [7.20], but up to year 2000, the presented numerical studies comparing these three kinds of coding are still not sufficiently definitive. Some experimental results support the NRZ coding [7.21] and they show the NRZ is still beneficial even in very high-speed transmission, but more experimental results support the CRZ and RZ coding [7.19 and 7.33]. The BER is much lower than that in the NRZ coding condition at higher than 40Gbit/s.

Table 7.5 Transmitted Lengths of RZ and NRZ Coding in Some Conditions.

	Which effect is dominant	NRZ coding	RZ coding
1	Dispersion $L_D < L_{eff}$, $L_D < L_{NL}$	Always further than RZ by 20 to 30 %	
2	Nonlinear SPM $L_D \cong L_{NL} < L_{eff}$		Always further than NRZ by a factor of 2–3
3	Fiber loss $L_{eff} < L_D$, $L_{eff} < L_{NL}$	To be better, if power compensation is not enough (Sect. 7.3.2)	To be better, if power compensation is a little excessive (Sect. 7.3.2)

Recently, a study compared the CRZ, RZ and NRZ modulation formats in a 64 × 12.3Gbit/s WDM transmission experiment up to 9000km [7.33]. Although the distinctions are fairly small for the band-center channels, the the CRZ provides notably better long-haul transmission performance for the band-edge channels, where the accumulated dispersion and the nonlinear inter-symbol interference are both large. Another study shows the results on the automated compensation for the NRZ and RZ coding data, using clock regeneration and fading without adding signaling [7.34] (Table 7.5).

Many improvements, including some basic techniques, are still needed for applying the CRZ and RZ modulation formats to commercial systems. Even for very short pulse and long-haul transmission, the CRZ and RZ modulation formats have to face the challenges from others on cost, easy of use, and compatibility with existing facilities.

7.3.2 The Compensation of Energy Losses in the Actual Soliton Transmission

In the perfect first-order soliton transmission, the effect of dispersion is uniformly compensated by the fiber nonlinearity in the total soliton transmission. Consequently, the chirp and deformation of the transmitted pulses can be completely offset. Such a perfect soliton is an ideal transmission mode of both the ultra-high bit-rate (40–100Gbit/s) and long-distance transmission. These conditions of uniform dispersion compensations and the complete chirp offsetting in spectrum are both ideal and critical. The stiffness of this ideal soliton transmission is weak, and its stability is esayly distorted by the power disturbing and frequency chirp disturbing. In these conditions, this perfect soliton will breakdown if there is no further management.

The above discussion on the perfect soliton transmissions in Sect. 7.3.1 has the condition that the fiber loss is neglected, and only the interactions of the fiber dispersion and nonlinearities are considered. If the loss term is induced into the soliton equation (7.16), the resultant pulse-width will increase exponentially with z, but less than that in the absence of the SPM. At the same time, the soliton will lose energy. When the energy loss exceeds the extent to which the nonlinear effect weakened is unable to keep the soliton, the soliton transmission becomes unstable. This is why we have to supplement the soliton energy. There are two kinds of energy supplements.

7.3.2.1 Distributed Energy Supplements

This is an ideal approach conceptually. If the whole transmitted line is distributively amplified, the solitons can get supplements everywhere and the balance between the GVD and SPM effects will occur everywhere. At present, the distributed SRS amplification of the conventional fiber and the distributed doped-fiber amplification of the low-doped erbium fiber are both unsatisfactory. Actually, since the pump sources are placed periodically, these two are both periodically-distributed energy supplements.

If the distance between the two adjacent pump sources $L_{\text{p-p}} >>$ the soliton period $z_{\text{period}} = \pi L_D/2$, the soliton amplification is an adiabatic amplification, in which the pulses are compressed and restored in the time domain, while the amplitudes heightened in the space domain. But if $L_{\text{p-p}} << z_{\text{period}}$, the soliton amplification is a nonadiabatic amplification in which no pulse is compressed in the time domain, and only their amplitudes increase. If $L_{\text{p-p}} \cong z_{\text{period}}$, the pulse compression and the increase of amplitude are about equal.

These three conditions are all very experimental. Some new approaches will soon be observed. The required SRS pump power is too high because of the very low gain, about 0.05dB/mW, while the cost of the low-doped EDFA is presently too great.

7.3.2.2 Lumped Energy Supplements

This is a practical approach for nonamplified traditional fibers. If we supplement the broadened solitons after some distance by lumped amplifiers, the solitons will be intensified. Then the nonlinear effect (SPM chirp) is enhanced and exceeds the dispersion effect (GVD chirp), due to the more intense soliton. The soliton pulse-width is narrowed and restored, but after a certain distance, the nonlinear effect is again weaker than the dispersion, and the solitons broaden once more, again requiring energy supplement. Thus it is a periodically balanced process, and its stability should be considered. There are two approaches for implementing the lumped supplement of the soliton energy: the routing-averaged soliton and the dynamic soliton.

A. Routing averaged soliton

The routing-averaged soliton was proposed by Mollenauer [7.9]. In the fiber transmission system with perturbations of pulse power and fiber dispersion, when the perturbation distance (averaged routing) $L_{\text{p-p}}$ is much less than z_{period}, the soliton transmission is stable if the pulse routing-averaged pump-power $P_{\text{RAP-P}}$ equals the first-order soliton peak power $P_{1\text{st}}$ determined by the dispersion [7.9],

$$P_{\text{RAP-P}} = P_{1\text{st}} = 0.776 \cdot \frac{\lambda^3 |D| A_{\text{eff}}}{\pi^2 c n_2 T_{\text{FWHM}}^2} \tag{7.20}$$

where $T_{\text{FWHM}} = 1.76 T_0$.

If the averaged dispersion on $L_{\text{p-p}}$ is roughly equal throughout L_{total}, and if the pulse-averaged peak power P_{RAPP} in $L_{\text{p-p}}$ after each EDFA equals the first-order soliton peak-power $P_{1\text{st}}$ (loss neglected), and if the accumulated nonlinear effect is nearly equal to the total effect in first-order soliton stable transmission, then in this routing-averaged approach, the lumped supplements of soliton energy will retain the pulse shape over long distances.

Practically, $L_{\text{p-p}}$ is approximately 25–50km, since the condition $L_{\text{p-p}} \ll z_{\text{period}}$ must remain. The z_{period} is hence several hundred km. For such a z_{period}, the soliton pulsewidth is about 30–50ps, with 4–7Gbit/s. A GVD as low as 1ps/nm.km (DSF 1.55μm window) is needed.

B. Dynamic fiber soliton

The dynamic fiber soliton approach was proposed by Nakazawa [7.10]. In the dynamic fiber soliton system, an initial soliton with a strong initial peak-power of a higher-order soliton of $N = 1.2$–2.0 is injected into the transmission line and narrows at the earliest stage. Then the SPM effect is weakened gradually by the fiber loss. After a certain distance with narrowing, the solitons are broadened by the GVD. When the soliton pulse-width returns to the initial condition, an EDFA for implementing the lumped supplement of the soliton energy is used. The injected dynamic pump-power $P_{\text{DP-P}}$ should be 1.4–4.0 times that for the first-order soliton. Thus [7.10],

$$\begin{aligned} P_{\text{DP-P}} &= (1.4 \sim 4.0) P_{1\text{st}} \\ &= (1.09 \sim 3.1) \cdot \frac{\lambda^3 |D| A_{\text{eff}}}{\pi^2 c n_2 T_{\text{FWHM}}^2}. \end{aligned} \tag{7.21}$$

Due to the high supplement power $P_{\text{DP-P}}$, the soliton can remain stable over 50 to 70km, which is comparatively long. The long-distance soliton dynamic transmission can be stably implemented if the process is repeated.

Compared with the conditions in averaged-routing, one advantage is the fiber length between the two pump sources $\cong z_{period}$. Thus, for the same $L_{p\text{-}p}$, z_{period} can be 60 to 100km, less than in the averaged-routing soliton transmission. Since $\zeta_{period} = z_{period}/L_D = \pi/2$, the periodic distance $z_{period} = \pi L_D/2 = \pi T_0^2/2|\beta_2| = 0.322\pi T_{FWHM}^2/2|\beta_2|$. Its pulse-width is less than about 20ps when the DSF dispersion is 2–3ps/km.nm. Accordingly, solitons with bit-rates above 10Gbit/s can be transmitted.

7.3.3 The Jitter-Controlled Solition and the Gordon-Haus Limit

When the EDFA is used for implementing the lumped supplement of the soliton energy, the amplified spontaneous emission (ASE) noise is induced unavoidably. As long as the random noise intensity is overlapped on the signal field, the additional random phase-shift and random frequency-shift (the chirp) are generated. Then the arrival time of pulses at the receiver is also random-jittering due to the GVD effects and this is the chief source of error bits [7.8]. In addition, the ASE causes the energy jittering of the soliton and degrades the figure of BER. The left-upper corner of Fig. 7.12 illustrates the limited condition of the system length L (km) ~ bit rate B (Gbit/s). For a certain SNR permitted, the system capacity product $BL <$ a critical constant (Gordon-Haus jitter). At BER = 10^{-9}, the Gordon-Haus limit is [7.8]:

$$(BL)^3 = 0.1372 T_{FWHM} \cdot (\Delta\tau_{window})^2 \cdot \frac{B^3 \cdot A_{eff}}{\hbar \alpha n_2 D}, \tag{7.22}$$

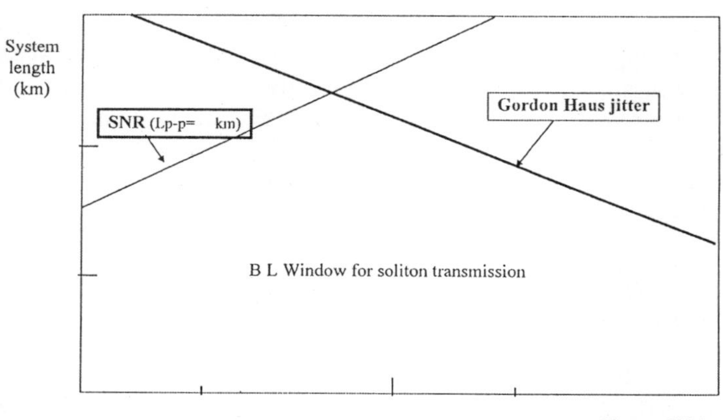

Fig. 7.12 The BL window for soliton transmission determined by the Gordon-Haus jitter and SNR permitted.

where $T_{FWHM} = 1.76T_0$ is the soliton pulse-width, $\Delta\tau_{window} = 1/(3B)$ is the half-width of the detection window, A_{eff} is the fiber core's effective area, α is the fiber intrinsic loss per km in (3.66), $\alpha = \alpha_{dB}/4.34 = 0.055\text{km}^{-1}$ (corresponding to $\alpha_{dB} = 0.25\text{dB/km}$), n_2 is the nonlinear refractive index coefficient (such as $n_2 = 3.2\times10^{-16}\text{cm}^2/\text{W}$), D is the GVD coefficient of the fiber.

Example

If $\lambda = 1.55\mu\text{m}$, $D = 1\text{ps/nm-km}$, $\alpha = 0.055\text{km}^{-1}$ corresponding to 0.25dB/km, $T_{FWHM} = 20\text{ps}$, $L_{P\text{-}P} = 30\text{km}$, $B = 10\text{Gbit/s}$, $A_{eff} = 35\mu\text{m}^2$, and $n_2 = 3.2\times10^{-16}\ \text{cm}^2/\text{W}$, then the soliton transmission distance L, limited by the Gordon-Haus effect (7.22), is 4,200km.

In addition, the conventional SNR at the receiver output settles simultaneously on the figure of BER. Here, the noises include both the receiver thermal noise and the cascaded ASE of the EDFAs. For a certain transmitted distance $L_{P\text{-}P}$ (km), the system capacity product $BL <$ a critical constant due to SNR permitted, is illustrated at the left-upper corner in Fig.7.12. Then, the total BL window of operation is revealed [7.23]. Compared with the limits caused by the effects of mutual soliton interaction, the Gordon-Haus effect is dominant. There are some existing methods to decrease the time-jitters while maintaining a significant uniformed dispersion by controlling the frequency or time domain.

7.3.3.1 The Synchronous Modulations — a Routine for Controlling Time Jitter

At present, this effect can be restrained in two ways. One is the soliton synchronized reforming and retiming. In this procedure, solitons pass through the loop with an inserted high-speed modulator which reforms and retimes the solitons synchronously. As a result, the accumulated ASE noises can be neglected. It is now conceivable to implement the soliton transmission of $10\text{Gbit/s}\times10^6\text{km}$, but precise synchronization is not yet sufficiently reliable.

7.3.3.2 The Filtered Controlling — a Routine for Reducing Time Jitter

The other method is the filtered controlling. In this method, the filter is added after each EDFA to restrain the ASE, since the noises out of the soliton band are filtered. Then the limit considering the filtered controlling can be relaxed, and becomes [7.11],

$$B^3 L \cdot L_0^2 = 4.573 \times 10^{-2} \cdot T_{\text{FWHM}} (\Delta \tau_{\text{window}})^2 \cdot \frac{B^3 A_{\text{eff}}}{\hbar \alpha n_2 D} \tag{7.23}$$

where $L_0 = (3\Omega_f^2 L_{\text{P-P}} T_0^2)/4$, $T_0 = T_{\text{FWHM}}/1.76$, $\Delta\tau_{\text{window}} = 1/(3B)$, $\Omega_f = 2\pi\Delta f_f$ is the filter's bandwidth, and $L_{\text{P-P}}$ is the fiber length between the two pump sources. Because $L/L_0 >> 1$, the limit may be relaxed by (7.23).

When considering the effect of the self-frequency-shifting, a pattern of the "slide-frequency guiding filters" is proposed [7.12]. In this case, the center filtering frequency ω_{0f} gradually slides a little, as shown in Fig. 7.13, and the ASE noises' accumulation can be avoided. In this way, the Gordon-Haus limit by (7.22) is exceeded and the communication capacity BL is greatly increased.

7.3.4 The Dispersion-Managed Soliton and the Enhanced System Stiffness

7.3.4.1 Stiffness Enhanced Approaches for Modified Soliton Systems

In the dispersion-uniformed soliton, the dispersion holds up the timing-jitters and hence makes more BER in the soliton receiving. From the view of time-jitters, the lower the dispersion, the better the jitter eliminating, but an enough-anomalous uniform-dispersion D known from (7.19) is indispensable for the shape-SNR kept power. If the uniform dispersion is too low, it is unfavorable for either dispersion-nonlinearity compensating (DNC) or shape-SNR keeping.

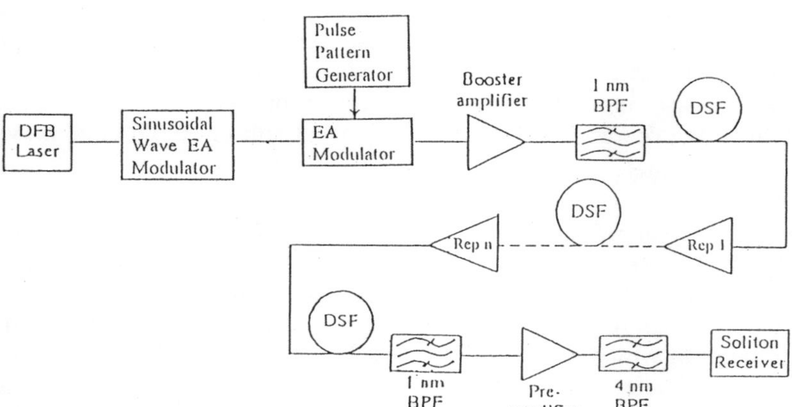

Fig. 7.13 The schematic diagram of a soliton transmission experiment on the "filtered-controled" approach (the slide-frequency guiding filter).

In the recent years, some approaches are proposed for the applied soliton system. The tendency is to relax the critically uniform demand of the conventional first-order soliton, spatially lessen and strengthen the controlling of the chirp and shape in the different distributed fiber sections. The system stiffness in the sequential sections is comparatively high, and of course the whole communication-line is more stable.

7.3.4.2 Dispersion-Managed Soliton and Short-Period Dispersion-Map WDM Systems

The dispersion-managed (DM) soliton is a new paradigm proposed by N.J. Doran [7.14] and others. The simplest of a DM soliton is a two-stage map with alternating sections of opposite sign dispersion.

In these proposals, the dispersion-managed soliton (now called conventional dispersion-managed soliton, or CDM soliton) occurs in the system with spatially varying (usually periodic) dispersion. The periods of dispersion management Z_p and the amplifier spacing Z_a are such that $Z_p \geq Z_a$. The total (net) dispersion may be kept very low, thus eliminating the long-distance temporal Gorgon-Haus jitter and the increasing signal-to-noise ratio. These solitons have been discovered to have enhanced power such that the stable pulse power is not simply reduced on reducing the net dispersion, but can be increased by increasing the depth (amplitude) of the dispersion variations. The local dispersion (of either sign changing-rate) may be relative large.

The relatively high dispersion in each local section fiber makes the soliton pulses in relatively high power, which is required the by shape-SNR kept power of first-order soliton. At the same time, its low path average-dispersion (such as $-0.1\text{ps}^2/\text{km}$) of the total soliton fiber makes the jitter low, which is required by the low BER kept dispersion. The length of the soliton-soliton interaction increases with the depth of the dispersion-map up to a maximum after which the interaction length decreases. Thus, the conventional dispersion-managed (CDM) soliton provides an extra degree of freedom in the design compared to the perfect soliton mentioned above in Sect. 7.3.1, namely the map strength, $S = |\beta_1 L_1 - \beta_2 L_2| / \tau^2$ [7.35 and 7.36].

However, at very high bit-rate (such as 160Gbit/s) transmission, extremely short pulses are required. The CDM soliton system then becomes incapable to restrict the soliton interaction and the four-wave mixing.

An alternative approach is a short-period dispersion map WDM (SPDM-WDM) system, in which $Z_p << Z_a$, which is just the opposite of that in the CDM soliton system. In this SPDM-WDM system, the lengths of the dispersion segments L_1 and L_2 can be much reduced, making it possible to keep extremely short pulses in the dispersion maps with high local dispersion at map strengths that will lessen the short pulse interaction and the four-wave mixing [7.35 and 7.36].

7.4 Nonlinear Optical Fiber Loop and Light-Leading-Light (L^3)

The light intensity-induced nonlinear phase-shift guides the loop transmission features. This approach may produce a series of all-optical high-speed controlling devices. One of them is the technique of light-leading-light (L^3).

7.4.1 The Phase Features in the Nonlinear Fiber Loop

7.4.1.1 The Transmission Features of a Twin Port Nonlinear Fiber Loop

The most fundamental model of the nonlinear fiber loop (NFL) is a fiber couple of which two output ports are connected. Thus, a four-port coupler becomes a twin-port (ports 1 and 2) fiber loop (Fig. 7.14).

First, we discuss the linear case in which the light-intensity-induced refractive index change in the loop can be neglected. We suppose only one input E_1 is injected into the Port 1. It is divided by the coupler C into two opposite-moving parts E_3 and E_4 which are then introduced into the loop. These two meet again and mutually interfere at C after travelling through the loop. Then the reflective signal E_{r1} from Port 1 and the transparent signal E_{t2} from Port 2 are displayed. This proves that the fiber loop action is like the arms in a Mach-Zehnder interferometer. In this case, the paths and hence the phase delays are the same except in the opposite directions.

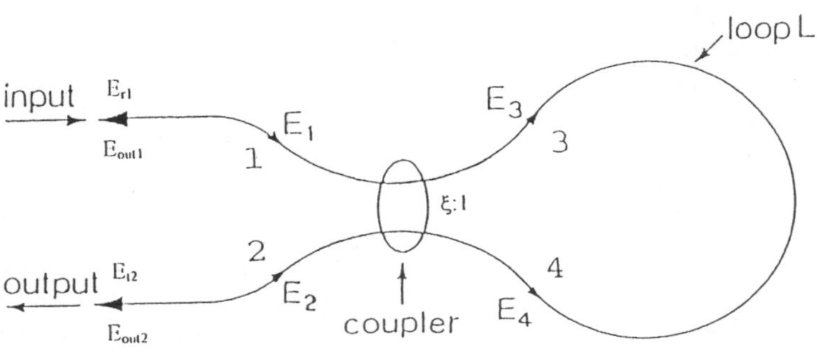

Fig. 7.14 A twin port nonlinear fiber loop.

If the coupling ratio $\xi = 1/2$, the injection $E_{in} = E_1$ is totally reflected at Port 1.

If the injection $E_{in} = E_1$ is strong enough, the light-intensity-induced refractive index change in the loop has to be considered. The phase delay should include the intensity induced nonlinear parts caused by the third-order nonlinear refractive index change. If the pulse-width is much less than the loop length, the interaction between E_3 and E_4 is disregarded. The only factor considered is the additional SPM-induced phase delay $\delta\varphi$. Since the phase is

$$\varphi = \omega t = \omega n\rho/v = kn\rho = 2\pi n\rho/\lambda. \tag{7.24}$$

The phase change $\delta\varphi$ due to intensity-induced refractive index change is

$$\delta\varphi = k\cdot\delta n\cdot\rho = 2\pi n_2|E_{in}|^2\rho/\lambda \tag{7.25}$$

where δn is defined in (2.108) and (2.109), and

$$n_2 = (3/8)[Re\chi^{(3)}(-\omega; +\omega, -\omega + \omega)/n_0(\omega)] \tag{2.102}$$

is the nonlinear refractive index coefficient.

The SPM-induced phase delay $\delta\varphi$, from (7.25), is increased by a shorter λ, longer L, stronger $|E|^2$, and larger n_2. In practice, λ is set at 1.55μm for the OFC due to the fiber loss, and it is definitely impossible to select 0.155μm or shorter. The commonly used loop length L is already rather long, and is difficult to extend from 1km to 50km or more, since the dispersion effects will be more serious. $|E|^2$ cannot increase in injection peak power, as from 100 mW to 10W, even if an amplifier is inserted. For a larger n_2, or susceptibility $\chi^{(3)}{}_{\text{fiber}}$, some semiconductor components or nonlinear fibers may be used. $|E|^2$ and n_2 will be discussed in this section.

The fundamental analysis of the nonlinear fiber loop (NFL) as a selected transparent device or a total reflector here is contributed by N.L. Doran and his colleagues [7.13].

If the coupling coefficient ratio $\xi \neq 1/2$, the amplitudes of E_3 and E_4 are unequal. Then the nonlinear phase delays are different, causing the phase difference $\delta\varphi$, and total transparency $|E_{out2}|^2 = |E_{in}|^2$ can exist. Here $E_{in} = E_1$ enters from port 1. Suppose the power-coupling ratio of the coupler is ξ : $1-\xi$, then the transmission feature of the coupler is [7.13]

$$\begin{aligned} E_3 &= \xi^{1/2}\cdot E_1 + i(1-\xi)^{1/2}\cdot E_2 \\ E_4 &= i(1-\xi)^{1/2}\cdot E_1 + \xi^{1/2}\cdot E_2. \end{aligned} \tag{7.26}$$

When $E_2 = 0$, only $E_1 = E_{in}$ enters from port 1. Then E_3 and E_4 meet again at C after traversing the loop and have the phase delays $\delta\varphi_3$ and $\delta\varphi_4$ obtained from (7.25), then we have

$$E_3 = \xi^{1/2} E_{in} \exp(i\delta\varphi_3)$$

$$E_4 = i(1-\xi)^{1/2} E_{in} \exp(i\delta\varphi_4). \qquad (7.27)$$

To calculate the outputs E_{out1} and E_{out2}, we make the transformations $E_3 = E_4^*$ and $E_4 = E_3^*$ and invert (7.26) to find E_{out2} at Port 2,

$$|E_{out2}|^2 = |E_4 E_4^*| = |E_{in}|^2[1 - 2\xi(1-\xi)]\{1 + \cos[(1-2\xi)|E_{in}|^2 2\pi n_2 L/\lambda]\}.$$

In the total transparency condition, the power of input terminal transfers totally to output port, $|E_{out2}|^2{}_{max} = |E_{in}|^2$ and $E_{out1}|^2{}_{min} = 0$ can occur at $\xi \neq 1/2$ and, for m odd in,

$$|E_{in}|^2 2\pi n_2 \rho/\lambda = m\pi/(1-2\xi). \qquad (7.28)$$

In this condition, the power of one input terminal transfers totally to the other output port, which is totally transparent. Accordingly, the minimum output power at out2 point is,

$$|Et_2| = |E_{out2}|^2{}_{min} = |E_{in}|^2 [1 - 4\xi\cdot(1-\xi)] \qquad (7.29)$$

where the corresponding reflection coefficient $R = 4\xi(1-\xi)$. From (7.29)

$$|E_{out2}|_{min} = |E_{in}|\cdot(2\xi - 1)$$

When $\xi \to 1/2$, $|E_{out2}| \to 0$, then the fiber loop is a total reflector.

$$|E_{r1}| = |E_{out1}|_{max} \cong |E_{in}| = |E_1|. \qquad (7.30)$$

The on-off power is determined by (7.28). If $|E_{in}|^2 = 0$, then $|E_{out1}|^2 = |E_{out2}|^2 = 0$. If $|E_{in}|^2$ increases from 0 to $\lambda/[2(1-2\xi)n_2 L]$, then from (7.28), the transparent $|E_{out2}|^2$ returns to 0 and the reflected power $|E_{out1}|^2 = |E_{out1}|^2{}_{max}$. The on-off power of a nonlinear fiber loop (NFL), needed to switch from on to off or vice versa, is given by [7.13],

$$P_{NFL-off\to on} \propto \tfrac{1}{2}|E_{in}|^2_{NFL-off\to on} = \frac{\lambda}{4(1-2\xi)n_2\rho}. \qquad (7.31)$$

From (7.31), the lower on-off power needs the higher nonlinear susceptibility $\chi^{(3)}$ and the longer fiber.

7.4.1.2 Nonlinear Fiber Loop with Amplifiers

To enhance the loop function and to decrease both the driving light power and the loop length, the effective methods are by enhancing the susceptibility in the loop or inserting the amplification. We discuss here first the amplifier insertion. Fig. 7.15 shows this type of NFL. An optical amplifier is inserted with spatial symmetry. If $\xi = 1/2$, the linear reflectivity is total, but when the amplifier is inserted asymmetrically, near

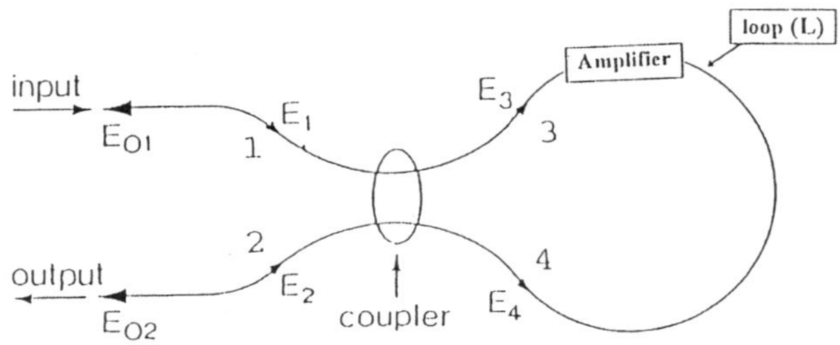

Fig. 7.15 The nonlinear fiber loop with amplifiers.

Port 3 as shown in Fig. 7.15, the clockwise wave propagates after amplification while the counter-clockwise wave amplifies after propagation.

When they again meet at C, the nonlinear phase delays of the clockwise and counter-clockwise waves are $\delta\varphi_c$ and $\delta\varphi_{cc}$ respectively. If the pulse-width is much shorter than the loop length, and the amplifier is unsaturated with amplification (gain) g, then the solutions are (7.32)

$$\delta\varphi_c = k\delta nL = 2\pi n_2 g|E_{in}|^2 L/\lambda$$

$$\delta\varphi_{cc} = k\delta nL = 2\pi n_2|E_{in}|^2 L/\lambda. \tag{7.32}$$

When $|E_{in}|^2 = \lambda/2n_2\rho(g-1)$, the transparency of NFLA is maximum [7.15], and the on-off power is:

$$P_{\text{NFLA-off}\to\text{on}} \propto (1/2)\cdot|E_{in}|^2{}_{\text{NFLA-off}\to\text{on}} = \lambda/[4(g-1)n_2\rho]. \tag{7.33}$$

Compared with the NFL without an amplifier, the factor $(1-2\xi)$ in (7.31) is just replaced by the factor $(g-1)$ in (7.33). Generally, $(g-1) >> 1 > (1-2\xi)$, so the on-off power and the loop length may be greatly decreased in NFLA, but it is best to avoid saturating the optical amplifiers.

Since the phase change $\delta\varphi \propto \omega$, n_2, $|E_{in}|^2$, and L in the nonlinear fiber loop should be greatly enhanced by the susceptibility, e.g. a very high n_2, but $\delta\varphi$ is governed by the medium materials. The method using the time feature of the resonant susceptibility, such as the refractive index delay due to the long lifetime of the resonant level, is one example of the NFL with susceptibility control. Unfortunately, it is impossible to discuss so many specific approaches in nonlinear fiber loops in this chapter.

7.4.2 Light-Leading-Light in High-Speed OFC

The nonlinear fiber loop's significance is that it is a light-leading-light, all-optical device. This approach might produce a series of all-optical controlling devices. Light-leading-light is a noteworthy example. It is possible to control certain light pulses by leading or selecting them from a series of pulses, by using the beacon or controlling light pulses. This leading, or selection, may be for a certain sequence in time, polarization states, frequencies, and so on. Light-leading-light is a candidate for demultiplexing at OTDM of 40–100 Gbit/s.

7.4.2.1 Nonlinear Optical Loop Mirror DEMUX by Induced Phase Delay

In a twin-port nonlinear fiber loop (described in Sect. 7.4.1), two other ports are added at points 3 and 4 respectively (as shown in Fig. 7.16). This forms the nonlinear optical loop mirror (NOLM) demultiplexer for very high-speed OFC [7.16]. When a certain channel needs to be demultiplexed, the high-speed time-division multiplexing (TDM) signal pulses are injected into the NOLM-DEMUX at Port 1. If a controlling series of pulses is introduced clockwise into the loop at Port 3 for demultiplexing, only the clockwise pulses of the demultiplexed channel, overlapped with the controlling pulses, change their phase delay. When these clockwise pulses of the channel interfere with the other counterclockwise pulses, the clockwise pulse are demultiplexed from output Port 2, which is the normally transparent port. The controlling pulses exit from Port 4 due to their perpendicular polarization, hence this approach is polarization-dependent. The signal pulses of the other channels have the same phases

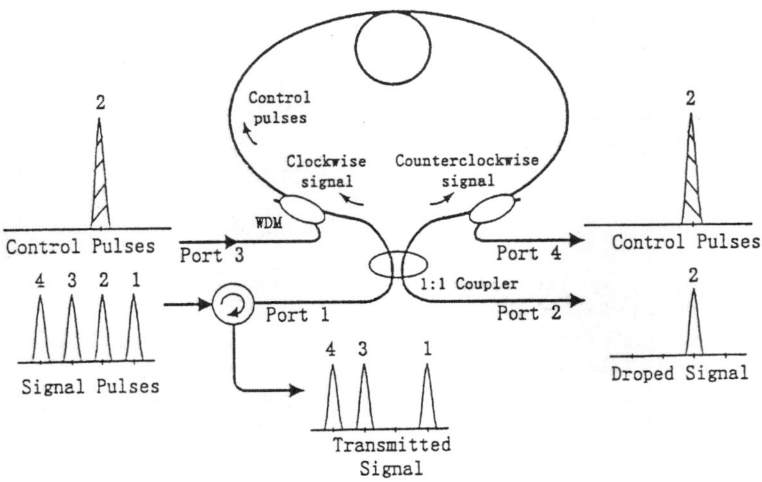

Fig. 7.16 The nonlinear optical loop mirror (NOLM) demultiplexer for very high-seed OFC.

for both halves, hence after interference, they reflect back to Port 1 but are arranged to a new routing. Dispersion-shifted polarization maintenance (DS-PM) fiber of 3 km length is used to make the NOLM-DEMUX. After demultiplexing, the 40Gbit/s signal TDM channel is demultiplexed to 16 channels of 2.5Gbit/s signals. In this case, the power of the controlling pulse series is about several hundreds of mW. Now, NOLM can demultiplex the OTDM up to hundreds of Gbit/s.

7.4.2.2 NOLM-DEMUX by XPM-Induced Frequency Changes

This approach combines the physical processes mentioned in this chapter. The four TDM signals in one frame accompanied by the pump pulse are injected into the nonlinear fiber loop. The loop is polarization-independent due to a PBS at the entrance [7.15]. As discussed in Sect. 7.2.2 (Fig. 7.10), there is a frequency-chirp $\delta\omega$ profile corresponding to the pump-intensity profile (the same induced phase profile as in Fig. 7.10) of $\delta\omega$ versus T, where T is the time domain abscissa from the leading edge to the trailing edge. The four TDM signals in one frame correspond to the different positions at the profile of $\delta\omega$ versus T. Because of the cross-phase modulation (XPM) effect expressed by (7.14–7.15), the frequency of these TDM signals has different shifts due to the XPM. After exit from the NOLM, they are separated (demultiplexed) by a wavelength selection device such as a grating.

References

[7.1] Charles K. Kao and George H. Hockham: Dielectric-fiber surface waveguides for optical frequencies, *Proc. IEE* **113**, 1151 (1966)

[7.2] G.P. Agrawal: *Nonlinear Fiber Optics,* Academic Press, New York (1989)

[7.3] A.W. Snyder and J.D. Love: *Optical Waveguide Theory*, Chapman and Hall, London (1983)

[7.4] D.N. Payne and W.A. Gambling: New silica-based low-loss optical fiber, *Electron. Lett.* **10**, 289 (1974)

[7.5] J.E. Midwinter and Yili Guo: *Optoelectronics and Lightwave Technology*, John Wiley, Chichester (1992)

[7.6] R.H. Stolen and Chinlon Lin: Self-phase-modulation in silica optical fibers, *Phys. Rev. A* **17**, 1448–1453 (1978)

[7.7] A. Hasegawa and F. Tappert: Transmission of stationary nonlinear optical pulses in dipersive dielectric fibers, I. Anomalous dispersion, *Appl. Phys. Lett.* **23**, 142 (1973)

[7.8] J.P. Gordon and H.A. Haus: Random walk of coherently amplified solitons in optical fiber transmission, *Opt. Lett.* **11**, 665–667 (1986)

[7.9] L.F. Mollenauer, et al.: Demonstration of error-free soliton trans-mission at 2.5 Gbit/s over more than 14000 km, *Electron. Lett.* **27**, 2055–2057 (1991)

[7.10] M. Nakazawa, et al.: Dynamic optical soliton communication, *IEEE J. Quantum Electron* **26**, 2095–2102 (1990)

[7.11] A. Mecozzi, H.A. Haus, et al.: Soliton transmission control, *Opt. Lett.* **16**, 1841–1843 (1991)

[7.12] L.F. Mollenauer, J.P. Gordon, et al.: The sliding-frequency guiding filter: an improved form of soliton jitter control, *Opt. Lett.* **17**, 1575 (1992)

[7.13] N.J. Doran and D. Wood: Nonlinear-optical loop mirror, *Opt. Lett.* **13**, 56–58 (1988)

[7.14] N.J. Smith, N.J. Doran, et al.: *Electron. Lett.* **32**, 54–55 (1996)

[7.15] M.E. Fermann, et al.: Nonlinear amplifying loop mirror, *Opt. Lett.* **15**, 752–754 (1990)

[7.16] S. Shimada, et al.: Very high speed optical signal processing. *Proceedings of the IEEE* **81**, 1633–1646 (1993)

[7.17] Yili Guo, et al.: A study on stimulated Raman scattering in doped optical fibers, *J. of Tsinghua University* **21**(1), 25–33 (1981)

[7.18] Jinghui Li, Yili Guo, Yizhi Gao, et al.: An all fiber filter and its application in FDM technology, *Acta Optica Sinica* (English Edition) **12**, 175–178 (1992)

[7.19] D. Breuer and K. Petermann: Comparison of NRZ- and RZ-modulation format for 40-Gbit/s TDM standard fiber system, *IEEE Photo. Tech. Lett.* **9**, 398–400 (1997)

[7.20] K. Ennser and K. Petermann: Performance of RZ versus NRZ transmission on standard single-mode fibers, *IEEE Photo. Tech. Lett.* **8**, 443–445 (1996)

[7.21] T. Matsuda, A. Naka and S. Saito: 10 Gbit/s, 6000 km NRZ and 4400 km RZ signal transmission experiments at zero-dispersion wavelength, *Electron. Lett.* **32**, 229–231 (1996)

[7.22] D. Marcuse: Pulse distortion in single-mode fibers, 3: Chirped pulses, *Appl.Opt.* **20**, 3573–3579 (1981)

[7.23] M.J. O'Mahony: Nonlinear optical transmission. *European Trans. on Telecommunications and Related Technologies* **4**, 629–640 (1993)

[7.24] T.K. Chiang, L.G. Kazovsky, et al.: Cross-phase modulation in fiber links with multiple optical amplifiers and dispersion compensators, *J. of Lightwave Technology* **14**, 249–260 (1996)

[7.25] R. Noe, D. Sandel, et. al.: Polarization mode dispersion compensation at 10, 20, and 40 Gb/s with various optical equalizers, *J. of Lightwave Technology* **17**, 1602–1616 (1999)

[7.26] Haocheng Zheng, Shizhong Xie, et al.: A simple formula on allowable fiber input power for unequal channel spaced WDM transmission systems, *J. of Optical Communication* **20**(4), 153–155 (1999)

[7.27] E.A. Golovchenko, V.J. Mazurczyk, et al.: Four-wave mixing penalties in long-haul WDM transmission links, *IEEE Photo. Tech. Lett.* **11**, 821–823 (1999)

[7.28] Grüner-nielsen Lars, et. al.: New dispersion compensation fibers for simultaneous compensation of dispersion and dispersion slop of non-zero dispersion shifted fibers in the C or L band, *OFC'2000,* Paper Tu G6 (2000)

[7.29] K. Mukasa and T. Yagi: Dispersion flat and low nonlinear optical link with new type of reverse dicpersion fiber (RDF-60), *OFC'2001,* Paper TuH7 (2001)

[7.30] H. Takenouchi, et al.: 2 × 40-channel dispersion-slop compensator for 40 Gbit/s WDM transmission systems covering entire C- and L- bands, *OFC'2001,* Paper TuS2 (2001)
[7.31] D. Penninckx and S. Lanne: Reducing PMD impairment, *OFC'2001,* Paper TuP1 (2001)
[7.32] H. Rosenfeldt, U. Feiste, et al.: Automatic PMD compensation at Gbit/s and 80 Gbit/s using a 3-dimensional DOP evaluation for feedback, *OFC'2001,* Paper PD27 (2001)
[7.33] B. Bakhshi, M. Vaa, et al.: Comparison of CRZ, RZ and NRZ modulation formats in a 64 × 12.3 Db/s WDM transmission experiment over 9000 km, *OFC'2001,* Paper WF4 (2001)
[7.34] Pan Zongqi, Qian Yu, et al.: Chromatic dispersion monitoring and automated compensation for NRZ and RZ data using clock regeneration and fading without adding signaling, *OFC'2001,* Paper WH5 (2001)
[7.35] L.J. Richardson, W. Forysiak and N.J. Doran: Dispersion-managed soliton propagation in short-period dispersion maps, *Opt. Lett.,* **25**, 1010 (2000)
[7.36] T. Hirooka, T. Nakada and A. Hasegawa: Feasibility of densely dispersion managed soliton transmission at 160 Gb/s, *IEEE Photo. Tech. Lett.* **12**, 633–635 (2000)

8. Photonic Nonlinearities in Fiber Communications

8.1 Nonlinear Photonics in Fiber Transmission and Optical Networks, Accumulation of Undesirable Effects in Transparent Networks

Ultra-broadband and high-capacity DWDM (or WDM/OTDM combined) transmission systems, high-compatibility optical networks with OADM, OXCs and optical switching, as well as the IP/WDM data optical networks with IP routers, optical wavelength routers and smart control plane are some of the most meaningful topics of OFC development in recent years.

Since 1966, the fiber has become the information society's mainstay and an indispensable component of the infrastructure of the whole world. Fibers underground are a permanent carrier of information. Professor Charles K. Kao has discussed fiber's great importance, present and future, in Chap. 1. The OFC's next focuses are its large-capacity transmission and high-transparency optical network. At the end of the 20th century, the communication traffic had become an important component in the infrastructure of the whole human society. Today, a growth of more than 15% a month in communication traffic is reported. On the carrier network, this 15% translates to a 80% growth in data traffic each year. Up to 2001, 10 Gbit/s facilities have already become the main force of commercial systems in the market and the 40 Gbit/s/channel technologies with high signal spectral efficiency have been successful in the laboratory. Some demonstrations of the large capacity fiber transmission systems are listed in Table 8.1.

A new proposal of the optical networks, the all optical network (AON), with high-compatibility in bit-rate and modulation formats at the optical layer nodes, will be the very urgently needed for the coming decades. The optical nodes of the add/drop multiplexing (OADM) and the optical cross-connect (OXC) as well as the novel functional devices for AON (some using optically nonlinear effects) will be promoted in the market soon.

Table 8.1 Some demonstrations of the large capacity fiber transmission systems.

	Demonstrated Capacity	*Comment*	*Reference*
Single wave-length (long-haul)	10Gbit/s × 9,000km	DSF + EDFA.	*OFC*'93, Paper PD1, 1993.
	10Gbit/s × 20,000km	3R regenerator of SOA (spacing of 140km) + DSF.	*OFC*'98, Paper PD15, 1998.
WDM	2.6Terabit/s	Optical duobinary coding.	*ECOC*'96, Paper ThB.3.1, 1996.
	100 channel × 10Gbit/s × 400km (50 or 100Ghz channel spacing)	True Wave™ fiber using large dynamic range, gain-flattened ultra wide band EDFAs.	*OFC*'98, Paper PD10, 1998.
	10.2Tbits/s (256 channel × 42.7Gbit/s) PDM/WDM, 100km	Tera Light™ fiber, in C and L band, 1.28bit/s/Hz spectral efficiency.	*OFC*'2001, Paper PD25, 2001.
	10.92Tbit/s (273 channel × 40Gbit/s), 117 km	Gain-shifted thulium-DFA, 50GHz spacing channel, 3 bands of S, C and L, very small PMD.	*OFC*'2001, Paper PD24, 2001.
Soliton	10Gbit/s × 106km	Synchronous retiming, loop.	*Electronics Letters*, Vol. 29, No. 9, 1993, p. 729.
	40Gbit/s × 5,000km	Periodic dispersion compensation, stright line.	*ECOC*'96, Paper TuD3.1, 1996.

Table 8.1 (continued)

	Demonstrated Capacity	*Comment*	*Reference*
WDM + Soliton	4 channel WDM × 10Gbit/s × 10,000km	Synchronous modulation, narrow band filter.	*Electronics Letters*, Vol. 32, 1996, p. 828.
	5 channel WDM × 20Gbit/s × 10,000km	Dispersion decreasing configuration with DCF.	*OFC*'97, Paper PD21, 1997.
OTDM (Single wave-length)	100Gbit/s × 500km	OTDM/PDM, NOLM Demux transform-limited optical pulses, optimizing wavelength.	*OFC*'95, Paper ThL2, 1995.
	640Gbit/s × 60km	Mode-locked fiber laser, pulse compressor, walk-off free dispersion-flattened NOLM.	*OFC*'98, Paper PD14, 1998.
WDM + OTDM	7 channel WDM × 200Gbit/s OTDM × 50km	Supercontinuum pulse source from ML-EDFL, PM-AWG, PM-PLC, PM-EDFA, DSF.	*Electronics Letters*, Vol. 33, September 1997, pp. 1716– 1717.
	30 channel WDM × 40Gbit/s NRZ-OTDM × 85km	Non-zero dispersion (True Wave) fiber.	*OFC*'98, Paper PD21, 1998.
WDM + Raman	3.08Tbit/s (77 channel × 42.7Gbit/s), 1200 km	Distributed Raman amplification, NZ-DSF, C and L bands with 100km spans in a re-circulating loop.	*OFC*'2001, Paper PD23, 2001.
	2.56 Tbit/s (64 channel × 40Gbit/s), 230km	Raman amplifiers, 0.8bit/s/Hz spectral efficiency, low-noise fiber, 170 μm^2 A_{eff} fiber.	*OFC*'2001, Paper PD26, 2001.

Accumulation of Undesirable Effects in Transparent Networks

At the end of the 20th century, the existing conventional fiber networks faced extensive pressures from various different needs: the increased needs of huge throughput, the multi-media services in the same channel, the different bit-rats per channel, the conventional but different formats (such as SDH/SONET, ATM), the different service formats (analogue and digital services) in the same fiber, etc. It was impossible to deal with all of these in the mixed O/E or optical local networks or access networks, and a revolution of networks was required. Recently, the Internet Protocol (IP) digital communication, which is based on home computers and is different from the existing communication formats, has shown very strong challenges by its field-trails or proposals of IP over fibers and WDM.

The all optical networking, or AON, is one of the important measures. Many pioneer backbone trunk-AONs in USA and Europe have shown distinctive characteristics. The laboratory's network test-beds, loop set-ups and field trails, with several tens of Gbit/s per fiber and over 1000km of distance, exhibit exceptional BER and Q-factor. In these cases, the photonic nonlinearities accumulated after cascaded optical nodes are of concern. Furthermore, the photonic nonlinearities in one channel will extend by crosstalks and cause the quality of another channel to deteriorate. The experimental results of the loop set-ups, including the OADM and OXC, with semiconductor wavelength converters, show the accumulated effects. There are obvious accumulations after five or ten turns, but it is not so awful because of the careful design and management. Some characteristics of optical networking are shown in Table 8.2.

Table 8.2 Characteristics of optical backbone networking.

Add/drop	Only those λ_i needed locally are dropped; other λ_s pass through optically at the OADM nods.
Cross-connect	Routings needed to be cross-connected and are switched totally by optics.
Network management	A special Ch_{manag} of λ_{added}; or $\textit{pilot-tone}_{\text{manag}}$ with λ-routing.
Scalability and re-configuration	No influence when new optical nodes or routes added; being changed easily as mesh, ring, and star, etc.
Transparency	High-compatibility in bit-rate and modulation formats at the optical layer. Save money in facilities.
Accumulation of undesirable effects	Being decreased by careful design and management for cascaded nodes.

8.2 Photonic Nonlinearities in Long-Distance Fiber Transmission

The SPM-GVD effects limit dominantly the long-distance transmission with the cascaded EDFAs, hence the dispersion compensation and dispersion management are of much concerned in most cases.

The OFC has no strict definition on what is long-range; it is the distance at which when troubles and peculiarities, which are dependent on length, become obvious. Of course, repeaters are certainly needed for long-distance communications because of the signal power exhaustion caused by the fiber loss. For the conventional single mode fiber (G.652 standard fiber, 1.3μm zero-dispersion), the 0.22dB/km loss at the channel wavelength (λ_{ch} = 1.55μm) and the fiber dispersion (+17ps/nm·km) limited repetition span is about 60km for 10Gbit/s transmission without any compensation. Although the coherent properties of lightwave are potentially useful for communication, it has not been accepted because of technological difficulties. After the EDFA was produced commercially in the 1990s, the financial expenses and manpower of facility maintenance for compensating the fiber loss have been decreased remarkably. In the real long-haul OFC, several tens and even several hundreds of the EDFAs are cascaded with the repetition span of about one hundred kilometers. Figure 8.1 shows the schematic diagram of the experimental setup.

There are two examples of single-channel long-range transmission. One is the PDP-1, with BL = 10Gbit/s × 9000km = 90Tbit·km/s (DSF + EDFA), presented at the OFC'93. It was implemented by using the IM-DD straight-line transmission and 274 EDFAs cascaded repeaters with a spacing of 33km. There are some pulse shape deformation after long distance transmission in the resultant peaky and compressed eye patterns, caused both by noise accumulation (especially the ASE of the EDFAs) and nonlinear

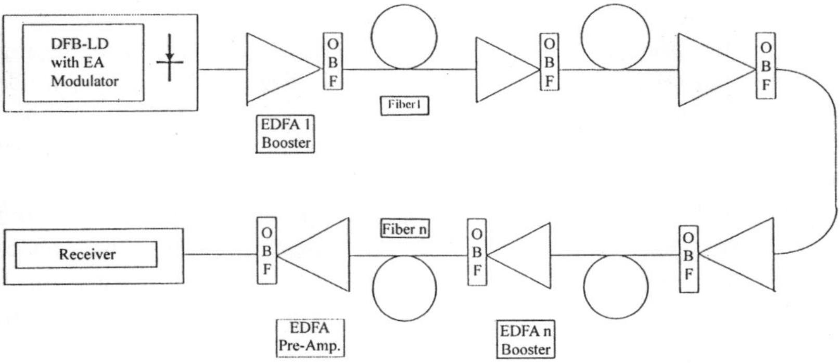

Fig. 8.1 The schematic diagram of the single wavelength channel long-haul transmission with the EDFAs.

effects (especially the SPM and frequency mixing). The other is the PDP-15 presented at the OFC'98 with BL = 10Gbit/s × 20,000km = 200Tbit·km/s. An original optical regenerator based on the semiconductor optical amplifier (SOA) has been used to demonstrate the error-free re-circulation loop transmission with a very large amplifier spacing of 140km DSF.

8.2.1 Noise Spectrum Broadening, Eye-Pattern Deterioration, Power Penalty Caused by the FWM, SBS, SRS

8.2.1.1 FWM Effects in Single Channel at Zero Dispersion Wavelength

If the EDFA is absent and the second-order GVD (β_3) can be neglected, then for GVD-induced pulse-broadening, the zero dispersion wavelength of the dispersion-shifted fiber will be the best for long-range transmission. In practice, cascaded EDFAs are indispensable for long-range transmission. If $\lambda_{\mathrm{channel}}$ is exactly at λ_0, then the EDFA-ASE noise λ_{noise} (at least a part of the ASE spectrum) and the FWM resultant $\lambda'_{\mathrm{noise}}$ will be symmetrically distributed around λ_0.

The relation between these refractive indices is

$$n(\lambda_{\mathrm{noise}}) + n(\lambda'_{\mathrm{noise}}) = 2n(\lambda_0), \tag{8.1}$$

and the phase-matching condition in the same direction of transmission,

$$\Delta k = k(\lambda_{\mathrm{noise}}) + k(\lambda'_{\mathrm{noise}}) - 2k(\lambda_0) = 0, \tag{8.2}$$

are always satisfied (as shown in Sects. 4.1 and 5.3) for parametric FWM processes. $D(\lambda)$, which is much near λ_0, could be approximated as a short line segment since there is an inflection point at λ_0, as shown in the upper curves in Fig. 7.2 in Chap. 7.

As a result, the nonlinear FWM induces the noise spectrum broadening and power penalty, as shown in the upper curve in Fig. 8.2. This ASE noise spectrum broadening will be accumulated and amplified by the cascaded EDFAs transmission and will cause the transmission quality decline on both the output spectrum broadening and eye-pattern deterioration.

If λ_{ch} deviates from λ_0, $\delta\lambda = \lambda_0 - \lambda_{ch} \cong 8$–10nm, this will induce the GVD coefficient $D = -1.0$ps/nm·km. In the dispersion case, the phase-matchings of λ_{noise}, λ_{ch} and $\lambda''_{\mathrm{noise}}$ are difficult as shown in the lower curve in Fig. 8.2. In the long-distance systems, there are many approaches to settle for the dispersion compensation or the dispersion management. Table 7.4 in Chap. 7 shows some of them, e.g. transmission at the normal dispersion region and compensated by a short segment fiber of anomalous

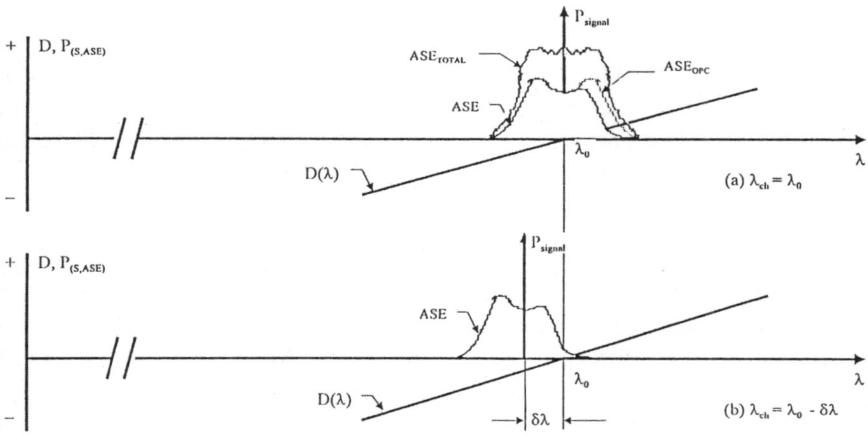

Fig. 8.2 a, b The noise spctrum broadenings caused by the FWM in a single channel at zero dispersion wavelength.

dispersion, or the alternative arrangement of equal-length fiber sections with opposite signs of fiber dispersion.

In the normal dispersion region,

$$n(\lambda_{\text{noise}}) + n(\lambda'_{\text{noise}}) > 2n(\lambda_{\text{ch}}) \tag{8.3}$$

and the phase-mismatching parameter

$$\Delta k = k(\lambda_{\text{noise}}) + k(\lambda'_{\text{noise}}) - 2k(\lambda_{\text{ch}}) \neq 0. \tag{8.4}$$

As shown in Fig. 8.2, the phase-mismatching is enough to restrain the FWM processes.

The theoretical analysis on the FWM effects at the zero-dispersion wavelength λ_0 for the cascaded EDFAs long-haul OFC is a very typical and significant condition [8.7]. It is easy to consider that the zero dispersion wavelength is ideal for the single channel, but actually at the zero-dispersion wavelength λ_0, the FWM caused by the signal and ASE induce the broadening of the ASE spectrum. Then the cascaded EDFAs will amplify this broadened spectrum. The transmitted eye-pattern has much disturbances due to the noises and the resultant noise spectrum is about several hundreds of GHz. If a little departure from the zero-dispersion wavelength λ_0 is settled, e.g., about D = 0.6ps/nm·km, the FWM will be difficult to be built, because of the dispersion-caused phase-mismatching. Then both the resultant output power spectrum and the eye-pattern are much better than that in the previous zero-dispersion wavelength case.

As a result, we settle the signal channel wavelenth λ_{ch} to be about several nm departure from the wavelength λ_0 . It will be enough to avoid the FWM between the signal and ASE. Then the domain nonlinearities in the single channel long distance case are the SPM and SBS.

In the WDM long-haul systems, the FWM-induced crosstalk between the multi-channels will deteriorate the communication quality heavily at the near-zero dispersion wavelength region. Some of the proposals to reduce the FWM in WDM systems will be considered in Sect. 8.4.

8.2.1.2 SPM Effect and Stimulated Scattering in Long-Distance OFC

Other factors are the interactions of nonlinear SPM effects and GVD, already discussed in Sects. 7.1 and 7.2. The bit-rate of the long-range OFC is limited by the SPM-produced pulse-width broadening. In the single channel long-haul communication, when the power is about mWs and the bit rate is about 10Gbit/s (the related pulse width is as short as several tens of ps), the SPM is a serious effect which will induce the pulse broadening and spectrum broadening and limit heavily the long-haul transmission.

Also, the nonlinear SBS and SRS (see Sect. 5.4) will cause power penalties. The interaction length of the light and the medium in the optical fiber is extremely long and the power density is relatively high, with the transmitted power bounded in an effective area of about $50\mu m^2$ in the standard non-zero dispersion shifted fiber (standard NZDSF). Recently, a new kind of NZDSF with a large effective area of $70–80\mu m^2$ is proposed to decrease partially the effects of almost all kinds of fiber nonlinearities. All of these proposals will be discussed in Sect. 8.4 of this chapter.

8.2.2 Dispersion Compensation in the Conventional Single Mode Fiber (CSMF) by the FWM-OPC

In the 1980s, the communication bit rate was not so high. The dominant limit in the long-haul transmission was the fiber loss. The wavelength of $\lambda = 1.55\mu m$ was selected for use because of both the minimum loss coefficient $\alpha_{min} = 0.22$dB/km for the conventional single mode fiber (CSMF) and the high quality of laser diodes. At the end of the 1980s, the EDFA was invented and it seemed to be the solution for the loss, hence the communication wavelength was fixed further at $\lambda = 1.55\mu m$, since it was also the operating wavelength of the EDFAs.

Nearly at the same time, the communication bit rate was also extended to 10Gbit/s and the linear dispersion coefficient $D(\lambda) = +17$ps/nm·km of the CSMF was then too high for a trunk. Numerous methods were proposed and valuable results were obtained. The near-zero dispersion fiber at $1.55\mu m$ using the dispersion shifted or dispersion flattened technologies was first proposed at that time.

The total CSMF length installed underground at present is hundreds of million kilometers in the world. Therefore, the dispersion compensation of the CSMF should be seriously considered. The method of combining the +17ps/nm·km CSMF fiber with the dispersion compensation fiber (DCF) of −125ps/nm·km is first proposed for the single channel long distance OFC. The fiber grating with chirp corrugation may be a better one since it is compact, but the long time stability (with temperature, stress, and so on) should be satisfied first.

In the dispersion managed fiber (DMF) systems, the standard NZDSF has been developed rapidly and installed for the long distance trunk since 1994. Just as we mentioned in Chap. 7, a lot of ideas have been proposed in recent years. The dispersion compensating fiber, the fiber Bragg gratings (broadband compensating and tunable channel compensating), virtual imaged phase array (by interference mode) are some promising means of compensating both the dispersion and dispersion slop for NZDSF. However, most of them are not within the scope of nonlinear photonics of this book.

The spectrum inversion by FWM-OPC is also a candidate in which dispersion compensation is total and thorough. It is discussed here since FWM-OPC is a very standard process in nonlinear photonics. It is shown that the pulse dispersion compensation in conventional SMF can be achieved by FWM-OPC, the principles of which have been discussed in Sect. 5.3.

For system installation, the FWM-OPC mirror should be a segment of the nonlinear fiber, a piece of the waveguide, or a compact semiconductor amplifier device. Since both the high susceptibility $\chi^{(3)}{}_{\mathrm{FWM}}$ and amplification are required, a compact device of semiconductor optical amplifier (SOA) will be one of the best candidates.

Figure 8.3 shows the schematic sketch of the chromatic dispersion compensation by the FWM-OPC in a hundred kilometer fiber span and the

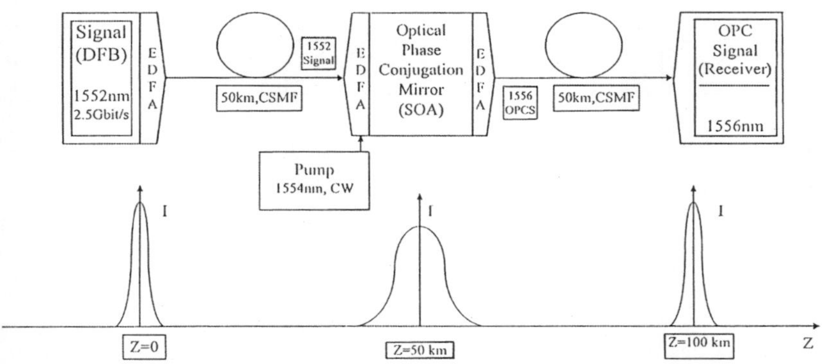

Fig. 8.3 A schematic sketch of the dispersion compensation by the FWM-OPC.

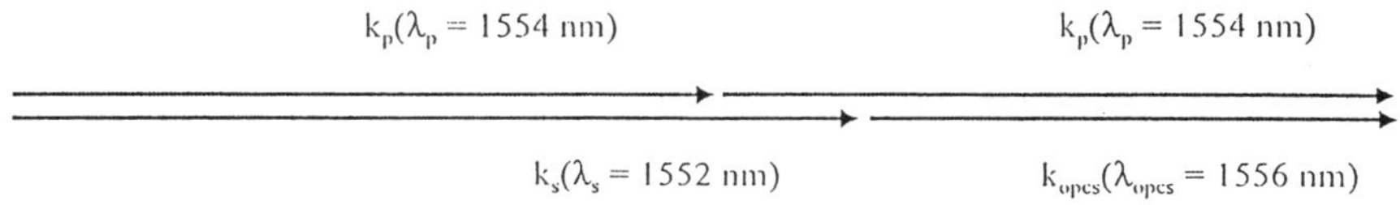

Fig. 8.4 The phase-matching condition of the dispersion compensation in the CSMF using the optical phase conjugation.

transmitted signals are 2.5Gbit/s ($2^{15}-1$) NRZ pseudo-random sequences. The OPC mirror is a semiconductor optical amplifier (SOA). Many benefits in these kinds of arrangements have been acquired from experiments. The bit error rate (BER) curves indicate that there are improvements of several dB (6–8dB or more) for the 100km transmission with the FWM-OPC dispersion compensation. The BER curve exhibits only a little difference (less than 1dB) from that received back to back.

The physical processes of the fiber dispersion compensation by the FWM-OPC are shown in Figs. 8.4 and 8.5. Figure 8.4 shows the diagram of momentum conservation by wavevector-matching or phase-matching (see Sect. 4.3). In this, the conjugate wavevector, as well as the two pump photons, are collinear and in the same direction as the signal wavevector. This is a forward FWM-OPC.

The physical patterns in the frequency domain are shown in the upper part of Fig 8.5. A transmitted short pulse with frequency ω_S is accompanied by a certain spectral distribution, caused by an initial chirp or linewidth of the laser diode. A higher frequency component ($\omega_S + \Delta\omega$) is selected to compare with the ω_S. An optical phase-conjugation medium(OPCM), such

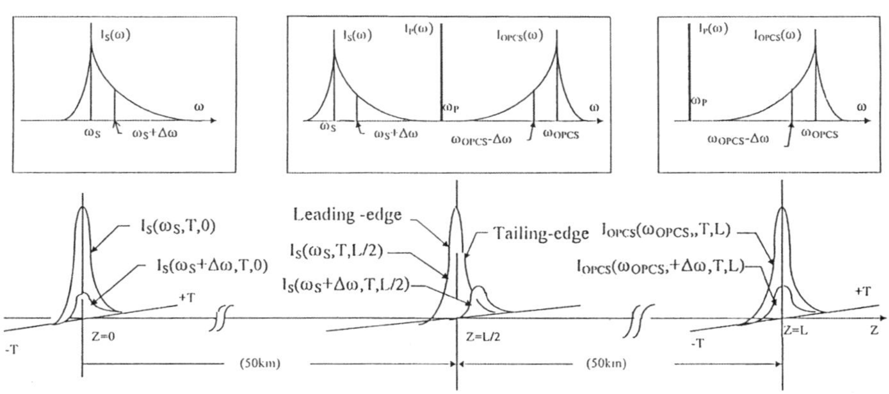

Fig. 8.5 The physical processes of the fiber dispersion compensation by the FWM-OPC in the domains of frequency, time and space.

as a SOA, is located at the midpoint $z = L/2$ (after 50km transmission). The pump wave $I_P(\omega_P)$ and the signal wave $I_S(\omega)$ were injected into the OPCM simultaneously and the OPC-signal wave is created. The spectrum inverses between the original signal spectrum and the OPC signal spectrum. The energy conservation is $\omega_P + \omega_P = \omega_S + \omega_{\text{opcs}}$ and the wave-vector matching is $k_P + k_P = k_S + k_{\text{opcs}}$. It is noticed that the spectrum distributions of the original signal and its OPC signal are symmetrical relatively to the pump wavelength, and this is really the characteristics of the optical phase conjugation, as discussed in Sect. 4.2. Consequently, there are the relations of $\omega_P + \omega_P = (\omega_S + \Delta\omega) + (\omega_{\text{opcs}} - \Delta\omega)$ and $k_P + k_P = k_{\omega_S+\Delta\omega} + k_{\omega_{\text{opcs}}-\Delta\omega}$ for the higher frequency component $(\omega_S + \Delta\omega)$ mentioned above. After the OPCM, the pulse with this inversed spectrum transports along the other 50km of the fiber.

The correspondingly physical patterns in the time domain (along the pulse-width direction from the leading-edge to the tailing-edge) and in the space domain (along the transmission) are shown in the lower part of Fig 8.5. At the beginning $z = 0$, the pulse profiles of the frequency component ω_S and of the component $(\omega_S + \Delta\omega)$ are at the same position. As long as in the positive dispersion region ($D(\lambda)$ = +17ps/nm·km), the refractive index $n(\omega)$ is larger and the group velocity v_g is lower for the high frequency component. Therefore, in the transmission process, the profiles of the higher frequency component $(\omega_S + \Delta\omega)$ will be postponed to the rear of the tailing-edge as shown in Fig. 8.5 (the midpoint of the lower part). At the same time, the profiles of the lower frequency component $(\omega_S - \Delta\omega)$ will lead to the front of the leading-edge. As a result, at the 50km point the pulse shows apparent broadening. After the action of the FWM in the OPCM, the frequency ω_S transfers into the frequency ω_{opcs}. In addition, the original higher frequency component $(\omega_S + \Delta\omega)$ transfers into the lower frequency component $(\omega_{\text{opcs}} - \Delta\omega)$ in the OPC spectrum, and its profile will catch up that of the center ω_S, since it is in the positive dispersion region. Simultaneously, the originally lower frequency component $(\omega_S - \Delta\omega)$ transfers into the higher frequency component $(\omega_{\text{opcs}} + \Delta\omega)$, and its profile will be pulled back to that of ω_S. As a result, the whole pulse profile returns to the initial profile at the end of a 100km transmission span, as shown in Fig. 8.5.

The improvement on the beam quality in fibers by FWM-OPC has already been successful in the 1970s, but the improvement on dispersion compensation was not so urgent at that time, since the bit rate was comparatively low. In that decade, the effects of the FWM-OPC was used to improve the beam qualities of a solid-state laser and that of the disturbed transmissions in the air. Now the FWM-OPC becomes a possible candidate for the dispersion compensation of the long-haul fiber communication. Its success will be determined by the efficiency and fitting of the FWM-OPC mirror. The key is to find an efficient and fitting medium for the "OPC mirror". For the efficient medium, it means at least 20–40% or more efficiency is needed for commercial use, and for the fitting medium, it means the medium should be easy to use and is able to match with the whole engineering environment. The FWM-OPC compensation devices should be compact and simple, with economical components.

8.3 High-Speed and OTDM/Soliton Communication Using Nonlinear Effects

The application of some inline fast processes of optical nonlinearities to achieve novel techniques is of great significance in high-speed and soliton communications.

In the OFC, there is no strict definition of high speed. In practical trunk transmission engineering or metropolitan networks, 20Gbit/s is considered high, but the all-optical experiments in laboratories have gone to as high as 100–640Gbit/s. Some high-speed single-channel experiments are listed in Table 8.3.

8.3.1 Pulse Width Broadening and BER Penalty Caused by the SPM and FWM

Limits of photonic nonlinearities in the single-channel high-speed OFC are the same as in the long-range OFC. Generally, the higher the speed, the shorter the pulse and pulse-spacing, hence the more significant the nonlinear effects. More attention should be paid to the pulse broadening deriving from the SPM-GVD and to the BER penalty caused by the nonlinear FWM between the signal and the ASE noise of the EDFAs.

8.3.2 Fiber Soliton and All Optical Demultiplexing in the OTDM Using the SPM, XPM and FWM

8.3.2.1 Fiber Soliton Transmissions Based on the Joint Action of SPM and GVD

Fiber soliton transmission is a typical nonlinear photonic process, based on controlling the balance of the SPM and GVD, which preserves the pulse shape for a very great distance. As discussed in Sect. 7.3, the positive chirping of the SPM on a short pulse can be exactly balanced by the

Table 8.3 High-speed single-wavelength channel experiments.

Bit-rate Gbit/s	Length km	B·L Tbit·km/s	Year	Technologies
640	60	38.4	1998	Mode-locked fiber laser, pulse compressor, walk-off free dispersion-flattened NOLM.
400	40	16	1996	Supercontinuum pulse, FWM demultiplexing, dispersion-slop compensated fiber.
200	100	20	1995	Supercontinuum pulse, FWM demultiplexing, PLL timing extraction.
100	500	50	1995	OTDM/PDM, NOLM Demux.
40	750	30	1993	OTDM/PDM, Soliton.
20	11,500	230	1994	Soliton/AAGF/loop.
20	125	2.5	1995	DCF, OTDM.
17	150	2.6	1990	LN mod.
15	25,000	375	1994	Soliton/SFF/loop.
10	1000,000	10,000	1993	Soliton/Syn. retime/loop.
10	9,000	90	1993	EA mod.
10	204	2	1993	DST.

negative chirping of the GVD in the anomalous dispersion region ($\beta_2 < 0$). The chirp-free pulse shape can remain unchanged for 10,000km, or even a million km in a fiber loop. Figure 8.6 shows a solitary wave and the solution of the KdV equation in an image. Although it is an old painting, it vividly portrays a soliton standing high and majestic in the raging waves.

Timing jitter-reduction or timing jitter-control is one of the solicitudes in the soliton transmission. There are some proposals in the practical demonstrations of the inline control of filtering in the frequency domain and regeneration in the time domain, as mentioned in Sect. 7.3. Superior proposals are those expected to match easily with the conventional photonics methods and easy to be accepted by system engineers.

When should a soliton be applied to OFC engineering systems or is it possible to be used at the beginning of this new century? This is an interesting question since the soliton transmission has demonstrated its unique properties in many aspects, but there are still some practical problems to be solved. One is the transmission stabilities: The system stiffness, the timing jitter, the soliton-soliton interaction, and the polarization mode dispersion of fibers are all severe, especially when the bit rate is over 20Gbit/s. Another problem is the transmission cost-effectiveness (including the use of less expenses and manpower for management) due to the competition with other non-soliton formats.

8.3.2.2 All-Optical Demultiplexing in OTDM Communication by Nonlinear Fiber Loops

The rapid development of the global information network requires high capacity communications. Both in the WDM and OTDM, the potential of having bit rates up to several hundreds of Gbit/s or even several Terabit/s has been demonstrated in recent years.

There are some difficulties in demultiplexing high-speed TDM signals (e.g. higher than 40Gbit/s) by using conventional electronic or opto-electronic demultiplexers. Although some types of optical switch demultiplexers are optical devices, their drivers are still electronic. If the driving voltage is about 5V, the multiplexing/demultiplexing speed will be limited to about 10 Gbit/s due to the electronics. Hence, all-optical methods are needed.

The principles and technologies of the nonlinear fiber loop and its applications to all-optical demultiplexing or the light-leading-light (L^3) in high-speed OFC, based on the intensity-induced refractive index (the real part of $\chi^{(3)}(\omega)$) changes, are developed in Sect. 7.4. Since the nonlinear fiber loops are light-leading-light, they may yield many all-optical controlling devices. It is possible to control, lead, and select light pulses from a series of pulses by using beacon light pulses. This control may select for time sequences, polarization states, and frequencies, and is a candidate for the demultiplexing high-speed OFC.

The physical phenomena in the light-leading-light (L^3) include light-intensity-induced refractive index changes caused by the intense controlling pulses, the XPM-induced phase changes of the signal pulses, the interference function of the fiber loop interferometer, and the leading of some pulses from the total pulse train to a certain destination.

Figure 8.6 shows an example of the schematic arrangement of the all-optical high-speed demultiplexing by the nonlinear fiber loop in the OTDM-OFC. In high-speed communications, the bit rate of the separate single time-division channel (B_{STDC}) my be as high as 10Gbit/s, the channel number N may be as much as 16 ($B_{\mathrm{STDC}} \times N$ = 160Gbit/s), and the transmission distance as long as several hundreds of kilometers (the EDFAs cascaded). Low jitters PLL timing extractor offer the clock pulse (frequency f_{clock} $(Hz) = B_{\mathrm{STDC}}$) which acts simultaneously as the controlling pulse in the all-optical demultiplexer.

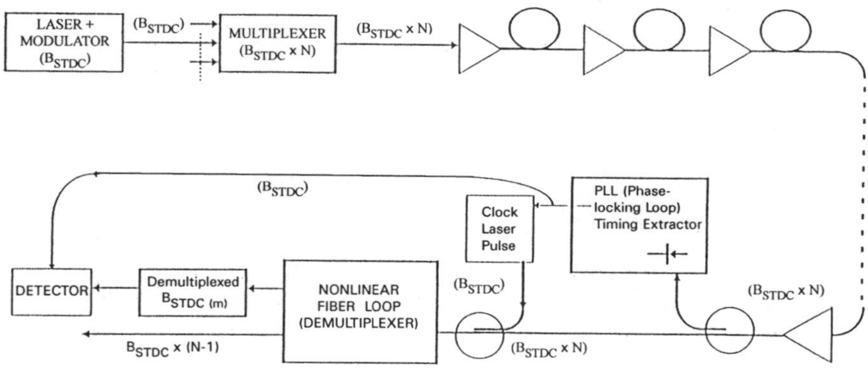

Fig. 8.6 The schematic arrangement of the all optical high-speed demultiplexing by a nonlinear fiber loop.

The nonlinear fiber loop (NFL) used as the all-optical demultiplexer is based on the polarization-insensitive FWM or XPM. In Fig. 8.6, $B_{STDC}(m)$ is one of the time-division channels in the sequence of $B_{STDC}(1)$, $B_{STDC}(2)$, ····, B_{STDC}(N) and is demultiplexed by the nonlinear fiber loop. In the proposal of the forward nearly-degenerated FWM, the energy conservation condition is $\omega_{DEMUX} - \omega_{clock} = \omega_{clock} - \omega_{signal}$, and the phase-matching condition is $\Delta k = 2k_{clock} - k_{signal} - k_{DEMUX} = 0$. Other types of the nonlinear fiber loop used as a very high speed demultiplexer have been discussed in Sect. 7.4.2.

8.4 Nonlinear Effects on the WDM/DWDM OFC

The influences of photonic nonlinearities are significant chiefly in the DWDM. The FWM, SPM, XPM, SBS and SRS are important in this case. The large effective area fiber (LEAF), true-wave dispersion-controlled fiber (TWDCF) and other new types of fibers are investigated for improving the trunk fiber. In the WDM and DWDM transmission fields, very meaningful works have been contributed by Tingye Li and his colleagues at the Bell Labs.

The FWM, SRS and XPM processes generate possibly the crosstalk between channels, and in the case of the WDM, several tens of high bit-rate channels. Power penalty due to the SBS is also important in this case. When the channel number in the DWDM is as large as 64 or 128 (spacing about 25–50GHz), these nonlinear effects will really be the final limitations of the high-capacity OFC. The difficulties for the 25GHz spacing at present are the nonlinearities and the filters with very narrow and step spectrum.

It is well known that the DWDM is limited by the fiber FWM and XPM. In contrast, the considerable influences of the FWM and XPM in the EDFA and SOA have not received enough concern.

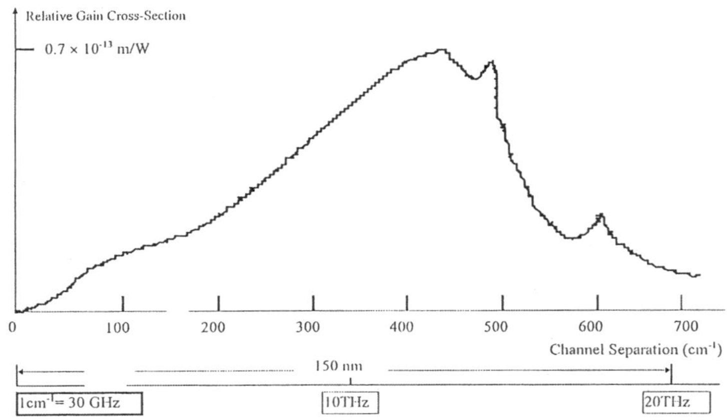

Fig. 8.7 Raman gain in fused silica fibers.

8.4.1 Stimulated Light Scattering in WDM/DWDM OFC, Crosstalk by SBS Near 11GHz Channel-Spacing

8.4.1.1 Stimulated Raman Scattering (SRS) in Single Channel

The principles of the SRS have been discussed in Sect. 5.4.2. Figure 8.7 shows that the Raman gain spectrum, g_{Raman} versus $\Delta(1/\lambda)_{\text{frequency-shift}}\lambda$, is as broad as 500 to 600cm^{-1} (the maximum shift is about 470cm^{-1}, e.g. 13THz). The fused silica is a kind of non-crystal material of which the vibration energy levels of the molecules are overlapped and broadened into an energy-band. Hence the Raman gain exists continuously in a wide range of about 15,000GHz, by which the silica fiber can be a wide-band amplifier. Of course, the comparatively high and flat gain range is only 15–20nm (about 2–2.5THz). When a detective wave of ω_S is injected together with the pump of ω_P, and if $\omega_P - \omega_S$ (> 0) is located in the wide Raman gain spectrum-range, the longer-wavelength ω_S wave will be amplified. The power can then be transferred from the shorter wavelength channels to the longer wavelength channels, and the laser line-width has no influences on the Raman gain.

From [8.11], the critical power of the SRS in the fiber is $P_{\text{crit}} = 16 (A\alpha P/\gamma_0)$. We consider here $A = A_{\text{eff}}$, $\alpha P = 1/L_{\text{eff}}$ and $\gamma_0 = 1/2\ G_{\text{Raman}}$ for a conventional SMF, then in the case of the obvious power degradation due to the SRS threshold being over, the permitted signal power of the single wavelength channel is

$$P_{\text{limited-SRS}} = \frac{32}{G_{\text{Raman}}} \cdot \frac{A_{\text{eff}}}{L_{\text{eff}}} \cong 45.72\times 10^{13}\left(\tfrac{W}{m}\right)\cdot \frac{A_{\text{eff}}\left(m^2\right)}{L_{\text{eff}}\left(m\right)} \tag{8.5}$$

where the gain coefficient $G_{\text{Raman}} \cong 7\times 10^{-14}$m/W.

Example

Find the power needed at λ = 1.55μm, if α_{dB} = 0.2dB/km, L = 30km for DSF, where $\alpha_{\text{PUMP}} = \alpha_{dB}/4.34 = 0.046$/km and $A_{\text{eff}} \cong 50\mu\text{m}^2$ at λ = 1.55μm.

The permitted power of the signal power degradation due to the SRS is

$$P_{\text{limited-SRS}} = \frac{32 \cdot A_{\text{eff}}}{G_{\text{Raman}} \cdot L_{\text{eff}}} \tag{8.5'}$$

If L = 30km, then $-\alpha_{\text{PUMP}}L = -1.38$, and
$L_{\text{eff}} = (1/\alpha_{\text{PUMP}})[1 - \exp(-\alpha_{\text{PUMP}}L)] = 16.27$km.
Thus from (8.1), $P_{\text{limited-SRS}}$ = 1.4W.

In the existing signal wavelength channel of the OFC, the standard injected power is only 1–2mW. Even in the signal wavelength soliton systems, the injected signal power is only about 50mW. Therefore, the obvious power penalty due to the SRS is unrealistic. But in long distance WDM/DWDM systems with over a hundred channels, all these channels are located in the same wide Raman gain spectrum-range. For this reason, the induced crosstalks, caused by the power transferred from the shorter-wavelength channels to the longer-wavelength channels, have to be considered.

8.4.1.2 Stimulated Raman Scattering (SRS) in the Multi-channel Case

Figure 8.8 shows the bit pattern with SRS in the simplest two-channel case. Both channels' BERs increase.

In the multi-channel case, all the channels are in the same Raman gain spectrum range since it is so extensive. All the channels join the mutual Raman gain spectrum, and the power in the longer wavelength channels is amplified, but is attenuated in the shorter wavelength channels. The multi-channel theory [8.2] (originally from A.R.Chraplyvy [8.10]) shows that in a N channel system with spacing of $\Delta f_{\text{spacing}}$ (GHz) and the degradation due to SRS of each channel not more than 1dB, the permitted channel power $P_{\text{ch-1dB-limited-SRS}}$ is

$$N \cdot P_{\text{ch-1dB-limited-SRS}} \cdot [(N-1)\cdot \Delta f_{\text{spacing}}] \cdot L_{\text{eff}} \le \zeta \cdot 10^4 \cdot W\ (\text{GHz}\cdot\text{km})\,. \tag{8.6}$$

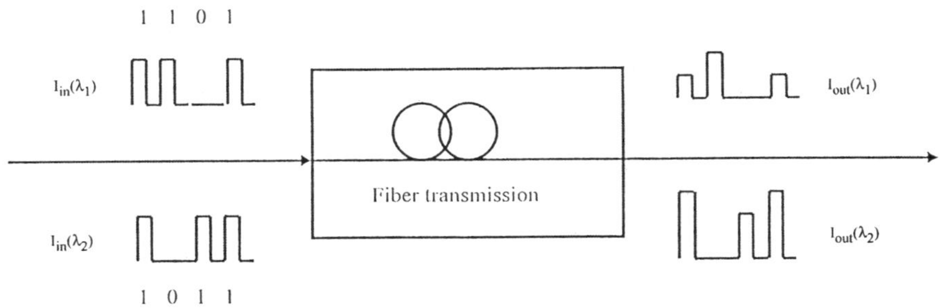

Fig. 8.8 The bit pattern with SRS crosstalk in the simplest two-channel case.

The value of ζ is roughly 1–2, which depends on some given conditions. Among them are the core effective area A_{eff} (about 50μm^2 (DSF) to 80μm^2 (CSMF or LEA fiber)), the actual value of G_{Raman} (about $6–8\times10^{-14}$ m/W, dependent on the Raman differential scattering cross-section of the doped material in the silica fiber, and the distinct position in the gain profile) and other existent settings. When $L_{\text{eff}} = nl_{\text{eff}}$ in the cascaded EDFA systems. $\Delta f_{\text{spacing}}$ is always 50–200GHz. When N is 16–64, $P_{\text{ch-1dB-limited-SRS}}$ is always about 4–1.2mW.

If the above condition is satisfied, the deterioration of signal quality caused by the SRS can be neglected. The SRS effects are notable only when light (for example the code "1" in the ASK) exists in two interactive channels. For this reason, the statistical results of the amplitude-shift keying (ASK) modulation are strongly influenced by this SRS effect.

8.4.1.3 Virtual State Hypotheses in Some Nonlinear Processes

From physics, virtual level states are introduced into the nonlinear photonic processes, such as the SRS (the dash-line in Fig. 8.5), the two-photon absorption (TPA, the dash-line always at the middle half), and others. A virtual level state, presented to complete the process, does not really exist in the molecules or atoms. Consequently, this intermediary state, of which no lifetime is needed to be considered, does not influence the speed of the processes. In the case of the SRS, only at the moment when the injection light beam (and also the virtual state) acts with the vibration-state (the real state) of the medium, the nonlinear processing exists. Once the injection light pulse vanishes, the virtual state also vanishes at the same time. Similar conditions are always linked with some other nonlinear phenomena in fibers and semiconductor materials. The virtual state hypothesis is possibly significant for the high-speed photonic processes and optical communication devices.

8.4.1.4 Stimulated Brillouin Scattering (SBS) in the WDM OFC

The principles of the SBS are shown in Sect. 5.4.1. The SBS has some differences from the SRS:

(1) In single-mode fibers, the gain coefficient $G_{\text{Brillouin}} \cong 5\times10^{-11}$m/W, about 10^3 times that of the SRS ($G_{\text{Raman}} \cong 0.7\times10^{-13}$m/W). Hence the SBS is a considerable photonic nonlinearity in fibers. It is noticed that the $G_{\text{Brillouin}}$ is decreased when the modulating dynamic-linewidth of the pump laser or the modulating bit rate of the 2PSK or ASK exists. The SBS will be reduced as

$$G_{\text{Brillouin-Mod}} \cong G_{\text{Brillouin}} \cdot[\Delta\nu_{\text{linewidth-Br}} /(B_{\text{Mod LW-Pump}} + \Delta\nu_{\text{linewidth-Br}})] \tag{8.7}$$

where $B_{\text{Mod LW-Pump}}$ is the dynamic-linewidth of the pump laser or the modulating bit rate of the 2PSK or ASK, hence $B_{\text{Mod LW-Mod}}$ is decreased obviously with the higher modulation rate.

(2) In silica fibers, the product ($G_{\text{Brillouin}}$ × bandwidth × $\Delta\nu_{\text{linewidth-B}}$) is as narrow as only 20–50MHz at λ =1.55μm. In some directly-modulated lasers, the linewidths (e.g. 400MHz) are much higher than this product. The gain $G_{\text{Brillouin}}$ will then be decreased by ($\Delta\nu_{\text{linewidth-B}}/\Delta\nu_{\text{linewidth-laser}}$) (e.g. 1/20), but the gain G_{Raman} × bandwidth is so wide (15,000GHz) that even the large linewidth laser used has no influence on it.

(3) The SRS propagates collinearly with the pump (here it is the signal) in the fiber with both the forward and backward directions, but the SBS propagates only the backward direction.

(4) The frequency down shift $\Delta f_{\text{Bri-shift}}$ is

$$\Delta f_{\text{Bri-shift}} = \omega_{\text{acoustic}}/2\pi \,. \tag{8.8}$$

From (5.50) in Sect. 5.4.1,

since $\quad \omega_{\text{acoustic}} = |k_{\text{acoustic}}|\cdot v_{\text{acoustic}} \cong 2|k_{\text{pump}}|\cdot v_{\text{acoustic}}$,

thus $\quad |k_{\text{acoustic}}| \cong 2|k_{\text{SBS}}| \cong 2|k_{\text{pump}}|$.

Substituting into (8.8),

$$\Delta f_{\text{Bri-shift}} = \omega_{\text{acoustic}}/2\pi \cong 2|k_{\text{pump}}|v_{\text{acoustic}}/2\pi,$$

where $|k_{\text{pump}}| = 2\pi/\lambda_{\text{pump}} = 2\pi n/\lambda_{\text{pump-vacuum}}$.

Then $\quad \Delta f_{\text{Bri-shift}} = 2nv_{\text{acoustic}}/\lambda_{\text{pump-vacuum}}. \quad$ (8.9)

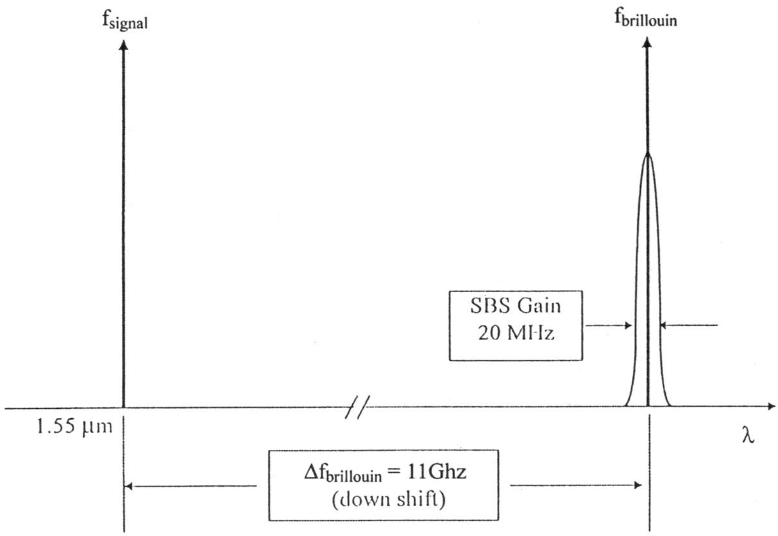

Fig. 8.9 The SBS down-shift $\Delta f_{\text{Brillouin}}$ and the gain spectrum in the single mode fiber.

Example

If the acoustic wave speed v_{acoustic} is 5.95km/s, n = 1.45, and $\lambda_{\text{pump-vacuum}}$ = 1.55μm, then the frequency shift

$$\Delta f_{\text{Bri-shift}} = (\omega_{\text{acoustic}}/2\pi) = (2nv_{\text{acoustic}}/\lambda_{\text{pump-vacuum}}) = 11.1 \text{ GHz}.$$

Figure 8.9 shows the SBS frequency downshift $\Delta f_{\text{Bri-shift}}$ in the single-mode silica fiber, about 13GHz at λ_{signal} = 1.3μm; and 11GHz at 1.55μm. As a result, the DWDM or FDM, if with a spacing of 11GHz, will be jeopardized by the SBS disturbance. From (5.51), the threshold pump power density of SBS is

$$\begin{aligned} I_{\text{SBS-threshold}} &\geq \frac{c\delta_s \delta\varepsilon_1 \varepsilon_2}{2\pi^2 \alpha S^2} \cdot \frac{\lambda_{1\text{Brillouin}}^2}{n_{1\text{Brillouin}}^2 \cdot n_{2\text{pump}}} \\ &= \varsigma \cdot \frac{\lambda_{1\text{Brillouin}}^2}{n_{1\text{Brillouin}}^2 \cdot n_{2\text{pump}}} \end{aligned} \tag{5.51}$$

In fiber, the limited power of the obvious signal degradation due to SBS can be referred from [8.11], in which $P_{crit} = 21\ (A\alpha_p/\gamma_0)$. We consider here $A = A_{\text{eff}}$, $\alpha_P = 1/L_{\text{eff}}$ and $\gamma_0 = 1/2\ G_{\text{Raman}}$ for the SMF. Then the permitted signal power in the single wavelength channel is

$$P_{\text{limited-SBS}} = \frac{42}{G_{\text{Brillouin}}} \cdot \frac{A_{\text{eff}}}{L_{\text{eff}}} \cong 8.4 \times 10^{11}\left(\frac{W}{m}\right) \cdot \frac{A_{\text{eff}}\left(m^2\right)}{L_{\text{eff}}\left(m\right)} \tag{8.10}$$

where the gain coefficient $G_{\text{Brillouin}} \cong 5\times10^{-11}$m/W.

Example

Find the power needed at λ = 1.55μm, if α_{dB} = 0.2dB/km, L = 30km, $\alpha_{\text{pump}} = \alpha_{\text{dB}}/4.34$ = 0.046/km, and $A_{\text{eff}} \cong 50\mu\text{m}^2$.

$n\ \alpha_{\text{pump}}L = -1.38$, $L_{\text{eff}} = (1/\alpha_{\text{pump}})[1-\exp(-\alpha_{\text{pump}}L)]$ = 16.27km.

From (8.10), $P_{\text{limited-SBS}}$ = 2.58mW. So the SBS occurs more easily and sooner in the fiber than does the SRS, due to the low $P_{\text{limited-SBS}}$.

8.4.1.5 Crosstalk by the SBS near 11 GHz Channel-Spacing

The downshift spacing $\Delta f_{\text{Bri-shift}}$ is about 11 GHz at λ = 1.55μm, while the $G_{\text{Brillouin}}$ gain bandwidth $\Delta\nu_{\text{linewidth-B}}$ is as narrow as only 20MHz. The WDM channel spacing was proposed as the integer of 10GHz at 1.55μm by ITU-T 1997. Consequently, the crosstalks generated by the SBS can be avoided and so the SBS-caused power limitation is independent of the number of WDM channels. In addition, because the lifetime of an acoustic phonon is rather long, the SBS is not a transient process and depends on the signal modulation rate. The higher the rate the broader the signal spectrum and the lower the $G_{\text{Brillouin}}$. Thus the SBS can be suppressed in this way.

The FDM channel spacing will probably be reduced to about 11GHz, if the number of FDM channels is further increased in the coming years. In 1998, the channel spacing of the Lucent Wavestar OLS-400 system is only 50 GHz, which is already less than the mentioned 100GHz. If the condition of $\Delta f_{\text{channel-spacing}} = \Delta f_{\text{Bri-shift}}$ (the Brillouin downshift of 11GHz) is satisfied, the SBS of this channel will drop on the adjacent channel and cause the crosstalk.

Table 8.4 shows the comparison of the SRS and SBS characteristics in the OFC.

8.4.2 Four-Wave Mixing (FWM) in WDM/DWDM OFC

Generally, FWM effect of the large-spcing(several nms) WDM communication in the conventional fiber is insignificant since the DWDM larger channel spacing and higher dispersion (larger phase mismatching in wave mixing) causes lower η_{FWM}.

Table 8.4 Comparison of the SRS and SBS characteristics in OFC fibers.

	SRS	SBS
Mechanism	Light acts with optical phonon.	Light acts with acoustic phonon.
Effects	Channel power from shorter transfer to longer wavelength.	Power penalty and intensity noise of signal channel.
Gain spectrum	See Fig. 8.7. Wide Raman gain G_{Raman} spectrum of 15THz.	See Fig. 8.9. Narrow Brillouin gain spectrum of $\Delta\nu_{\text{linewidth-B}} \cong$ 20MHz.
Gain coefficient	7×10^{-14} m/W(max).	5×10^{-11} m/W(1.55μm).
Directions of scattering	Both forward and backward, but $I_{\text{threshold-forward}} < I_{\text{threshold-backward}}$	Only backward, the forward-pump is used to amplify the backward SBS.
Scattering threshold	$P_{\text{SRS-threshold}}$ $= P_{\text{limited-SRS}}$ = 1–1.4W.	$P_{\text{SBS-threshold}} = P_{\text{limited-SBS}}$ = 2–3mW. Frequency shift $\Delta f_{\text{Brillouin-shift}}$ $\cong$ 11GHz at 1.55μm
Multi-channel case	Channel number is limited by the SRS, since the Raman gain spectrum is terribly broad.	$\Delta\nu_{\text{linewidth-B}} \cong$ 20MHz, no channel number limited by the SBS. Crosstalk will be of concern only if the spacing is near 11 GHz.
Response time	Broad band $G_R \Leftrightarrow$ short T_{response} optical phonon $\Rightarrow$ G_R is independent with modulation rate.	Narrow band $G_B \Leftrightarrow$ long life acoustic phonon $\Rightarrow G_B$ is decreased with higher modulation rate B of the signal.
Way to weaken it	No appropriate way for power WDM systems except the large effective area fiber.	$G_{\text{Brillouin-Mod}} \cong G_{\text{Brillouin}} \cdot \left[\Delta\nu_{\text{linewidth-Br}} / \left(B_{\text{ModLW-Pump}} + \Delta\nu_{\text{linewidth-Br}}\right)\right]$

But the channel BER and power penalties due to fiber FWM crosstalk are important effects in WDM/FDM systems. The spatial distribution of channels is dense, and so it is very easy to form many combinations of nearly degenerate FWM in adjacent dispersion-free zones, a peculiarity of FWM in FDM systems. The principles of FWM are discussed in Sects. 3.3 and 5.3.

For the same reason as in Fig. 8.2, if one WDM wavelength λ_n is located just at the zero-dispersion wavelength λ_0, then the four-wave mixing and thus the channel crosstalks occur among it and the symmetrically adjacent wavelengths λ_{n-1} and λ_{n+1}. Generally, the dispersion curve is treated as a short line segment, and so the phase-matching condition $k_{n-1} + k_{n+1} = 2k_n$ is naturally satisfied. For that reason, no channel should be located exactly at the zero-dispersion wavelength λ_0 in multichannel WDM systems. The usual, simple example is the many combinations of nearly degenerate four-wave parametric processes originating from only three injected channels. The profile is shown in Fig. 8.10.

The energy conservation and momentum conservation conditions, that is, phase-matching conditions, on the nonlinear parametric processes are

$$\sum_{i=1}^{m} \hbar\omega_i = \sum_{j=1}^{n} \hbar\omega_j \tag{4.1}$$

$$\Delta k = \sum_{r=1}^{m+n} k_r = 0 . \tag{4.2}$$

Equation (4.2) can be satisfied perfectly when the multi-wave interactive parameters are arranged carefully.

When the injection frequencies shown in Fig. 8.10 are $\nu_{I=1}$, $\nu_{j=2}$ and $\nu_{k=3}$, the generated frequencies are $\nu_{123,213}$, $\nu_{231,321}$, and $\nu_{132,312}$, the energy conservation and phase-matching conditions are

$$\nu_{ijk} = \nu_i + \nu_j - \nu_k \tag{8.11}$$

$$k_{ijk} = k_i + k_j - k_k . \tag{8.12}$$

These two conditions have to be satisfied in each combined group produced from FWM, as shown in Fig. 8.10.

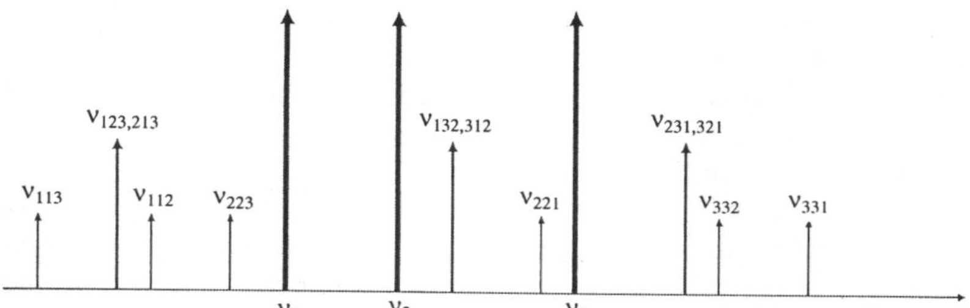

Fig. 8.10 The manifold combinations of the nearly-degenerated FWM originally from three wavelength channels injected.

If the spacing of frequencies are equal, the new frequencies generated by the FWM are also uniformly distributed. Hence the channel power penalties and crosstalks will cause the system quality to deteriorate. The power of the generated ω_m is

$$P(\omega_m) = \xi_{\text{FWM}} \cdot (\chi^{(3)}{}_{xxxx})^2 \cdot (L_{\text{eff}}/A_{\text{eff}})^2 \times P_i P_j P_k \exp(-\alpha L) \tag{8.13}$$

where ξ_{FWM} is a factor linked with the signal wavelength, the refractive index n, etc. $\chi^{(3)}{}_{xxxx}$ is the third-order nonlinear susceptibility at the same direction x in the fiber.

The FWM and XPM in EDFAs

Recent report shows that the impact of the FWM and XPM in EDFAs can be comparable with and even greater than that in a transmission fiber for the broadband DWDM systems.

Firstly, the FWM crosstalk in the L-band can considerably exceed that in the C-band since the doped-fiber length of the L-band EDFA is longer than that of the C-band EDFA. Secondly, the total FWM noise of a long-haul transmission is obtained by the summation over all cascaded EDFAs. Hence the total FWM crosstalk caused by all cascaded EDFAs can be stronger than that caused by the transmission fiber [8.20].

The XPM in EDFAs can be comparable with and even exceed that in a transmission fiber for the broadband DWDM systems [8.21].

8.4.3 Reducing the Nonlinearities of FWM/XPM Crosstalk, SPM-GVD Broadening and Stimulated Scattering

In the several (or several tens of) Gbit/s-DWDM transmission, except SPM-GVD and polarization mode dispersion(PMD), broadening is serious and the FWM/XPM crosstalk and stimulated scattering become dominant, since the channel spacing becomes smaller and the power intensity of the signal channels gets higher.

The FWM and XPM are dependent strongly on the channel spacing, fiber dispersion and fiber effective area size. Generally, when the channel spacing narrows to about 50GHz, or even 25GHz, the FWM and XPM crosstalk will be more serious. In these cases, the XPM crosstalk is somewhat important as the FWM effects.

We shall discuss in two physical ways. One is to enhance the fiber's dispersion managing, that is to keep the adapted small dispersion (2–4ps/nm·km), and the other is to lower the signal power intensity by the large core area fiber as well as by mixing the Raman distributed amplification (RDA) with EDFAs.

8.4.3.1 Small Dispersion Fiber Trunk and Dispersion Management

The FWM crosstalk in fiber transmission is chiefly stemmed from the third-order of nonlinearities in optical fibers. There are many methods to alleviate or to avoid it, such as to make the phase-mismatching by dispersion management, or to lower the light power intensity, as well as sometimes to tolerate the FWM by unequal spacing. Both the broadened bandwidth and complicated demultiplexing are needed in this unequal spacing proposal and this makes it less competitive.

The FWM/XPM crosstalk of the low dispersion fiber trunk, such as in the low positive dispersion non-zero DSF, is obviously lower than that in the dispersion-shifted fiber (almost zero dispersion) trunk. Actually, some people will select to install the trunk fiber which has $D = +(2\text{–}4)\text{ps/nm}\cdot\text{km}$ and the large effective core area $A_{\text{eff}} = (70\text{–}80)\mu\text{m}^2$. Also, people will use the high negative dispersion ($D = -(75\text{–}100)\text{ps/nm}\cdot\text{km}$) fiber with small negative dispersion slop ($-(0.1\text{–}0.3)\text{ps/nm}^2\cdot\text{km}$) as the management fiber.

Table 8.5 shows some of the ways to alleviate or to avoid FWM crosstalk. Fibers with large effective core area and small dispersion are two physical approaches. The combination of these two is a practical and efficient fiber pattern.

Table 8.5 Some modes to alleviate or to avoid the FWM crosstalk.

Modes	*Physical Mechanism*	*Illustration Profile*	*Fibers*
Dispersion management	FWM is suppressed by dipersion-caused phase-mismatching.	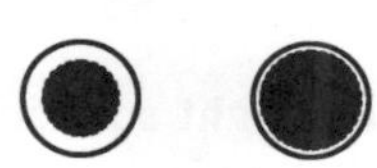	New fibers
Large A_{eff}	Lower optical nonlinearities by decreasing the light power intensity.	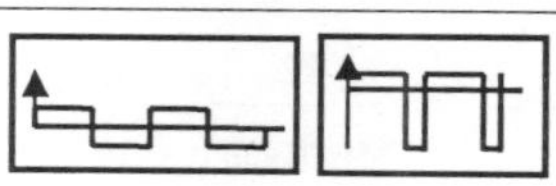	New fibers
Unequal spacing	The resultant wave to be departed from the transmission channels.	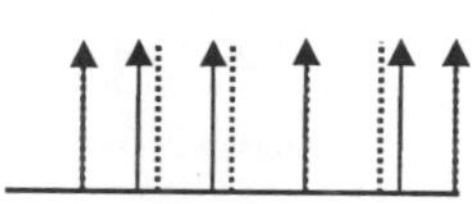	

8.4.3.2 Large Effective Area (LEA) Fiber for Lowering the Photon Intensity of the Signal Channel

Large effective area fiber (LEAF) is an approach to lower both the total signal photon intensity and the signal intensity of the separate channel of DWDM. Fiber nonlinearities are the limitation factors in the long distance, high speed, and WDM fiber communications as mentioned above. From the discussions in Sect. 3.5 of Chap. 3 and Sect. 7.2 of Chap. 7, as well as Sect. 8.4 of Chap. 8, we have the impression that the effective area A_{eff} of the fiber is very important for fiber nonlinearities and the larger effective area fiber can lessen most types of nonlinear effects.

From Sect. 3.5, the nonlinear fiber parameter is

$$\delta_{\mathrm{fiber}} = \frac{n_2}{A_{\mathrm{eff}}} . \tag{3.84}$$

The nonlinearity length is

$$L_{\mathrm{NL}} = \frac{1}{\gamma} \cdot \frac{1}{P_0} = \frac{1}{k_0} \cdot \frac{1}{P_0} \cdot \frac{A_{\mathrm{eff}}}{n_2} = \frac{1}{\delta_{\mathrm{photon}}} \cdot \frac{1}{\delta_{\mathrm{fiber}}} . \tag{3.82}$$

From Sect. 7.2, the SPM-induced nonlinear phase shift is

$$\begin{aligned} \phi_{\mathrm{SPM}} &= |U(0,\mathrm{T})|^2 \cdot \frac{L_{\mathrm{eff}}}{L_{\mathrm{NL}}} = \left[|U(0,\mathrm{T})|^2 \cdot L_{\mathrm{eff}} \cdot k_0 \cdot P_0\right] \cdot \frac{n_2}{A_{\mathrm{eff}}} \\ &= \left[|U(0,\mathrm{T})|^2 \cdot L_{\mathrm{eff}} \cdot \delta_{\mathrm{photon}}\right] \cdot \delta_{\mathrm{fiber}} \end{aligned} \tag{7.8}$$

and the SPM-induced nonlinear frequency chirp is

$$\delta\omega_{\mathrm{SPM}} = -\frac{\partial |U(0,\mathrm{T})|^2}{\partial T} \cdot \frac{L_{\mathrm{eff}}}{L_{\mathrm{NL}}} = -\left[\frac{\partial |U(0,\mathrm{T})|^2}{\partial T} \cdot L_{\mathrm{eff}} \cdot \delta_{\mathrm{photon}}\right] \cdot \delta_{\mathrm{fiber}} . \tag{7.10}$$

In the nonlinear light scattering, from Sect. 8.4.1, the permitted channel power in the SBS case is

$$P_{\mathrm{3db-limited-SBS}} = \frac{42}{G_{\mathrm{Brillouin}}} \cdot \frac{A_{\mathrm{eff}}}{L_{\mathrm{eff}}} \tag{8.10}$$

and the permitted channel power in the SRS case is

$$P_{\mathrm{3db-limited-SRS}} = \frac{32}{G_{\mathrm{Raman}}} \cdot \frac{A_{\mathrm{eff}}}{L_{\mathrm{eff}}} \tag{8.5}$$

Therefore, to decrease the nonlinear fiber parameter $\delta_{\mathrm{fiber}} = n_2/A_{\mathrm{eff}}$ will lessen the fiber nonlinearities, including the effects of the SPM, XPM, FWM, SBS and SRS, and so on. It is noticeable that the nonlinear fiber parameter δ_{fiber} is inversely propotional to the effective area of fiber A_{eff}. In the usual dispersion-shifted fiber, the effective area is always about 50μm^2.

Some designs and fabrication technologies of the large effective area fiber, which has quite different refractive index profiles from the conventional fiber, have been proposed and demonstrated [8.12]. To get a larger A_{eff}, such as increasing from 50–55μm^2 to 70–80μm^2, the increase of the MFD from 8.5μm to 10.5μm is required. In general, these cause an increase of bending loss. On that account, we have to have a large A_{eff} and to keep the bending loss low. It is effective to design and fabricate a fiber of a certain refractive index profile to make the fundamental mode strongly confined in the core region and to maintain the single-mode operation simultaneously.

Some types of the special index profile design have been proposed in recent years to produce a flat-top light-field distribution feeding the fiber core cross-section. For the same transmitted power, the light intensity in this type of fiber is substantially lower than the intensity at the cross-section center of the usual dispersion-shifted fiber. In this case, the large effective area fiber (LEAF) is about 70–80μm^2, and the nonlinear fiber parameter $\delta_{\text{fiber-LEA-DSF}} = (50/70)\cdot\delta_{\text{fiber-conven.DSF}} = 0.7\cdot\delta_{\text{fiber-conven.DSF}}$. The effects of the SPM, XPM, FWM, SBS and SRS mentioned above will be lessened.

In the long distance fiber communication, for example, the transmission-line repeater spacing $L_{\text{rept.s-LEAF}} \cong 1.2\cdot L_{\text{rept.s- conven.DSF}}$. In addition, the large effective area fiber can carry more wavelength channels in WDM systems at a higher bite rate.

At the recent OFC'2001, a kind of A_{eff} enlarged fiber with 170μm^2 was successfully exploited in a 2.56Tbit/s transmission demonstration, as mentioned in Table 8.2 above [8.28]. In this case, a 170μm^2 A_{eff} positive-dispersion fiber reduces the intense nonlinearities due to the 50 GHz channel spacing. In addition, a couple of NZDSF and A_{eff} positive-dispersion fiber segment is to play the role of a low-noise Raman pre-amplifier.

8.4.4 Extremely Broadband Raman Amplifiers

The subject of using the fiber Raman amplifiers (FRA) in the OFC has been discussed enthusiatically in the whole 1990s. A series of problems have been overcome just at the end of the 20th century and at the very beginning of this century. The FRA was a very hot topic in the recent OFC'2001, and also has become a promoting device to long-haul 10Gbit/s and 40Gbit/s transmission.

8.4.4.1 Extremely Broadband Fiber Raman Amplifier (RA) with Low Noise and High Gain. Proposal for an Optimized Located-Profile of the Multi-Pump Intensities to Flatten the Raman Gain Spectrum

The study of the extremely broadband (1480–1610nm) transmission in a single fiber, simultaneously with the S-band (1480–1530nm), C-band

(1530–1565nm) and L-band (1565–1610nm), as well as the XL-band (1610–1645nm) has progressed rapidly in recent years. The Fiber Raman amplifiers (FRA) can greatly improve the optical signal to noise ratio (OSNR) and give efficient amplification in this extremely extended broadband (such as 1480–1610nm) transmission. This implies that an extremely broadband Raman amplifier with multi-wavelength high power laser diode pumping (e.g. 10 pumps of different wavelengths come together by a WDM coupler) distributed within the range of 1395–1510nm can fit this broadband transmission [8.18 and 8.29]. Some advantages of the FRA are listed in Table 8.6.

As mentioned in Sect. 8.4.1, a single pump can contribute a comparatively high and flat gain range of only 15–20nm (about 2–2.5THz). This is much less than the gain range needed by the C plus L bands, of which the range is about 80nm (almost 10THz). The current approach is to arrange a pump-troop, so the overall gain profile is a superposition of the contributions from all the pumps.

We use here the C plus L bands as an example, the pumping band of which is about 1410–1510nm. The pump-to-pump power transferred from the long wavelength to the short wavelength due to the SRS will impact the gain-spectrum's distribution of the signal channel band of 1530–1610nm. Therefore, the optimized design of the intensity-location profile of the pump-troop will be extremely valuable. In addition, the dynamical tuning of the pump intensities in this profile is possibly one of the methods to achieve the gain controlling in the future, although it is extremely difficult at present.

Table 8.6 Some advantages of an ideal Raman amplifier proposal.

	Advantages	*Physical basis*
1	The lower trunk nonlinearities	RA supplies the distributed energy and lower the trunk power, then lower the nonlinearities
2	Lesser ASE in fiber trunk	RA shares part of the needed gain of the whole trunk and EDFAs offer only half of the gain, then ASE is lessened.
3	Higher bit rate	40Gbit/s or more higher.
4	Longer span length	Due to an improvement of the OSNR budget by the distributed Raman amplification
5	Broader gain bandwidth	RA has a broad Raman gain (3 dB) bandwidth of 50–60nm, e.g. 1515–1570nm at the 1.55μm region.
6	Extremely broad-band RA trunk	Multiple pumps with arranged wavelength-spacing density and intensity to extend the RA bandwidth of 60–100nm.

One can propose the intensity of pumps to be successively lowered or propose some other optimized located-profiles of the multi-pump intensities in a broad pump band, such as from 1395–1510nm, to offset the minimum influence of pump-to-pump power transfer via Raman scattering or Four-Wave Mixing (FWM).

In order to achieve the gain-flattened Raman amplifying, someone put 24 pumps and obtained a gain of 6.5dB with a low fluctuation of ± 0.5dB, but this is complicated. Recently there is a proposal of using a high dual wavelength pumped with long period fiber Bragg grating (LPG) to flatten the Raman gain spectrum. It achieved a gain of 16dB with a gain ripper of 1.27dB [8.21].

8.4.4.2 Raman Distributed Amplification (RDA)

In this scheme, the distributed Raman amplifier (DRA) holds both the posts of the trunk fiber and the amplifier fiber and always makes following distinctive benefits.

First, the gain bandwidth of the signal channel by the multiple-pumped DRA can stretch 75–80nm for the C plus L band, and the average gain of signal is about 8–10dB. Meanwhile, the gain variation is about 0.5–1dB and no gain equalizer for the DRA is needed. However, between different pumps, the SRS and FWM can severely degrade the performance of the transmission systems.

Second, mixing the DRA with the EDFAs makes the channel power very low. In this case, a channel power can be less than 0.1mW and the total signal power are only of a few mW in the fiber transmission. Consequently, it lessens evidently the fiber nonlinearities. In the usual pure EDFA WDM trunk, it may be several tens of mW for several tens of channels.

Third, the DRA shares the total gain with the EDFA. Even though the total gain is 30dB, the DFRA can share 15dB and the EDFA shares only 15dB. The ASE of the EDFA is much weakened and hence the system Q factor is much enhanced.

8.4.4.3 Separated Raman Amplifier (SRA) and Achieving the Flattened Gain

As different from the above DRA case, in this separated Raman amplifier (SRA) scheme, the Raman amplification fiber and the trunk fiber are separated in location.

Since the research on the SRA are much less than that on the DRA, the Rayleigh scattering impairs severely (at least a grade of magnitude) the Raman pumping efficiency for a long haul fiber span in this separated Raman amplifier (SRA) scheme (also called the discrete scheme) than in the distributed RA case.

But recently, by using a highly nonlinear new fiber and a broadened pump source, a broadband (30nm) discrete fiber Raman amplifier (FRA) with high differential gain (0.08dB/mW), low noise (< 5dB) and flat gain (< ± 1dB) has been demonstrated [8.30]. By using a broadband noise pump scheme, it eliminates the gain ripples that are often observed under the multi-wavelength laser-pumping scheme.

8.4.4.4 Problems of FRA and Challenges from Other New Amplifiers

Some of the problems needed to be solved for extremely extended broadband are listed in Table 8.7.

Furthermore, it is worthwhile to construct new kinds of Raman fiber with larger Raman scattering differential gain cross-section $\partial\sigma/\partial\Omega$, which means a higher Raman gain in (5.69) of Sect. 5.4.2. In the engineering case, the RA has to face the challenges from existing and novel amplifier components.

Table 8.7 Some problems of the fiber Raman amplifier.

	Problems	*Causes*
1	Amplifier spontaneous noise.	Spontaneous emission caused by the thermal population of photons in the ground state.
2	Signal-spontaneous beat noise.	The position of the signal wavelength related to the pump wavelength.
3	The OSNR of signal-channels degraded.	By FWM between the multi-wavelength pumps.
4	Gain limit.	Rayleigh backscattering induced by the forward ASE. Double Rayleigh backscattering induced by the signal power.
5	Signal-to-signal power transfer (even pump-to-pump power transfer in multiple pumps from the shorter λ to the longer one).	By SRS
6	Pump power loss.	By SBS

References

[8.1] Charles K. Kao and George H. Hockham: Dielectric-fiber surface waveguides for optical frequencies, *Proc. IEEE* **113**, 1151 (1966)

[8.2] Tingye Li: The impact of optical amplifiers on long-distance lightwve telecommunications, *Proc. IEEE* **81**, 1568–1579 (1993)
[8.3] I.P. Kaminow and S.E. Miller: *Optical Fiber Communication,* Academic Press, New York (1988)
[8.4] G.P. Agrawal: *Fiber-optic Communication Systems*, John Wiley, New York (1992)
[8.5] D. Marcuse and Chinlon Lin: Low dispersion single-mode fiber transmission — the question of practical versus theoretical maximum transmission bandwidth, *IEEE J. Quantum Electron* **17,** 867–877 (1981)
[8.6] A.R. Chraplyvy: Limitation on lightwave communication imposed by optical fiber nonlinearities, *J. Lightwave Technol.* **8,** 1548–1557 (1990)
[8.7] D. Marcuse: Effect of fiber nonlinearity on long distance transmission, *J. Lightwave Technol.* **9,** 121–128 (1991)
[8.8] H. Taga, et al.: In: *Postdeadline Paper,* PD-1, OFC/IOOC'93, San Jose, CA (1993)
[8.9] L.F. Mollenauer, et al.: Demonstration, using sliding-frequency guiding filters, of error-free soliton transmission over more than 20 Mm at 10Gbit/s, single channel, and over more than 13 Mm at 20 Gbit/s in a two channel WDM, *Electron. Lett.* **29,** 910–911 (1993)
[8.10] A.R. Chraplyvy: Fiber nonlinearities. In: *Tutorial Sections, OFC'93,* Paper WE (1993)
[8.11] R.G. Smith: Optical power handling capacity of low loss optical fibers as determined by stimulated Raman and Brillouin scattering, *Appl. Opt.* **11**, 2489–2494 (1972)
[8.12] Y. Liu, et al.: In: *Technical Digest, Postdeadline Paper,* PD-19, IOOC'95 (1995)
[8.13] Yili Guo, et al.: A study on stimulated Raman scattering in doped optical fibers, *J. of Tsinghua University* **21**(1), 25–33 (1981)
[8.14] Zhijian Zhang and Yili Guo: Optical buffer in an asynchronous transfer mode photonic switching system, *Optical Engineering* (USA) **35**, 1151–1152 (1996)
[8.15] Wenlu Chen, Hanyi Zhang, Yili Guo, et al.: Configuration optimization and wavelength resources distribution algorithm in WDM network, *Journal of Tsinghua University* **38**, 43–46 (1998)
[8.16] Yunsi Fei, Xiaoping Zheng, Yili Guo, et al.: A novel scheme of power equalization and power management in WDM all optical networks, *IEEE Photo. Tech. Lett.* **11**, 1189–1191 (1999)
[8.17] J.E. Midwinter and Yili Guo: *Optoelectronics and Lightwave Technology,* John Wiely, Chichester (1992)
[8.18] H. Kidorf, et al.: Pump interactions in a 100-nm bandwidth Raman amplifier, *IEEE Photo. Tech. Lett.* **11**, 530–532 (1999)
[8.19] Yongqian Liu, Sergey Burtsev, et al.: Four wave mixing in EDFAs. In: *OFC'2000*, Paper WG4 (2000)
[8.20] M. Shtaif, M. Eiselt, et al.: Crosstalk in WDM systems caused by cross-phase modulation in erbium-doped fiber amplifiers, *IEEE Photo. Tech. Lett.* **10**, 1796–1798 (1998)

[8.21] F. Koch, et al.: Broadband gain flattened Raman amplifier to extend operation in the third telecommunication window. In: *OFC'2000,* Paper FF3 (2000)

[8.22] E. Ciaramella and F. Curti: Experimental assessment of node Crosstalk limitations enhanced by nonlinear effects in all optical transport network, *IEEE Photo. Tech. Lett.* **11**, 751–753 (1999)

[8.23] M. Eiselt, L.D. Garrett and R.W. Tkach: Experimental comparison of WDM system capacity in conventional and nonzero dispersion shifted fiber, *IEEE Photo. Tech. Lett.* **11**, 281–283 (1999)

[8.24] B.T. Doshi, et al.: Optical network design and restortion, *Bell Labs Technical Journal* **4**, 58–84 (1999)

[8.25] J. Bannister, J. Touch and A. Willner: How many wavelengths do we really need? A study of the performance limits of packet over wavelengths, *Optical Network Magazine* **1**, 17–27 (2000)

[8.26] N. Ghani: Lambda-labeling: a framework for IP-over-WDM using MPLS, *Optical Network Magazine* **1**, 45–58 (2000)

[8.27] Yabin Ye, Hanyi Zhang, Yili Guo, et al.: Statistics study of routing and wavelength assignment algorithms in WDM all optical network, *Optics Communications* **185**, 315–320 (2000)

[8.28] T. Miyakawa, et al.: 2.56Tbit/s (40Gbit/s × 64WDM) unrepeatered 230km transmission with 0.8bit/s/Hz spectral efficiency using low-noise fiber Raman amplifier and 170μm^2-A_{eff} fiber, *OFC'2001*, Paper PD26 (2001)

[8.29] J. Bromage, et al.: S-band all-Raman amplifiers for 40 × 10Gbit/s transmission over 6 × 100km of non-zero dispersion fiber, *OFC'2001*, Paper PD4 (2001)

[8.30] T. Tsuzaki, M. Kakui, et al.: Broadband discrete fiber Raman amplifier with high differential gain operating over 1.65μm-band, *OFC'2001*, Paper MA3 (2001)

9. Nonlinear Optical Waveguides

In Chaps. 7 and 8, some important nonlinear effects in optical fibers and fiber-based systems have been discussed. As clearly shown in these chapters, it is the waveguiding property of an optical fiber that makes a rich variety of nonlinear phenomena become prominently observable with relatively weak optical intensity. Apart from optical fibers, there is another class of optical waveguides, which are fabricated on flat substrates, such as slab waveguides, channel waveguides, and rib waveguides. Because of their geometries, these waveguides are called planar waveguides. The study of planar waveguides and devices forms a distinct discipline, known as integrated optics [9.1–9.3]. In this chapter, some nonlinear effects in planar waveguides are discussed.

As in the fiber, a planar waveguide can confine a high intensity of light within an area comparable to the wavelength of light, over a distance that is much longer than that with a bulk material. Compared with optical fibers, planar waveguides are much shorter. However, the thin-film and micro-fabrication technology that planar waveguides are based upon is much more flexible. In addition to glasses, a wide range of crystals, semiconductors, and polymers can be used as waveguide materials, and various kinds of devices, such as X-junctions, Y-junctions, directional couplers, and Mach-Zehnder interferometers, can be fabricated with these materials. Because of these advantages, nonlinear effects that do not normally exist in silica fibers, such as the second-harmonic generation, can be observed and, in principle, be optimized relatively easily with planar waveguides. Furthermore, the application of nonlinear effects to various waveguide devices can open many new opportunities such as the all-optical switching, which has already generated much interest.

In most practical situations, the light confinement in a nonlinear waveguide in the cross-sectional plane can be assumed unchanged, as the nonlinear change in refractive index is usually small. Recently, there have been extensive studies on the effects arising from a large nonlinear change in refractive index. When the nonlinear change in refractive index is sufficiently large, the light confinement in the waveguide can be significantly modified and new classes of guided waves can be generated. Under appropriate conditions, stable spatially-invariant waves in unbound media, known as "spatial solitons", can be formed by balancing the effects of nonlinearity and diffraction. As spatial solitons can be regarded as light-induced waveguides, they can trap weak optical signals, producing the phenomenon of light-guiding-light. The interactions among spatial solitons can also generate many interesting phenomena, which can find applications in all-optical signal processing.

This chapter consists of four sections. The first two sections concentrate on nonlinear effects that do not involve modification of light confinement in the waveguide. Second-harmonic generation and nonlinear switching in directional couplers are chosen for discussion as examples of nonlinear change in frequency and phase. The third section considers effects due to a large nonlinear change in the refractive index. Nonlinear guided modes and spatial solitons are introduced, together with a description of some pioneering experiments. The concluding remarks are given in the last section.

9.1 Second-Harmonic Generation

9.1.1 Principles

The second-harmonic generation (SHG) can be regarded as a wave-mixing process, where an optical wave is mixed with itself and generates a new wave at twice the frequency through the second-order nonlinear susceptibility. The principle of the SHG in an unbound medium was discussed in Chap. 5. The same principle applies to the SHG in a waveguide except that the waves involved in a waveguide are usually guided modes of the waveguide. A review of the SHG and other second-order nonlinear effects in optical waveguides is in [9.4].

The dynamics of the SHG is, in general, governed by two coupled-mode equations, which couple the optical fields of the fundamental and second-harmonic waves. In most practical cases, the power converted from the fundamental wave to the second-harmonic wave is small and the coupled-mode equations can be reduced to a single equation, which can be solved analytically. The conversion efficiency based on this no-depletion approximation can be expressed by

$$P_2/P_1 \cong [(2\omega^2 L^2 P_1 f^2)/(c^3 \varepsilon_0 N_1^{\,2} N_2)]\ [\sin^2(\Delta\beta L/2)]/(\Delta\beta L/2)^2 \tag{9.1}$$

where

$$f = \int_{-\infty}^{+\infty}\int_{-\infty}^{+\infty} d(x,y)\psi_1^2(x,y)\psi_2(x,y)dxdy\,. \tag{9.2}$$

P_1 and P_2 are the input optical power at the fundamental frequency and the output optical power at the second-harmonic frequency, respectively, ω is the fundamental (angular) frequency, L is the interaction length, d is the nonlinear optical coefficient, $\Delta\beta = (2\beta_1 - \beta_2)$ (β_1 and β_2 are the propagation constants for the fundamental wave and the second-harmonic wave respectively), ψ_1 and ψ_2 are the corresponding normalized mode fields, and $N_1 = \beta_1 c/\omega$ and $N_2 = \beta_2 c/2\omega$ are the corresponding effective indices. The fundamental wave and the second-harmonic wave can be in different modes of the waveguide. It is clear from (9.1) and (9.2) that, to obtain a high conversion efficiency, it is necessary to have a long interaction length, use a material with a large nonlinear coefficient, optimize as much as possible the overlap integral f, and maintain the phase-matching condition (i.e., $\Delta\beta = 0$) over the interaction length.

The length of a planar waveguide can be of the order of cm, which provides a much longer interaction length than that available with a bulk material. The physical length of a waveguide is therefore not much of a problem. The key is to maintain the uniformity of the waveguide and, hence the phase-matching condition over the entire length. The attenuation in the waveguide, 0.1 to 1.0 dB/cm, may also set a limit on the useful length for the SHG and, indeed, for any nonlinear interactions.

9.1.2 Materials

Many nonlinear materials are available for use in the SHG, but not all of them can form high-quality waveguides. Fortunately, electro-optic materials such as $LiNbO_3$, $LiTaO_3$, and KTP, which have been well developed for waveguide applications, possess large second-order nonlinear coefficients. Most promising results of the SHG in waveguides have been obtained with these materials [9.5–9.12].

SHG has also been demonstrated with waveguides fabricated from semiconductors [9.13] and polymers [9.14 and 9.15]. The advantage of using semiconductor waveguides is their compatibility with AlGaAs/GaAs laser diodes, essential for making monolithic frequency doublers. Organic polymers in bulk form exhibit much larger optical nonlinearity, but it is difficult to fabricate a high-quality polymer waveguide while maintaining the optical nonlinearity. As a result, polymer waveguides often suffer from stability problems. Nevertheless, progress in polymer waveguides is rapid and looks promising for the future.

We see from (9.1) that d^2/n^3, where n is the refractive index, is a figure of merit for materials used in the second-harmonic generation. The nonlinear optical coefficients, as well as the relative figures of merit for some typical materials, are given in Table 9.1 [9.4].

Table 9.1 Figures of merit for second-harmonic generation in waveguide materials.

Material	D 1/9 × 10^{-22} (MKS)	Figure of merit d^2/n^3
$LiNbO_3$	$d_{13} = 4.6$	1
	$d_{33} = 32$	50
$LiTaO_3$	$d_{33} = 15$	12
KTP	$d_{13} \cong 4$	1.5
metanitroaniline (MNA)	$d \cong 30$	75
N-(4-nitrophenyl)-L-prolinol (NPP)	$d \cong 80$	600
DCV in PMMA	$d \cong 30$	85

9.1.3 Phase-Matching Techniques

The phase-matching condition $\Delta\beta = 0$ implies that the effective indices for the fundamental and second-harmonic waves are equal, i.e., $n_1 = n_2$. Because of the presence of dispersion in a waveguide, the above condition cannot be satisfied if the fundamental and second-harmonic waves are in the same mode.

If the waveguide supports many modes, phase-matching conditions can be easily found between two different modes [9.16]. Figure 9.1 shows the situation where the TE_m mode at the fundamental frequency in a slab waveguide is phase-matched with the TE_n mode at the second-harmonic frequency ($n > m$). To reduce the sensitivity of phase-matching to waveguide dimensions, the crossing dispersion curves as shown in Fig. 9.1 should subtend a small angle, which requires a small chromatic dispersion. Typical tolerances of waveguide dimensions are of the order of 5 nm, which is not easy to achieve [9.17]. It is possible, however, to choose the waveguide parameters so that the phase-matching condition becomes insensitive to the waveguide dimensions. Noncritical phase-matching waveguides in the form of channel waveguides have been analyzed and demonstrated [9.18].

The problem with phase-matching between guided modes of different orders is that the interference between the mode fields usually results in a small overlap integral (the factor f), and hence, a low conversion efficiency. A solution to this problem is to reduce the interference effect by using a guiding region consisting of layers of linear and nonlinear materials [9.19].

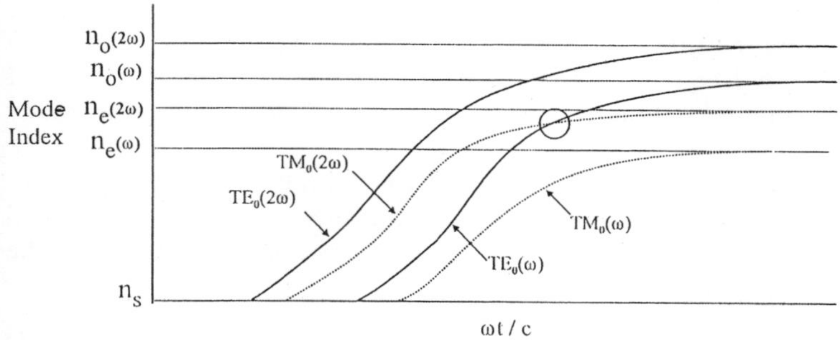

Fig. 9.1 Dispersion curves showing the phase-matching between the TE_m and TE_n modes of a slab waveguide with the film refraction index n_f, substrate refraction index n_s, and film thickness t.

The phase-matching can also be obtained between two orthogonally-polarized modes of the same order in an anisotropic waveguide [9.5–9.7]. An example is shown schematically in Fig. 9.2, where the TE_0 mode is phase-matched with the TM_0 mode in a slab waveguide. To reduce the sensitivity to waveguide parameters, it is desirable that the material birefringence is comparable to the material dispersion. The advantage of this scheme is that a large overlap between the mode fields of the fundamental and second-harmonic waves can be achieved. The problem is that the nonlinear coefficient responsible for the TE-TM mode coupling is usually significantly smaller than that responsible for the coupling with the same polarization. For example, in $LiNbO_3$, d_{31} is seven times smaller than d_{33}, as shown in Table 9.1.

It is also possible to obtain the phase-matching between a guided mode and a radiation mode (Cerenkov radiation) [9.8]. As the propagation

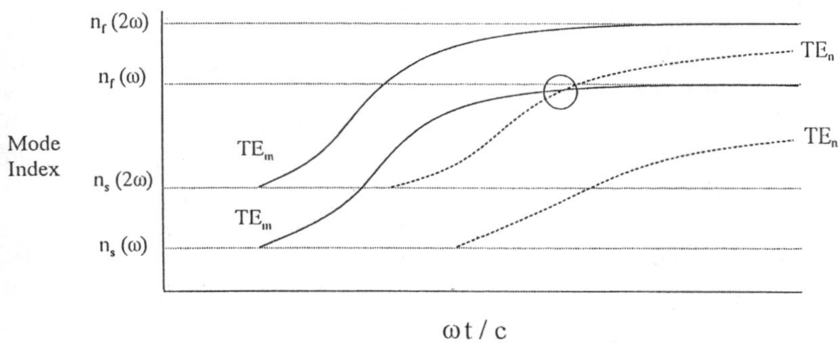

Fig. 9.2 Dispersion curves showing the phase-matching between the TE_0 and TM_0 modes of a slab waveguide with an anisotropic film. n_o and n_e denote the ordinary and extraordinary indices of the film.

constants for radiation modes spread over a continuous spectrum, the phase-matching condition is less sensitive to the waveguide parameters. It can also allow any frequency to be doubled. However, the spatial pattern of the second-harmonic output radiation is a cone, undesirable for many applications.

The most popular scheme for the phase-matching is the quasi-phase-matching (QPM) scheme, based on a periodic modulation of the optical nonlinearity (nonlinearity grating) along the interaction length [9.20]. The QPM condition is given by

$$\beta_2 - 2\beta_1 = 2\pi m/\Lambda, \tag{9.3}$$

where Λ is the period of the nonlinearity grating and m is an integer characterizing the order of the grating. The above condition guarantees that the second-harmonic waves generated in all the grating periods are in phase. As the fundamental and second-harmonic waves can be in the same mode, the QPM scheme can provide a large field overlap and make use of the largest nonlinear coefficient in the d tensor. Furthermore, by choosing the period Λ properly, the QPM condition can be satisfied at any wavelength. A comprehensive review of the technology of frequency conversion in nonlinear waveguides by QPM is in [9.21].

The performance of a waveguide frequency doubler is usually characterized by a normalized efficiency η, in % $W^{-1}cm^{-2}$, defined by,

$$\eta = 100\,|a_2(L)|^2 / |a_1(L)|^2 L^2 \tag{9.4}$$

where $|a_1(L)|$ and $|a_2(L)|$ are the amplitudes of the fundamental and second-harmonic waves at the exit port of the device. This definition measures the device's intrinsic efficiency, which is independent of length. When the factor L^2 is removed, the efficiency of a particular device is measured.

There have been numerous experimental demonstrations of the SHG in waveguides. For example, an efficiency of 4% $W^{-1}cm^{-2}$ was obtained by phase-matching the two polarized modes of the lowest order in a $LiNbO_3$ channel waveguide [9.5]. The efficiency was raised to 12% $W^{-1}cm^{-2}$ by forming a resonator for the fundamental wave [9.6], and further to 45% $W^{-1}cm^{-2}$ by forming a resonator for the second-harmonic wave as well [9.7]. By using Cerenkov phase-matching in a $LiNbO_3$ waveguide, an efficiency of 40% $W^{-1}cm^{-2}$ was demonstrated [9.8]. Comparable efficiencies were also obtained by using QPM in $LiNbO_3$ [9.9] and $LiTaO_3$ [9.10] waveguides. Particularly efficient frequency doublers in the blue region, with efficiencies of several hundred % $W^{-1}cm^{-2}$, were demonstrated with periodically-segmented KTP waveguides [9.11 and 9.12].

Despite the progress, the conversion efficiencies demonstrated so far are still considerably lower than the predicted values. The present limitation appears to be due to waveguide fabrication rather than nonlinearity.

9.2 Nonlinear Optical Switching in Directional Couplers

There is a wide range of all-optical switching devices which rely on the effect of intensity-dependent refractive index. Although in most practical situations the optical intensity in a waveguide can induce only a minute change in the refractive index, the corresponding change in the phase of the guided wave can be accumulated to an appreciable value over a long distance. It is therefore possible to achieve the all-optical switching by the control of the optical phase with the light intensity in phase-sensitive devices, such as directional couplers and various kinds of interferometers. In this section, the behavior of a nonlinear directional coupler is described, which serves as an example to illustrate the general principles and problems of nonlinear optical switching. Comprehensive reviews of various kinds of nonlinear optical switching devices are in [9.4, 9.22 and 9.23].

9.2.1 Switching of Continuous Waves

For simplicity, a directional coupler consisting of two parallel identical single-mode waveguides is considered. When the separation of the two waveguides is large, the electric field in the coupler can be expressed as a superposition of the mode fields in the individual waveguides:

$$E(x,y,z) = a_1(z)\ \psi_1(x,y) \exp(i\beta z) + a_2(z)\ \psi_2(x,y) \exp(i\beta z) \tag{9.5}$$

where ψ_1 and ψ_2 are the normalized mode fields in the two waveguides when separated, a_1 and a_2 are the slowly varying amplitudes associated with the two modes, and β is the propagation constant. Assuming uniform Kerr-law nonlinearity, the amplitudes satisfy the following coupled-mode equations [9.24–9.26]:

$$-i\, da_1/dz = Ca_2 + \gamma |a_1|^2 a_1 = 0, \tag{9.6}$$

$$-i\, da_2/dz = Ca_1 + \gamma |a_2|^2 a_2 = 0, \tag{9.7}$$

where C is the conventional coupling coefficient which measures the spatial overlap between the mode fields ψ_1 and ψ_2.

The parameter measuring the nonlinearity strength γ can be generally expressed as

$$\gamma = 2\pi n_2 \alpha / \lambda, \tag{9.8}$$

where n_2 is the nonlinear refractive-index coefficient of the medium, λ is the free-space optical wavelength, and α is the reduction factor due to the inhomogeneous cross-sectional intensity distribution. For single-mode optical fibers, α is about 0.5 [9.27]. Only the most significant nonlinear terms are retained in (9.7) and (9.8) and these equations are accurate only for well-separated waveguides [9.24–9.26].

The behavior of the nonlinear directional coupler can be characterized by a critical power P_C, given by [9.24]

$$P_C = 4C/\gamma\pi. \tag{9.9}$$

When the optical power launched into one waveguide is lower than P_C, the power can be completely transferred to the other waveguide. The coupling length, which is the shortest distance required for the complete power transfer and which has a value of $\pi/2C$ in the low-power limit, increases with the input power until the critical power is reached, at which the coupling length becomes infinite and the input power is evenly distributed in both waveguides over a long distance. When the input power exceeds the critical power, the propagation constant in the input waveguide becomes so different from that in the other waveguide that the two waveguides can be regarded as dissimilar. The behavior of the nonlinear coupler resembles that of a linear coupler consisting of two dissimilar waveguides and, therefore, a complete power transfer becomes impossible [9.28]. The response of the nonlinear coupler at different input power levels is summarized in Fig. 9.3 [9.24]. For the coupler to operate as an effective switch, the nonlinear phase-shift induced in each waveguide must exceed 4π over a coupling length [9.25].

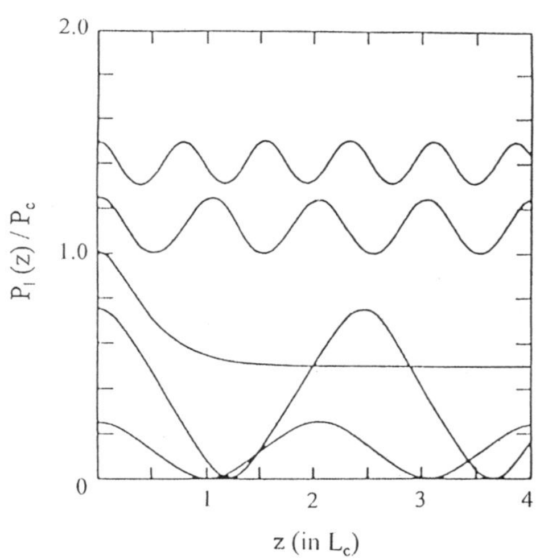

Fig. 9.3 Switching behavior of a nonlinear directional coupler when optical power is initially launched into one waveguide. $P_1(z) = |a_1(z)|^2$ is the optical power in the waveguide, P_C is the critical power, and $L = \pi/2C$ is the coupling length in the low-power limit [9.24].

The function of a nonlinear coupler can also be described with the normal-mode theory [9.29]. Although the individual waveguides that form the coupler are single-moded, the coupler as a composite waveguide structure supports two normal modes: the even mode with a symmetric field distribution and the odd mode with an antisymmetric field distribution. The electric field in the coupler can be expressed as

$$E(x,y,z) = a_+(z)\ \psi_+(x,y)\exp(i\beta_+ z) + a_-(z)\ \psi_-(x,y)\exp(i\beta_- z), \qquad (9.10)$$

where ψ_+ and ψ_- are the fields of the even and odd modes, respectively, a_+ and a_- are the slowly varying amplitudes associated with the modes, and β_+ and β_- are the corresponding propagation constants. The normal-mode equations with the nonlinearity-induced polarizations included [9.29] are given by

$$i\ da_+/dz + \gamma\,(|a_+|^2 + 2|a_-|^2)a_+ + a_-^2 a_+^* \exp(-4Cz) = 0 \qquad (9.11)$$

and

$$i\ da_-/dz + \gamma\,(|a_-|^2 + 2|a_+|^2)a_- + a_+^2 a_-^* \exp(4Cz) = 0, \qquad (9.12)$$

where $C = (\beta_+ - \beta_-)/2$ and γ is given by (9.8). It is obvious from these equations that the propagation of one mode is affected by the propagation of the other mode through the nonlinearity.

The normal-mode equations (9.11) and (9.12) are equivalent to the coupled-mode equations (9.6) and (9.7) for well-separated waveguides, when the following relations are satisfied:

$$\psi_+(x,y) = \psi_1(x,y) + \psi_2(x,y), \qquad (9.13)$$

$$\psi_-(x,y) = \psi_1(x,y) - \psi_2(x,y). \qquad (9.14)$$

In this case, the coupling coefficient C in (9.11) and (9.12) is equal to that in (9.6) and (9.7); and the solution of (9.11) and (9.12) leads essentially to the same results as shown in Fig. 9.3.

With the normal-mode theory, the switching behavior of the coupler can be explained in the following way. When the optical power is launched into only one waveguide, both normal modes are excited equally. As these two modes have different propagation constants, their phase difference varies along the coupler. Whether the power is confined in one waveguide or the other depends on whether the two modes are in phase or out of phase. The beating between the two modes thus gives rise to the optical power transfer between the two waveguides. In the presence of nonlinearity, the propagation constants of the two modes, and hence, the coupling length of the coupler, change with the input light intensity. When the input power is larger than the critical power, the induced phase-mismatch becomes so large that the two waveguides can be regarded as dissimilar and the beating between the two normal modes can no longer give rise to complete power transfer between the two waveguides. Both the coupled-mode theory and the normal-mode theory can give only qualitative results at very large power levels, where the light confinement is inevitably modified by the light intensity.

9.2.2 Switching of Short Pulses

The coupled-mode equations (9.7) and (9.8), and the normal-mode equations (9.12) and (9.13) describe the switching phenomenon only for continuous waves. In most practical applications, it is the optical pulses, rather than the continuous waves, which must be switched. A problem with the nonlinear pulse switching is that the low and high intensity portions of the same pulse are switched differently [9.30 and 9.31]. Such an incomplete switching can cause the pulse distortion and pulse breakup. By using rectangular pulses, distortion of this kind can be minimized [9.31]. The use of soliton pulses has also been proposed to overcome this difficulty [9.32], because a soliton pulse should always propagate as a single unit and, therefore, switches completely.

There is another mechanism that can cause problems in the switching of short pulses. As mentioned earlier, a directional coupler supports two normal modes. As the group velocities of the two modes are in general different, the pulses carried respectively by the two modes may "walk off" in the time domain as they propagate down the coupler [9.33–9.35]. When the group-delay difference between the two modes, i.e., the intermodal dispersion, is larger than the pulse width, the two modes are completely displaced in time and can no longer beat against each other to produce the effect of pulse switching. The intermodal dispersion in the coupler thus sets a limit on the shortest duration of the optical pulse that can be switched effectively. The intermodal dispersion in a typical two-core step-index fiber is 1 to 10ps/m [9.33]. Although it has been proposed to use a long two-core fiber to reduce the optical power to achieve nonlinear switching of ultra-short pulses [9.30], it is clear that a trade-off between the optical power and the pulse duration exists because of the presence of the intermodal dispersion.

The soliton switching in a coupler is also affected by the intermodal dispersion. In a single-mode fiber, a soliton pulse is formed by balancing the effects of the group-velocity dispersion and the Kerr nonlinearity. In a coupler, however, a self-confined soliton-like pulse may only be formed by balancing the effects of the intermodal dispersion and the Kerr nonlinearity [9.35]. Because the magnitude of the intermodal dispersion can be much larger than that of the group-velocity dispersion, the soliton power required in a coupler can be much larger than the soliton power required in a single-mode fiber [9.35]. In that case, a single-mode fiber soliton pulse is just like a low-power linear pulse, and therefore, subject to distortion by the intermodal dispersion in the coupler. Previous theoretical studies of the soliton switching in couplers [9.32, 9.36–9.39] fail to consider the effect of the intermodal dispersion, and may give unrealistic results.

9.2.3 Materials and Experiments

The theory just outlined is valid for a lossless coupler with uniform Kerr nonlinearity. In a realistic material, there always exists saturation in the nonlinear index change, as well as absorption and scattering that account for optical attenuation. To analyze nonlinear couplers with saturable index changes, numerical techniques such as the beam propagation method must be used [9.40]. The key parameter that determines the effectiveness of switching in a coupler with saturable nonlinearity is [9.41]

$$w = |\Delta n_{\text{sat}}| L/\lambda, \tag{9.15}$$

where Δn_{sat} is the saturated index change, i.e., the maximum index change that can be induced optically, and L is the length of the coupler. For a coupler with a length equal to the coupling length in the low-power limit to switch effectively, the phase induced in each waveguide of the coupler must exceed 4π. This leads to the condition $2\pi|\Delta n_{\text{sat}}|L/\lambda > 4\pi$, or $w > 2$. The switching behavior of the coupler at different values of w is shown in Fig. 9.4 [9.41]. It is clear from Fig. 9.4 that the larger the saturated index change, the lower is the required switching power, relative to the critical power.

For a given saturated index change, it is possible to increase the value of w by increasing the length of the coupler. The useful length of the device is

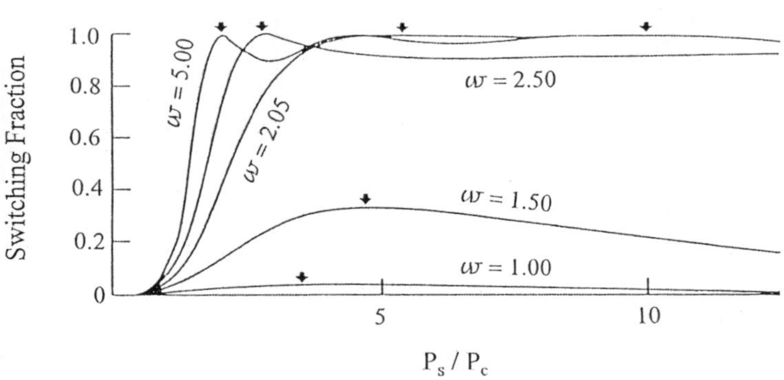

Fig. 9.4 A fraction of the switching as a function of the switching power P_S, relative to P_C, at different values of w [9.41].

limited by the optical attenuation. An excessively long device reduces not only the throughput but also the switching contrast. With the optical attenuation taken into account, a material figure of merit can be defined [9.4, 9.22 and 9.23]:

$$W = w/\alpha L = |\Delta n_{sat}|/\alpha\lambda \tag{9.16}$$

where α is the low-power attenuation coefficient. The values of W, as well as the other relevant properties for the different kinds of nonlinear materials, are compared in Table 9.2 [9.4 and 9.22]. Although silica glass has the smallest nonlinear coefficient, it has the largest W because of its exceedingly low attenuation coefficient. Low switching power and sharp switching contrast should be obtainable with optical fiber couplers. The drawback is that a very long length is usually required. On the other hand, semiconductors have the largest nonlinear coefficients. If the operating wavelength is near the material resonance, the nonlinearity can be further enhanced by several orders of magnitude, at the expense of a corresponding increase in the optical attenuation. The major problem with semiconductors is that the response is slow. Organic materials can potentially offer a good compromise between nonlinearity and optical attenuation, but the technology for the device fabrication with organic materials is still far from mature. Significant progress in the all-optical switching with organic materials has yet to be made.

Table 9.2 Figures of merit for materials used for nonlinear switching at λ = 1.0μm.

Material	n_2 m^2/W	Response Time sec	α cm^{-1}	Δn_{sat}	W
GaAlAs					
resonance	10^{-8}	10^{-8}	10^4	0.1	0.01
off-resonance (theory)	10^{-13}	10^{-8}	10	0.01	10
PTS (organic)					
resonance	2×10^{-15}	2×10^{-12}	10^5	0.1	0.01
off-resonance	10^{-16}		0.1	$>10^{-3}$	>100
SiO_2					
off-resonance	10^{-20}	10^{-14}	10^{-5}	$>10^{-6}$	$>10^3$

There have been many demonstrations of nonlinear switching with two-core fibers and semiconductor-based directional couplers. In one experiment, effective switching of 100fs pulses was observed with a two-core fused-quartz fiber as short as 5mm, but the switching power required was 32 kW [9.30]. In another experiment with pico-second pulses, a switching power of several watts was demonstrated with a 200m long two-core silica fiber [9.42]. Recently, a large nonlinear index change in an erbium-doped fiber was discovered [9.43], and a nonlinear coupler using a 2m long two-core erbium-doped fiber with a switching power as low as a fraction of 1 mW (continuous waves) was demonstrated [9.44]. On the other hand, a semiconductor-based (mainly GaAs) nonlinear coupler has a typical length of several mm, and can be fabricated with different kinds of waveguide structure, such as strip-loaded waveguides [9.45] and rib waveguides [9.46], which may incorporate quantum-well structures [9.45 and 9.47]. In many experimental devices, the semiconductors were operated near their half-band gaps, so that the nonlinear index changes were greatly enhanced without excessive increase of the absorption [9.45–9.47]. Typical switching power levels were of the order of tens of watts for pico-second pulses. The throughputs of these devices were relatively low because of the high losses. Nevertheless, more sophisticated devices or functions can be readily realized with such couplers. For example, an all-optical femto-second multiplexer [9.48] and a single-chip optical switch with two cascaded nonlinear couplers [9.49] were demonstrated recently.

9.3 Nonlinear Guided Waves

The theories outlined in Sect. 9.2 assume that the mode field distribution in the waveguide does not depend on the optical power carried by the mode. However, nonlinear changes in the mode field distribution can have significant effects on the switching behavior of a directional coupler [9.50]. A more accurate analysis of a nonlinear directional coupler should be based on finding the nonlinear guided modes of the composite waveguide structure [9.51].

In general, it is difficult to find the guided modes of a nonlinear waveguide. Exact analytical solutions have been obtained only for relatively simple waveguide structures, such as self-focusing [9.51–9.60] and self-defocusing [9.61] slab waveguides. For self-focusing waveguides, analytical solutions have been found for step [9.51–9.58], exponential [9.59], and power-law [9.60] index profiles. To handle more general nonlinearities and waveguide geometries, various numerical methods have been developed, which include, the variational method [9.62], the finite-element method [9.63 and 9.64], the inverse iteration method [9.65], the beam propagation method [9.66], the modal-expansion method [9.67], and the integration method [9.68–9.70], etc. Reviews of nonlinear guided waves in self-focusing media are in [9.71] and [9.72]. In this section, the TE modes of nonlinear step-index slab waveguides are used as examples to illustrate some general features of nonlinear guided waves. A discussion of a special class of nonlinear waves, known as the spatial solitons, is also given.

9.3.1 The Nonlinear Scalar Wave Equation

A dielectric waveguide with a refractive-index profile $n(x,y)$ is considered. We assume that the modes of the waveguide are linearly polarized and satisfy the scalar wave equation

$$\partial^2 \psi/\partial x^2 + \partial^2 \psi/\partial y^2 + [n^2(x,y)k^2 - \beta^2] = 0, \tag{9.17}$$

where ψ is the transverse electric field of the mode, assumed real, $k = 2\pi/\lambda$ is the free-space wavenumber, and β is the propagation constant. Obviously, the solution of (9.17) represents a stationary wave characterized by a transverse field distribution and a propagation constant which are independent of the time and distance propagated.

In a nonlinear waveguide, the dielectric constant can be decomposed into a linear component and a nonlinear component,

$$n^2(x,y) = n_1^2(x,y) + \alpha\varepsilon(x,y)f(\psi^2) \tag{9.18}$$

with

$$\alpha(x,y) \equiv c\varepsilon_0 n_1(x,y)n_2(x,y), \tag{9.19}$$

where n_1 is the linear index, c is the speed of light in vacuum, ε_0 is the permittivity for vacuum, n_2 is the nonlinear refractive-index coefficient, and $f(\psi^2)$ characterizes the form of the nonlinearity. In a Kerr-law or self-focusing medium, where $f(\psi^2) = \psi^2$, the nonlinear part of $n^2(x,y)$ is proportional to the local intensity ψ^2. The wave that satisfies (9.17) with the refractive-index profile given by (9.18), is a nonlinear guided mode, which carries an optical power level P given by

$$P = (\beta c\varepsilon_0 / 2k) \int_{-\infty}^{+\infty} \int_{-\infty}^{+\infty} \psi^2 (x, y) dxdy\,. \tag{9.20}$$

The wave equation (9.17) is exact only for the TE modes of a slab waveguide. Nevertheless, it provides an accurate description of the modes of a weakly-guiding waveguide, in which the refractive index of the guiding region is only slightly larger than that of a cladding region. An exact analysis of the TM modes of a slab waveguide, or the hybrid modes of a general waveguide structure, should allow for the vector nature of the modes [9.71–9.75].

According to the above formulation, a nonlinear guided mode can be understood as a guided mode of the linear waveguide induced by the mode itself. This self-consistency principle leads to an iterative process for finding the nonlinear guided modes of a general waveguide structure [9.76]. The process can be started by calculating a refractive-index profile from (9.19), making appropriate guesses for the propagation constant and the mode field, subject to a given optical power level. New approximations to the propagation constant and the mode field are calculated numerically by solving (9.18) with the refractive-index profile just calculated. The updated field and propagation constant are then used to calculate an updated refractive-index profile, and so on. The process continues until it converges to a final value for the propagation constant for a given convergence criterion. In each step only a linear inhomogeneous waveguide is analyzed and, therefore, many of the modal methods developed for linear waveguides can be applied. The iterative process just described is general and versatile but may fail to converge for unstable solutions [9.76].

9.3.2 Examples of Nonlinear Guided Modes

Many features of the nonlinear guided modes can be illustrated by studying a specific self-focusing slab waveguide [9.52–9.57 and 9.76]. The geometry of the waveguide is shown in Fig. 9.5, where a thin film with index $n_f = 1.57$ is sandwiched between two infinite cladding layers with identical indices $n_C = 1.55$. The waveguide operates at 0.515μm. Assuming that the mode field does not vary in the direction parallel to the waveguide surfaces, (9.17) reduces to the TE wave equation, and the power level P given in (9.20) reduces to the power density carried by the wave (watts per unit width of the waveguide). Three cases are considered. First, the film is linear and both claddings have the same nonlinear coefficient n_2. Second, the film is linear and the two claddings have nonlinear coefficients n_2 and $2n_2$, respectively. Third, the same as the first, except that the nonlinearity is saturable. In all cases, $n_2 = 10^{-9}\text{m}^2/\text{W}$, the value for the liquid crystal MBBA, is assumed.

For the first case where the film is linear and the claddings have identical nonlinearity, the dispersion curves for the TE_0 mode are shown in Fig. 9.6, where the mode index β/k is plotted as a function of the power density P. The results are presented for a film thickness of 1.25μm.

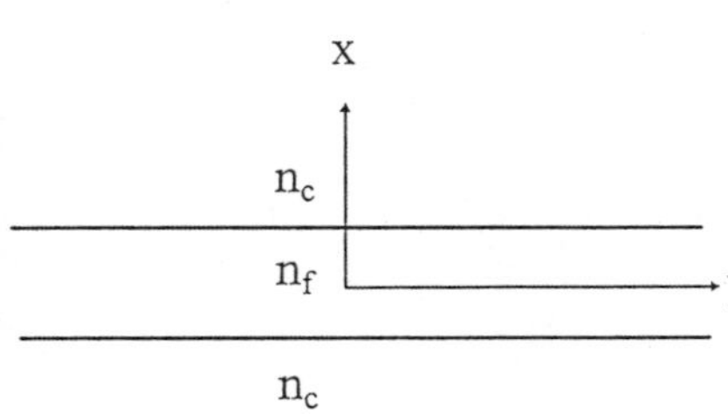

Fig. 9.5 The cross-section of a slab waveguide with the film index n_f and cladding index n_C.

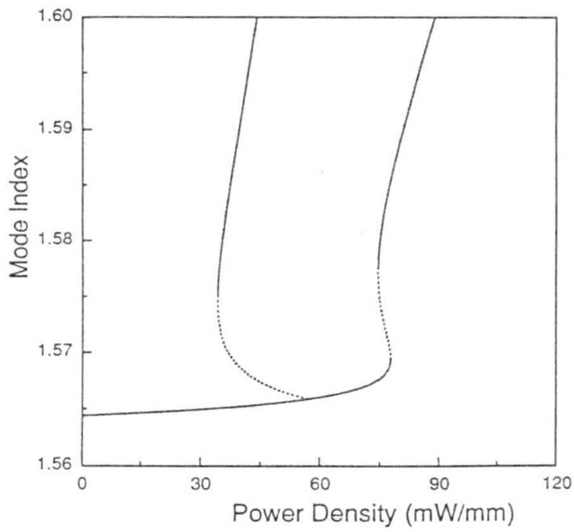

Fig. 9.6. Dispersion curves for the waveguide shown in Fig. 9.5, for the case where the film is linear and the claddings have identical nonlinear coefficients.

As shown in Fig. 9.6, there are three branches. The low-index branch extends from zero power density to a peak power density and is almost horizontal, indicating that the mode index varies little from its linear value, at $P = 0$, over the whole range of power densities. On this branch, as the power density increases, the induced refractive indices in the claddings increase and cause the peak of the field to split into two symmetric peaks, which move away from the center of the film towards the claddings. As the power density keeps increasing, the field eventually becomes localized symmetrically in the claddings. When this happens, the solution is described by the vertical branch at higher power densities. This branch is therefore a natural extension of the horizontal branch. There exists another vertical branch at the lower power densities, which bifurcates from the horizontal branch. This branch represents two degenerate solutions, each of which corresponds to an asymmetric field localized in a cladding, near the film boundary. With the same mode index, the field distribution of the asymmetric mode looks very much like the field distribution of the symmetric mode in one of the claddings. The asymmetric mode therefore carries approximately half of the power density that the corresponding symmetric mode carries. The field distributions along the various branches are sketched in Fig. 9.7. The transition regions between the horizontal branch and the vertical branches, as shown by the dotted lines with negative slopes in Fig. 9.6, become narrower as the film thickness decreases and eventually disappear completely when the film becomes thin enough.

When the film is linear and the claddings have different nonlinear coefficients, the dispersion curves for the TE_0 mode are as shown in Fig.

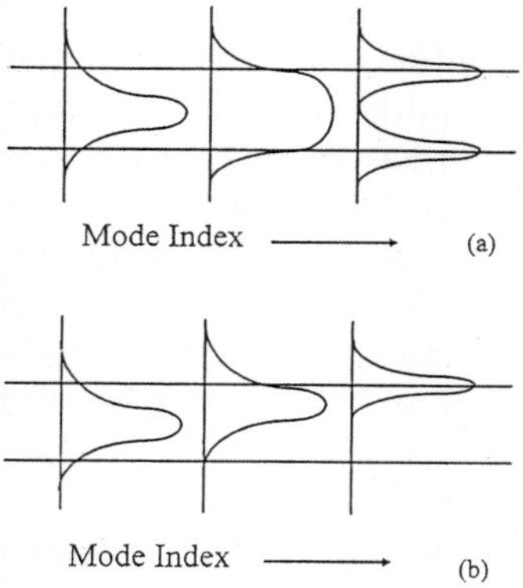

Fig. 9.7 a, b The field distributions in the waveguide with the dispersion curves shown in Fig. 9.6. **(a)** Transition from the horizontal branch to the high-power vertical branch (symmetric-field solutions). **(b)** Transition from the horizontal branch to the low-power vertical branch (asymmetric-field solutions).

9.8 for a film thickness of 2μm. Four branches exist, one horizontal and three vertical. Because the claddings have different nonlinear coefficients, the symmetry of the waveguide is broken at a finite power density and the mode field has a symmetric distribution only in the linear case, i.e., at zero power density. In any other case, the field distribution is asymmetric. The horizontal branch which extends from zero power density corresponds to a field localized in the film. On this branch, the peak of the field is shifted towards the cladding that has the stronger nonlinearity, as the power density increases. This branch therefore joins naturally to the first vertical

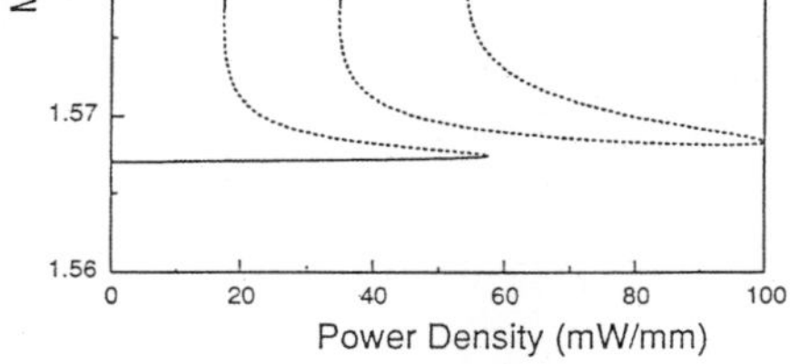

Fig. 9.8 Dispersion curves for the waveguide shown in Fig. 9.5 for the case where the film is linear and the claddings have different nonlinear coefficients.

branch which has the lowest power density and corresponds to a field localized in the cladding that has the stronger nonlinearity. The two other vertical branches are not connected to the horizontal branch. Instead, they are connected together. The second vertical branch has the medium power density and corresponds to a field localized in the cladding that has weaker nonlinearity, while the third vertical branch has the highest power density and corresponds to a field localized in both claddings. As one cladding is twice as nonlinear as the other, for a given mode index, the power density on the first vertical branch is approximately half of that on the second vertical branch, and one-third of that on the third vertical branch.

In the last case, the film is linear and has a thickness of 1.25μm, as in the first case, but the nonlinearity in the claddings is saturable and expressed by

$$n^2 = n_c^{\,2} + \Delta\varepsilon\,\{1 - \exp[(-2kn_c^{\,2}n_2P\psi_0^{\,2})/(\Delta\varepsilon\beta)]\}, \tag{9.21}$$

where $\Delta\varepsilon$ is the saturation value for the dielectric constant induced by the mode field, and ψ_0 is the normalized mode field in the sense that

$$\int_{-\infty}^{+\infty} \psi_0^2\,dx = 1$$

The dispersion curves for the TE_0 mode are presented in Fig. 9.9 for $\Delta\varepsilon$ = 0.1, 0.3, and $+\infty$. The results for $\Delta\varepsilon = +\infty$ are simply those shown in Fig. 9.6 for self-focusing nonlinearity. It is clear from Fig. 9.9 that the saturable nonlinearity limits the growth of the mode index with power density. As $\Delta\varepsilon$ decreases, the transition from a field localized symmetrically in the film to a field localized symmetrically in the claddings becomes smoother, and the

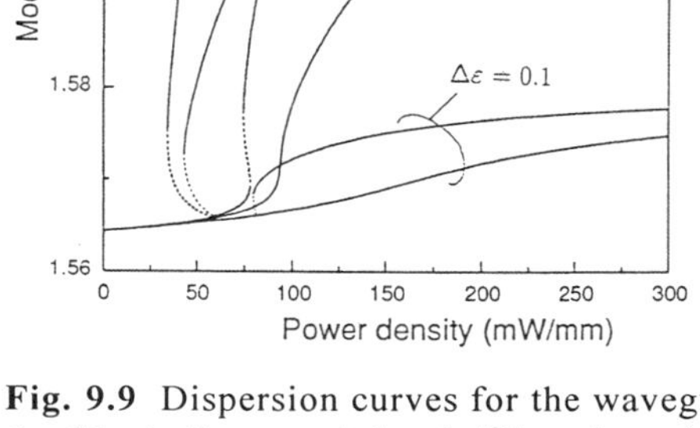

Fig. 9.9 Dispersion curves for the waveguide shown in Fig. 9.5 for the case where the film is linear and the claddings have identical saturable nonlinearities.

bifurcation point on the horizontal branch shifts to a higher power density. More detailed studies of the saturation effects can be found in [9.68–9.70].

It is clear from the examples given above that both the mode index and the field distribution depend on the power carried by the mode. When the nonlinearity becomes sufficiently large, the dispersion characteristics of even a simple slab waveguide can become quite complicated, and asymmetric field distributions can exist in a perfectly symmetric waveguide. Above a certain power level, multiple solutions, which correspond to different field distributions, may exist. These features could provide many opportunities for optical switching and bistability. Although examples are given only for the TE_0 mode of a step-index waveguide, similar features occur in higher-order TE modes [9.57], as well as in graded-index waveguides [9.59 and 9.60]. The properties of nonlinear TM modes [9.73–9.75] are like those of nonlinear TE modes.

In spite of the very extensive theoretical studies in this area, few experimental results have been reported. Nevertheless, in an experiment using a thin-film glass waveguide with liquid-crystal MBBA on the surface, nonlinear TE_0 and TE_1 modes were observed and a good agreement with the theory was obtained [9.77].

9.3.3 Stability of Nonlinear Guided Modes

It is known that in a uniform linear (lossless) waveguide, once a mode is excited, it will propagate along the waveguide and emerge at the output of the waveguide, regardless of the distance propagated. In a nonlinear waveguide, whether a particular mode can be sustained in the waveguide depends on both the launching condition and the stability of the mode. Loosely speaking, stability is a measure of how sensitive the mode is to the perturbations in the mode field. Perturbations can be due to any factors that can modify the field in the waveguide, such as the launching condition, the waveguide geometry, and the material properties. If a mode is highly unstable, it cannot be excited even under the ideal launching condition where the incident field matches exactly with the mode field. On the other hand, a highly stable mode can be excited with an incident field that deviates significantly from the mode field. The degree of stability is therefore related to the degree of perturbation that can be tolerated for the mode to be excited. Apparently, only stable modes can possibly play a significant role in practical applications.

There are two main methods for investigating stability, both of which are based on solving the propagation equation

$$2iN\,\partial\psi/\partial z + \partial^2\psi/\partial x^2 + \partial^2\psi/\partial y^2 + [n^2(x,y)k^2 - N^2]\,k^2\,\psi = 0 \qquad (9.22)$$

where N is the effective index β / k of the wave, $n(x,y)$ is the refractive-index profile given by (9.18), and $\psi(x,y,z)$ is the field. The propagation equation (9.22) describes the evolution of the field in the waveguide when the parabolic approximation is valid, i.e., $|\partial^2 \psi / \partial z^2| << 2N\, |\partial \psi / \partial z|$.

In the first method [9.78–9.80], the field is written as the sum of the exact mode field and a perturbation term in the form of an eigenfunction. The substitution of the field into (9.22) can lead to a linear equation for the perturbation term. The stability of the mode can be determined by investigating the solution for the perturbation term. If the perturbation has a solution that grows with distance, the mode is unstable. As not every form of perturbation can be represented in this way, modes cannot be definitely identified as being stable. Nevertheless, this method is particularly useful for identifying unstable modes.

In the second method, (9.22) is solved directly by a numerical propagation method [9.81–9.85] with a perturbed field as the initial condition. The stability of the mode can be decided by investigating the evolution of the field over a long distance. Again, this method is more useful for identifying unstable modes, in which case, the exact mode field is used as the initial field and the numerical errors in the calculations serve as the perturbation. As not every form of perturbation can be tested in practice, definite stability cannot be established. This method has the advantage of providing a full picture of field evolution for a given incident field, and therefore, is convenient to simulate experiments.

There are other methods for determining the stability, which are not based on the propagation equation, such as the variational method [9.86] and the iteration method [9.76], both of which are based on the stationary wave equation (9.17).

The stability of the nonlinear TE_0 modes described in the previous section has been thoroughly investigated. Solutions on all transition curves with negative slopes ($dP/d\beta < 0$), as shown in Figs. 9.6, 9.8, and 9.9, are unstable. In a symmetric waveguide, the symmetric-field solution, as represented by the horizontal branch in Fig. 9.6, is stable only up to the bifurcation point at which the vertical branch for the asymmetric mode starts. Beyond the bifurcation point, the symmetric-field solution is unstable. There is a range of power levels within which both symmetric-field and asymmetric-field solutions are present and stable. Which type of mode can evolve in the waveguide depends on the degree of asymmetry introduced to the incident wave [9.76]. The same stability criterion applies to the modes of the symmetric waveguide with saturable nonlinearity, whose dispersion curves are shown in Fig. 9.9. When the symmetry of the waveguide is broken, as shown by the dispersion curves in Fig. 9.8, there is no symmetric-field solution. The solution that corresponds to a field localized in the linear film is stable, and so is the solution that corresponds to a field localized in either cladding. The solution that corresponds to a field localized in both claddings, the vertical branch with the highest power in Fig. 9.8, is, however, unstable. This is consistent with the stability criterion for the symmetric waveguide. The stability criterion described here is applicable to the TE modes of waveguides with self-focusing nonlinearity. In a self-defocusing nonlinearity, the dispersion curves for the TE modes always have negative slopes, and yet the modes are stable [9.87 and 9.88].

Recently, a universal method of reading the stability of a solution directly from the dispersion diagram has been developed [9.80]. This method can be applied to the fundamental scalar modes of an arbitrary waveguide with an arbitrary form of nonlinearity. The stability problem of nonlinear vector modes, e.g., the TM modes of a slab waveguide, has yet to be addressed.

9.3.4 Spatial Solitons

While the discussion of nonlinear guided waves in Sect. 9.3.2 is based on a three-layer slab waveguide, it is in general not necessary to assume a waveguide structure for a nonlinear guided wave to exist. Solutions can still be sought from (9.18), even if the linear refractive-index profile $n_l(x,y)$ in (9.18) does not represent a guiding structure. This means that waveguiding can be solely induced by the wave itself through the nonlinearity. Such a wave, however, ceases to be a guided wave in the low-power limit because of diffraction. Its existence as a guided wave can be understood as the result of balancing the effects of diffraction and nonlinearity. For this reason, nonlinear waves of this kind are commonly referred to as spatial solitons. A spatial soliton is therefore a self-guided or self-trapped wave, i.e., a wave that is guided by the waveguide induced by the wave itself.

Spatial solitons of circular symmetry in a uniform self-focusing medium were identified long ago [9.89]. The effects of diffraction and nonlinearity balance out exactly at the critical power P_C given by [9.89]

$$P_C = 1.86\pi / k^2 n_l n_2. \tag{9.23}$$

Unfortunately, all two-dimensional spatial solitons in self-focusing media are unstable [9.90]. If the power carried by the wave is slightly smaller than the critical power, the wave will be diffracted away. But if the wave's power exceeds the critical power, self-focusing will occur and the wave will usually break up into filaments [9.91]. To obtain stable spatial solitons, either the dimension of the spatial solitons is reduced to one, or some other form of nonlinearity is used.

The one-dimensional spatial soliton in a uniform self-focusing medium has simple analytical solutions for the field and the mode index [9.71]:

$$\psi(x) = [\sqrt{2}(\beta^2 - n_1^2 k^2)/k\sqrt{\alpha}]\mathrm{sech}(x\sqrt{\beta^2 - n_1^2 k^2}), \tag{9.24}$$

$$\beta / k = (n_1 / \sqrt{2})\sqrt{[1 + \sqrt{1 + (n_2 k P)^2}]} \tag{9.25}$$

where P is the power density given by

$$P = \frac{2\beta\sqrt{\beta^2 - n_1^2 k^2}}{n_1^2 n_2 k^3} \tag{9.26}$$

Clearly, the field of the one-dimensional spatial solition has a form similar to that of the temporal soliton. The one-dimensional spatial soliton at a single interface between a linear medium and a self-focusing medium, known as the nonlinear surface polariton, has also been studied [9.71 and 9.72]. The field of the wave is trapped in the nonlinear medium near the interface. Since the three-layer slab waveguide discussed in Sect. 9.3.2 can be considered as a structure that consists of two such interfaces, the solution that corresponds to a field localized in either cladding, as shown in Fig. 9.7, is in fact a nonlinear surface polariton. When the thin film of a three-layer structure as shown in Fig. 9.5 has a linear index smaller than that of the claddings, it does not support any guided waves in the low-power limit. However, when the power of the wave exceeds a certain level, spatial solitons can be formed [9.55 and 9.56].

All the one-dimensional spatial solitons described above, though stable, are based on the unrealistic assumption that the field in the other transverse direction does not vary. When this plane-wave assumption is removed in analyzing a thin-film waveguide, the spatial-soliton solutions can be found in the direction parallel to the film-cladding interfaces [9.65, and 9.92–9.94]. In that case, the field in the direction normal to the film-cladding interfaces is confined by the waveguide structure in the way described in Sect. 9.3.2, while the field in the other direction, which is unbound in the low-power limit, is confined solely by the nonlinearly-induced refractive index. A combination of these two waveguiding effects produces an effective two-dimensional waveguide for the wave. Although the wave is confined in both transverse dimensions, self-guiding takes place only in one dimension. It is therefore a one-dimensional spatial soliton. In fact, the dispersion characteristics of this particular type of spatial soliton resemble those of one-dimensional nonlinear guided waves, such as those described in Sect. 9.3.2 [9.94]. Spatial solitons of this kind were observed first in a CS_2 thin-film waveguide [9.95], and later in a slab glass waveguide, using several hundred kilowatts of peak power [9.92].

The other method of obtaining stable spatial solitons uses a nonlinearity other than the self-focusing nonlinearity. Saturable nonlinearity can produce stable two-dimensional solitons with circular symmetry [9.96 and 9.97]. Under some ideal situations, such "self-induced optical fibers" can be analyzed quite simply with the well-known linear theory of optical fibers [9.98 and 9.99]. The concept of self-induced optical fibers in a uniform medium with saturable nonlinearity is illustrated in Fig. 9.10 [9.96]. It has been shown theoretically that the effective diameter of the self-induced optical fiber can be controlled by introducing the gain [9.100] or absorption [9.101] in the medium, resulting in the phenomenon of self-tapering, which could find applications in connecting incompatible optical components.

Because the spatial solitons can be regarded as light-induced waveguides, they can be used to guide weak optical signals of different wavelengths [9.102–9.104]. The spatial-soliton-induced guiding has been demonstrated with CS_2 liquid [9.102 and 9.103] and AlGaAs waveguides [9.104]. Like the temporal solitons, the spatial solitons can interact with each other and produce interesting effects. For example, two identical spatial solitons propagating in the same direction can attract each other if they are in phase, repel if they are π out of phase, or coalesce if they are $\pi/2$ out of phase, as confirmed by [9.105–9.107]. The theory of interactions of spatial solitons with different widths [9.108] or different wavelengths [9.109 and 9.110] has also been studied.

The spatial solitons described so far are "bright" solitons, which have a region of high intensity in a dark background. Figure 9.10 shows that a bright spatial soliton can only be formed in a medium whose refractive index increases with the light intensity. One can also imagine "dark" or "black" spatial solitons, which have a region of low intensity in a bright

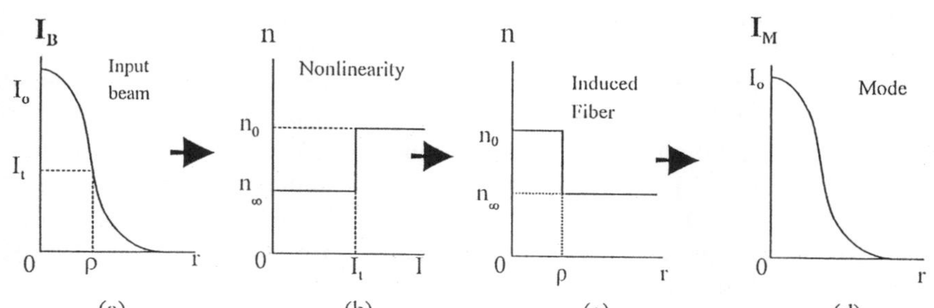

Fig. 9.10 a–d Self-induced optical fiber. An input beam (a) in a nonlinear medium with idealized saturation (b) induces a step-index optical fiber (c), which supports a mode (d). The beam is self-guided only when its intensity profile in (a) is identical to that of the mode in (d) [9.96].

background. In fact, a stable dark spatial soliton can exist in a uniform self-defocusing medium [9.111]. Dark spatial solitons have been observed in the bulk semiconductor ZnSe [9.112], and thermally defocusing liquid (chlorophyll in ethanol solvent) [9.113 and 9.114]. The formation of a Y-junction with the dark solitons [9.113] and a method of steering the dark solitons [9.114] have also been demonstrated experimentally.

The discussion so far has been restricted to the spatial solitons in the media with Kerr-like nonlinearity. In the last few years, spatial solitons in photorefractive crystals have received considerable attention [9.115–9.118]. Unlike Kerr-like solitons, the photorefractive solitons are transient by nature. It takes time for the soliton beam to induce the waveguide, and the induced waveguide persists for some time after the soliton beam is removed. Photorefractive solitons can be formed with weak visible light at a few microwatts, and can guide the intense near-infrared light on the order of watts [9.118]. More extensive discussions of this subject are in [9.115–9.118] and their references.

9.4 Concluding Remarks

In this chapter, some examples of nonlinear phenomena in optical waveguides have been described. For the second-harmonic generation and nonlinear switching with couplers, only weak nonlinearity in the waveguide is required. The effectiveness of these phenomena can usually be enhanced by increasing the interaction length. The progress in the second-harmonic generation with waveguides is expected to continue steadily, along with the continuous improvement in waveguide fabrication, and may eventually lead to practical devices. Whether the waveguide second-harmonic generators can ultimately become commercially viable also depends on the competition with solid-state visible lasers. The outlook for the all-optical switching with nonlinear couplers or similar devices is uncertain, as it is difficult to overcome the fundamental difficulties related to pulse switching. So far the most promising results, in terms of switching power, contrast, and throughput, have been obtained with the long two-core fibers, which, however, do not lend themselves to the construction of compact switching circuits. The spatial solitons do provide an innovative approach to realizing all-optical processing. To generate the spatial solitons, however, very strong nonlinearity is required, which can only be achieved with impractically-high optical power in glasses and semiconductors. With photorefractive crystals, the spatial solitons can be formed with low power, but the response is inherently slow. Polymers could offer new possibilities in all of these areas, but at this stage it is not clear whether the polymer-based waveguides could possibly reach the quality required for practical applications. For practical purposes, the materials will remain the key issue.

References

[9.1] T. Tamir (ed.): *Integrated Optics,* Springer-Verlag, New York (1985)

[9.2] D.L. Lee: *Electromagnetic Principles of Integrated Optics,* John Wiley, New York (1986)

[9.3] H. Nishihara, M. Haruna and T. Suhara. *Optical Integrated Circuits,* McGraw-Hill, New York (1989)

[9.4] G.I. Stegeman: In: A.D. Boardman, et al. (eds.): *Nonlinear Waves in Solid State Physics,* Plenum Press, New York, p. 463 (1990)

[9.5] N. Uesugi, K. Daikoku and M. Fukuma: *J. Appl. Phys.* **49**, 4945 (1978)

[9.6] W. Sohler and H. Suche. In: L.D. Hucheson and D.G. Hall (eds.): *Integrated Optics III, Proc. SPIE* **408**, 163 (1983)

[9.7] R. Regener and W. Sohler: *J. Opt. Soc. Amer.* **B5**, 267 (1988)

[9.8] T. Taniuchi and K. Yamamoto: *Digest of CLEO'86,* Paper WR3, Optical Society of America, Washington (1986)

[9.9] E.J. Lim, M.M. Fejer, et al.: *Electron. Lett.* **25**, 731 (1989)

[9.10] K. Yamamoto, K. Mizuuchi and T. Taniuchi: *Digest of CLEO'91*, Paper CPDP23, Optical Society of America, Washington (1991)

[9.11] C.J. Van der Poel, J.D. Bierlein, et al.: *Appl. Phys. Lett.* **57**, 2074 (1990)

[9.12] Z. Weissman, A. Hardy, et al.: *Opt. Lett.* **20**, 674 (1995)

[9.13] R. Lodenkamper, M.L. Bortz, et al.: *Opt. Lett.* **18**, 1798 (1993)

[9.14] G.H. Hewig and J. Jain: *Opt. Commun.* **47**, 347 (1983)

[9.15] K.D. Singer, J.E. Sohn and S.J. Lalama: *Appl. Phys. Lett.* **49**, 248 (1986)

[9.16] D.B. Anderson and J.T. Boyd: *Appl. Phys. Lett.* **19**, 266 (1971)

[9.17] Y. Suematsu, Y. Sasaki, et al.: *Quantum Electron* **10**, 222 (1974)

[9.18] X.F. Cao, J. Natour, et al. *Appl. Phys. Lett.*, **58**(1991), 2331)

[9.19] H. Ito and H. Inaba: *Opt. Lett.* **2**, 139 (1978)

[9.20] N. Bloembergen and A.J. Sievers: *Appl. Phys. Lett.* **17**, 483 (1970)

[9.21] R.V. Ramaswamy and X. Cao: *Mol. Cryst. Liq. Cryst. Sci. Technol. — Sec. B (Nonlinear Optics* **14**, 115 (1993)

[9.22] G.I. Stegeman, E.M. Wright, et al.: *J. Lightwave Technol.* **6**, 953 (1988)

[9.23] G.I. Stegeman and E.M. Wright: *Opt. Quantum Electron* **22**, 95 (1990)

[9.24] S.M. Jensen: *IEEE J. Quantum Electron* **18**, 1580 (1982)

[9.25] A. Maier: *Soviet J. Quantum Electron* **12**, 1490 (1982)

[9.26] F.J. Fraile-Pelaez and G. Assanto: *Appl. Opt.* **29**, 2216 (1990)

[9.27] A. Hasegawa: *Opt. Lett.* **5**, 416 (1980)

[9.28] A.W. Snyder, D.J. Mitchell, et al.: *J. Opt. Soc. Amer. B* **8**, 2102 (1991)

[9.29] Y. Silberberg and G.I. Stegeman: *Appl. Phys. Lett.* **50**, 801 (1987)

[9.30] S.R. Friberg, A.M. Weiner, et al.: *Opt. Lett.* **13**, 904 (1988)

[9.31] A.M. Weiner, Y. Silberberg, et al.: *IEEE J. Quantum Electron* **25**, 2648 (1989)

[9.32] S. Trillo, S. Wabnitz, et al.: *Opt. Lett.* **13**, 672 (1988)

[9.33] K.S. Chiang: *Opt. Lett.* **20**, 997 (1995)

[9.34] K.S. Chiang: *IEEE J. Quantum Electron* **33**, 950 (1997)
[9.35] K.S. Chiang: *J. Opt. Soc. Amer. B* **14**, 1437 (1997)
[9.36] J.M. Soto-Crespo and E.M. Wright: *J. Appl. Phys*. **70**, 7240 (1991)
[9.37] M. Romagnoli, S. Trillo and S. Wabnitz: *Opt. Quantum Electron* **24**, S1237 (1992)
[9.38] P.L. Chu, B.A. Malomed and G.D. Peng: *Opt. Lett.* **18**, 328 (1993)
[9.39] Y.S. Kivshar: *Opt. Lett*. **18**, 7 (1993)
[9.40] L. Thylen, E.M. Wright and G.I. Stegeman: *J. Opt. Soc. Amer. B* **5**, 467 (1988)
[9.41] S. Trillo, S. Wabnitz, et al.: *Opt. Commun*. **63**, 281 (1987)
[9.42] B.K. Nayar, N. Finlayson, et al.: *Opt. Lett.* **16**, 408 (1991)
[9.43] R.A. Betts, T. Tjugiarto, et al.: *IEEE J. Quantum Electron* **27**, 908 (1991)
[9.44] P.L. Chu and B. Wu: *Opt. Lett*. **17**, 255 (1992)
[9.45] R.Jin, C.L. Chuang, et al.: *Appl. Phys. Lett*. **53**, 1791 (1988)
[9.46] J.S. Aitchison, A.H. Kean, et al.: *Electron. Lett*. **27**, 1709 (1991)
[9.47] A.Villeneuve, C.C. Yang, et al.: *Appl. Phys. Lett*. **61**, 147 (1992)
[9.48] J.U. Kang, G.I. Stegeman and JS. Aitchison: *Electron. Lett*. **31**, 118 (1995)
[9.49] J.S. Aitchison, A. Villeneuve and G.I. Stegeman: *Opt. Lett*. **20**, 698 (1995)
[9.50] A. Ankiewicz and G.D. Peng: *J. Opt. Soc. Amer. B* **9**, 1341 (1992)
[9.51] S.Y. Shin, E.M. Wright and G.I. Stegeman: *J. Lightwave Technol*. **6**, 977 (1988)
[9.52] N.N. Akhmediev: *Sov. Phys. — JETP* **56**, 299 (1982)
[9.53] D.J. Robbins: *Opt. Commun*. **47**, 309 (1983)
[9.54] U. Langbein, F. Lederer and H.E. Ponath: *Opt. Commun.* **46**, 167 (1983)
[9.55] A.D. Boardman and P. Egan: *IEEE J. Quantum Electron* **21**, 1701 (1985)
[9.56] A.D. Boardman and P. Egan: *IEEE J. Quantum Electron* **22**, 319 (1986)
[9.57] C.T. Seaton, J.D. Valera, et al.: *IEEE J. Quantum Electron* **21**, 774 (1985)
[9.58] J.S. Jeong, C.H. Kwak and E.H. Lee: *Opt. Commun.* **116**, 351 (1995)
[9.59] R.K. Varshney, M.A. Nehme, et al.: *Appl. Opt*. **25**, 3899 (1986)
[9.60] R.A. Sammut: *Opt. Quantum Electron* **26**, S301 (1994)
[9.61] P.M. Lambkin and K.A. Shore: *IEEE J. Quantum Electron* **24**, 2046 (1988)
[9.62] Q.Y. Li, R.A. Sammut and C. Pask. *Opt. Commun.,* **94**(1992), 37)
[9.63] B.M.A. Rahman, J.R. Souza and J.B. Davies: *IEEE Photon. Technol. Lett*. **2**, 265 (1990)
[9.64] K. Hayata, A. Misawa and M. Koshiba: *J. Opt. Soc. Amer. B* **7**, 1772 (1990)
[9.65] N.N. Akhmediev, R.F. Nabiev and Y.M. Popov: *Opt. Commun.* **69**, 247 (1989)
[9.66] T.T. Shi and S. Chi: *J. Opt. Soc. Amer. B* **8**, 2318 (1991)
[9.67] T. Rozzi and L. Zappelli: *J. Lightwave Technol.* **14**, 229 (1996)
[9.68] U. Langbein, F. Lederer, et al.: *Opt. Lett*. **10**, 571 (1985)
[9.69] G.I. Stegeman, E.M. Wright, et al.: *IEEE J. Quantum Electron* **22**, 977 (1986)

[9.70] S.J. Al-Bader and H.A. Jamid: *IEEE J. Quantum Electron* **24**, 2052 (1988)

[9.71] A.D. Boardman, P. Egan, et al. In: A.D. Boardman, et al. (eds.): *Nonlinear Waves in Solid State Physics,* Plenum, New York, p. 1 (1990)

[9.72] A.C. Newell and J.V. Moloney: *Nonlinear Optics*, Addison-Wesley, Redwood City, Calif., Chap. 3 (1992)

[9.73] K. Hayata, M. Nagai and M. Koshiba: *Electron. Lett.* **23**, 1305 (1987)

[9.74] W. Biehlig, F. Lederer, et al.: *IEEE J. Quantum Electron* **27**, 238 (1991)

[9.75] Q. Chen and Z.H. Wang: *Opt. Lett.* **18**, 260 (1993)

[9.76] K.S. Chiang and R.A. Sammut: *Opt. Commun.* **109**, 59 (1994)

[9.77] H. Vach, C.T. Seaton, et al.: *Opt. Lett.* **9**, 238 (1984)

[9.78] N.N. Akhmediev and N.V. Ostrovskaya: *Sov. Phys. Tech. Phys.* **33**, 1333 (1988)

[9.79] H.T. Tran and A. Ankiewicz: *IEEE J. Quantum Electron* **28**, 488 (1992)

[9.80] D.J. Michell and A.W. Snyder: *J. Opt. Soc. Amer. B* **10**, 1572 (1993)

[9.81] J.V. Moloney, J. Ariyasu, et al.: *Appl. Phys. Lett.* **48**, 826 (1986)

[9.82] L. Leine, C. Wcchter, et al.: *J. Opt. Soc. Amer. B* **5**, 547 (1988)

[9.83] J. Ariyasu, C.T. Seaton, et al.: *IEEE J. Quantum Electron* **22**, 984 (1986)

[9.84] P. Varatharajah, A.B. Aceves, et al.: *J. Opt. Soc. Amer. B* **7**, 220 (1990)

[9.85] C.K.R.T. Jones and J.V. Moloney: *Phys. Lett. A* **117**, 175 (1986)

[9.86] R.A. Sammut, Q.Y. Li and C. Pask: *J. Opt. Soc. Amer. B* **9**, 884 (1992)

[9.87] D. Hart and E.M. Wright: *Opt. Lett.* **17**, 121 (1992)

[9.88] H.T. Tran: *Opt. Lett.* **17**, 1767 (1992)

[9.89] R.Y. Chiao, E. Garmire and C.H. Townes: *Phys. Rev. Lett.* **13**, 479 (1964)

[9.90] P.L. Kelly: *Phys. Rev. Lett.* **15**, 1005 (1965)

[9.91] V.I. Bespalov and V.I. Talanov: *JETP Lett.* **3**, 471 (1966)

[9.92] J.S. Aitchison, A.M. Weiner, et al.: *Opt. Lett.* **15**, 471 (1990)

[9.93] R.A. Sammut, C. Pask and Q.Y. Li: *J. Opt. Soc. Amer. B* **10**, 485 (1993)

[9.94] K.S. Chiang and R.A. Sammut: *J. Opt. Soc. Amer. B* **10**, 704 (1993)

[9.95] S. Maneuf, R. Desailly and C. Froehly: *Opt. Commun.* **65**, 193 (1988)

[9.96] A.W. Snyder, D.J. Mitchell, et al.: *Opt. Lett.* **16**, 21 (1991)

[9.97] Y. Chen: *J. Lightwave Technol.* **9**, 1208 (1991)

[9.98] A.W. Snyder, D.J. Mitchell, et al.: *J. Opt. Soc. Amer. B* **8**, 1618 (1991)

[9.99] R.A. Sammut and C. Pask: *Electron. Lett.* **26**, 1131 (1990)

[9.100] A.W. Snyder, L. Poladian and D.J. Mitchell: *Opt. Lett.* **17**, 267 (1992)

[9.101] M.S. Sodha, S. Medhekar, et al.: *Opt. Lett.* **19**, 1110 (1994)

[9.102] R. De la Fuente, A. Barthelemy and C. Froehly: *Opt. Lett.* **16**, 793 (1991)

[9.103] R. De la Fuente and A. Barthelemy: *IEEE J. Quantum Electron* **28**, 547 (1992)

[9.104] J.U. Kang, G.I. Stegeman and J.A. Aitchison. *Opt. Lett.,* **20**(1995), 2069)

[9.105] J.S. Aitchison, Y. Silberberg, et al.: *J. Opt. Soc. Amer. B* **8**, 1290 (1991)

[9.106] F. Reynaud and A. Barthelemy. In: D.B.Ostrowsky and R.Reinisch (eds.): *Guided-wave Nonlinear Optics,* Kluwer Academic, Dordrecht (1992)

[9.107] M. Shalaby, F. Reynaud and A. Barthelemy: *Opt. Lett.* **17**, 778 (1992)

[9.108] B. Crosignani and P. Di Porto: *Opt. Lett.* **18**, 1394 (1993)

[9.109] H.T. Tran, R.A. Sammut and W. Samir: *Electron. Lett.* **30**, 1080 (1994)

[9.110] H.T. Tran, R.A. Sammut and W. Samir: *Opt. Lett.* **19**, 945 (1994)

[9.111] A.W. Snyder, L. Poladian and D.J. Mitchell: *Opt. Lett.* **17**, 789 (1992)
[9.112] G.R. Allan, S.R. Skinner, et al.: *Opt. Lett.* **16**, 156 (1991)
[9.113] B. Luther-Davies and X. Yang: *Opt. Lett.* **17**, 496 (1992)
[9.114] B. Luther-Davies and X. Yang: *Opt. Lett.* **17**, 1755 (1992)
[9.115] M. Segev, B. Crosignani, et al.: *Phys. Rev. Lett.* **68**, 923 (1992)
[9.116] G. Duree, J.L. Shultz, et al.: *Phys. Rev. Lett.* **71**, 533 (1993)
[9.117] G. Duree, G. Salamo, et al.: *Opt. Lett.* **19**, 1195 (1994)
[9.118] M. Morin, G. Duree, et al.: *Opt. Lett.* **20**, 2066 (1995)

10. Interdiffused Quantum Wells — Optical Properties and Device Applications Part I

10.1 Introduction

Over the past decade there has been a rapid and intense development of quantum well (QW) devices for many applications including optoelectronics. Extensive reviews of the properties of QW structures extending from the basic physical principles to device applications have been published. They include topics such as the electronic and optical properties of QWs and superlattices (SLs) [10.1–10.4], physical properties of QWs for optoelectronics [10.5], digital optical devices [10.6], optical modulators [10.7], optical switches [10.8], and electroabsorption devices [10.9 and 10.10].

To-date most of the developments in QW devices have been based on symmetric square (rectangular) single wells, and in structures containing a number of such periods, i.e. multiquantum wells (MQW), also known as superlattices (SL), which lead to the formation of a large number of narrow bands within the conduction band (minibands). In this chapter, these structures are referred to as "square-wells", whilst wells whose band-edge potential varies along the growth direction are called "non-square quantum wells" (NSQW). These carrier confinement potential profiles (or simply well shapes) may be linearly or nonlinearly varied, for instance, due to a variation of the alloy composition (denoted by x in the system such as AlxGa1-xAs/GaAs) of III-V semiconductor along the growth axis, and are known as linear or nonlinear profiles. It is now accepted that square QWs represent an improvement over epitaxial materials which do not utilize the quantum confined size effect, and these QWs are promising for some electro-optic applications [10.8]. Although QWs do present the possibility of varying the carrier confinement properties by varying the well width, in many applications this parameter alone is not sufficient to produce a structure with the desired properties. Consequently more complex structures which vary the properties of the QWs are required to allow more flexibility in determining the properties of the confined states. In the past, most of the NSQWs have been produced by the application of an external electric field to a square well, which results in an asymmetric well with a parallelogram-shaped potential profile. However, recent advances in MBE growth techniques, such as atomic layer epitaxy [10.11], and pulse beam techniques, such as pulsed molecular beam epitaxy [10.12], have enabled the growth of single atomic layers of good quality. In turn, these developments have enabled NSQWs to be fabricated, where a graded composition of the alloy is created along the growth direction. A selection of such shapes are shown in Fig. 10.1. It can be shown [10.13] that, irrespective of

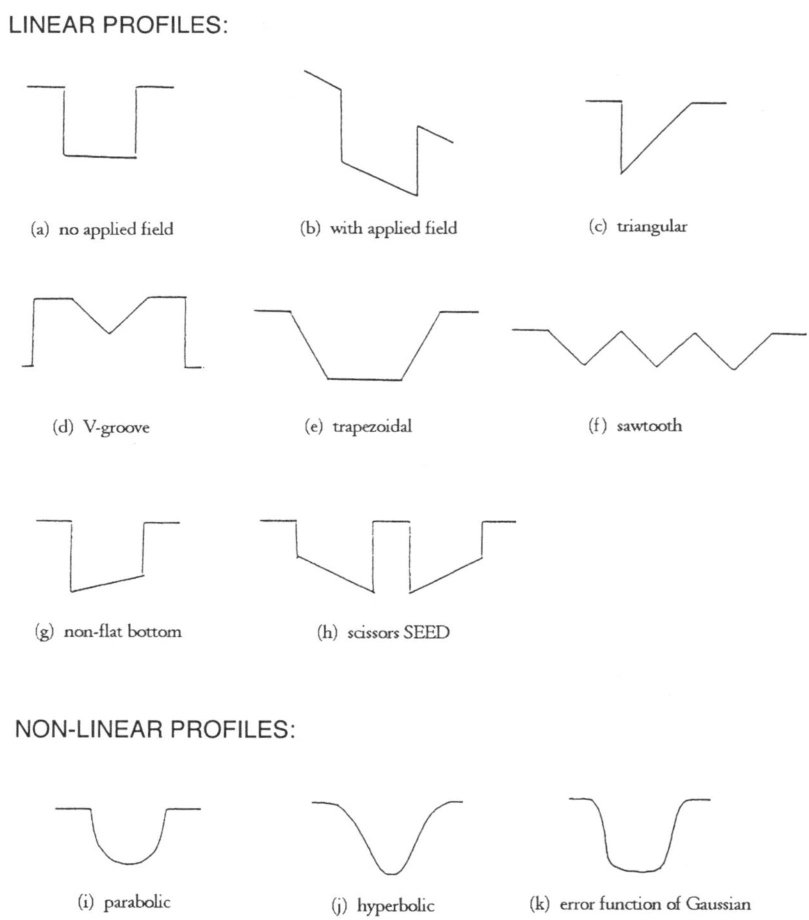

Fig. 10.1 a-k All the profiles shown are represented by the conduction band-edge. A similar band-edge will result for the valence band.

the exact spatial variations of the potential profile and the effective mass at the graded well-barrier interface, the bound state energies of the NSQW may coincide with that of a square well having an effective thickness L_{eff} which is slightly larger than the nominal QW thickness L. It is also true that the well shape does not play an important role in determining the exciton binding and emission energies so that an NSQW can effectively be replaced with a square QW of width L_{eff} and still be able to obtain a reasonably accurate value for the above energies [10.14]. However, for the NSQW, the physical properties are quite different and thus it permits the ability to modify the electrical and optical properties of the QW. For example, it is expected that the positions of the quantized energy levels of the QW are more sensitive to the profile of the NSQW than to its width and depth in the case of a square well [10.12], and therefore is a potential candidate for fast switching devices. In the presence of an electric field, the tunneling rates of the carriers into and out of the well are also extremely sensitive to the shape of the QW. In optical modulators, the tunneling rate for electrons and holes needs to be kept approximately equal in order to avoid space charge build-up and this can be accomplished by the use of an NSQW with asymmetric barrier compositions [10.14]. Consequently, the properties of these NSQWs are clearly of interest due to their ability to modify the positions of the quantised energy levels, the effective quasi-two dimensional band structure, the properties of the confined carriers, and therefore the properties of the QW structure, and may lead to interesting new device concepts.

This chapter presents a simple technique of fabricating integrated optoelectronic devices using spatially selective modification of the QW shape after standard growth, which in turn modifies the QW bandgap energy [10.15–10.17]. This diffused quantum well (DFQW) is a non-square QW produced by the interdiffusion of constituent atoms through the hetero-interface (Fig. 10.2). In some literature, DFQW is also referred to as quantum well mixing or quantum well intermixing (QWI). It is based on the fact that a QW is an inherently metastable system, due to the large concentration gradient of atomic species across the well/barrier interface. For example, in InGaAs/InGaAsP QWs, the phosphorous concentration changes from 43% to 0% in a distance of less than 1nm. At a temperature above 750°C, significant diffusion of atomic species will occur resulting in an interdiffusion of the well and barrier alloys. This process causes an increase in the bandgap energy. It can also be greatly enhanced by the presence of impurities or defects or vacancies in the vicinity of the interfaces of the QW, allowing interdiffusion to occur at temperatures that are substantially lower than that normally required. All these techniques are spatially selective, permitting interdiffusion enhancement only in the regions requiring a large bandgap, leaving other regions unmodified.

This bandgap modification is a powerful technique for monolithically integrating optoelectronic devices of varying functionalities on a single

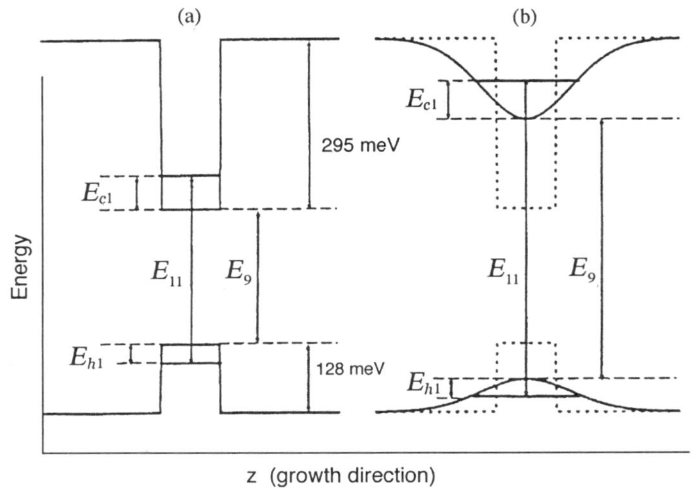

Fig. 10.2 a, b The conduction band and valence band potential energy profiles for an as-grown rectangular SQW structure and those of its interdiffused structure, where $L_d \approx ½L_z$. The tilde denotes the respective parameter in an interdiffused SQW and E_{cf} is the sum of the electron E_{c1} and the hole E_{h1} confinement energies [10.18].

wafer. It is simple and has no deleterious effects on the optical, electrical or lifetime properties of the optoelectronic devices. In fact, it possesses surprising advantages such as reducing the polarization sensitivity and enhancing the reflectance change. Since the process is easy to implement, it will simplify the fabrication of many complex photonic ICs and may finally render them practical enough to be adopted by industry. The advantage of this technology is that through interdiffusion (intermixing), the bandgap, frequency response and optical properties of the QW materials (AlGaAs/GaAs, InGaAs/GaAs and InGaAs(P)/InP) can be accruately modified, and this tuning ability can then be used to integrate several device structures (such as waveguide-detector, laser-modulator and laser-laser) to form photonic ICs [10.19]. In particular, this allows a multi-section integrated structure to be fabricated for wide band-width and multi-wavelength applications, such as the wavelength division multiplexing (WDM) in high bit rate communication systems.

Impurity-free vacancy intermixing using SiO_2 and SrF_2 dielectric caps (Fig. 10.3) to induce selective QW intermixing in the GaAs-AlGaAs system has also been developed. The intermixing rate of IFVD was found to be higher in intrinsic n-i-p than in p-i-n structures, which suggests that the diffusion of the Group III vacancy is not supported in the p-type material.

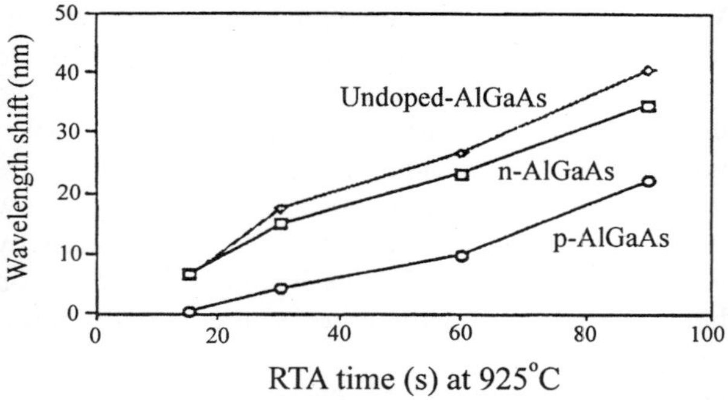

Fig. 10.3 Wavelength shifts as a function of annealing time, obtained from SiO_2 capped intrinsic n-i-o and p-i-n samples annealed at 925°C [10.20].

Single-mode waveguides have been fabricated from both as-grown and bandgap-tuned double-quantum-well (DQW) laser samples. Propagation losses as low as $8.5 dB \cdot cm^{-1}$ were measured from the bandgap-tuned waveguides at the lasing wavelength of the unintermixed material, i.e., 860nm. Simulation was also carried out to study the contribution of the free-carrier absorption from the cladding layers, and the leakage loss induced by the heavily p-doped GaAs contact layer. It was found that the leakage loss contributed by the GaAs cap layer is significant and increases with wavelength. Based on IFVD, the fabrication of multiple-wavelength lasers and multichannel wavelength division multiplexers were also demonstrated using the one-step "selective intermixing in selected area" technique. This technique enables one to control the degree of intermixing across a wafer. Lasers with bandgaps tuned to five different positions have been fabricated on a single chip. These lasers showed only small changes in threshold current, internal quantum efficiency, or internal propagation loss, which indicates that the quality of the material remains high after being intermixed. Four-channel wavelength demultiplexers based on a waveguide photodetector design have also been fabricated. The photocurrent and the spontaneous emission spectra from the individual diodes showed that the absorption edge was shifted by different degrees due to the selective degree of QW intermixing. The results obtained also imply that the technique can be used in the fabrication of broad-wavelength emission superluminenscent diodes.

Optoelectronic modulators are fundamental elements of optical communications systems, optical computers and other systems in which the electronic control of guided light waves is required. Since the mid-1980s, modulators based on semiconductor QWs have been studied extensively [10.21–10.38]. In these devices, light is coupled into an absorbing region which consists of QW layers sandwiched between barrier layers with a larger bandgap. The QW region is in turn sandwiched between p- and n-type cladding layers to which ohmic contacts are made to enable a field to be applied across the QW. The application of a reverse-bias control voltage to the modulator structure modifies the absorption properties of the QWs, hence the amount of light transmitted through the QW region, therefore providing voltage control of the output intensity. The small size, low switching power, high switching speed, and design flexibility offered by these QW devices makes them attractive candidates for optical modulation applications, as does their compatibility with other semiconductor opto-electronic devices, such as laser diodes.

The operation of QW modulators is based on an electroabsorption mechanism called the "quantum-confined Stark effect" (QCSE), which is unique to QW structures [10.39]. Electroabsorption is primarily a consequence of the fact that the application of an electric field across a QW shifts the single particle confinement energies of electrons and holes by different amounts, although a field-induced modification of electron-hole Coulomb interaction contributes as well. The net effect is an overall decrease in the energy of the fundamental optical absorption edge of the QW with applied field. This effect is useful for optical modulation since the transmission of light through a QW waveguide may be turned on and off by Stark-shifting the QW absorption edge to an energy below that of the propagating photons. The QW Fabry-Perot modulators using the QCSE in this manner exhibit reflection changes as high as 47% for an applied voltage swing of 2V [10.40].

The electro-optic effect in QWs has attracted wide attention, because of its applicability to various optoelectronic devices. The QCSE is the most notable, since strong exciton absorption occurs at room-temperature [10.41], and this demonstrates that QWs are ideally suited for use in optical switching devices, such as the self-electro-optic effect devices (SEEDs) [10.42]. Demonstrations using these devices have included arrays of SEEDs as large as 128·256 (32K elements) [10.43]. One drawback of these devices is their switching power, and at present the 32K array requires about 1 picojoule to change state. Therefore, there is a requirement to improve the performance of the electro-optic modulators to make them compatible with VLSI electronic circuits for true optoelectronic integration. The nonlinear QW shape, such as the one produced by the interdiffusion process (i.e. QW disordering) [10.44], is a promising candidate for such modification. This process has been used in the past to produce optical confinement in lasers and waveguides [10.45]. Also it has the potential to improve modulator performance, such as a lower driving voltage, increasing the ON/OFF ratio, and the ability to form two-dimensional integration [10.46]. The motivation for analyzing electroabsorption in single QW structures stems from a desire to produce high speed optoelectronic integrated circuits, which include both lasers and modulators [10.47].

One of the major issues for the design of a modulator is the ability to produce a high ON/OFF ratio with the QCSE. Since the output of the modulated signal is fed into a system for further processing, to some extent the ON/OFF ratio requirement depends upon the system needs and configuration, such as high speed, small size, and ability to integrate with other optoelectronic devices. The intermixed QW, as discussed early in the last chapter, may provide a solution to the above requirement while at the same time being compatible with the existing Si planar technology for VLSI.

The quantum confinement of carriers in semiconductor non-square QWs, which leads to the quantization of energy levels and the formation of subbands, has been presented in the last chapter. The linear and nonlinear transitions due to these subbands are of interest, as they contribute to the absorption coefficient and refractive index of the structure. An important consequence of these two-dimensional carrier confinement in QW structures is an enhancement of the exciton binding energies to the degree that clear exciton absorption features can be observed at room temperature. The associated optical nonlinearlity has been exploited for a variety of novel device applications, and the ability to selectively interdiffuse QW structures has also found application in the design of novel optical device structures. In addition, the optical characteristics of these interdiffused QW structures consist of large dipole strengths and narrow bandwidth due to their intersubband transitions, which suggest that they have highly nonlinear optical absorption, although the intersubband absorption edge occurs at far-infra-red wavelengths. These effects have given rise to numerous device developments such as low-loss waveguides, self-electro-optic devices and photodetectors.

It is intuitive that the quantized energy levels are more sensitive to non-square QW shapes than to square QW shapes. Consequently the non-square QWs are clearly of interest as promising candidates for their ability to tailor the optical properties of the structure, and may give rise to novel optical device operations. The selective induced composition intermixing of QWs using masked ion implantation or impurity vacancy diffusion is a useful technique for providing lateral photon confinement in the fabrication of integrated structures, such as the planar integration of waveguides, modulators, and lasers on a single substrate for optical integration. In spite of a large interest in this technique, limited knowledge is available about the properties of the refractive index produced by this technique at wavelength around the QW band-edge for modulator applications and at long wavelengths ($< 1.6\mu m$) for waveguide applications, as well as the change of the lateral refractive index step between the intermixed and the unintentional intermixed regions. As this lateral index step is a key parameter for the design of integrated optoelectronic devices, there is a strong need to understand its variation with the extent of intermixing. Moreover, in reports of waveguiding nature, the wavelength range of interest is often characterized around or below the active QW layer band-edge. It is expected that, from a theoretical point of view, the refractive index of the intermixed QW should strongly depend on the structure of the subband (other than just the bandgap, which is only the lowest subband-edge). This gives rise to a strong variation of the refractive index at wavelengths around and below the band-edge of the QW and may affect the refractive index at longer wavelengths. These are the main themes of this chapter.

10.2 Techniques and Mechanism in QW Interdiffusion

The intermixing of QW together with a planar selectivity has received much interest in the field of integrated optoelectronic devices as it offers a planar technology capable of modifying the band structure of the material. [10.48]

The intrinsic interdiffusion rate of III-V material QWs, such as AlGaAs/GaAs QWs, is found to be extremely low. Photoluminescence spectroscopy has been employed to determine the temperature and composition dependence of the interdiffusion of Al and Ga in AlGaAs/GaAs superlattices [10.49].

Based on the above reason, a lot of work had been done to solve this problem. It is found that the rate of interdiffusion is increased by several orders of magnitude by introducing substitutional impurities during the process of crystal growth [10.50], by ion implantation [10.51], or by diffusion [10.52], and impurity-free induced intermixing by diffusion.

The kinetics of implantation enhanced interdiffusion has been investigated by cathodoluminescence and transmission electron microscopy [10.53]. There are also many experimental data presenting the dependence of intermixing on III-V material QW structures. Ion implantation enhanced intermixing of III-V material QW structures has been studied using low-temperature photoluminescence [10.54]. Some works indicate that the ion-implantation enhancement of the diffusion length depends on ion dose but not significantly on annealing time or temperature. The influence of ion mass and dose on the intermixing of III-V material QW structures using photoluminescence and secondary-ion-mass spectroscopy techniques has been studied [10.55].

Since the thermal annealing process modifies the Ga and Al composition profiles in an AlGaAs/GaAs single QW structure due to the interdiffusion of Al and Ga, the structure is transformed from an as-grown square QW (AG-SQW) profile to a graded nonlinear profile. This composition profile will determine the confinement potential profile which confines the carriers to the QW. The quantum confinement subband states in this interdiffusion induced QW potential profile have been studied theoretically using an error function profile (EQW) [10.48, 10.54, 10.56–10.64], a hyperbolic function profile (HQW) [10.65–10.69], and a Green's function model [10.70]. The advantage is that they rely on Fick's second law of diffusion, with a constant interdiffusion coefficient, for a one dimensional diffusion process. Numerical results reported for the EQW models have been obtained using variational methods, multistep transmission resonance tunneling methods, and finite difference methods.

In the ten years since the first demonstration of the phenomenon of impurity-induced disordering (IID) [10.52 and 10.71], considerable effort has been devoted to studying and utilizing this effect for the application to integrated optoelectronics [10.72]. It is becoming increasingly apparent that this effect is an almost universal one, being observed in a wide variety of III-V alloy systems as well as being induced by a wide variety of impurity species. Studies have been carried out using surface diffusion, ion implantation and doping during epitaxial growth. A specialized conference on QW mixing was first organized in 1991 [10.73].

This process of IID may be used to produce enhanced intermixing in the presence of suitable impurities. Since the rate for this interdiffusion process may be increased significantly by the introduction of specific impurities, such as Al [10.54 and 10.74], Ga [10.53 and 10.75], or O [10.76 and 10.77] in AlGaAs/GaAs QWs. If the impurities are introduced selectively at different locations in the wafer, a refractive index step is obtained due to the different interdiffusion rates. The planar process of masked implantation of impurities is a very useful technique for the production of selected area QW intermixing and thus it may be used to fabricate two- and three-dimensional integrated optoelectronic devices, such as waveguides, lasers and transistors, on a single substrate [10.78].

IID was first used to produce an index-guided laser [10.79] by using zinc diffusion to intermix a single QW. However, in this case the major interest was in the change of the bandgap of the material when it is mixed, so that the current is injected into the active region only, thus lowering the threshold current. The change in refractive index and the index guiding was an added advantage. The first transverse mode guided laser [10.80] was reported in the same year, again, using selective impurity (zinc) diffusion-induced intermixing into the MQW structure. In this case the capping layer above the area to be intermixed was removed by dry etching using a Si_3N_4 mask. Then by using the same mask, zinc could be selectively diffused at 650°C for 40 minutes, the widths of the mask was changed to obtain different stripe widths of the laser active region. It was found that a single transverse mode was obtained for a stripe width of 3.3μm, and the refractive difference between the laser active region and the GaAlAs-buried active layer was estimated to be 0.06 (approx. 2%). This change, though small, is sufficient to produce no undulation in the transverse mode profiles. However, in using this system, the MQW layers to be intermixed have to be relatively close to the surface, since the source of zinc for the diffusion mechanism comes from the surface. More recently, an integrated external cavity single QW laser has been fabricated by selective IID [10.81]. An active and a passive (intermixed) section were formed to achieve low resonant losses in the cavity due to a blue shift of the bandgap in the passive section, which is transparent at the lasing wavelength of the active section.

The first buried 3-D waveguide was produced by ion implantation of the impurities at high energy (keV to MeV) using masks [10.82]. Waveguide stripe bends were produced by implanted zinc impurity-induced intermixing during thermal annealing. Buried stripe waveguide bends with angles varying from 0° to 20° in increments of 2.5° were fabricated, and by studying the modes guided by the structures, the refractive index difference was estimated to be approximately 0.9%. It was not possible to determine the true index difference because of the complicated (non-rectangular) shape of the guide produced by the mixing process. The loss in the guide was also found to be low. It is described as "semi-transparent". This reduction was thought to be due to mainly the increased 3-D optical confinement within the waveguide.

There is some dispute about the number of modes that can be supported by a 3-D waveguide produced by IID. It was found that only TE modes were supported in a 100 period 100Å GaAs/ 100Å $Al_{0.3}Ga_{0.7}As$ MQW structure [10.83]. TEM was used to find that the intermixing process did not produce simple rectangular waveguides, where intermixing was found not to occur throughout all the layers in the QW structure, but only at a depth between 0.1 to 0.5μm. This structure was found not to support the TM modes at 1.15μm irrespective of the stripe width used, although the TE modes were found to guide light for stripe widths down to 3μm. In contrast, a 29 period 180Å $Al_{0.24}Ga_{0.76}As$ and 120Å GaAs MQW intermixed by Si ion implantation and thermal annealing was found to guide for both TE and TM fundamental modes at a wavelength of 1.15μm [10.84]. However, the differences in the guiding properties may be due to the variations in the many different process and material parameters. They include the fundamental differences in the layer structure, the thickness and aluminium concentration in the barriers (possibly resulting in the more complete mixing in the case of [10.84]), and the different annealing conditions or the different dopant levels.

Another structure, which contains a 75Å GaAs/80Å $Al_{0.35}Ga_{0.65}As$ MQW with six wells and five barriers between two 650Å $Al_{0.35}Ga_{0.65}As$ cladding layers, was used to measure the refractive index stop produced by mixing [10.85]. This structure was intermixed using silicon diffusion from the surface. The lateral index step was calculated by studying the lateral mode cut-off for higher order modes versus waveguide width at 1.06μm and at a small range of wavelengths center ed at 875nm. From this study, the refractive index step was calculated, using a rectangular waveguide model, and it was found that the lateral refractive index step depended strongly on the wavelength, increasing with energy towards the bandgap. Also the birefringence of the MQW layer can lead to a situation that produces lateral guiding only, and the index step is always larger for the TE modes than the TM modes. During the anneal cycle, if an As overpressure was used in intermixing the QWs, a lower refractive index step was produced which can be attributed to the absence of impurity dopants or the incomplete intermixing of the QWs.

Selective intermixing can also be achieved by impurity-free vacancy diffusion (IFVD) [10.86 and 10.87], using, for example, an SiO_2 cap onto the sample of an AlGaAs/GaAs QW structure during rapid thermal annealing [10.86–10.93].

The basic mechanism is to have Ga diffused out into the SiO_2 cap, generating vacancies which diffuse through the structure and enhance the intermixing at the well barrier interface. The advantage provided by IFVD is the absence of impurity, which is more ideal in optical devices. The process also creates a blue shift in the band edge for an AlGaAs/GaAs single QW, while the room temperature exciton remain resolvable. The IFVD process has been applied to a reflection modulator [10.93], which enables the production of both normally-off and normally-on asymmetric Fabry-Perot modulators from the same wafer for optoelectronic integration.

10.2.1 Impurity Induced Disordering (IID) by Ion Implantation

Ion implantation, as it has evolved in practice for IC fabrication, involves the direct injection of energetic, charged atoms or molecules to form precisely delineated regions of selectively doped Si, GaAs or other substrate material. The practical values of ion energies extends from 20 to 200keV with implant doses ranging from 10^{10}ions/cm^2 to more than 10^{16}ions/cm^2.

The growing use of ion implantation processing, in the face of the complexities of the process and the cost of the operating equipment, is based on the need to exploit the unique capabilities of ion implantation for controlled doping levels in increasingly tighter spatial limits. The reliance on ion implantation for device doping will be virtually complete for the fabrication of VLSI geometries. The advantages of ion implantation processing has been discussed in many review articles [10.94–10.98]. These advantages include:

(1) *Accurate doping over a concentration range form 10^{14} to more than 10^{21}atoms/cm^3*. This doping can be achieved for concentrations far in excess of solid solubility limits. Implants can be made through barrier

layers (such as thin oxides and nitrides) which block the diffusion from surface layers, thus protecting the device from contamination by unwanted impurities. In addition, direct monitoring of the accumulated charge delivered to the wafer provides a real-time feedback on the dopant levels. This is essential for accurate and reproducible high-volume IC processing.

(2) *A direct influence on the dopant profile and the junction depths through the control of ion energies, molecular species and dielectric screen layers over the Si material.* With the evolution of IC circuitry to VLSI dimensions and correspondingly shallower junction depths, the control on the dopant profiles becomes a paramount advantage of implantation over traditional chemical doping methods. The parallel development of techniques for "rapid annealing" with minimal dopant diffusion further enhances our ability to obtain well-controlled, steeply-graded dopant profiles [10.99].

Ion implantation processing has been, therefore, gradually replacing the traditional methods of diffusion doping in all areas of IC fabrication. During the process of ion implantation, atoms or molecules are ionized, accelerated in an electrostatic field, and implanted into a solid. In this way, a great many combinations of ions and substrates are possible, such as p-type ion (Zn, Be), n-type ion (Si), neutral-type ion (O), or constituent ion (Al, Ga, As) on AlGaAs/GaAs. The acceleration energy can be from a few thousand electron volts (keV) to several million electron volts (MeV). The penetration depth of the ions depends not only upon the energy, but also upon the mass of the ions and the atomic mass of the solid.

On the other hand, it has been shown that the presence of impurities, such as Si, O, B, and F, induces the intermixing of QW structures by the interdiffusion of the layers at temperatures below which they are thermally stable. This opens up new possibilities in bandgap engineering and device fabrication when photolithographic masking techniques are used for selected-area processing. Depending on which atom diffuses during intermixing, the spectral absorption curve will move to either longer or shorter wavelengths. This technique may be used to modify the effective bandgap and refractive index of the QW structures [10.100–10.102]. For example, selected-area intermixing may be used to provide the lateral confinement in a stripe optical waveguide, as shown in Fig. 10.4, where the QW regions form the guiding layer and the region for electroabsorption modulators.

This process also retains the planar structure and the advantages of the planar processing technology for the fabrication of 3-D devices, such as buried-heterostructure QWL diodes and optical waveguides. The intermixing process also allows the perpendicular penetration to the QW layers, so that devices with three-dimensional carrier and photon confinement can be produced using photolithography. Consequently it is one of the few present

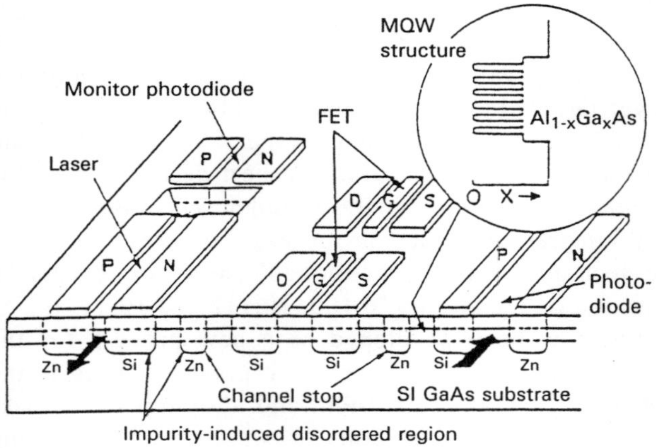

Fig. 10.4 Proposed planar compatible OEIC structure with MQW commonly used for constituent devices [10.45].

day processes that has the potential to fabricate QWW or QWD structures [10.103].

In the early 1960s, the theoretical bases regarding the range distribution of low-energy ions in solids were established in Denmark by Lindhard, Scharff and Schiott [10.104 and 10.105], proceeding from the work of Bohr [10.106].

In the theory of Lindhard, Scharff and Schiott [10.107], a statistical model has been used to arrive at the integrals for moments of the range distribution. In many practical calculations [10.108], the first and the second moment of the range are taken into account, and the Gaussian form is assumed as the range distribution. The concentration along the depth of the structure is found to be a exponential function.

Because of the recent developments of lithography techniques, allowing the fabrication of implantation masks in the submicrometer and even in the nanometer range, it is necessary to estimate the lateral spread of implanted ions [10.109]. The mask's edge is assumed to be infinitely steep and the mask is assumed infinitely high to stop all ions from penetrating. It has been shown that the lateral spread of the implanted ions leads to a considerable lateral extension of the distribution of the implanted ions beneath the mask. The actual distribution function along the lateral direction is expressed approximately by the complementary error function at the window edge, in case the ions are implanted through the mask-window.

An improvement of this method was later proposed by Runge [10.107] by taking into account the effect of the shape of the mask edge on the two-dimensional distribution of implanted ions. Barthel, Lorentz, and Ryssel [10.110] consider the possibility of having different materials form the substrate and the masking layer. Models for the description of implantation profiles in the multilayer targets have been compared [10.111]. It is found that all analytical models have severe limitations, especially if the different layers possess very different stopping powers. The best description of the multilayer targets is obtained by a combination model. This uses density scaling in the range and integration to account for mass conservation.

In addition to the technical reports mentioned, there are textbooks dealing exclusively with ion implantation [10.100, 10.112 and 10.113].

Ion implantation have now been developed as a new process for the doping of semiconductor crystals. The most common processes used in the past have been doping during epitaxy, diffusion, and alloying. In the case of epitaxy, layers having the desired doping are grown on a wafer. During the diffusion process, doping atoms are diffused from the surface into the semiconductor. In the case of alloying, the semiconductor layer nearest to the surface is dissolved and doped during recrystallization.

However, the use of ion implantation is not restricted to the field of semiconductors. It is also being applied to other areas of physics and chemistry of solids, though efforts here are still in the research stage. With the development of high current implantation machines, it has become possible not only to dope semiconductors, but also to alter the chemical, mechanical and optical properties of solids. Using implantation with high doses, a transformation of the layer near the surface into another chemical compound is possible. Electrochemical surface passivation by ion bombardment is possible for several metals. Other areas of application are superconduction, hardening of metals, fabrication of light guides by implantation into quartz and GaAs, and investigations in nuclear physics. Thus, a wide new field for the application of ion implantation has been opened.

10.2.2 Impurity Diffusion-Induced Disordering (DID)

Studies of the diffusion of zinc into superlattices have yielded quite remarkable results on the impurity-induced disordering by diffusion. It was found that the presence of the diffusing zinc increased the interdiffusion between AlAs and GaAs by several orders of magnitude [10.114–116]. By masking the whole surface of a sample except for a strip, the zinc could enter the material through this 10μm strip. After a ten minute anneal at 575°C, the region under the strip became quite different from the rest of the sample. Under the mask the layers remained distinct, while directly under the strip where zinc diffusion had occurred, the sample became a homogeneous $Al_wGa_{1-w}As$.

Investigation on the effect of varying the layer thickness have been conducted [10.117]. AlAs/GaAs superlattices were produced by MBE with periods 320Å, 630Å and 1100Å. A period is the combined width of one AlAs layer and one GaAs layer. In all cases, the total superlattice thickness was 1.8μm. Zinc diffusion was carried out at 550°C, 575°C and 600°C. The intermix was determined using AES and the zinc diffusion profile was plotted by SIMS. It was convenient to grade the degree of intermix into three categories: "total", "intermediate" and "slight". As expected, the intermix is greater at greater diffusion temperatures. It is interesting, however, that the specimens of the smaller period showed a greater degree of intermix at the end of a diffusion. This suggests that the layer interfaces play an important role in the effect. Some basic techniques were used to extract the approximate values of interdiffusion coefficient from the AES data. The coefficients in the region of $10^{-16}cm^2s^{-1}$ were obtained in the temperature range employed, and the values were a little larger in the specimens of the smaller period. Similarly, the diffusion coefficients for the zinc were calculated from the SIMS results. Again the evidence suggested that the diffusion was slightly greater in the finer structures. Comparing the two sets of experimental data indicated that the intermixing of the superlattice sets in when the zinc concentration rises to about $10^{18}cm^{-3}$ [10.118].

A variation on the above experiment has been carried out with unexpected results [10.119]. Using MBE specimens of alternate layers of GaP and $GaAs_{0.4}P_{0.6}$, a strained-layer structure was created. The samples all consisted of 40 periods of layer width 120Å. Zinc was diffused through a narrow opening in a Si_3N_4 cap for 14 hours at 825°C. Mixing occurred under the strips but not elsewhere, just as in the AlAs/GaAs samples described above.

Zinc occupies the group III site in the III-V compounds and it is perhaps not too surprising to find its diffusion interacting in some way with the self-diffusion of group III atoms. However, it is not easy to see why it should interfere with the self-diffusion of group V atoms. Matters were further complicated by research [10.120], suggesting that the intermixing effect does not occur if the superlattice layers are doped with certain atoms. Prepared structures consisting of 100Å layers of beryllium-doped GaAs were alternated with 90Å layers of tin-doped $Al_{0.3}Ga_{0.7}As$. The doping level in each case was about $2\times10^{17}cm^{-3}$. Zinc was diffused into the superlattice for 30 minutes at 600°C. Observation under an electron microscope revealed that the layers remained quite distinct after the diffusion.

A number of groups have reported that silicon acts in much the same way as zinc in initiating DID. Silicon was diffused into a superlattice structure consisting of 40 periods of 280Å GaAs and 320Å $Al_{0.6}Ga_{0.4}As$ layers [10.121 and 10.56]. Si_3N_4 was used to mask the surface leaving 15μm strips, through which silicon could enter the sample. The diffusion was at 850°C for ten hours. At the end of this time, the layers were mixed under the strip but unaffected elsewhere, as with the analogous zinc experiment.

A slightly different experiment has been performed in which the layers were uniformly doped with silicon during growth [10.50 and 10.122]. After growth, the superlattice structure was perfect, but heat treatment for two hours at 850°C induced the mixing of the layers if the silicon concentration was greater than about 4×10^{18}cm^{-3}. This experiment established that there is no need for a concentration gradient to exist for the intermixing effect to be observed. Silicon was also grown into the specimens [10.123] producing a fairly complicated structure consisting of an AlAs/GaAs superlattice grown on a relatively thick silicon-doped layer of $Al_{0.3}Ga_{0.7}As$. In this case, the silicon did diffuse during growth and intermix the superlattice. The growth temperature of 700°C was rather higher than is normally used in MBE, however.

As mentioned, there is evidence that beryllium can inhibit intermixing induced by zinc diffusion. There is even stronger evidence that it inhibits the silicon-induced effect. Kawabe, et al. [10.124] grew AlAs/GaAs superlattices using MBE with all layers 150Å in thickness. They found that doping the structures with beryllium during growth and then heat-treating did not bring about mixing. Further structures were grown with undoped, silicon-doped and silicon-plus-beryllium-doped sections. These were then annealed at 780°C for two hours. The layer structure was monitored after the anneal using AES profiling. It was found that the undoped section did not intermix and the silicon-doped section did, in agreement with the previous work. The section doped with both elements intermixed if the beryllium doping was less than about 10^{19}cm^{-3}. For higher concentrations of beryllium, the mixing was suppressed. Similar results were obtained [10.125] when a model was put forward for the inhibiting action based on the mechanism for silicon diffusion in GaAs [10.126]. According to this model, the diffusion of silicon occurs via the substitutional exchange of $(Si_{III}Si_{V})$ pairs with vacancies. It was suggested that beryllium, an acceptor occupying the group III site, forms donor-acceptor pairs with silicon of the form $(Be_{III}Si_{III})$. This prevents the formation of the silicon pairs and reduces the normal silicon diffusion, thereby inhibiting the silicon-induced mixing effect.

At present there is conflicting evidence as to whether donors occupying group V sites in the superlattice can cause group III intermixing. Researchers have diffused sulphur into $Al_wGa_{1-w}As$/GaAs structures at 850°C from a vapour source and reported mixing of the lattices [10.127]. The intermix was detected by photoluminescence. A reduction in luminescence wavelength after diffusion was taken to indicate mixing. The same method has been employed by a number of workers but, in general, it has been used in combination with other techniques, such as AES, TEM, etc.

Both silicon and sulphur have been ion-implanted into AlAs/GaAs superlattices and then heat-treated the samples for two hours at 800°C [10.128]. DID was found in the silicon-implanted specimens, but not in those containing sulphur. On the other hand, when a wide range of atoms were implanted, including sulphur, into the same structure, DID was reported in all specimens after annealing [10.129]. The problem with using ion-implantation rather than simple diffusion to study DID, of course, is that it is difficult to separate diffusion effects from those due to damage.

When DID is used to introduce selectively intermixing into a superlattice structure by partially masking the surface of a sample against diffusion, the region of superlattice under the masked area remains ordered, and the rest become intermixed. The procedure of producing strips of ordered material into an intermixed matrix has considerable potential for use in opto-electronics. The difference in refractive index between the ordered and the intermixed material, for instance, gives rise to the use of the strips as waveguides and as the active regions in QW lasers [10.80, 10.120 and 10.130] In the latter case, the electrical barrier between the ordered and the intermixed portions also acts to confine the carriers.

An example of the use of DID in forming the active region of a multi-quantum-well (MQW) laser is shown in Fig. 10.5. The layers, grown by MBE, were a 3μm $Al_{0.35}Ga_{0.65}As$ cladding layer, doped with beryllium, a 1μm $Al_{0.3}Ga_{0.7}As$ optical guide layer (also beryllium-doped), an undoped MQW active layer (composed of five 80Å GaAs wells separated by four 120Å $Al_{0.3}Ga_{0.7}As$ barriers), a 1μm $Al_{0.35}Ga_{0.65}As$ cladding layer (silicon doped) and a 1μm GaAs cap layer. Zinc was diffused into this structure, leaving a 4μm stripe unaffected by using a Si_3N_4 mask. The active MQW is simply and controllably defined by the edge of the zinc diffused region. Recently, researchers [10.85] have speculated on the possibility of creating three-dimensional structures in superlattices by creating some desired pattern using DID and then subjecting the sample to further layer growth, followed by more DID.

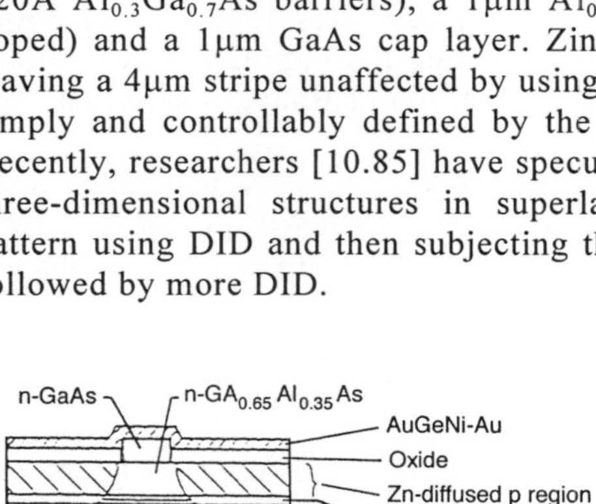

Fig. 10.5 Multi-quantum-well laser fabricated using diffusion-induced disorder (from Fukuzawa, et al., 1984, reproduced by permission of the American Institute of Physics).

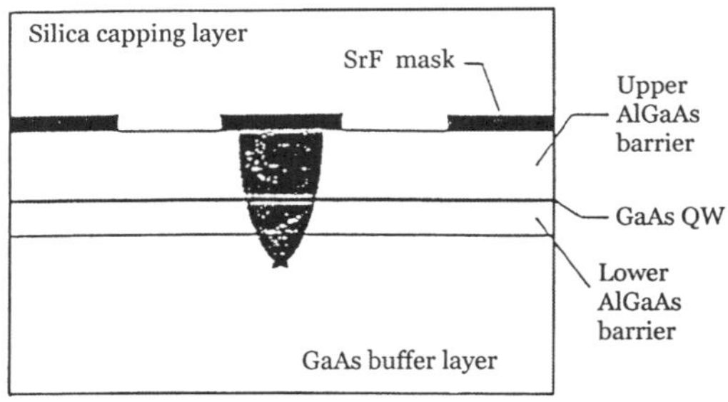

Fig. 10.6 Schematic diagram of the vacancy diffusion in SrF_2 masked single quantum well material [10.161].

10.2.3 Impurity-Free Induced Disordering by Diffusion

An alternative method that can produce the selective intermixing is impurity-free vacancy diffusion (IFVD) [10.131]. Figure 10.6 is a schematic representation of vacancy diffusion in SrF_2 masked single QW material.

This method involves the use of an electron-beam to evaporate the deposition of a SiO_2. Gallium outdiffuses into the vacancies which is generated by the SiO_2 cap. The vacancies diffuse through the structure and enhance the intermixing at the well/barrier interface. There is a patterned SrF_2 (strontium fluoride) mask for preventing the outdiffusion of gallium between the QW material and the SiO_2 cap, via photolithography with wet chemical etching.

The advantage of using IFVD is that while it causes a blue shift in the band-edge, it retains the inherent optical and electrical properties of the MQW structures [10.97], such as resolvable room temperature excitons and QCSE [10.132].

For the Al-Ga interdiffusion mediated by group III vacancies,

$$D = fD_V\,[V] \tag{10.1}$$

where D_V and $[V]$ are the group III-vacancy diffusion coefficient and concentration respectively, and f is the correlation factor [10.133]. The thermal equilibrium between the As overpressure and group III-vacancies with valence –m is described by the reaction

$$V^{-m} + mp \leftrightarrow \tfrac{1}{4}As_4\,. \tag{10.2}$$

Applying the law of mass action, the group III-vacancy concentration is given by

$$\left[V^{-m}\right] = K(T) p_{\mathrm{As_4}}^{\frac{1}{4}} \left(\frac{n_i}{p}\right)^m \tag{10.3}$$

where $K(T)$ is a temperature dependent equilibrium constant, and p and n_i are the hole and intrinsic carrier concentrations respectively. The dominant charge state in intrinsic and n-type GaAs has been theoretically determined to be $m = 3$ [10.134–10.137].

For Al-Ga interdiffusion mediated by group III vacancies, the activation energy is equal to the sum of the group III-vacancy formation and migration energies. The formation energy of ionized group III vacancies is generally calculated to be between 2.6 and 3.0eV in intrinsic and stoichiometric GaAs [10.134, 10.135 and 10.137]. Others have calculated it to be 1.6eV [10.138] and approximately 4eV [10.136]. The migration energy of group III vacancies has been estimated to be greater than 4eV for the first nearest-neighbour hopping and approximately 2.5eV for the second nearest-neighbour hopping [10.138 and 10.139]. The larger migration energy required for the first nearest-neighbour hopping is due to the creation and annihilation of antisites, whereas for the second nearest-neighbour hopping the group III vacancy migrates solely within the group III sublattice. (The group III-vacancy migration energy is equivalent to the activation energy of group III-vacancy diffusion, which has been experimentally determined in GaAs and AlGaAs heterostructures to be 2.1 or 2.7eV [10.140 and 10.141].) Therefore, the activation energy for Al-Ga interdiffusion should be at least 7eV when mediated by group III-vacancy first nearest-neighbour hopping in intrinsic and stoichionetric, but only 5eV is needed when due to the second nearest-neighbour hopping.

10.2.4 Interdiffusion Mechanism

The interdiffusion process modifies singles QW structure from an as-grown square QW profile to a graded nonlinear profile. The composition profile determines the potential profile which confines the carriers to the QW. The quantum confinement subband states in this interdiffusion induced QW potential profile have been studied theoretically using an error function profile (EQW) [10.142], [10.143], [10.144], [10.145], [10.146] and a Green's function model [10.147]. The advantage is that they rely on Fick's second law of diffusion, with a constant interdiffusion coefficient, for a one-dimensional diffusion process [10.148].

The following shows Fick's second law of diffusion. In Fig. 10.7, assume the concentration profile be

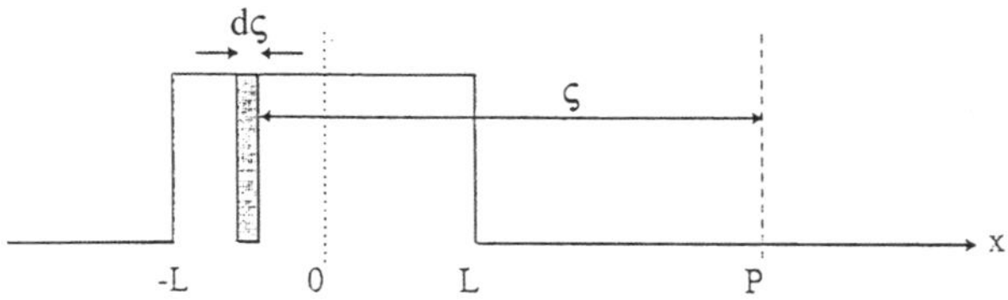

Fig. 10.7 Extended initial distribution.

$$C = \begin{cases} W & -L < x < L \\ 0 & \text{elsewhere} \end{cases} \quad \text{at } t = 0.$$

The concentration at P after diffusion time t is equal to:

$$\frac{W d\varsigma}{2\sqrt{\pi D t}} \exp\left(\frac{-\varsigma^2}{4Dt}\right). \tag{10.4}$$

The concentration at P due to all of the elemental thin film is equal to the sum from $\varsigma = z + L$ to $\varsigma = z - \mathrm{L}$.

$$C = \int_{z-L}^{z+L} \frac{W}{2\sqrt{\pi D t}} \exp\left(\frac{-\varsigma^2}{4Dt}\right) d\varsigma$$

$$= \frac{W}{2}\left(erf\,\frac{L+z}{2\sqrt{Dt}} + erf\,\frac{L-z}{2\sqrt{Dt}} \right), \tag{10.5}$$

where $erf\ y = \frac{2}{\sqrt{\pi}} \int_0^y \exp\left(-\theta^2\right) d\theta$ is the error function.

10.2.4.1 AlGaAs / GaAs

Error function profile (EQW)

The QW structure modeled here is an $Al_wGa_{1-w}As$/GaAs single QW, which is assumed to be undoped, and the effect of unintentional impurities is ignored. The barrier is also assumed to be infinitely thick and is approximated by a large finite barrier width for numerical convenience and efficiency.

The extent of the intermixing process is characterized by a diffusion length L_d, which is defined here as $L_d = (Dt)^{1/2}$, where D and t are the diffusion coefficient and annealing time respectively. It should be noted that the aim of the present study is to determine the effect of the interdiffusion process, which is characterized by L_d, on the QW bound states rather than to determine the magnitude of D or t. In the model presented here we assume an isotropic interdiffusion of Al and Ga with the concentrations, which are also assumed to be equal and constant, i.e., according to Fick's second law of diffusion.

The diffused Al composition profile $\tilde{w}(z)$ across the QW structure is given by

$$\tilde{w}(z) = w\left\{1 - \frac{1}{2}\left[erf\left(\frac{L_z + 2z}{4L_d}\right) + erf\left(\frac{L_z - 2z}{4L_d}\right)\right]\right\} \tag{10.6}$$

where w_0 is the as-grown Al mole fraction in the barrier, L_Z is the as-grown width of the QW, z is both the quantization and the growth axis (QW center ed at $z = 0$), and *erf* denotes the error function [10.149]. Figure 10.8 shows $\tilde{w}(z)$ for different values of L_d. The AGSQW is defined by $L_d = 0$ and a small value of L_d corresponds to an extensively intermixed QW, and $L_d \rightarrow \infty$ corresponds to a fully intermixed QW, i.e., a bulk material.

However, both error function profile and Green's function model are not able to determine the subband energies and wave functions analytically, which makes it difficult to determine the relationship between some of the physical parameters of the structure, which may be of use in the designing of devices.

In contrast, the hyperbolic quantum well (HQW) model can be solved exactly by using a modified hyperbolic function profile and, since it closely resembles the error function profiles in shape, it is a simple and realistic

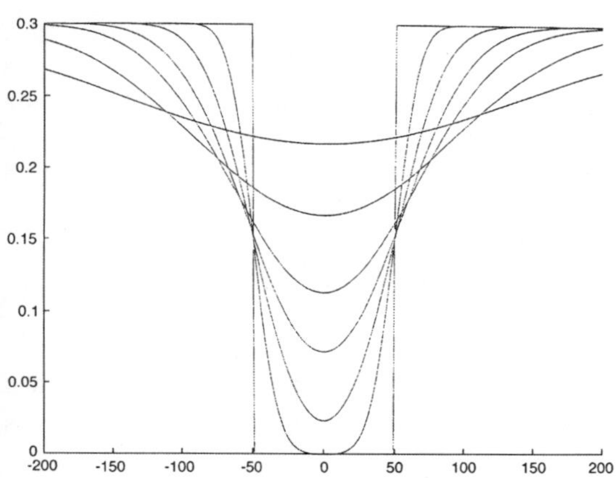

Fig. 10.8 The Al concentration profile as a function of position.

model for the calculation of optical parameters, such as absorption coefficient and refractive index of interdiffusion-induced non-square QWs. The advantage of such an analytical solution is that it may give a better physical insight into the problem, as shown by its application to the calculation of the photon confinement criteria for intermix delineated waveguides.

Hyperbolic quantum well model (HQW)

The interdiffusion process is characterized by a diffusion length L_d which is defined as $L_d = (Dt)^{1/2}$, where D and t are the diffusion coefficient and annealing time respectively. It should be noted here that some authors use the alternative definition of $L_d = 2(Dt)^{1/2}$ [10.145]. A small value of L_d corresponds to a partially intermixed well, while a large value of L_d corresponds to an extensively intermixed well. The diffused Al composition profile $\tilde{w}$ across the QW structure, given by (10.6) [10.143], is then evaluated at the center of the QW ($z = 0$) to obtain the interdiffusion-induced bandgap $\tilde{E}_g$ and the depth $\Delta\tilde{E}_r$ of the QW, where the subscript r denotes either the electron in the conduction band C, or the heavy and light holes in the valence band ($V = HH$ or LH). The cross-over point (COP) potential $U_{COP,r}$ is defined here as the potential at the intersections of the as-grown and the diffused potential profiles at $z = \pm \frac{1}{2} L_z$, measured positively from the bottom of the interdiffused HQW. These parameters are defined as follows:

$$\tilde{E}_g(z) = \tilde{E}_g(w = w(z)) \tag{10.7a}$$

$$\Delta\tilde{E}_g = E_g(w_0) - \tilde{E}_g(z = 0) \tag{10.7b}$$

$$\Delta E_r = Q_r \Delta\tilde{E}_g \tag{10.7c}$$

$$U_{COP,r} = Q_r[\tilde{E}_g(z = \tfrac{1}{2} L_z) - \tilde{E}_g(z = 0)]. \tag{10.7d}$$

The band offset splitting ratio is taken as 60:40, i.e., $Q_C = 0.6$ and $Q_V = 0.4$, and the AlGaAs bulk bandgap $E_g(w)$ and spin-orbit splitting $\Delta_0(w)$ are based on the cubic [10.150] and linear [10.151] Al composition dependence at room temperature (see Appendix 1).

The interdiffusion-induced HQW non-square potential profile is defined here by:

$$U_r(z) = U_{eff,r}\left[1 - \frac{1}{\cosh^2(\beta z)}\right] \tag{10.8}$$

where $U_{eff,r} = \Delta\tilde{E}_g$ is the effective barrier depth after interdiffusion, and β is determined by fitting $U_r(z)$ to the $U_{COP,r}$ at the COPs in (10.7d), and is expressed as:

$$\beta = \frac{2}{L_z}\left[\cosh^{-1}\sqrt{\frac{1}{1\text{-}\hat{U}_{COP}}}\right] \tag{10.9}$$

where $\hat{I}U_{COP} = U_{COP}/U_{eff}$ is the potential of the COP normalized to the well depth and the subscript r is removed from the Us since the Q_rs are canceled in the present unstrained QW case. However, for a strained QW structure, such as in an InGaAs/GaAs QW structure, the ratio of U_{COP}/U_{eff} should be dependent upon r [10.152]. The hyperbolic profile modeled here by the fitting of the parameter β is a realistic approximation to the interdiffusion-induced confinement profile, and differs from that used in [10.153], where the COP is 1/2 $U_{eff,r}$.

10.2.4.2 InGaAs / GaAs

The intermixing process in InGaAs/GaAs QW structures is assumed to result from the interdiffusion of Group III (In and Ga) atoms. The interdiffusion is also characterized by the diffusion length L_d. A small L_d corresponds to the initial stages of intermixing with the QW compositional profile remaining almost rectangular, while a larger L_d corresponds to a more extensively intermixed QW structure.

After interdiffusion, the In concentration $\tilde{w}(z)$ across the InGaAs/GaAs QW structure is assumed to have an error function profile, given by:

$$\tilde{w}(z) = \frac{w}{2}\left[erf\left(\frac{L_z + 2z}{4L_d}\right) + erf\left(\frac{L_z - 2z}{4L_d}\right)\right] \tag{10.10}$$

where z is the growth direction, L_z is the well width, with the QW center ed at $z = 0$. The compositional profile in the QW structure after interdiffusion implies that the carrier effective mass, the bulk bandgap, and the hydrostatic and shear strain vary continuously across the QW.

10.3 Band Structure of Interdiffused QWs

The use of non-square QWs with a nonlinear graded bandgap profile along the growth direction, such as those produced by interdiffusion, are of growing interest for the controlled modification of the properties of QW structure. The one-dimensional quantum confinement of electron and holes in semiconductor QWs leads to the formation of discrete energy levels within the QW, which form a subband of bound states, and may be modified by its confinement profile. The subband states of an inter-diffusion-induced QW have been studied theoretically, using an error function profile (EQW) [10.154], a Green's function model [10.155], and a hyperbolic function model [10.156 and 10.157]. Numerical results which have been reported for the EQW model are limited to the subband ground states using a variational method [10.154], or a multistep transmission resonance tunneling method [10.158], while recently a finite difference method has been used to study an AlGaAs/GaAs multiple EQW [10.159] and a strained InGaAs/GaAs EQW [10.160]. However, in the first variational method, it should be noted that, although the ground-state energy can be obtained accurately, the error produced for the corresponding wave function is an order of magnitude higher and the absorption coefficient peak can be over-estimated by more than 20% [10.161]. Also in this method, the excited states, which are essential for the calculation of optical parameters of the QW structure, are difficult to obtain. In practice, all the bound states and a limited number of unbound continuum states are required to enable the dielectric function, and thus the optical parameter, such as the refractive index and absorption coefficient, to be calculated. For the case of a nonlinear profile, such as the error function, the second method of tunneling requires the multiplication of a large number of matrices and thus it suffers from the accumulation of numerical errors and can be numerically inefficient and unstable (especially for the high energy states) [10.162]. Also, the resonant states are different between a smooth profile and a graded superlattice [10.163], thus extreme care should be taken in implementing this method. In comparison to the first two methods, the finite difference method is better for the determination of a number of states in the EQW.

10.3.1 Confined Eigenstates of EQW

To calculate the electron and hole wave function in QW, we use the multiband effective mass theory. For most III-V semiconductors, such as GaAs-based materials, it is a good approximation that the conduction and valence bands are decoupled. A simple isotropic parabolic band is used in the conduction, and we use a 4×4 Luttinger-Kohn Hamiltonian [10.164] with the strain-component in a block diagonal (called the Pikus-bir Hamiltonian) in the valence band. The electron states near the conduction subband-edge are assumed to be almost purely s-like and non-degenerate (excluding spin), while the hole states near the valence subband-edge are almost purely p-like and four-fold degenerate (including spin). The envelope function scheme is adopted to describe the slowly varying (spatially extended) part of the subband states. The non-interacting electron and hole states are thus described by

$$\Psi_C(k_{\|}, z) = \psi_C(z) u_{0C}(r) e^{ik_{\|} \bullet \rho} \tag{10.11a}$$

and

$$\Psi_V(k_{\|},z) = \psi_V(k_{\|},z)u_{0V}(r)e^{ik_{\|}\bullet\rho} \tag{10.11b}$$

respectively. The u_0s are zone-center Block functions for the bulk materials; ψs are the envelope functions for the QW confinement; $k_{\|} = (k_x,k_y)$ is the transverse (in-plane) wave vector which is in the direction parallel to the plane of the QW layer and perpendicular to the z axis; $r = r(x,y,z)$, and $\rho = \rho(x,y)$. (A note on the direction notation: $\|$ denotes the x-y plane and $\perp$ denotes the z axis.)

The electron and hole subband-edge ($k_{\|} = 0$) structure at the zone center of Γ_6 valley symmetry can be calculated in the envelope function separately for the electron and holes, using the Ben-Daniel and Duke model [10.165] by the one-dimensional Schrödinger-like equation, which is written as follows:

$$-\frac{\hbar^2}{2}\frac{d}{dz}\left[\frac{1}{m_{\perp r}{}^*(z)}\frac{d\psi_{rl}(z)}{dz}\right] + U_r(z)\cdot\psi_{rl}(z) = E_{rl}\psi_{rl}(z) \tag{10.12}$$

where $\psi_{rl}(z)$ is the envelope wave function, $l = 1, 2, \ldots$ are the QW subband levels for either the electrons (Cl) or holes (Vl) respectively; $m_{\perp r}{}^*(z)$ is the carrier effective mass in the z direction; $E_{rl} = E_{rl}(k_{\|} = 0)$ is the subband-edge energy level. (10.12) is solved numerically, using a finite difference method with confinement profile, and the boundary condition is taken to be zero at the end of a finite and large barrier [$\psi_{rl}(\pm z_f) = 0$]. In other words, the QW is embedded in a well of width $L_f = 2z_f$ with an infinitely high barrier to approximate the thick barrier.

10.3.2 Valence Subband-Mixing

For valence band-mixing, it is much more complicated to solve in comparison with parabolic band model. This band-mixing prevents the diagonalization of the fourfold degenerate Luttinger Hamiltonian, which can be solved using a standard finite difference method. However, it can be solved in the neighborhood of a high-symmetry zone center (i.e. $k_{\|} = 0$), using approximation methods such as the one developed by Broido and Sham using series expansions [10.166], or the one by Chan using an effective Hamiltonian [10.167]. The latter approach is adopted here and is briefly described below.

This approach is based on the analogous application of the $k{\cdot}p$ method to the subbands. The coupled envelope function $\psi_V(k_{\|},z)$ in (10.11b) at any finite $k_{\|}$ not too far away from the highest symmetry point ($k_{\|} = 0$) can be expressed as a linear combination of the decoupled envelope functions $\psi_V(z)$ at $k_{\|} = 0$, which are obtained *a priori* from the solution of (10.12).

The validity of the approximation is dependent upon a limited range of $k_{\|}$, since only a finite dimension basis set for the expansion is used in practice. However, a small range of $k_{\|}$ near 0 is sufficient for this study of the QW properties, as the QW effect only manifest themselves near the zone-center subband-edge. The basis for the expansion is defined as follows:

$$\Psi(k_{\|}, z) = \sum_{l=1}^{N} \sum_{v=-3/2}^{3/2} d_{v,l}(k_{\|}) \cdot \psi_{v,l}(z) \tag{10.13}$$

where V is running over the lifted fourfold degenerate hole states and N is the maximum number of gases taken in each V, and the effective Hamiltonian is

$$\begin{bmatrix} E_{3/2} + \xi & C & B & 0 \\ C^* & E_{-1/2} - \xi & 0 & B^T \\ B^* & 0 & E_{+1/2} - \xi & C^T \\ 0 & B^* & C^* & E_{-3/2} + \xi \end{bmatrix} \tag{10.14}$$

where

$$C_{\|} = \left[\frac{3}{4}\right]^{1/2} \frac{\hbar^2}{m_o} \gamma_2 (k_x - ik_y)^2 \int_0^{L_f} dz \psi_{-3/2,l}(z) \psi_{1/2,'}(z) \tag{10.15a}$$

$$B_{\|} = 3^{1/2} \frac{\hbar^2}{m_o} \gamma_2 (-k_x - ik_y) \int_0^{L_f} dz \psi_{3/2,l}(z) \frac{\partial}{\partial z} \psi_{-1/2,'}(z) \tag{10.15b}$$

$$E_{\pm 3/2,ss'} = \delta_{ss'} E_{Hs} - \frac{\hbar^2}{2m_{\|H}} k_{\|}^2 \tag{10.15c}$$

$$E_{\pm 1/2,ss'} = \delta_{ss'} E_{Ls} - \frac{\hbar^2}{2m_{\|L}} k_{\|}^2 \tag{10.15d}$$

$\gamma_2(w)$ is the Luttinger parameter, and L_f is the width of a square QW with infinitely high barriers which contains the QW, as defined earlier. The introduction of this larger well is to numerically determine the continuum levels above the top of the QW. ξ is the perturbative strain parameter, which will be described later. It should be noted that, in the absence of an applied electric field or any other mechanism that will distort the inversion symmetry assumed here, the symmetric QW confinement potential profile causes the valence subband to exhibit a twofold Kramer's spin degeneracy.

The valence subband-mixing allows the "forbidden" transitions (lC + lV = odd), which contribute to the finite absorption and negative hole mass. It also increases the reduced mass and the exciton absorption strength as well as the saturation limit. Figure 10.9 shows the dispersion curves for only the six topmost valence subbands, in the cases of L_d = 1 and 10Å, and an additional number of subbands including the continuum hole state in the cases of L_d = 40 and 100Å. These four cases provide some information on the characteristics of subband-mixing due to interdiffusion. The first six valence subbands can be labeled as "heavy" or "light" according to their band-edge line-up at $k_{||} = 0$ which, starting from the top down (well bottom), are H1, L1, H2, H3, L2, H4, and these states are now denoted by numbers in ascending order (V = 1,..., 6, respectively), as for $k_{||} \neq 0$, the admixture of heavy- and light-hole states makes it difficult to distinguish between them. For an as-grown QW, as shown in Fig. 10.9(a), non-parabolicities give rise to an electron-like upward dispersion which results in a negative effective mass at the zone center for V2(L1) and V4(H3). This negative mass implies that not only heavy- and light-hole masses are

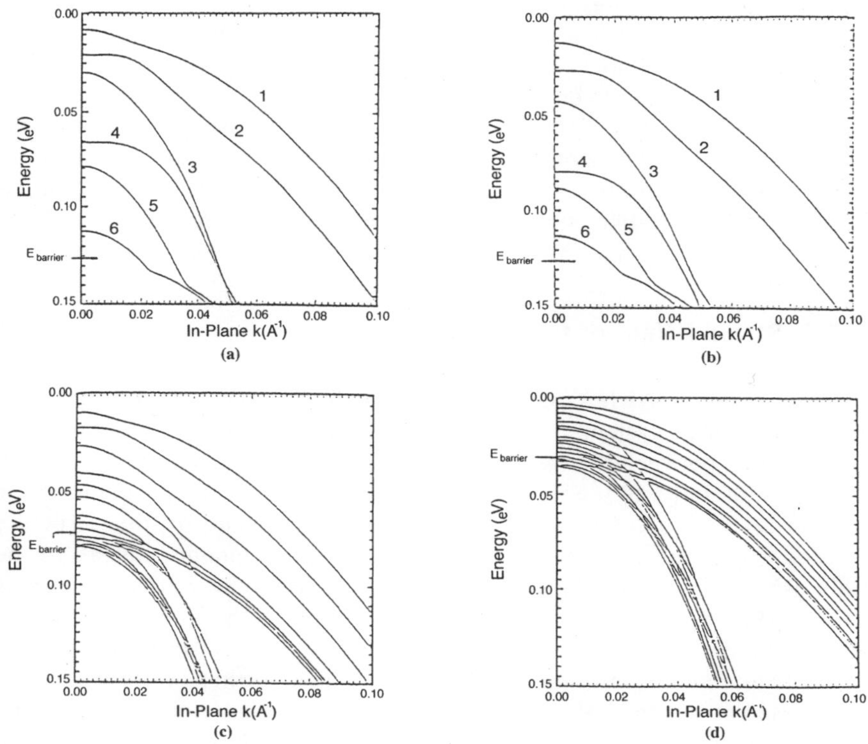

Fig. 10.9 a–d The valence subband potential (zero taken at the respective well bottom) as a function of the transverse direction wave vector $k_{||}$ for the EQW (W_0 = 0.3, L_z = 100Å) with various Ld. The barrierr is noted by an arrow. **(a)** L_d = 1Å, **(b)** L_d = 10Å, **(c)** L_d = 40Å, and **(d)** L_d = 100Å.

reversed in the transverse direction, they also depend on the type of subband-mixing. As interdiffusion proceeds (see Fig. 10.9(b)), the separation between subbands is initially reduced and the strong mixing shown between V5 and V6 is transferred to V4 and V5. This implies an interaction of order $k_{\|}^2$ ($l_L + l_H$ = even) between L2 and H4 is reduced and the interaction of order $k_{\|}$ ($l_L + l_H$ = odd) between L2 and H3 is increased. In Fig. 10.9, it can be seen that the V4(H3) curve is coupled to and pulled down by V5(L2) and results in a less negative mass.

In Fig. 10.9(c), strong band-mixing still persists and the lower subbands, V1 to V4, are still clearly discretized with good 2-D behaviour. For states near $k_{\|} = 0$, much less distinction between the heavy- and light-holes states is observed although the V2 edge still remain negative in mass. However, it can be observed that below the barrier (energy less than 73meV) the subbands (V5 and below) start to become densely packed, forming a bulk-like continuum with two sets of lines representing the heavy-hole-like and light-hole-like bulk states.

In Fig. 10.9(d), which is a latter stage of the interdiffusion process, most of the subbands become almost parabolic and densely packed. The band-mixing effect remains strong with fine details only at around $k_{\|} = 0.02\text{A}^{-1}$. For the states near $k_{\|} = 0$, their mass is indistinguishable between the heavier and lighter parts of the bulk material, while at larger $k_{\|}$ the states converge into two groups. These two groups, referred to as the heavy-hole-like and light-hole-like states, are the result of the restoration of the fourfold degeneracy, and the bulk-like nature of the material is restored.

10.3.3 Strain Effect of the InGaAs/GaAs QW

Different from the AlGaAs/GaAs system, the InGaAs/GaAs is a lattice mismatched material system. In the lattice-mismatched QW heterostructures, the mismatch between the well and barrier materials is taken up by uniform elastic strain only if the layers are sufficiently thin. In pseudomorphic strained-layer structures, the in-plane lattice constant of the structure matches that of the substrate so that the QW layer accommodates all of the strain while the barrier layer is unstrained. Within the critical thickness regime [10.168], the QW lattice constant parallel to the interface matches the in-plane lattice constant by compression or expansion in the interfacial plane. A tetragonal deformation also takes place so that the QW layer is compressed or expanded in the direction perpendicular to the interface. The QW will be coherently strained with a biaxial hydrostatic strain parallel to the interfacial plane and a uniaxial shear strain perpendicular to the interfacial plane. A compressive (tensile) hydrostatic strain causes an increase (decrease) in the bandgap energy by an amount determined by the hydrostatic deformation potential, defined as the shift in energy of the band-edge per unit hydrostatic strain. The uniaxial (shear) strain disrupts the cubic symmetry of the semiconductor and lifts the degeneracy of the heavy hole (HH) and the light hole (LH) band-edges at the Brillouin zone center Γ. The heavy hole band moves towards (away from) the conduction band and the light hole band moves away from (towards) the conduction band. Furthermore, the presence of the strain induces the coupling between the LH valence band and the split-off band.

The unstrained bandgap in the well $E_g(\tilde{x})$ is also a function of the indiffused compositional profile $\tilde{x}(z)$, so that the unstrained potential profile after interdiffusion $\Delta E_r(\tilde{x})$ varies across the well (see Fig. 10.10) and is given by:

$$\Delta E_r(\tilde{x}) = Q_r \Delta E_g(x = \tilde{x}) \tag{10.16}$$

where Q_r is the band offset ratio and ΔE_g is the unstrained bandgap offset.

The in-plane strain across the well, $\varepsilon(\tilde{x})$ will also vary so that the strain effects are also z-dependent. Assuming that the growth direction z is along (001), then for the biaxial stress parallel to the interface the strain components after interdiffusion are given by:

$$\begin{aligned} \varepsilon_{xx} &= \varepsilon_{yy} = \varepsilon(\tilde{x}) \\ \varepsilon_{xx} &= -2\left[C_{12}(\tilde{x})/C_{11}(\tilde{x})\right]\varepsilon(\tilde{x}) \\ \varepsilon_{xy} &= \varepsilon_{yz} = \varepsilon_{zx} = 0 \end{aligned} \tag{10.17}$$

where $\varepsilon(\tilde{x})$ is defined to be negative for the compressive strain and $C_{ij}(\tilde{x})$ are the elastic stiffness constants.

The change in the bulk bandgap, $S_\perp(\tilde{x})$ due to the biaxial component of the strain is given by:

$$S_\perp(\tilde{x}) = -2a(\tilde{x})\left[1 - C_{12}(\tilde{x})/C_{11}(\tilde{x})\right]\varepsilon(\tilde{x}) \tag{10.18}$$

where $a(\tilde{x})$ is the hydrostatic deformation potential calculated from:

$$a(\tilde{x}) = -\frac{1}{3}\left[C_{11}(\tilde{x}) + 2C_{12}(\tilde{x})\right]\frac{dE_g(\tilde{x})}{dP} \tag{10.19}$$

where dE_g/dP is the hydrostatic pressure coefficient of the lowest direct energy gap E_g.

The splitting energy $S_H(\tilde{x})$ between the HH and LH edges induced by the uniaxial component of strain is given by:

$$S_H(\tilde{x}) = -b(\tilde{x})\left[1 + 2C_{12}(\tilde{x})/C_{11}(\tilde{x})\right]\varepsilon(\tilde{x}) \tag{10.20}$$

where $b(\tilde{x})$ is the shear deformation potential. The parameters a, b, C_{ij}, and dE_g/dP in the above equations are assumed to obey Vegard's Law so that

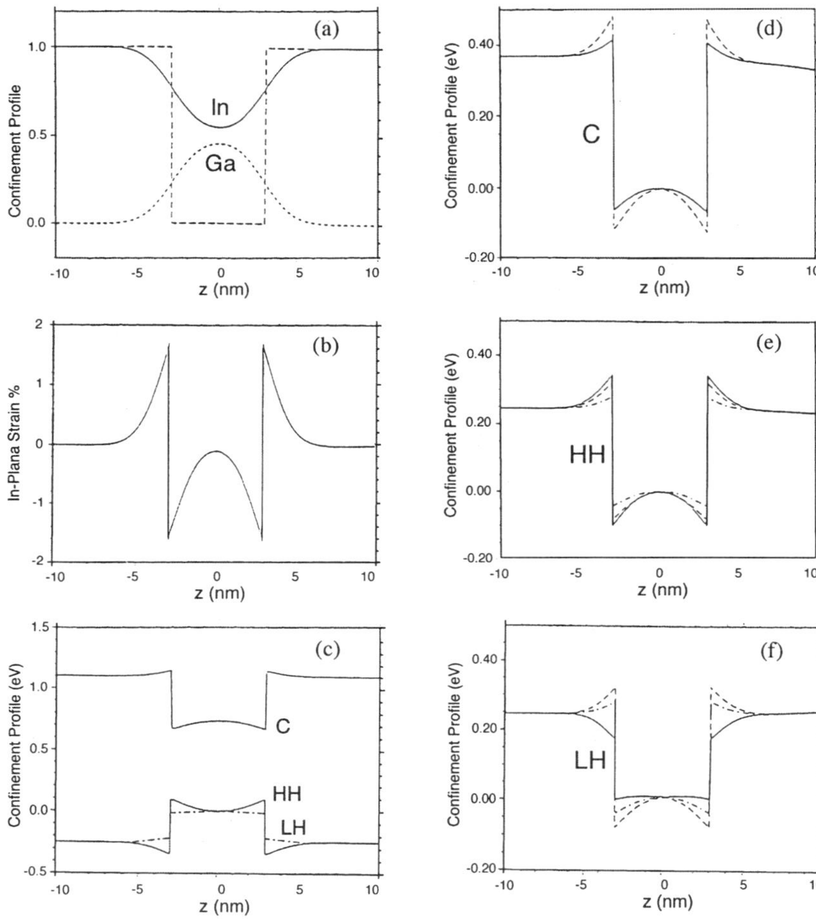

Fig. 10.10 a–f Schematic diagram showing the effects of strain and intermixing on the carrier confinement profiles in an intermixed $In_{0.2}As_{0.8}As$/GaAs single QW for L_d = 1.5nm. The dashed lines represent the unstrained bandgap after intermixing, and the dot-dash line shows how the unstrained bandgap is modified by the effect of the hydrostatic strain. The solid line represents the final electron (C), HH and LH confinement profile. Note that for the electron confinement profile, the dot-dash and solid lines coincide. All terms refer to the well center z = 0 [10.163].

their respective values depend directly on the compositional profiles across the QW.

Moreover, for the case of the InGaAs/GaAs system, the LH and the spin-orbit bands are assumed to be coupled due to the presence of the stress while the HH state remains uncoupled.

The valence band splitting at Γ for the HH band and for the LH band are given by:

$$S_{\parallel \, HH}(x_{df}) = S_{\parallel}(x_{df}) \tag{10.21}$$

$$S_{\parallel \, LH}(x_{df}) = -\tfrac{1}{2}[\, S_{\parallel}(x_{df}) + \Delta_o(x_{df})] + \tfrac{1}{2}[9\{ S_{\parallel}(x_{df})\}^2 + \{\Delta_o(x_{df})\}^2$$

$$S_{\parallel \, LH}(x_{df}) = 2\, S_{\parallel}(x_{df})\Delta_o(x_{df})]^{\frac{1}{2}} \tag{10.22}$$

respectively, where $\Delta_o(x_{df})$ is the spin-orbit splitting.

The QW confinement potential after the intermixing process, $U_r(z)$ is obtained by modifying the unstrained potential profile after processing $\Delta E_r(\tilde{x})$ by the variable strain effects, and is given by:

$$U_r(z) = \Delta E_r(\tilde{x}) - S_{\perp r}(\tilde{x}) \pm S_{Hr}(\tilde{x}) \tag{10.23}$$

where $S_{\perp r}(\tilde{x}) = Q_r S_{\perp}(\tilde{x})$, the '+' and '–' signs represent the confined HH and LH profiles respectively and $S_{HC}(\tilde{x}) = 0$.

Thus, due to the presence of the strain, different confinement potential profiles result for the HH and LH (as well as the spin-orbit) bands, while the intermixing produces non-square, graded profiles (see Fig. 10.10). Since the hole occupancy of the spin-orbit split-off band is negligibly small, its details are not considered.

10.4 Intersubband Optical Transition in DFQW

10.4.1 Introduction

The quantum confinement of carriers in semiconductor quantum wells (QW) and superlattice structures, which lead to the quantization of energy levels and the formation of subbands, have been discussed in Sect. 10.3 [10.169–10.176]. The QW structures have been used to improve the performance of conventional optolelectronic devices. The characteristics of these devices are determined by the optical properties of the QW structure, which are derived directly from the confinement subband structure, and their performance may be further improved by the modification of the QW confinement profile, such as the thermal modification of the Al(In) composition profile in an AlGaAs-GaAs (InGaAs-GaAs) QW structure to produce the various "nonsquare" QWs. There are several methods to produce the non-square confinement potential profile of the QW.

This quantum confinement of carriers in a semiconductor QW leads not only to the formation of discrete energy levels, but also the structural change of the optical-absorption spectra [10.177]. One of the most remarkable properties of these quasi-two dimensional electronic systems is that the optical transitions between the size-quantized subbands are tunable. The linear intersubband optical absorption have been studied experimentally with [10.178] and without [10.179 and 10.180] an electric field. A very large dipole strength and a narrow band width were observed. This suggest that the intersubband optical transitions in a QW may have very large dipole nonlinearities. The linear and third-order nonlinear absorption coefficients are then obtained [10.181] for the general asymmetric two-level QW systems. Since the behaviour of the linear and third-order absorption coefficients inside the interdiffused QW have not been discussed in detail, it is worthwhile to have a thorough analysis of them.

The linear and nonlinear electroabsorptions in modulation-doped QW have also been examined [10.182]. In doped QW, the larger the occupancy of the ground states, the higher the absorption strength. In addition, modulation doping offers higher mobility and low-noise characteristics such that the carriers are confined to areas where they are subjected to reduced impurity scattering. As the analysis of electroabsorption in modulation-doped QW is of vital importance, the examination of the doped QW inside interdiffused QWs is also a significant research area.

In this section, the linear and third-order absorption coefficient in undoped interdiffused QW and also the linear and nonlinear electroabsorption inside the modulation-doped interdiffused QW are discussed. Interdiffusion is very effective for tuning the intersubband transition energy and the results are found especially useful for photodetector in the mid-infra-red (10 to 30μm) region. Radiation detectors that work at the mid-infra-red and longer wavelengths have many applications, including night vision, navigation, weather monitoring, flight control and early warning radars. The absorption lines of many gas molecules such as ozone, water, carbon monoxide and nitrous oxide occur at these wavelengths (between 3 and 18μm), so infra-red detectors can be used to monitor pollution, humidity and the distribution of different molecular species in the atmosphere. Such detectors also have the potential to play a major role in various remote-sensing satellites due to be launched in the next decade, in particular those operating in the atmospheric window, which is a low-absorption region between 8 and 10μm.

10.4.1.1 Intersubband Absorption in Square QW

The quantum confinement of electrons in the conduction band of a one-dimensional semiconductor QW grown in the z-direction are quantized as E_1, E_2,...etc. If light polarized in the z-direction is incident on the QW, photons whose energy corresponds to one of the transition energies ($E_{n+1} - E_n$) will make the transition.

Intersubband transitions in QWs and modulation-doped high-electron-mobility-transistor structures have been extensively studied in the last few years by the measurement of absorption [10.183–10.195], photoconductivity [10.188, 10.196 and 10.197], Raman scattering [10.198–10.200], and photoluminescence excitation spectroscopy [10.199]. The theory for linear and nonlinear intersuubband transitions has also been discussed [10.202–10.221]. Based on this kind of transitions, several types of infra-red detectors and modulators have been proposed and demonstrated [10.222 and 10.223].

Fig. 10.11 shows the room-temperature absorption spectra of two (001)-grown $Al_{0.3}Ga_{0.7}As/GaAs$ QWs measured by West and Eglash [10.183]. The samples, each containing 50 GaAs QWs (L_Z = 65Å and 82Å) with each of the wells cladded with 150Å of undoped $Al_{0.3}Ga_{0.7}As$ on both sides, were grown by molecular-beam epitaxy. Between these buffered wells were placed 100Å layers of $Al_{0.3}Ga_{0.7}As$ doped with $4\times10^{17}cm^{-3}$ Si donors. The light was incident at Brewster's angle of 73° and polarized to eliminate the surface reflection. Note that the intersubband transitions in n-type QWs are forbidden for normally incident radiation, i.e., for light polarization perpendicular to the growth direction. Because of the high n (refractive index) value of the GaAs at these infra-red wavelengths, the light beam is refracted strongly. Thus, only a small component of the electric-field vector is parallel to the growth direction, resulting in a greatly reduced coupling to the dipole of the intersubband transitions [10.183].

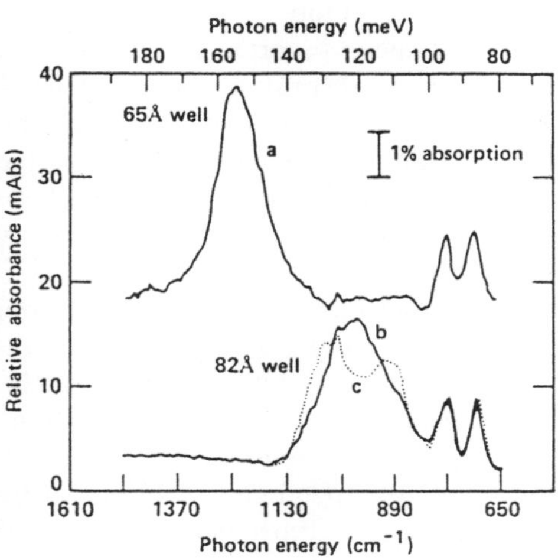

Fig. 10.11 Absorption spectra of two quantum well samples. $mAbs = -10^{-3}\log_{10}$ (transmission). The spectra are with $\theta = 73°$ and at room temperature. Case (a) is the spectrum of a 65Å quantum well sample and was independent of rotation about ϕ. Cases (b) and (c) are for the same 82Å sample with $\phi = 0°$ (b) and $\phi = 90°$ (c), showing sample inhomogeneity with the scan direction [10.183].

The intersubband-absorption-resonance peak energy can be recognized to shift from 152 to 121meV as the well size is increased from 65 to 82Å. A computer solution for the wave functions taken account of the band deviation from the quadratic and the finite barrier height gave transition energies of 139meV and 111meV for the 65Å and 82Å wells respectively, for the lowest-energy ($E_1(n = 1) \rightarrow E_2(n = 2)$) transition [10.183]. In addition, two small resonances at 88 and 96meV have been observed which are originating from the high-order phonon absorption.

Jogai, et al. [10.194] have observed the intersubband and optical absorption not only between the ground and the first excited states ($E_1(n = 1) \rightarrow E_2(n = 2)$) but also between the first and second excited states ($E_2(n = 2) \rightarrow E_3(n = 3)$) in an n-type $Al_wGa_{1-w}As$/GaAs multiple QW structure. The sample consisted of 100 periods of 75Å Si-doped ($\sim 8\times10^{18}cm^{-3}$ GaAs wells separated by 100 undoped $Al_{0.3}Ga_{0.7}As$ barriers. The Fermi level for this sample was estimated to be ~171meV above E_1 and 44meV above E_2. The absorption peak energies measured were 144.8meV ($E_1 \rightarrow E_2$) and 112.7meV ($E_2 \rightarrow E_3$). To verify that these peaks were not due to processes other than intersubband transitions, for example, neutral-donar-to-subband transitions, they also measured the spectra for normal incidence and no peaks were observed. Also, if the dopant density was high enough to raise the Fermi level above two subbands, two peaks were always observed.

As the intersubband absorption is proportional to the occupancy of the lowest intersubband level, it was investigated mainly in doped QW. The photo-induced intersubband absorpiton in undoped $Al_wGa_{1-w}As$/GaAs multiple QWs with different well widths has been reported by Elbert, et al. [10.193]. They had observed two different absorption bands. One is polarized along the growth direction, is linear with the pump intensity, and is characteristic of an intersubband transition of (intrinsic) free carriers confined within the QW. The other is not polarized, depends sublinearly on the pump intensity, and originates from the bound carriers whose recombination is probably controlled by (extrinsic) defects.

In order to achieve the structures which can intrinsically achieve strong absorption of normally incident light, QWs consisting of indirect-gap semiconductors are required [10.195]. In these indirect-gap materials, the conduction electrons occupy indirect valleys with ellipsoidal constant-energy surfaces. The effective-mass anisotropy of electrons in these valleys can provide coupling between the parallel and perpendicular motions of the electrons when the principal axes of one of the ellipsoids are tilted with respect to the growth direction. As a result of this coupling, interconduction subband transitions at normal incidence in indirect-gap semiconductor QWs are allowed. Katz, et al. [10.195] have reported the first observation of normal incidence infra-red absorption due to the interconduction subband transitions in $Al_{0.4}Ga_{0.6}As$/AlAs X-valley multiple-QWs. In an X-valley material, the six equivalent minima of the conduction

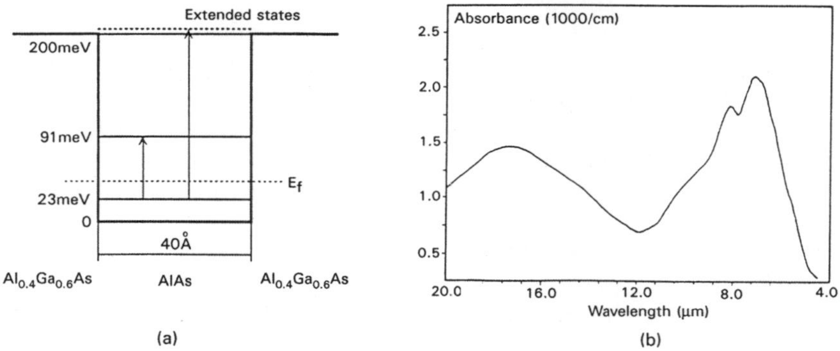

Fig. 10.12 a, b **(a)** The conduction-subband structure of a $AlAs/Al_{0.4}Ga_{0.6}As$ *X*-valley QW and the intersubband transitions, and **(b)** the intersubband absorption spectra of a $AlAs/Al_{0.4}Ga_{0.6}As$ *X*-valley MQW at 77°K [10.195].

band are located at the *X* point of the *BZ*. The values of the effective mass components in such valleys can be obtained in terms of the longitudinal and transverse effective masses m_t and m_l, which are defined along the principal axes of the conduction-band valleys.

The intersubband absorption spectrum of a (113)-grown $Al_{0.4}Ga_{0.6}As$/AlAs *X*-valley multiple QWs has been measured by Katz at 77°K [10.195]. In this structure, the wells are AlAs and the barrier is $Al_{0.4}Ga_{0.6}As$. The conduction band offset is about 220meV (Fig. 10.12(a)).

The observed spectrum shows two clear absorption peaks at wavelengths of 7 and 17μm. They considered that the 17μm peak is a result of transitions between the ground and the first excited states, while the 7μm peak is due to the transitions between the ground and the second excited (or continuum) states. They also measured the absorption of the same QWs grown on a (115)-substrate orientation and achieved similar results. No notable absorption for the samples grown on the (001)-substrate orientation has, however, been observed. These facts indicate the complicated orientation dependence of the effective-mass tensor in the indirect-gap semiconductor QWs, in good agreement with their theoretical calculations [10.195 and 11.196].

10.4.1.2 Square QW Infrared Photodetector

As discussed in the introduction section, the tendency for the use of the QW material in the design of long-wavelength infra-red photodetector have increased in the recent years. Major steps have been made toward a new approach for designing semiconductor structures with tailored electronic and optical properties for a new generation of long-wavelength infrared quantum detectors. This approach has come to be known as the bandgap engineering [10.224 and 10.225]. Central to the bandgap engineering is the concept that, by spatially varying the composition and doping of a semiconductor over distances ranging from a few micrometers down to ~2.5Å (~1 monolayer), we can tailor the electronic band structure in a nearly arbitrary and continuous way. Thus, semiconductor structures with new electronic and optical properties can be custom-designed for specific applications.

To implement the bandgap engineering of infra-red detectors, precise control of the composition, doping and thickness of the semiconductor layers is required. Modern thin film growth techniques such as molecular beam epitaxy and metalorganic chemical vapour deposition have made it possible to deposit ultra-thin (10~1000Å) epitaxial layers of precisely controlled alloy compositions on semiconductor substrates on an atomic layer-by-layer basis [10.226]. High-quality single crystal epitaxial layers composed of a variety of semiconductor alloy combinations (such as GaAs, $Si_{1-w}Ge_w$, $Al_wGa_{1-w}As$) have been grown by using these techniques. In addition, to the single composition epitaxial layers, these epitaxial growth techniques can be used to deposit multiple layers of two different semiconductor layers, which have precisely controlled compositions and are doped within the individual layers. These layers are only a few tens of angstroms thick and are so thin that the energy levels in them show the quantum effects associated with the electron confinement within the layers. The confinement, due to the heterojunctions at the interfaces, causes the continuous-allowed energy levels associated with the macroscopic thickness (a free electron) to become discrete-allowed energy levels [10.227].

When there is a conduction band offset in energy at the heterostructure interface due to a difference in the bandgap between the alternating compositions, the QWs are formed in the band structure. Typically, the QWs are created by sandwiching a very thin layer of a small bandgap semiconductor material between two thick layers of a wide bandgap material. The allowed electron energies are quantized by the presence of the two potential energy barriers formed by the wide bandgap material, that is, the classic electron in a finite potential well problem.

The actual values of the discrete energy levels in the QWs are determined by the thickness and the depth of the well. The depth of the well is the band discontinuity of the two materials. The allowed electronic transitions are then further tailored by the selection of the thickness of the QW. Wider wells contain more bound states that are closer in energy than the bound states in narrower wells.

If many QWs are grown on top of one another and the barrier layers are made thick enough so that the electrons wave functions in the adjacent QWs do not overlap, then the structure is called a multiple QW (see Fig.

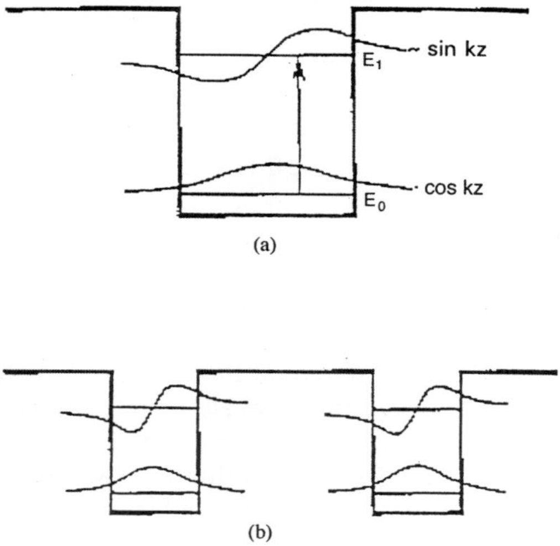

Fig. 10.13 a, b **(a)** A single quantum well, and **(b)** a double quantum well.

10.13(b)). If the barrier layers are so thin that the tunneling between wells is significant (thickness < 50Å), the result is a superlattice. One of the uses of the multiple QW and superlattice is the design of the novel long-wavelength ($\lambda > 12\mu m$) infra-red detectors [10.228]. In bulk semiconductor materials, the selection of compounds or alloys with the appropriate bandgaps and long-wavelength cutoffs for infra-red detectors is very limited.

It is readily apparent that only the InAsSb system has the wavelength cutoffs approaching 12μm. Therefore, long-wavelength absorption is not possible via the valence band to conduction band transitions in any of these semiconductor systems. In the II-VI materials, two alloy systems, HgCdTe and HgZnTe, are available for band-to-band infra-red absorption at long wavelengths. The main drawback of these alloys for infra-red detection is that they are difficult to grow with precisely controlled compositions, and this strongly affects the bandgap. On the other hand, long wavelength quantum detectors made of III-V semiconductor QWs and superlattices are expected to have several advantages over the bulk HgCdTe detectors as well as other bulk long-wavelength detectors for the following reasons:

(1) A higher degree of uniformity, which is of importance for detector arrays;
(2) Smaller leakage currents due to the suppression of tunneling available in superlattices;
(3) Technologically mature device fabrication and materials processing techniques;

(4) Tunability of the upper cutoff wavelengths (theoretically to arbitrarily high values);
(5) A large cross-section for photon absorption;
(6) Radiation hardness;
(7) Ease of crystal growth;
(8) Specific detectivity comparable to that of HgCdTe at temperatures as high as 80°K;
(9) Lower cost.

In the GaAs/$Al_wGa_{1-w}As$ multiple QW heterostructures, the mechanism for obtaining long-wavelength infra-red radiation absorption is by the intersubband transition between the ground state and the first excited state of the QWs. The energy separation between these levels is determined by the size of the QW, and theoretically goes to zero for sufficiently thick wells. A number of types of GaAs.$Al_wGa_{1-w}As$ multiple QW detector structures with different Al content in the barriers and various layer thicknesses can be selected to have intersubband transition energies of approximately 12.5μm [10.197, 10.229–10.231]. Several key experimental and theoretical milestones ranging from the first detection of infra-red radiation, fast-response QW detectors, 1% quantum efficiency, and high-detectivity infra-red detectors, to the origin of the blue shift in the inter-subband transitions have been achieved during the past several years in these types of infra-red detectors.

10.4.2 Linear and Third-Order Intersubband Absorption Coefficients

The mid-IR (λ = 10–100μm) technology has found its use in a number of applications. They include environmental monitoring, astronomy observations, military and medical imaging [10.232]. In order to detect this type of radiation, different type of materials have been studied and used. In the past, the limited choice of semiconductors (e.g., Si, Ge, InSb) had restricted the operation wavelength range of the IR radiation. As the semiconductor technology matures, narrowgap materials such as HgCdTe could be used for the λ = 10μm detector, but because of the difficulty in controlling its composition, it was ruled out. It should be noted that both tactical and strategic uses of the IR imaging systems require the availability of high performance detector arrays operating at wavelengths beyond the 10–18μm spectral band. In addition, several researchers have proposed a mode of IR detection based upon the excitation of carriers out of a QW, either by an intersubband-like transition [10.233] or through enhanced free-carrier absorption [10.234]. This type of detector might be preferable to the narrow bandgap or Schottky barrier detectors in the 10μm and above range, since its fabrication is simpler than the former and its detection threshold is sharper. The quantum efficiency is also larger than the latter.

In recent years, major steps have been made towards the use of III-V semiconductor QW structures, such as the fabrication of 10μm IR detectors. [10.235 and 10.236] The GaAs QW detector has a number of potential advantages over the conventional HgCdTe detectors operating in the 8–12μm spectrum region. This is a well-understood material with mature growth technologies, higher device uniformity, monolithic integration with GaAs electronics, high speed, and the ability to vary the operation wavelength by controlling the well parameters. The special feature of the QW material is that the electronic band structure can be tailored in a nearly arbitrary and continuous way by spatially varying the alloy composition, the well dimension and the doping of a semiconductor. This is known as the bandgap engineering [10.237 and 10.238]. The characteristics of these devices are determined by the optical properties of the QW structure, which are determined directly from the confinement subband structure of the QW and their performance may be further improved by the modification of the QW confinement profile, such as changing the Al(In) composition profile in an AlGaA/GaAs (InGaAs/GaAs) QW structure, to produce various "on-square" QWs. One such non-square confinement potential profile is due to the interdiffusion of the cation [10.239 and 10.240] across the well-barrier interface produced by the post-growth thermal processing [10.241]. This interdiffused QW (DFQW) has been demonstrated in a lattice-matched AlGaAs/GaAs [10.242] and strained InGaAs/GaAs [10.243] material systems. The InGaAs/GaAs is a compressive strained material system that offers the potential of photonic applications at longer optical wavelengths than lattice-matched AlGaAs/GaAs system [10.244], and can also be integrated with GaAs electronics. The InGaAs has aroused a growing interest of its use, due to the promising development in a strained InGaAs/(Al)GaAs (QW) lasers [10.245 and 10.246], which lase at a wavelength about 1μm, and which offer the potential of having lower threshold current than that of the GaAs/AlGaAs laser [10.247 and 10.248].

The intersubband transitions have large oscillator strengths which has been evidenced for the first time by West and Eglash [10.250]. These intersuband transitions have given birth to a wealth of new IR devices [10.235, 10.236, 10.249–10.252]. In order to facilitate the use of intersubband in the far-IR region, the linear and nonlinear intersubband optical absorption in the interdiffused QW have been analyzed [10.230]. On the other hand, the intersubband transition energy in a partially interdiffused multiple QW can be redshifted [10.254] and the postgrowth tuning of the peak detection wavelength [10.255] of the QW IR detectors using rapid thermal annealing have also been demonstrated.

As the DFQW has high flexibility in designing the required wavelengths for operation, it is worthwhile to have an in-depth analysis of the influence of interdiffusion to different QW structures. Although the effects of interdiffusion on the bandedge wavelength of the QW materials had been investigated, other influences on the DFQW intersubband transitions (such as applying the electric field, impurity concentration, and well width) have not been investigated fully. In this chapter, the single DFQW structures containing a maximum of three energy states inside the well are considered. Since the magnitude of intersubband absorption coefficient is mainly determined by the ground and the first excited state, the effect of the second excited state is neglected. In order to characterize the AlGaAs/GaAs and the InGaAs/GaAs DFQW material for IR photodetection, the intersubband transition and absorption as a function of diffusion length are analyzed for several as-grown QW structures. It is expected that these DFQW can have a large tuning range (from 5 to 25μm) while the optical resonance remain strong enough for device applications. In Sect. 10.2, the modeling of the DFQW is described. This includes the intersubband eigenstate calculation and the linear and third-order absorption of the optical transitions. In Sect. 10.3, the various effects of the intersubband transition properties are discussed as a function of the diffusion length, electric field, as-grown QW width, and alloy (Al and In) concentration. This is followed by a brief proposal of wide bandwidth device applications. Finally, the main results of this report is summarized in Sect. 10.4.

10.4.2.1 Modeling of Intersubband Optical Transition and Absorption

The thermal annealing process modifies the trivalent composition profile in the AlGaAs/GaAs (InGaAs/GaAs) QW structure due to the interdiffusion of Al(In) and Ga. This annealing results in the transition from an as-grown square QW composition profile to a graded nonlinear profile, which determines the confinement profile of the intermixed QW. The modeling of the subband states in the DFQW confinement profile have been presented in Sect. 10.3.

Consider a QW under a dc applied field in the z direction $F_{cd,z}$ and in the presence of an optical field $E(\omega t)$ where the polarization is along the z-direction of the well. the electronic polarization can be expanded as

$$P(t) = \in_0\chi^{(1)}E(i\omega t) + \in_0\chi^{(2)}E(2\ i\omega t) + \in_0\chi^{(2)}E(0)$$

$$P(t) = + \in_0\chi^{(3)}E(i\omega t) + \in_0\chi^{(3)}E(3\ i\omega t) + \ldots \qquad (10.24)$$

The dc optical rectification term $\in_0\chi^{(2)}E(0)$ will be neglected due to the small size of $\chi^{(2)}$ for symmetric QWs with dc electric field less than 30kV/cm. Contributions due to the higher order harmonic terms will also be neglected. These assumptions yield the approximate form for $P(t)$ as,

$$P(t) \approx \in_0\chi^{(1)}E(i\omega t) + \in_0\chi^{(3)}E(i\omega t). \qquad (10.25)$$

The analytical forms of the linear ($\chi^{(1)}$) and the nonlinear ($\chi^{(3)}$) susceptibilities have been calculated by modeling the QW as a two-level system [10.255 and 10.256].

The susceptibility χ is related to the absorption coefficient $\alpha(\omega,I)$ by

$$\alpha(\omega) = \omega\sqrt{\mu / \in_R}\, Im(\in_0\, \chi(\omega)) \tag{10.26}$$

where μ is the permeability of the material and $\in_R$ is the real part of the permittivity.

For first-order absorption coefficient,

$$\alpha_1 = \frac{\hbar\omega}{c_0 \hbar n_r \varepsilon_0} \left|M_{fi}\right|^2 \left(\frac{m^* k_B T}{L\pi\hbar^2}\right) \times \ln\left(\frac{1+\exp\left(\frac{E_F - E_i}{k_B T}\right)}{1+\exp\left(\frac{E_F - E_f}{k_B T}\right)}\right) \times \frac{\frac{i\hbar}{\tau_{if}}}{(E_f - E_i - \hbar\omega)^2 - \left(\frac{i\hbar}{\tau_{if}}\right)^2}. \tag{10.27}$$

For the third-order absorption coefficient,

$$\alpha_3 = -\omega\sqrt{\frac{\mu}{\varepsilon_R}} \frac{I}{2c_0 n_r \varepsilon_0} \left|M_{fi}\right|^2 \left(\frac{m^* k_B T}{L\pi\hbar^2}\right) \times \ln\left(\frac{1+\exp\left(\frac{E_F - E_i}{k_B T}\right)}{1+\exp\left(\frac{E_F - E_f}{k_B T}\right)}\right)$$

$$\times \frac{1}{\left(\left(E_f - E_i - \hbar\omega\right)^2 + \left(\frac{\hbar}{\tau_{if}}\right)^2\right)^2}$$

$$\times\left[\frac{2}{\tau_{if}^2}(\hbar)\left(\tau_{ff} + \tau_{ii}\right)\left|M_{fi}\right|^2 - \frac{\left(M_{ff} - M_{ii}\right)}{\left(\left(E_f - E_i\right)^2 + \left(\hbar^2 / \tau_{if}^2\right)\right)}\right.$$

$$\left.\times\left\{\frac{\hbar}{\tau_{if}}\left(\left(E_f - E_i - \hbar\omega\right)^2 - \frac{\hbar^2}{\tau_{if}^2}\right) + \frac{\hbar}{\tau_{if}} 2\left(E_f - E_i\right)\left(E_f - E_i - \hbar\omega\right)\right\}\right]. \tag{10.28}$$

The dipole matrix element between states approximated as

$$M_{if} = |q|\langle i|z|f\rangle = |q| \int_{-\infty}^{+\infty} \psi_i z \psi_f \, dz. \tag{10.29}$$

Here the subscripts i and f denote the initial and final states, E_f and E_i are the energies of the initial and final states, k_B is the Boltzman constant and T is the temperature in degree Kelvin. The value of I is the optical power per unit area, c is the speed of light in free space, τ is the intrasubband relaxation time (for simplicity, we assume $\tau = \tau_{12} = \tau_{21} = \tau_{22} = \tau_{11}$), and E_F(eV) is the Fermi level measured from the bottom of the well. The corresponding doping concentration of the QW is equal to 3×10^{16}cm^3, the refractive index is n_r ($\in_R = n^2{}_R \in_0$), and the value of the parameters used are listed in Appendix 1

10.4.2.2 Intersubband Transition and Optical Properties

In this section, a detailed analysis of the AlGaAs/GaAs (InGaAs/GaAs) QW with various parameters of different structures will be presented. We have varied the Al(In) mole fraction from 0.2 to 0.4 (0.15 to 0.25) with the well width varied from 50 to 150Å (50 to 100Å). For all the AlGaAs/GaAs and InGaAs/GaAS DFQW structures analyzed, the diffusion length has increased from zero to L_d=50Å. The intersubband absorptions for those DFQWs that contain two or three subband states are determined. It is assumed that the third subband state only contributes negligible energy to the total intersubband absorption.

(A) Interdiffused subband states

For a fixed as-grown well width AlGaAs/GaAs structure, both small and large values of L_d corresponds to the narrow-deep and wide-shallow square QWs. Since the interdiffusion can modify the entire depth, width and shape of the QW structure, the first and second energy states can also be modified [10.258]. As a result, the intersubband transition energy can be changed. A general trend for the different Al concentration fractions (0.2 to 0.4) of the distribution of absorption peaks for different diffusion lengths can be seen: for small L_z under 90Å, there is an increase of diffusion length from zero to L_d = 20Å causing a large decrease of the intersubband transition energy, while the drop of the energy of the absorption decreases gradually as the diffusion increases from L_d=30Å to L_d = 50Å.

In a more detailed analysis, the first and second subband energies decrease monotonically for well widths under 90Å. When the well width increases from 90 to 130Å, the energies of the first and second subband increases to a maximum when $L_d \approx$20Å and then decreases monotonically. This agrees with the result of reference [10.258]. However, when the well width is increased to 150Å, the energies of the first subband state increases to a maximum when L_d = 30Å while the maximum energy of the second subband remains at 20Å. It should be noted that the second subband energy varies over a larger range than the first subband energy as the diffusion length increases. However, this effect reduces as the well width increases. As the first and second subband states changes little when the diffusion length increases at large well width, the intersubband transition energies for diffusion lengths 10Å and 20Å may become approximately the same when the well width is large, and may even shifted to higher transition energies, as compared to the square well with $L_d = 0$.

For the InGaAs/GaAs, we have also analyzed the diffusion length effect from 10 to 50Å at well widths 50 to 100Å. The In concentrations are 0.15, 0.2 and 0.25. In this range of analysis, two subband states are found in the QW only when no field or a small field is applied and they are usually presented in a larger well width (approximately 100Å), a higher concentration ratio (0.2 or 0.25) and a larger diffusion length. The energies of the first and second subband states, however, decrease monotonically when the diffusion length is increased for a well width up to 90Å. When the well width is equal to 100Å, the first subband state energies increase to a maximum in the case of $L_d = 20$Å and then decrease montonically as L_d is increased to 50Å. This is quite similar to the AlGaAs/GaAs DFQW structure. However, the range of variation of the subband state energies is much smaller than that of the AlGaAs/GaAs, since for small well widths (50 to 70Å) and at small diffusion lengths (10 to 20Å), there is only one subband state. While at larger well widths (90 or 100Å), although two subband states are presented for different L_ds, the range of second subband energies change is less than 100meV.

(B) Effect of QW's Al and In concentration

The total absorption coefficients of QW analyzed consist of the first- and third-order absorption coefficients only. The effect of increasing the diffusion length to the first- and third-order absorption coefficients for the AlGaAs/GaAs are shown in Fig. 10.14(a). The effect of the diffusion length in three different aluminium concentration ratios 0.2, 0.3, 0.4 are shown in this figure with the well widths all equal to 70Å and with an applied field equals to 10kV/cm. The incident optical intensity is set at 0.5MW/cm^2. We can observe that the total positive total absorption α is mainly contributed by α_1 and it is significantly reduced by the α_3. This reduction of absorption coefficient due to the decrease of α_3 is called bleaching. Within each concentration ratio, there is a general decrease of α_1 but the α_3 on the other hand becomes more negative as the L_d increases from 10 to 30Å and then to 50Å, as also shown in Fig. 10.14(a).

These two effects together cause the absorption coefficient to decrease as the wavelength is increased to higher diffusion lengths (as mentioned in Sect. 3.3.1, the increase in diffusion length would generally causes the transition energy to decrease). The total absorption coefficient of Fig. 10.14(a) is shown in Fig. 10.14(b). The reduction of the absorption coeffi-

(a) in 02-025 lz100 ld0-30-50 f0

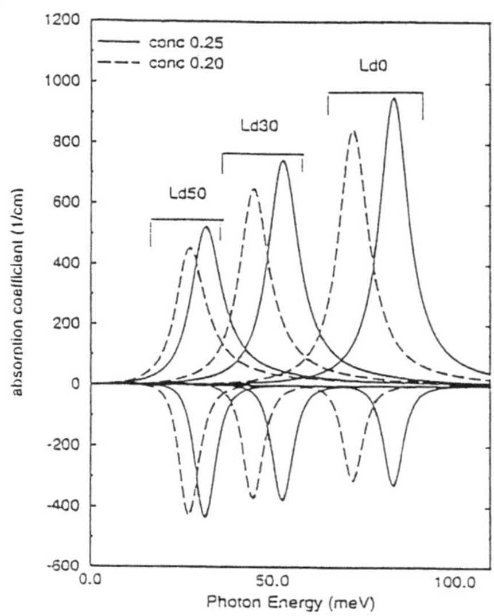

(b) in 02 025 lz100 ld0-30-50 f0

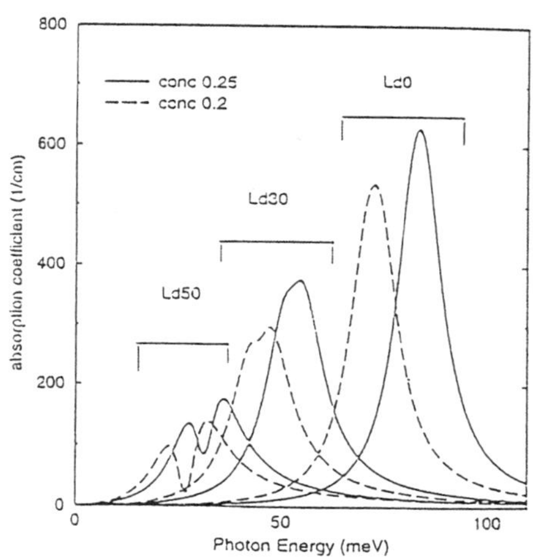

Fig. 10.14 a, b **(a)** The first- and third-order absorption coefficients at different L_ds (Al = 0.2, 0.3 and 0.4, L_z = 70Å, and F = 10kVcm^{-1}). **(b)** The total absorption coefficients at different L_ds (Al = 0.2, 0.3 and 0.4, L_z = 70Å, and F = 10keVcm^{-1}) [10.326].

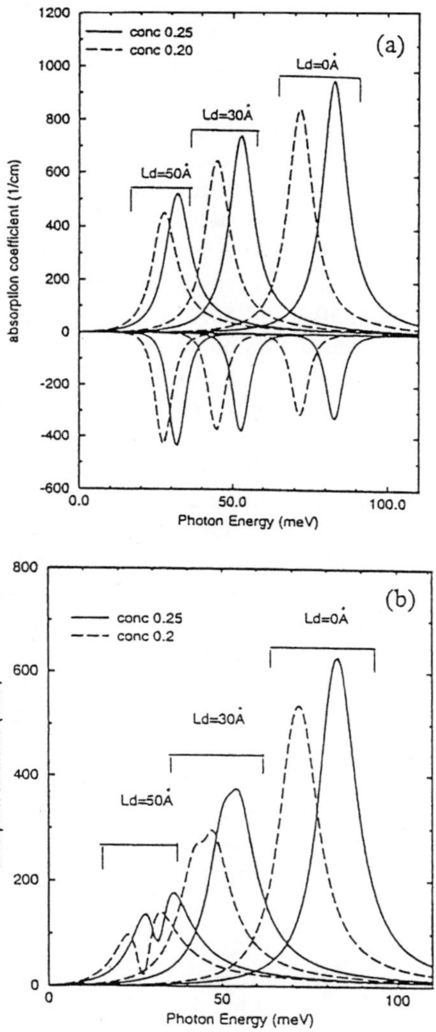

Fig. 10.15 a, b (a) The first- and third-order absorption coefficients at different L_ds (In = 0.2, 0.25, L_z = 100Å, and F = 0kVcm^{-1}). **(b)** The total absorption cofficients at different L_ds (In = 0.2, 0.25, L_z = 100Å, and F = 0kVcm^{-1}).

cient at high L_d (≈40Å) is mainly due to the significant drop of α_1 while α_3 remains approximately the same. In addition, the extent of the drop of absorption coefficients are different at various concentrations ratios. From Fig. 10.14(b), bleaching is observed when the diffusion length is increased to 40Å for an Al concentration ratio = 0.2. While the bleaching effect is obvious at L_d = 50Å for an Al concentration ratio = 0.3, it only begins to bleach at L_d = 50Å for an Al concentration ratio = 0.4. In addition to later bleaching, the absorption coefficient for higher concentration ratios is also larger in a small extent, too.

For the InGaAs/GaAs material system, a similar observation is seen. The absorption coefficient peaks are also mainly contributed by the first-order absorption coefficients and reduced by the third-order absorption coefficients. The case for a well width of 100Å, with no electric field applied, the incident optical power equals to 0.5 MW/cm^2, and the Indium concentration equals to 0.2 and also 0.25, are investigated. The diffusion lengths shown in Fig. 10.15(a) include 0Å, 30Å and 50Å. There is a general decrease of α_1 and an increase of α_3 as the diffusion length is increased. The extent of the decrease of α_1 is much more larger as compared with the increase of α_3. So, the total of absorption coefficients show a general decrease when the diffusion length increases gradually. In addition, the value of the absorption coefficient for larger concentration ratios (In = 0.25 in Fig. 10.15(a)) is higher than that of smaller Indium concentration ratios (In = 0.15 or 0.2). This result is similar to the AlGaAs/GaAs. However, the range of intersubband absorption coefficients that can exist without bleaching is much smaller than the AlGaAs/GaAs. With the optical absorption coefficient equals to 0.5MW/cm^2, for the well width at a maximum 100Å for an Indium concentration ratio = 0.25, we can only increase the diffusion length to 30Å. When the concentration ratio is increased, the extent of the diffusion length that can be increased becomes smaller. When the concentration ratio is 0.2 for a well width equals to 100Å, the maximum diffusion length that can be applied is 20Å. The effects of the well width and the incident optical power will be discussed in a later section. On the other hand, it should be noted that the transition energies becomes closer together for higher diffusion lengths as in Fig. 10.15(b).

(C) Effect of the QW well width

The analysis of the first and second subband state energies of the AlGaAs/GaAs have shown that when the well width is increased to 90Å, the extent of decrease of the second subband state energy reduces significantly as compared to those well widths lower than 90Å. In Fig. 10.16, the variaions of the absorption coefficients from 10 to 30Å are shown for two different well widths 70Å and 90Å. We can see that the transition energies of the absorption peaks in the case of the well width = 90Å is much closer together than that of L_z = 70Å. On the other hand, the reduction of the absorption coefficients for smaller well widths when the diffusion length is increased is larger than that of the larger well widths. So, if we want to obtain a flatter response of the absorption coefficient for different L_ds, a larger well width should be used.

For the InGaAs/GaAs, the effect of increasing the diffusion length is quite similar to the AlGaAs/GaAs material system. When the diffusion length is increased from 10 to 30Å in both of the QWs with well width equals to 70Å and 100Å respectively, the transition peaks for the 70Å case is smaller than that of the 100Å, as shown in Figs. 10.15(a) and 10.15(b). This general trend exists in the range of well widths investigated. However, the first-order absorption coefficient decreases as the well width is decreased while the third-order absorption coefficient remains more or less the same. The effect of bleaching occurs and this affects the optimal width bandwidth material selected for the InGaAs/GaAs.

(D) Wide bandwidth photodetection

From the DFQW intersubband results shown, it is noted that we can realized a wide bandwidth material for photodetection by the variation of the diffusion length, and thus for a DFQW, we can change its detection wavelength. In view of the different design parameters affecting the QW structures, as discussed in the previous three Sects. 10.3.1, 10.3.2 and 10.3.3, we need to consider basically the concentration ratio and the well width. A higher concentration ratio can generally have a higher absorption coefficient and the later occurrence of bleaching in view of the increase of diffusion length. On the other hand, a smaller well width can have a larger range of transition energy, and hence a larger bandwidth. However, a larger well width can have a higher absorption coefficient. In order to compromise, we need to consider the extent of each effect. It is observed that the change of transition energies is very large when the well width is small (see Fig. 10.13(b)) while the drop of absorption coefficient is not so serious as well width is increased (see Fig. 10.15). In Fig. 10.16, an optimized wide-bandwidth AlGaAs/GaAs material with the bandwidth equals to 13.578μm or 21.542μm (5.179μm (zero L_d) to 18.757μm (L_d = 30Å) or 26.721μm (L_d = 40Å)). For the L_d = 40Å QW structure, the absorption coeffiecent may have dropped so low that we shall only use it up to L_d = 30Å.

Similarly, the optimal parameters chosen for the wide-bandwidth InGaAs/GaAs is also dependent on the concentration ratio and well width used. To compromise for the same reasons as above, a small well width and high concentration material is needed. For the range of w = 0.15, 0.2 or 0.25 with the well width varies from 50 to 100Å, the largest bandwidth obtained is 11.57μm (from 14.7μm to 26.27μm). The parameters of the QW used is: In concentration ratio = 0.25, well width = 70Å, incident optical power = 0.5MW/cm^2 (see Fig. 10.17(a)).

The optical intensity used in the above analysis is equal to 0.5MW/cm^2. Since the optical intensity only contributes to α_3, the increase of optical intensity will cause bleaching to occur [10.182]. The effect of optical inten-

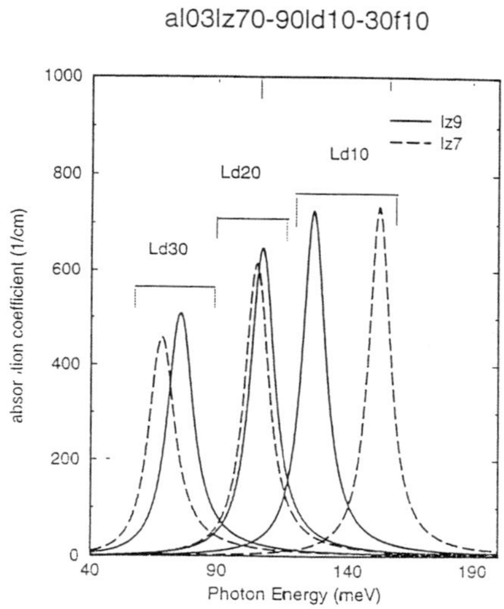

Fig. 10.16 Variation of the absorption coefficients at different well width DFQWs. (Al = 0.3, L_z = 70Å and 90Å, F = 10kVcm^{-1} [10.158].

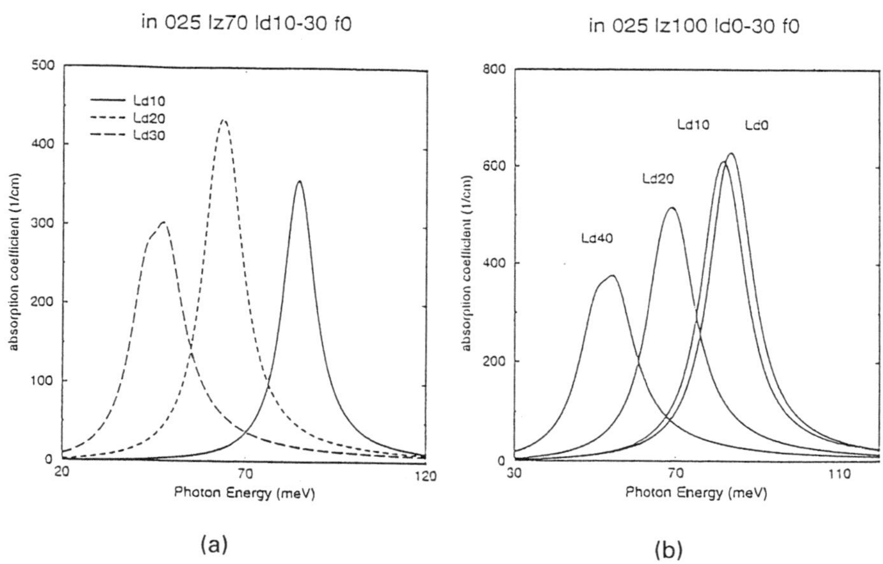

Fig. 10.17 a, b **(a)** Effect on the transition energies at the absorption peaks as L_d is increased in a DFQW with smaller L_z (In = 0.25, L_z = 70Å, and F = 0kVcm^{-1}). **(b)** Effect on the transition energies at the absorption peaks as L_d is increased in a DFQW with larger L_z (In=0.25, L_z = 100Å, and F = 0kVcm^{-1}).

sity for the AlGaAs/GaAs in the case corresponding to Fig. 10.18 is shown in Fig. 10.19. It is observed that when the optical intensity is increased to 0.75MW/cm^2, bleaching has already occurred for L_d = 30Å. So, an optical intensity of approximately 0.5MW/cm^2 is quite limiting for the use of this wide bandwidth material. While for the InGaAs/GaAs, it is even more limiting to use 0.5mW/cm^2 as the optical power. To consider the optimal case of the InGaAs/GaAs (Fig. 10.20), it is noted that when we increased the optical power intensity to 0.75MW/cm^2, bleaching is serious for L_d = 30Å and it begins to bleach for L_d = 20Å. So, we can see that when the optical power intensity is increased beyond 0.5MW/cm^2 for this optimal case, the bandwidth which we can use to detect light without bleaching is much more narrower.

10.4.2.3 Summary

The DFQW is very useful and flexible in tailoring the operation wavelength for the intersubband transitions. The variation of the transition energies which contributes to the intersubband absorption is found to be dependent on the well width of the QW. A wider bandwidth of operation can be found in the case of smaller well widths of 50 and 70Å. For a square well and slightly diffused QW, the linear absorption coefficient is much larger than the third-order absorption coefficient. However, when the diffusion length

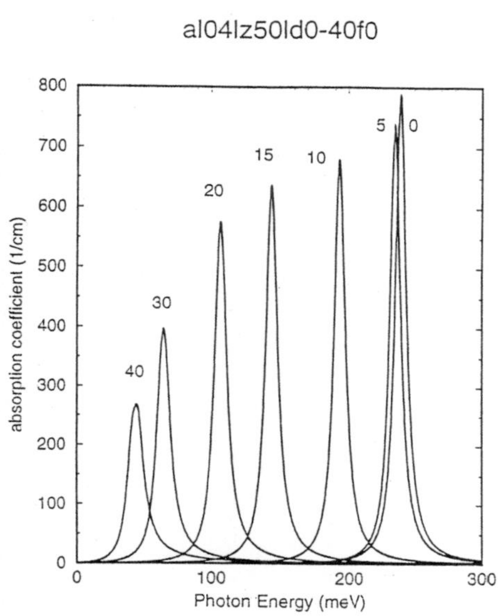

Fig. 10.18 Wide bandwidth operation for the AlGaAs/GaAs material by the variation of L_d (Al=0.4, L_z = 50Å, and F = 0kVcm^{-1}).

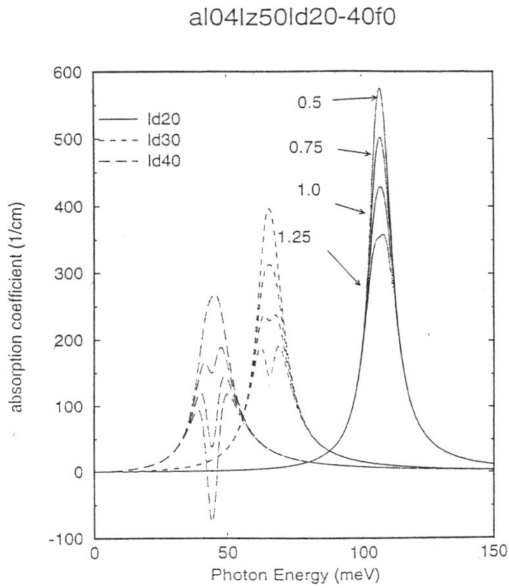

Fig. 10.19 Effect of increasing the optical intensities (in units of MW/cm^2) in DFQW with different L_d in the AlGaAs/GaAs material (Al = 0.4, L_z = 50Å, and F = $0kVcm^{-1}$).

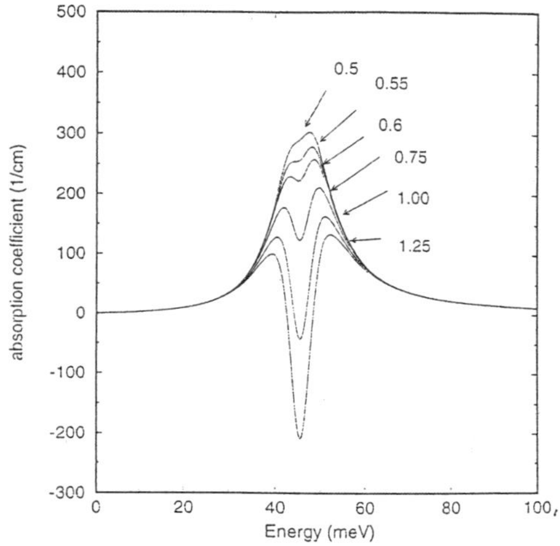

Fig. 10.20 Optimal optical intensities for (in units of MW/cm^2) the InGaAs/GaAs material for wide bandwidth operation (In = 0.25, Lz = 70Å, and Ld = 30Å).

is increased to 30 or 40Å (for optical intensity = 0.5MW/cm^2), the third-order absorption coefficients can reduce the absorption coefficients so that bleaching occurs. On the other hand, the absorption coefficient generally increases with the Al(In) concentration ratio used. A larger concentration ratio can increase the peaks of the absorption coefficient. With the concenration ratio and the well width appropriately chosen, a wide operation bandwidth and large absorption QW material for both the AlGaAs/GaAs and InGaAs/GaAs can be realized. The bandwidth of operation which equals to 18.8μm for the AlGaAs/GaAs and 11.6μm for the InGaAs/GaAs is possible while having the operation wavelength extended to the mid-IR region (~26μm) for both the AlGaAs/GaAs and InGaAs/GaAs material system. This is difficult to achieve by other metallic or bulk semiconductor material system or even a QW without diffusion. The large intersubband bsorption coefficients subjected to different diffusion lengths results in the possible use of the DFQW for photodetection. More research is needed for the DFQW with high carrier density, since the screening effect must be taken into account.

10.4.3 Linear and Nonlinear Intersubband Electroabsorptions in DFQW

Modulation-doped quantum-well structures have been widely used in studying interband and intersubband optical transitions. The spatial separation between the charged dopant atoms and the free electronic carriers creates an electric potential in a modulation-doped QW. This induced electric potential will modify the original potential profile which includes the built-in and external applied potentials. The modification of the overall potential profile leads to the changes in the electron energy levels and envelope wave functions, and subsequently the changes in the electronic and optical properties of the QW.

In this section, the intersubband optical processes in a modulation doped AlGaAs/GaAs QW are considered. The total width of the well plus the outer barrier region is equal to 100 nanometres. In order to investigate the influence of interdiffusion and the carrier effects in the QW thoroughly, DFQWs with different well widths and Aluminium concentration ratios are considered. The carrier-induced effects are taken into account by solving the Schrödinger and Poisson's equations self-consistently. The changes in the transition energy, linear absorption coefficient and refractive index in these modulation-doped QW are analysed for their suitable use in IR photo-etection. In addition, the nonlinear effect due to the optical intensity is examined and the saturation intensity is also included.

In the following sections, the formulation of the linear and nonlinear intersubband optical-absorption coefficients and refractive indices in the

DFQW and the procedures for solving the Schrödinger equation and Possion's equation self-consistently are discussed. The electronic properties, the carrier effects and the nonlinear effects due to optical field intensity and carrier effect are also analyzed. Furthermore, applications based on the effects of interdiffusion to the intersubband absorptions and the change of refractive indices in QWs with different parameters are also investigated.

10.4.3.1 Model of the Linear and Nonlinear Intersubband Optical-Absorption Coefficients and Change in Refractive Indices

Schrödinger-Poisson Solver

The interdiffusion-induced DFQW confinement profile $V_C(z)$ is defined here by (10.2) and using the Ben-Daniel and Duke model [10.261] with a z-position dependent effective mass on the interdiffused composition profile, by the one-dimensional Schrödinger-like equation shown in (10.14), which is solved self-consistently with the Poisson's Equation,

$$\frac{d^2}{dz^2}V_{He}(z) = -\frac{4\pi e}{\varepsilon}(n_{hh} + n_{lh} - n_e(z)) \tag{10.30}$$

where $V_{Hhh}(z) = V_{Hlh}(z) = nV_{He}(z)$, and n_e, n_{hh}, n_{lh} are the spatial distributions for the electron, heavy hole and light hole respectively. The Hartree terms V_{He}, V_{Hhh} and V_{Hlh} are obtained from the Poisson's equation and the boundary condition is taken with these Hartree terms equal to zero at $z = \pm 50$nm. The envelope wave functions inside the interdiffused QW can then be solved [10.262].

In order to calculate the change in the absorption coefficient and the refractive index for a given QW structure, we need to find out the Fermi level of the QW. In the QW, the electrons in each subband are considered as a quasi-two-dimensional electron gas. In each subband, the electrons occupy states in both the well region and the barrier region, and the ratio between the two regions depends on the probability of electron distribution in each region. For the *i*th subband, the surface concentration of electrons N_n, can be expressed as

$$N_n = \left(\frac{m_e^*}{\pi\hbar^2}\right)k_B T \ln\left\{1 + \exp\left[\frac{(E_f - E_n)}{k_B T}\right]\right\}, \tag{10.31}$$

m_e^* is the effective mass of the electrons inside the QW, w is the Al concentration ratio inside the QW, m_O is the free electron mass, E_f is the Fermi level energy of the QW, E_n is the energy of each subband where the subscript n refers to the number of that subband, and T is the temperature in degree Kelvin.

The total density of electrons in the QW N_S is the summation of the N_n of all the subbands in (10.12)

$$N_s = \sum N_n . \tag{10.32}$$

The quasi-Fermi level can then be obtained by solving (10.12) and (10.13) with the Schrödinger equation and Poisson's equation self-consistently [10.263].

Optical absorption coefficients and change in the refractive indices

With the Fermi level and the envelope wave functions calculated, the linear and nonlinear intersubband absorption coefficients and the change in refractive indices to the optical radiation will be [10.263]:

$$\alpha(\omega) = 2\,\mathrm{Im}(k) = \omega\sqrt{\frac{\mu}{\varepsilon}}\sum_f\sum_i \frac{\left|M_{fi}\right|^2 (n_i - n_f)(\hbar\gamma)}{(E_f - E_i - \hbar\omega)^2 + (\hbar\gamma)^2}\left(\frac{1}{1+\dfrac{I}{I_s}}\right) \tag{10.33}$$

$$\begin{aligned}\Delta n(\omega) &= \sqrt{\frac{\varepsilon}{\varepsilon_o}}\left(\frac{1}{2\varepsilon}\right)\mathrm{Re}[\varepsilon_o \chi_{zz}(\omega)] \\ &= \left(\frac{1}{2n_r\varepsilon_o}\right)\mathrm{Re}[\varepsilon_o \chi_{zz}(\omega)] \\ &= \left(\frac{1}{2n_r\varepsilon_o}\right)\sum_f\sum_i \frac{\left|M_{fi}^z\right|^2 (n_f - n_i)(E_f - E_i - \hbar\omega)}{(E_f - E_i - \hbar\omega)^2 + (\hbar\gamma)^2}\left(\frac{1}{1+\dfrac{I}{I_s}}\right)\end{aligned} \tag{10.34}$$

where

$$I = 2n_r|E|^2/\mu c,$$

$$I_s = \frac{n_r}{2\mu c}\frac{(E_f - E_i - \hbar\omega)^2 + (\hbar\omega)^2}{\left|M_{fi}^z\right|^2}.$$

n_i and n_f are the densities of the electrons in the initial (i) subband and the final (f) subband of the intersubband transition of the unperturbed system; γ, μ, ε and n_r are the relaxation time, the permeability, the real part of permitivity and the refractive index of the QW respectively. I is the intensities of the incident optical radiation and I_S is the saturation intensity of the intersubband optical transition between the two subbands i and f respectively.

10.4.3.2 Tunability of the Optical Properties

In this section, the theoretical calculation of the electron energy levels, linear and nonlinear intersubband absorption coefficients and changes in the refractive indices of the conduction band under the QW with different diffusion lengths are analysed.

We have considered earlier the modulation-doped structures of the QWs. In addition, a wide range of parameters of the QW are used. The aluminium concentration ratio used includes: 0.2, 0.3 and 0.4 with widths equal to 70, 90 and 130Å. The doping concentration of Si in the barrier includes 1×10^{16}, 1×10^{17} and $1\times10^{18}\text{cm}^{-3}$. The total surface concentration N_S in (10.9), is equal to the doping concentration times the total width of the doped region. Finally, the diffusion length L_d, which is the main concern here, is varied from 0 to 40Å in steps of 10Å. The parameter list is shown in Appendix 1.

w is the Al concentration ratio at the center of QW, m_0 is the free electron mass and ε_0 is the dielectric constant of free space. As interdiffusion occurs, the Al atoms diffuse into the QW while the Ga atoms inside the well diffuse out so, the value of w inside the QW varies as the potential profile changes with different L_d.

The results presented here are divided into three parts: in Sect. 10.4.3.1, the electronic properties of the interdiffused QW are discussed; in Sect. 10.4.3.2, the influences of the interdiffusion to the intersubband absorption coefficients and the changes in refractive indices of the QWs with the different parameters are considered; in Sect. 10.4.3.3, some applications of the effects due to the interdiffusion discussed in Sect. 10.4.3.2 are discussed.

(A) Intersubband transition energy

The QW structures, which are being considered in this section, contain at most five envelope wave functions inside the well. In Figs. 10.21–10.24, the envelope wave functions inside the QW with a well width of 70Å, doping concentrations of $1\times10^{16}\text{cm}^{-3}$ and $1\times10^{18}\text{cm}^{-3}$ and diffusion lengths of 0 and 30Å are plotted respectively. Since the intersubband absorption is mainly contributed by the intersubband transition between the ground and the first excited state, it is necessary to know the variation of the two subband energies in order to analyse the trend of the transition energies due to the change in L_d.

It is noted that, for a well width of 70Å, the ground state energies rise to a maximum at $L_d = 10$Å, and then decreases monotonically as the diffusion length increases. On the other hand, for well widths of 90 and 130Å, the energies of the ground state rise to a maximum at L_d = 20Å, and then decreases to its minimum when L_d = 40Å. For the first excited state, the trend for the variation of its energies is the same as the ground states for L_z = 70 and 130Å . However, the energy of the first excited state in the 90 Å wide QW increases to a maximum when L_d = 10Å instead of 20Å as for the ground state. This shifting of energy levels as discussed above can be explained by using a cross-over point (COP), which is defined as the potential, measured from the well bottom, of the nonlinear profile at the well barrier interface [10.241 and 10.242].

The variations of the transition energies with the diffusion length depends on the well width used but not the carrier density. For L_z = 70Å, the transition energy decreases with the increase in diffusion length. When L_z = 90Å, the transition energy increases to maximum when L_d = 10Å and then decreases. While for L_z = 130Å, the transition energy peaks at L_d = 20Å. It is also noted that when w increases, the transition energies have a blue shift for any diffusion length used.

The range of variation of the second subband is much larger than the first subband especially in a QW with smaller well width (< 90Å). Moreover, the number of energy states increases and they are packed more closely together for the QW with larger well width than with smaller

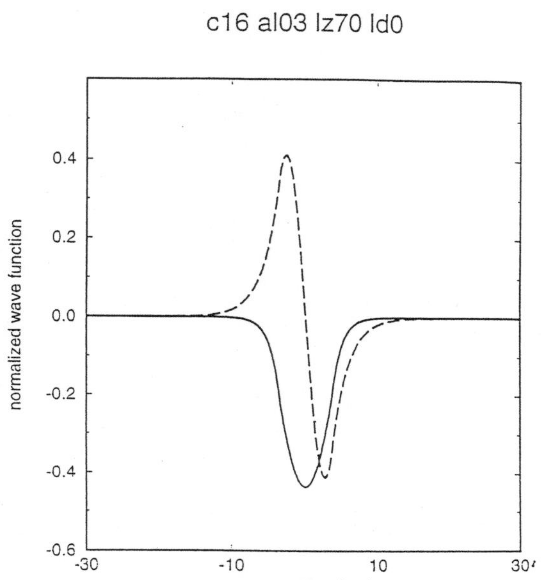

Fig. 10.21 The wave functions in an as-grown QW with doping concentration equal to 10^{16}cm^{-3} [10.182].

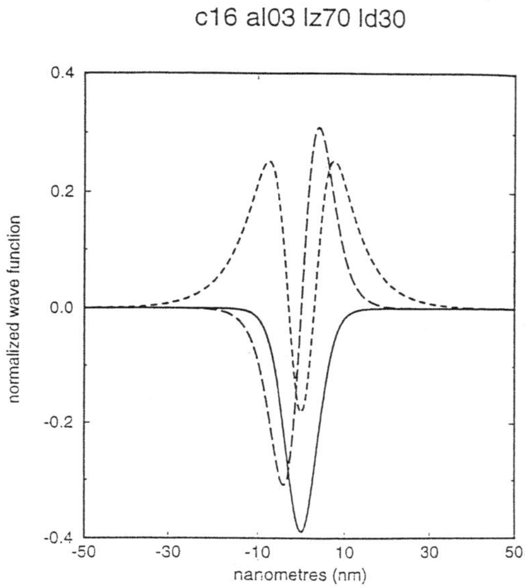

Fig. 10.22 The wave functions in a diffused QW with the doping concentration equals to $10^{16}cm^{-3}$ [10.182].

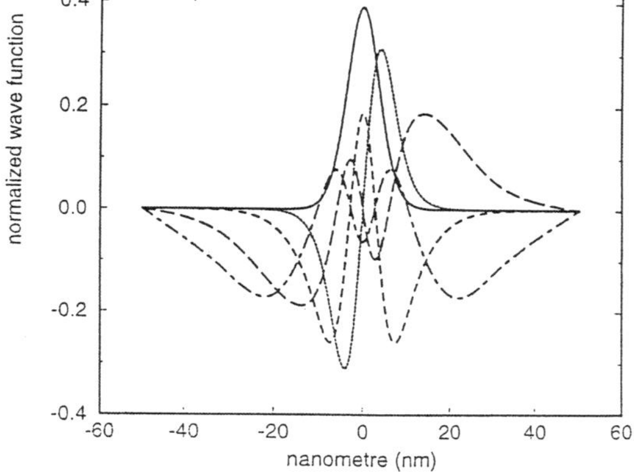

Fig. 10.23 The wave functions in a as-grown QW with the doping concentration equals to $10^{18}cm^{-3}$. The solid lines are for the ground state, the dashed lines are for the first excited state and the dot-dashed line is for third excited state [10.182].

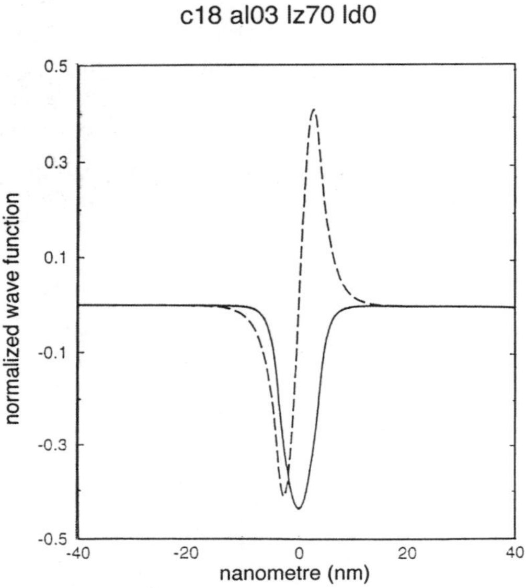

Fig. 10.24 The wave functions in a diffused QW with the doping concentration equals to 10^{18}cm^{-3} [10.182].

well width. As a consequence, the variation of the transition energies between the first two subbands are also affected as the interdiffusion proceeds. Since the intersubband absorption is mainly contributed from the first two subbands, the transition energies between them are very important for choosing the operation wavelength of the photodetector and modulator. So, in Table 10.1, the transition energies between the ground state and the first excited state for the aluminium mole fraction = 0.3 are listed and the corresponding dipole moments are shown in Table 10.2. The dipole moments decrease to a minimum when L_d = 10Å and then increases monotonically as the diffusion length increases for L_z = 70 and 90Å. For L_z = 130Å, the dipole moment decrease to a minimum when L_d = 20Å. The trend for L_z = 90 and 130Å are the same for w = 0.2, 0.3 and 0.4. However, when L_z = 70Å for w = 0.2 is considered, the values of the dipole moment increase with the increase in diffusion length.

(B) Linear and nonlinear optical properties

It has been discussed that the higher the densities of the electrons inside the GaAs well, the higher the absorption coefficients and the change of refractive indices [10.260 and 10.263]. This is because the increase in the densities of the electrons inside the QW will increase also the difference between the densities of the electrons in the lowest two subbands and also

Table 10.1 Transition Energies for the Intersubband Optical Transition between the Lowest Two Subbands in the AlGaAs/GaAs QW (L_Z) with Different Diffusion Length, Well Width and Doping Concentration with Al Concentration Ratio = 0.3 (the transition energies are in meV).

Well width L_Z (Doping concentration in cm^{-3})	L_d				
	0Å	10Å	20Å	30Å	40Å
70Å (1×10^{16})	155	153	104	68	47
70Å (1×10^{17})	155	153	104	68	48
70Å (1×10^{18})	130	119	82	55	39
90Å (1×10^{16})	113	128	107	75	54
70Å (1×10^{17})	113	128	107	76	54
70Å (1×10^{18})	100	105	86	61	45
130Å (1×10^{16})	66	80	90	78	61
170Å (1×10^{17})	66	80	90	78	61
170Å (1×10^{18})	61	70	74	64	50

Table 10.2 The Dipole Moments for the Intersubband Optical Transition between the Lowest Two Subbands in the GAAS Well with Different Diffusion Length, Well Width and Doping Concentrations (the dipole moments are in electron-angstroms (e-Å)).

Well width L_Z (Doping concentration in cm^{-3})	L_d				
	0Å	10Å	20Å	30Å	40Å
70Å (1×10^{16})	19	19	22	27	31
70Å (1×10^{17})	19	19	22	27	32
70Å (1×10^{18})	19	19	22	27	32
90Å (1×10^{16})	23	21	23	26	30
70Å (1×10^{17})	23	21	23	26	30
70Å (1×10^{18})	23	21	23	26	30
130Å (1×10^{16})	30	27	25	27	29
170Å (1×10^{17})	30	27	25	27	29
170Å (1×10^{18})	30	27	25	27	29

their dipole moments. The dipole moment decreases as the doping concentration increases for the case of small well width (70Å), but it increases for larger well widths (90Å and 130Å). The absorption coefficients, shown in Figs. 10.25–10.30 are for $L_d = 0$ and 40Å and the change of the refractive indices shown in Figs. 10.31–10.36 are for $L_d = 0$ and 30Å. The absorption coefficients generally increase with the doping concentration for any L_d from 0 to 40Å, but it is noted that when the doping concentration is increased to 1×10^{18}cm^{-3}, the ratio of the drop of the absorption coefficients when the diffusion length is raised to 40Å respective to 0Å is larger than when the doping concentration is equal to 1×10^{17}cm^{-3} and 1×10^{16}cm^{-3} in a QW with well width = 70Å. The peaks of the absorption coefficients are shown in Table 10.3. This may be due to the larger screening effect at higher doping concentrations and also at higher diffusion lengths (20–40Å), or due to the drop of the Fermi level, and this requires further analysis. For the refractive indices shown in Figs. 10.31–10.36 (well width = 70Å), it is noted that there are some sudden changes of the refractive indices besides the change of shape at the transition energies of the lowest two subbands. This is due to the fact that as the diffusion length increases, more electron states are found inside the well (as explained in Sect. A). So, more possible energy states are available for the electrons to transit which will then cause the sudden changes in the refractive indices. In QWs with larger well widths, the increase of the diffusion length has some special effect, and this will be discussed later.

Table 10.3 The Maximum Absorption Coefficient for the Intersubband Optical Transition at the Transition Energies of the Lowest Two Subbands for $x = 0.3$ (the unit of absorption coefficient is in cm^{-1}).

Well width L_Z (Doping concentration in cm^{-3})	L_d				
	0Å	10Å	20Å	30Å	40Å
70Å (1×10^{16})	413	391	355	306	252
70Å (1×10^{17})	4132	2082	3541	3061	2459
70Å (1×10^{18})	41071	38980	34286	27653	21429
90Å (1×10^{16})	315	320	296	259	219
90Å (1×10^{17})	3173	3194	2918	2582	2173
90Å (1×10^{18})	30510	31735	28265	23776	19286
130Å (1×10^{16})	191	209	214	196	174
130Å (1×10^{17})	1893	3918	2135	1959	1740
130Å (1×10^{18})	16633	19184	20510	18110	15612

The saturation intensity is defined as the intensity of the incident optical radiation that reduces the maximum absorption coefficient to half of that of the linear absorption. This is the nonlinear effect. In Figs. 10.23–10.27, the absorption coefficients when the optical intensity is equal to I_S are shown, while the change of the refractive indices are shown in Figs. 10.31–10.36. The values of I_S at different diffusion lengths for the Al concentration equals to 0.3 are shown in Table 10.4.

The values show that for L_Z = 70 and 90Å, The saturation intensities for all doping concentrations increase to a maximum when L_d = 10Å and then decrease as the diffusion length increases. When the well width is increased to 130Å, the maximum saturation intensity is found with L_d = 20Å. The range of optical radiation intensity for operation is then affected because of the variations of I_S caused by the interdiffusion. The above changes are basically the same for w = 0.2 and 0.4, except when w = 0.2 and L_Z = 70Å. The increase in the saturation intensity decreases with the increase of the diffusion length.

(C) Device applications

During the analysis of the linear absorption coefficients over a wide range of material parameter structures, it is found that the transition energies of α varies with the diffusion length and so this property can be used for wide

Table 10.4 The Saturation Intensities for the Intersubband Optical Transition between the Lowest Two Subbands in the AlGaAs/GaAs QW with Different Lengths, Doping Concentrations and Well Widths (the saturation intensities are in MW/cm^2, Al = 0.3).

Well width L_Z (Doping concentration in cm^{-3})	L_d				
	L_d = 0Å	L_d = 10Å	L_d = 20Å	L_d = 30Å	L_d = 40Å
70Å (1×10^{16})	3.21268	3.36358	2.44814	1.65150	1.18567
70Å (1×10^{17})	3.21311	3.36410	2.44911	1.65287	1.18172
70Å (1×10^{18})	3.21334	3.36714	2.45367	1.66172	1.18872
90Å (1×10^{16})	2.32754	2.64847	2.36148	1.74773	1.29459
70Å (1×10^{17})	2.32782	2.64903	2.36234	1.74886	1.29587
70Å (1×10^{18})	2.32546	2.64931	2.36476	1.75175	1.32024
130Å (1×10^{16})	1.36087	1.63955	1.86782	1.68823	1.37759
170Å (1×10^{17})	1.36076	1.63970	1.86838	1.68902	1.37850
170Å (1×10^{18})	1.35467	1.63477	1.86636	1.68775	1.37757

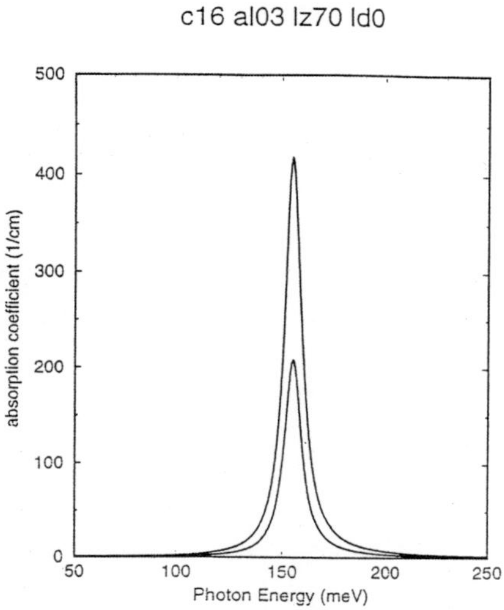

Fig. 10.25 The linear intersubband absorption coefficient for an as-grown and diffused QW with the doping concentration equals to 10^{16}cm^{-3} and L_d = 0Å. The reduced absorption is for the case with the optical intensity equals to the saturation intensity and the non-reduced one is for the case with the incident optical intensity equals to zero.

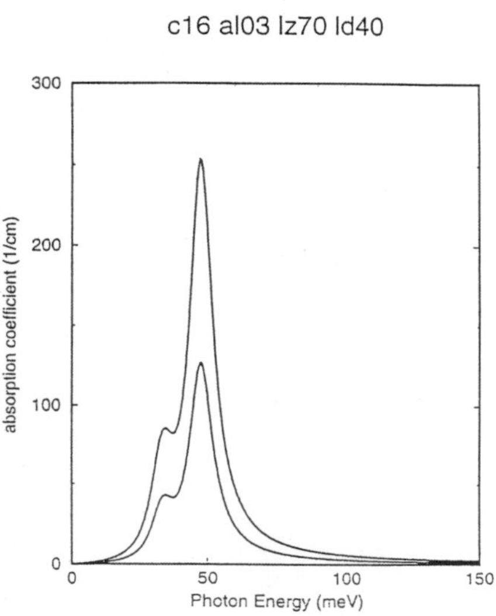

Fig. 10.26 The nonlinear intersubband absorption coefficient for an as-grown and diffused QW with the doping concentration equals to 10^{16}cm^{-3} and L_d = 40Å. The reduced absorption is for the case with the optical intensity equals to the saturation intensity and the non-reduced one is for the case with the incident optical intensity equals to zero.

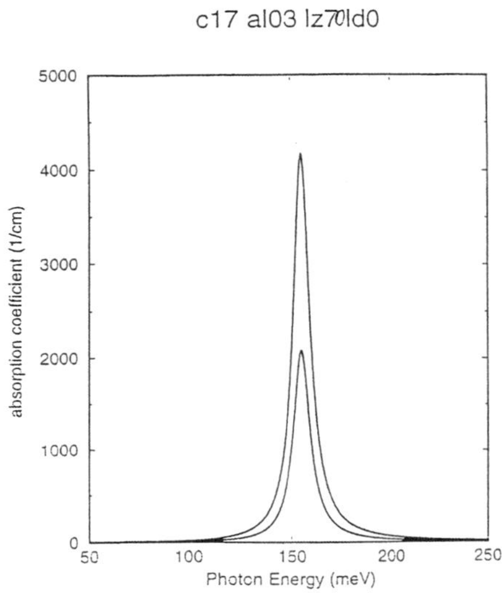

Fig. 10.27 The linear intersubband absorption coefficient for an as-grown and diffused QW with the doping concentration equals to 10^{17}cm^{-3} and $L_d = 0\text{Å}$. The reduced absorption is for the case with the optical intensity equals to the saturation intensity and the non-reduced one is for the case with the incident optical intensity equals to zero.

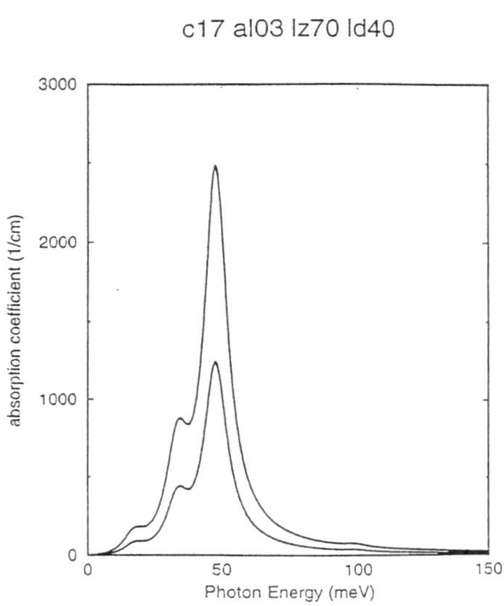

Fig. 10.28 The nonlinear intersubband absorption coefficient for an as-grown and diffused QW with the doping concentration equals to 10^{17}cm^{-3} and $L_d = 40\text{Å}$. The reduced absorption is for the case with the optical intensity equals to the saturation intensity and the non-reduced one is for the case with the incident optical intensity equals to zero.

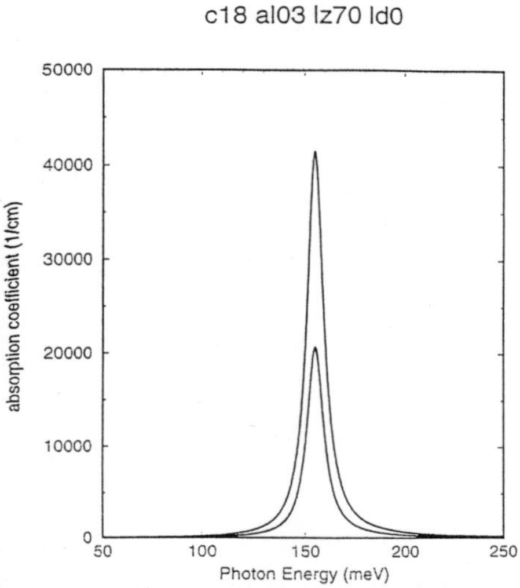

Fig. 10.29 The linear intersubband absorption coefficient for an as-grown and diffused QW with the doping concentration equals to 10^{18}cm^{-3} and $L_d = 0\text{Å}$. The reduced absorption is for the case with the optical intensity equals to the saturation intensity and the non-reduced one is for the case with the incident optical intensity equals to zero.

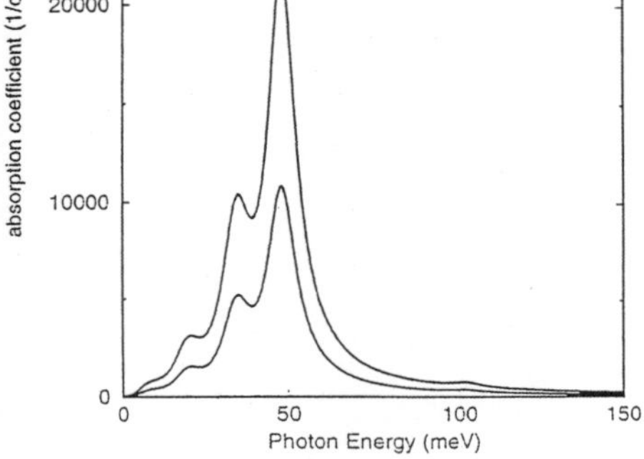

Fig. 10.30 The nonlinear intersubband absorption coefficient for an as-grown and diffused QW with the doping concentration equals to 10^{18}cm^{-3} and $L_d = 40\text{Å}$. The reduced absorption is for the case with the optical intensity equals to the saturation intensity and the non-reduced one is for the case with the incident optical intensity equals to zero [11.47].

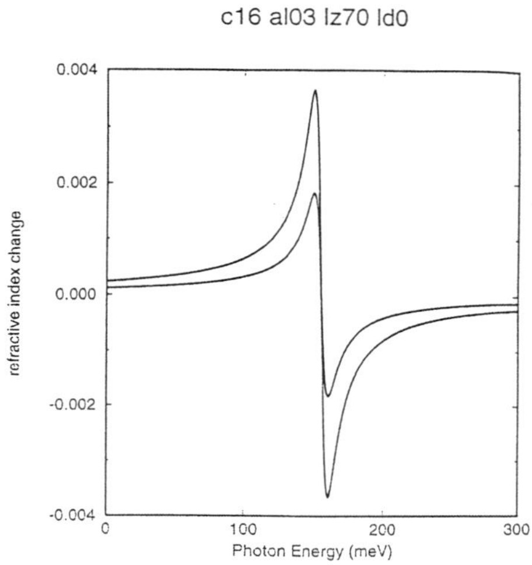

Fig. 10.31 The linear intersubband refractive index change for an as-grown and diffused QW with the doping concentration equals to 10^{16}cm^{-3} and L_d = 0Å. The reduced refractive index is for the case with the optical intensity equals to the saturation intensity and the non-reduced one is for the case with the incident optical intensity equals to zero [10.327].

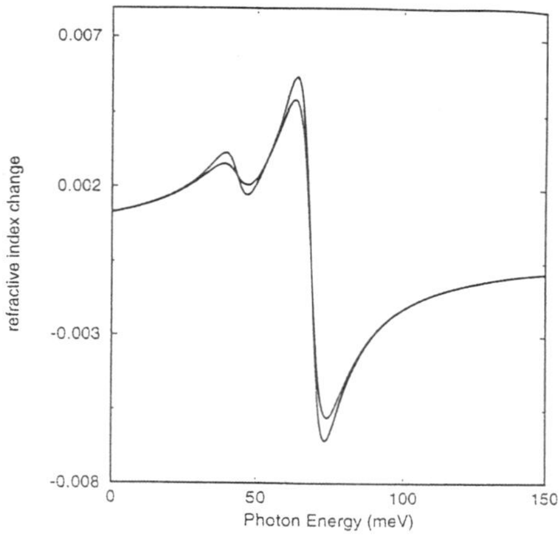

Fig. 10.32 The nonlinear intersubband refractive index change for an as-grown and diffused QW with the doping concentration equals to 10^{16}cm^{-3} and L_d = 30Å. The reduced refractive index is for the case with the optical intensity equals to the saturation intensity and the non-reduced one is for the case with the incident optical intensity equal to zero. [10.327]

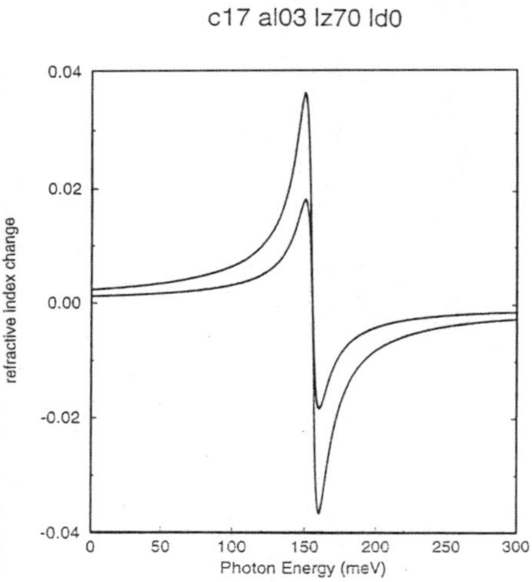

Fig. 10.33 The linear intersubband refractive index change for an as-grown and diffused QW with the doping concentration equals to 10^{17}cm^{-3} and L_d = 0Å. The reduced refractive index is for the case with the optical intensity equals to the saturation intensity and the non-reduced one is for the case with the incident optical intensity equals to zero [10.327].

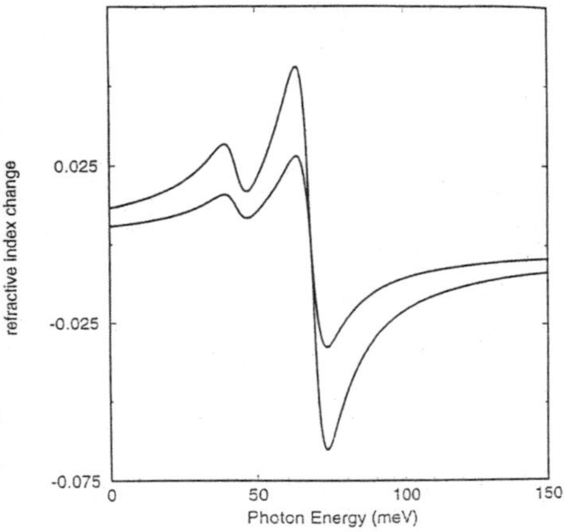

Fig. 10.34 The nonlinear intersubband refractive index change for an as-grown and diffused QW with the doping concentration = 10^{17}cm^{-3} and L_d = 30Å. The reduced refractive index is for the case with the optical intensity equals to the saturation intensity and the non-reduced one is for the case with the incident optical intensity equals to zero.

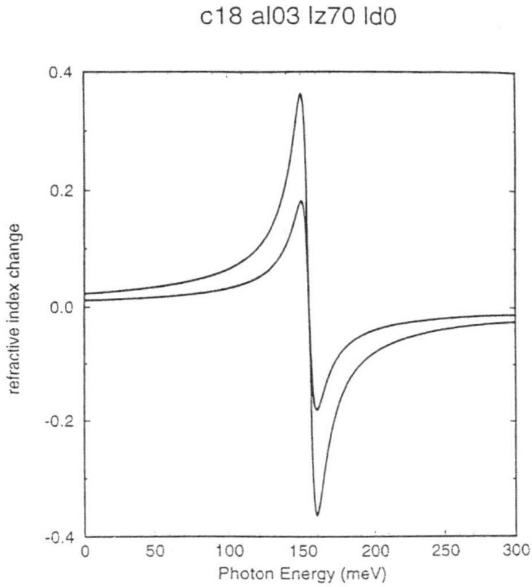

Fig. 10.35 The linear intersubband refractive index change for an as-grown and diffused QW with the doping concentration = 10^{18}cm^{-3} and $L_d = 0$Å. The reduced refractive index is for the case with the optical intensity equals to the saturation intensity and the non-reduced one is for the case with the incident optical intensity equals to zero.

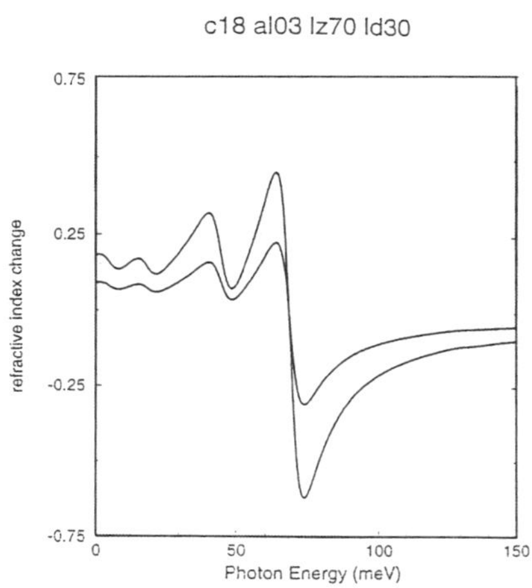

Fig. 10.36 The nonlinear intersubband refractive index change for an as-grown and diffused QW with the doping concentration = 10^{18}cm^{-3} and $L_d = 30$Å. The reduced refractive index is for the case with the optical intensity equals to the saturation intensity and the non-reduced one is for the case with the incident optical intensity equals to zero.

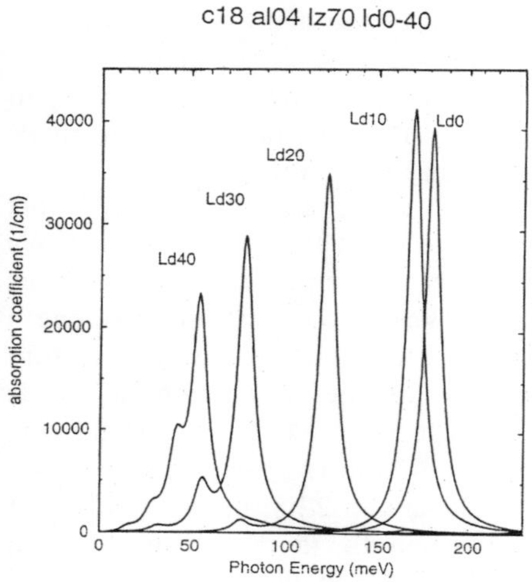

Fig. 10.37 The absorption coefficients in different diffused QWs for wide bandwidth photodetection.

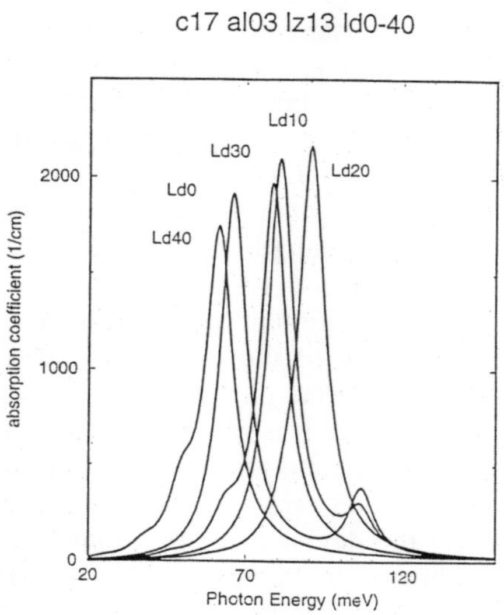

Fig. 10.38 The absorption coefficients in different diffused QWs for broadband photodetection [10.182].

bandwidth photodetection materials, since the bandwidth of operation increases with the rise of the aluminium concentration ratio and the reduction of well width. A wide bandwidth material of 15.9μm (from 6.9 to 22.8μm) with L_Z = 70Å and w = 0.4 is shown in Fig. 10.37. Besides the wide bandwidth operation, it can also be used as the material for multiple-colour QW infra-red photodetectors (QWIPs). As observed in Fig. 10.37, the overlapping between the adjacent absorption spectra corresponding to different L_ds (except for L_d = 0 and 10Å) is so small that a sharp discrepancy for the operation of the QWIP can be obtained.

Comparing with those two-colour QWIP using different well widths QWs [10.297], the DFQW has the advantage of fabricating several QWs with the same size but operating them at different temperatures. This makes fabrication easier. In addition, if we want to vary the operation wavelength of the QW infra-red photodetector, it is flexible to change the temperature applied to the DFQW. The even response of the absorption as L_d is increased is also a favourable feature for it to be used. In Fig. 10.37, the operating wavelengths of the DFQW include 6.9μm (L_d = 0), 7.3μm (L_d = 10Å), 10.2μm (L_d = 20Å), 16.3μm (L_d = 30Å) and 22.8μm (L_d = 40Å) .If we want to vary slightly the operation wavelength, then we can choose a DFQW with different L_z and w, or even apply an electric field to the DFQW.

A second type of application for the DFQW material for intersubband transitions is seen in the QW with a larger L_z of 130Å. In Fig. 10.38, a broadband operation of 6.5μm (13.7μm to 20.2μm) is shown. At larger well widths (> 10Å), the variation of L_d changes the transition energies of the absorption coefficients only a little. However, the peak values of the absorption remain approximately the same as L_d is increased (for smaller well width, L_z = 70Å, the peak of absorption reduces as L_d is increased). As a result, it is appropriate to design QWIPs with DFQWs for broadband operation at larger well widths. Nevertheless, it is noted that the operation range for the AlGaAs/GaAs is at the higher wavelength region, approximately larger than 12μm (100meV) for w = 0.2, 03 or 0.4.

In [10.260], when choosing the operation wavelength at the intersubband transition energies, the problem of the sensitivity of the refractive index change with the QW width was pointed out. During the physical fabrication to the desired value, if the variation of the well width is more than 10Å, it will cause a large change in the refractive indices and also the absorption values. It is found that the change of the transition energy is small for large well width DFQW, it can be used to correct the deviation of the transition energies due to the error in well width fabrication. Sample results of the variation of the absorption coefficients and the change in refractive indices at different L_d for this application are shown in Figs. 10.39 and 10.40. It is found that the increase of diffusion length to L_d = 10 and 20Å can vary the transition energies to about 5–15meV, which is sufficient to tackle the problem of the well width change, as pointed out in [10.260].

As shown in Fig. 10.39, we have found that the absorption coefficient at L_d = 0Å will usually have a minor peak for large well width (> 100Å) QWs (also shown as the sudden change of refractive index besides the large change at the transition energies of the lowest two subbands in Fig. 10.40). This means that the possibility for the electrons at the ground states to promote to higher energy states is larger for bigger well width QW structures. However, when we increase the diffusion length to 20Å as in Fig. 10.40, the sudden variations in refractive index disappears. So, it is a

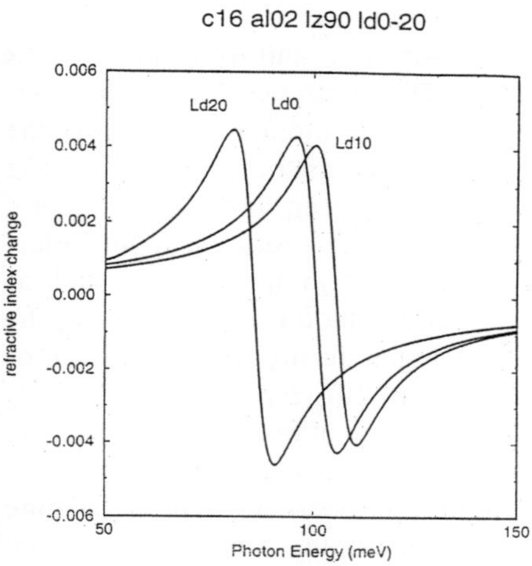

Fig. 10.39 Variation of the refractive index changes by interdiffusion in QWs with L_z = 90 Å, Al conc. = 0.2 and doping concentration = 10^{16}cm^{-3}.

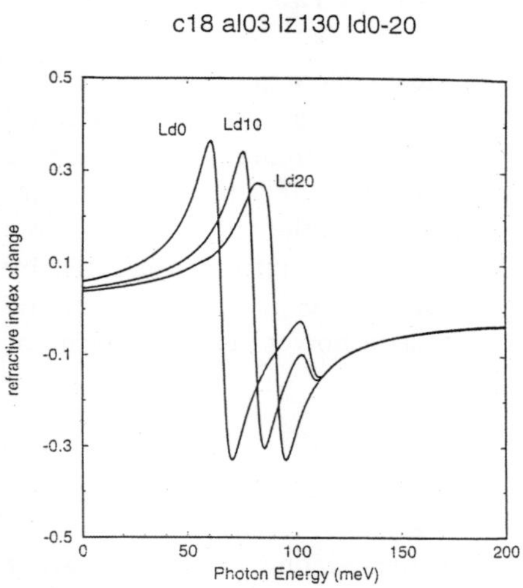

Fig. 10.40 Refractive index change varies as the diffusion increases in a QW with L_z = 130Å and doping [10.327].

probable application for using DFQW to remove this undesirable refractive index change (and also the absorption coefficient) which may affect the photodetection of light in some situations. Nevertheless, when the diffusion length is increased to a large value (40Å), these minor peaks of absorption and refractive index changes usually appear.

10.4.3.3 Summary

The linear and nonlinear intersubband absorption and refractive index changes in the DFQW have been considered. Different electronic properties including the transition energies and the dipole moments between the lowest two subbands and the optical properties such as the absorption peaks and the saturation intensities in the DFQW have been analysed in detail. The characteristic of the variation of the transition energies in the interdiffused QW with different well widths can be used in applications such as the multi-color QW and the broadband photodetector. The shift in the absorption peaks and the refractive index changes during the fabrication of the QW can also be corrected by applying a 10 to 20Å L_d.

The screening effect and nonlinear effect have been taken into account in the calculations. However, the many body effects due to the interactions among the electrons are not considered in this section. It will become important for heavier concentrations. For example, the depolarization effects and exciton-like effects [10.267] shift the energy levels and modify the absorption spectra. More in-depth analysis is required in these areas.

10.4.4 Enhancement of the Nonlinear Third-Order Susceptibility

Nonlinear optics [10.268] is the study of the phenomena that occur as a consequence of the modification of the optical properties of a material system by the presence of light. Typically, only higher power laser is sufficiently intense to modify the optical properties of a material system. Conventionally, nonlinear optics is often taken to be the study of intensive light generated second-harmonic frequency, as it is first demonstrated by Franken in 1961 [10.124]. The nonlinear optical response can often be described by expressing the polarization $P(t)$ as a power series in terms of the field strength, $E(t)$, and susceptibility, χ, as

$$P(t) = \sum_{k=1}^{N} \chi^{(k)} E^k(t) .$$

The quantities $\chi^{(2)}$ and $\chi^{(3)}$ are known as the second- and third-order nonlinear optical susceptibilities respectively. In the past decade, interest has been focused on the quantum confinement effect of the carriers in a semiconductor QW [10.270], including its nonlinear optical properties. Strong infra-red absorption features, such as the extremely large oscillator strength and relatively narrow line width, associated with the intraband transitions between the subbands of the QW have been observed [10.271]. This gives a very large optical nonlinearlity and for a conventional square QW, there are about six orders of magnitude larger than that of the bulk GaAs. However, this cannot be stretched too far due to the limitation of the single pair of the resonant subbands.

Recently, very strong third-harmonic susceptibility has been achieved in a superlattice structure [10.272] and a parabolic QW structure [10.273], due to the fourth-power dependence on the dipole moment. The equally-spaced eigenstate energy in a finite parabolic QW provides multiple pairs of resonant subbands, so that the enhancement in the third-order susceptibility is much more enhanced than that in the case of square QWs. Two fabrication techniques can be used to realize a parabolic QW, one is by changing the material composition of the well gradually [10.274] and the other is by using the alternate deposition of thin undoped layers of the GaAs and AlGaAs of varying thickness, which increase quadratically with position [10.275]. However, whether using a smooth or abrupt hetero-interface, there still remains difficulties in accurately generating the two types of parabolic structure, thus making it not cost-effective for mass production. On the other hand, there is some recent interest [10.276] in using the technique of interdiffusion to modify the confinement profile of a QW including its intersubband optical properties [10.277] and detector [10.278]. By adjusting carefully the parameters such as the composition, well width and diffusion length of the QW, the interdiffused QW (DFQW) with a parabolic-like confinement profile can be obtained. This is much simpler and easier than the methods suggested earlier for fabricating a parabolic QW, and by this a DFQW structure with a parabolic-like confinement profile is proposed to produce an enhanced third-order susceptibility similar to the case of a parabolic QW.

The thermal processing of a GaAs-$Al_wGa_{1-w}As$ QW, such as rapid thermal annealing around 1000°C, produces interdiffusion across the heterointerfaces of an as-grown square QW, which results in a modification of the composition and confinement profiles of the QW structure. The diffused Al composition profile $w(z)$ across the QW structure is given by (10.6). The Schrödinger equation is solved using this potential numerically by a finite difference method consistently with the Poisson's equation. Once the eigenstates have been obtained, the third-harmonic susceptibilities can now be readily computed via the density matrix formalism as follows [10.279]:

$$\chi^{(3)} = \frac{N}{\hbar^3} \times \sum_{abcd} \Bigg[(\Delta \rho_{bd}^{(0)}) \times \frac{M_{ba} M_{ac} M_{cd} M_{db}}{(\Delta \omega_{ab} - 3\omega - i\Gamma_{ab})(\Delta \omega_{cb} - 2\omega - i\Gamma_{cb})(\Delta \omega_{db} - \omega - i\Gamma_{db})}$$
$$- (\Delta \rho_{dc}^{(0)}) \times \frac{M_{ba} M_{ac} M_{db} M_{cd}}{(\Delta \omega_{ab} - 3\omega - i\Gamma_{ab})(\Delta \omega_{cb} - 2\omega - i\Gamma_{cb})(\Delta \omega_{cd} - \omega - i\Gamma_{cd})}$$
$$- (\Delta \rho_{cd}^{(0)}) \times \frac{M_{ba} M_{cb} M_{ad} M_{dc}}{(\Delta \omega_{ab} - 3\omega - i\Gamma_{ab})(\Delta \omega_{ac} - 2\omega - i\Gamma_{ac})(\Delta \omega_{dc} - \omega - i\Gamma_{dc})} \Bigg], \tag{10.35}$$

where N is the total electron density in the DFQW which is defined as N_{2DEG}/L_Z and the total surface 2-D electron gas is determined to be $1.2\times10^{12}\text{cm}^{-2}$ in this calculation, $\Delta\omega_{nm} = (E_n - E_m)/\hbar$, and $\Gamma_{nm}^{-1} = \tau_{nm}$ is the dephasing time between the DFQW states $|\phi_n\rangle$ and $|\phi_m\rangle$. The dipole moment matrix element M_{nm} is given by $M_{nm} = \int_{-5L_Z}^{5L_Z} \phi_n^*(z) q z \phi_m(z) dz$, where q is the electronic charge; $\Delta\rho_{mn}^{(0)} = \rho_{mm}^{(0)} - \rho_{nn}^{(0)}$ is the difference of unperturbed diagonal density matrix $\rho_{nn}^{(0)}$, which is equal to the thermal equilibrium occupation probability of the corresponding state and can be expressed as

$$\rho_{nn}^{(0)} = \{1 + \exp[(E_n - E_F)/(K_B T)]\}^{-1},$$

where E_F is the Fermi level, T is the temperature which is taken to be 300°K, and K_B is the Boltzmann constant. In (11.12), a, b, c, d are numbers representing the four lowest electron eigenstates and they are permutated in cyclic order in the calculation. Four iterations are required in the summation and their respective combinations are (a,b,c,d) = (1,2,3,4), (2,3,4,1), (3,4,1,2) and (4,1,2,3). The Fermi level is determined by a method briefly described as follows: The electrons in each subband are considered as a quasi 2-D electron gas, and their corresponding occupy states can be expressed as

$$N_n = \left(\frac{m_e^*}{\pi\hbar^2}\right) k_B T \ln\left\{1 + \exp\left[\frac{(E_f - E_n)}{k_B T}\right]\right\},$$

where $m_e^* = (0.0632 + 0.0856w + 0.0231w^2)$, m_e is the effective mass of the electron, and m_o is the free electron mass. Since $N_{2DEG} = \sum N_n$ which is already known, hence the quasi-Fermi level can be obtained.

In order to tailor-design the DFQW with an aim to achieve equally spaced eigenenergies, the structural parameters and the interdiffusion extent of the DFQW are chosen very carefully, so that its confinement profile is parabolic-like in shape. During the design, it is found that the eigenenergy spacing, and thus the magnitude and position of the third-harmonic susceptibility, is a strong function of the diffusion length. An optimized result shows that if the structural parameters is set at $w = 0.37$, $L_Z = 170$Å and $L_d = 22$Å in the AlGaAs/GaAs material system, this will give a good parabolic shaped profile. This is shown in Fig. 10.41 together with the parabolic QW ($L_Z = 220$Å, $w = 0.3$, the spring constant $K = 0.049$meV/Å^2) for comparison. It can be seen that the DFQW profile matches the parabolic profile exactly at the barrier, with only a slight mismatch in the center of the well. This should have a minor effect on the enhanced $\chi^{(3)}$ generated by the DFQW. The calculated eigenenergy differences for the DFQW are, $\Delta E_{21} = E_2 - E_1 = 74$meV, $\Delta E_{31} = E_3 - E_1 = 153$meV and $\Delta E_{41} = E_4 - E_1 = 226$meV. Thus it can be seen from these values that the four energy levels are nearly equally spaced.

The nonlinear $|\chi^{(3)}(3\omega)|$ of the optimized DFQW as a function of the photon wavelength λ is shown in Fig. 10.42 together with that of a parabolic QW ($L_Z = 170$Å, $w = 0.4$, K $= 0.125$meV/Å^2) for comparison purpose. It can be clearly seen that both spectra consist of only one peak and have about the same magnitude except in the case of the DFQW, the

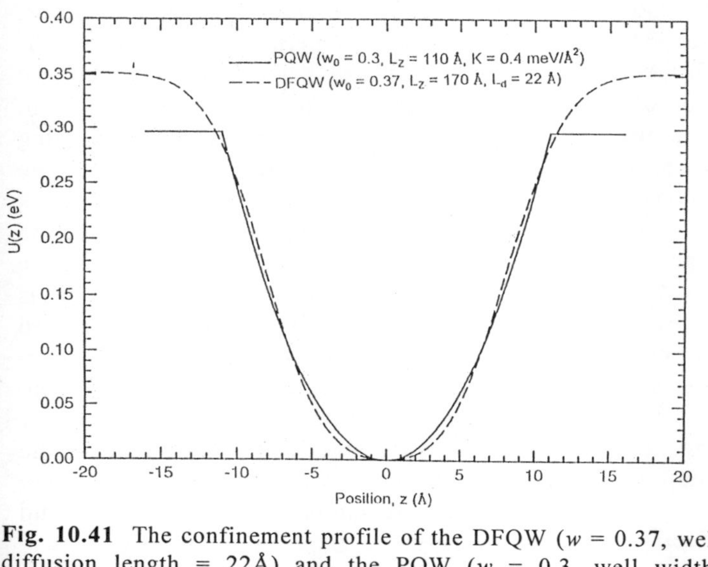

Fig. 10.41 The confinement profile of the DFQW ($w = 0.37$, well width = 170Å, diffusion length = 22Å) and the PQW ($w = 0.3$, well width = 220Å, K = 0.049meV/Å^2).

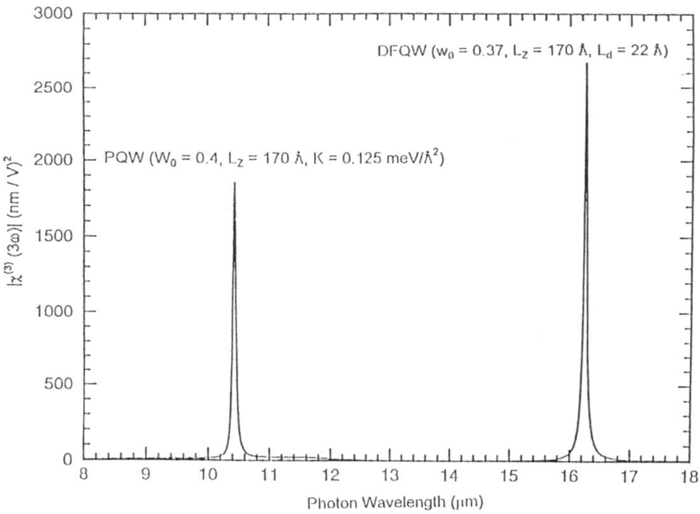

Fig. 10.42 Nonlinear $|\chi^{(3)}(3\omega)|$ of the DFQW and the PQW as a function of wavelength.

$\chi^{(3)}$ resonance is at a longer wavelength λ = 15.3μm. The magnitude of the DFQW peak is about 1000(nm/V)^2, more than that of the parabolic QW. This is because the nonlinear enhancement does not only depend on the energy ladder, but also on the overlapping of the eigenstate wavefunctions, a phenomena also observed in the case of a hyperbolic QW [10.280]. If both the as-grown QW structure and the diffusion length are adjusted, the wavelength position of the enhanced $\chi^{(3)}$ can be varied around λ = 15.3μm, thus providing a tuning range. This demonstrates the advantage of using DFQW to achieve an enhanced $\chi^{(3)}$ while at the same time simplifies the fabrication technology.

To summarize, a DFQW has been successfully employed to enhance the third-harmonic susceptibility to an extent that it is six times better than the case of a square QW. The obtained $\chi^{(3)}$ enhancement is as high as 2700(nm/V)^2. However, the operating wavelength is at around 15.3μm and not at the 10.6μm common pumped by CO_2 lasers. Although this enhancement can also be achieved by utilizing the parabolic QWs, technical difficulties still exist in their fabrication to achieve a high yield and low price. In addition, by carefully adjusting the DFQW structures, the operating wavelength of an enhanced $\chi^{(3)}$ can be varied to produce a tuning range. The technology of using the interdiffusion is thus useful, simple, and effective for producing large nonlinearlity.

References

[10.1] G. Bastard, J.A. Brum and R. Ferreira. In (H. Ehrenreich and D. Turnbull (eds.): *Solid State Physics*, **44**, Academic Press, New York, p. 232 (1991)

[10.2] S. Schmitt-Rink, D.S. Chemla and D.A.B. Miller: *Advances in Physics* **38**, 89 (1989)

[10.3] L. Esaki: *Int. J. Mod. Phys.* **B3**, 487 (1989)

[10.4] D.A.B. Miller, D.S. Chemla and S. Schmitt-Rink. In: H. Haug (ed.): *Optical Nonlinearities and Instabilities in Semiconductors*, Academic Press, San Diego, p. 325 (1988)

[10.5] H. Okamoto: *Jap. J. Appl. Phys.* **26**, 315 (1987)

[10.6] N.K. Streibl, H. Brenner, et al.: *Proc. IEEE* **77**, 1954 (1989)

[10.7] T.H. Wood: *J. Lightwave Technol.* **6**, 743 (1988)

[10.8] D.A.B. Miller: *Int. J. High Speed Electron.* **11**, 19 (1990)

[10.9] D.A.B. Miller: *Opt. Eng.* **26**, 368 (1987)

[10.10] D.A.B. Miller: *Opt. Quantum Electron.* **22**, 561 (1990)

[10.11] H. Watanabe, T. Mizutani and A. Usui. In: T. Ikoma (ed.): *Semiconductor and Semimetals*, **30**, Academic Press, New York, p. 1 (1990)

[10.12] A.C. Gossard, R.C. Miller and W. Wiegmann: *Surf. Sci.* **174**, 131 (1986)

[10.13] P.J. Price and F. Stern: *Surf. Sci.* **132**, 577 (1983)

[10.14] S. Hong and J. Singh. J.: *Appl. Phys.* **61**, 5346 (1987)

[10.15] L.L. Chang and A. Komai Interdiffusion between GaAs and AlAs, *Appl. Phys. Lett.* **29**(3), 138–141 (1976)

[10.16] R.M. Fleming, D.B. McWhan, et al.: X-ray diffraction study of inter-diffusion and growth. In: GaAs)n(AlAs)m multilayers, *J. Appl. Phys.* **51**(1), 357–363 (1980)

[10.17] W.D. Laidg, N. Holonyak Jr., et al.: Disorder of an AlAs-GaAs superlattice by impurity diffusion, *Appl. Phys. Lett.* **38**(10), 776–778 (1981)

[10.18] J.J. Coleman, P.D. Dapkus, et. al.: Disorder of an AlAs-GaAs superlattice by silicon implantation, *Appl. Phys. Lett.* **40**(10), 904–906 (1982)

[10.19] D.G. Deppe, L.J. Guido, et al. Stipe-geometry quantum well hetero-structure AlxGa1-xAs-GaAs lasers defined by defect diffusion, *Appl. Phys. Lett.* **49**(9), 510–512 (1986)

[10.20] Nick Holonyak, Jr. and Wyn D. Laidig: Method for producing integrated semiconductor light emitter, *United States Patent* **19**, No. 4378, 255 (1983)

[10.21] T.H. Wood: Multiple quantum well (MQW) waveguide modulators, *J. Lightwave Technol.* **6**, 743–757 (1988)

[10.22] M. Whitehead, G. Parry and P. Wheatly: Investigation of etalon effects in GaAs-AlGaAs multiple quantum well modulators, *Proc. Inst. Elect. Eng.* **136,** Pt. J, 52–58 (1989)

[10.23] R.H. Yan, R.J. Simes and L.A. Coldren: Analysis and design of surface-normal Fabry-Perot electrooptic modulators, *IEEE J. Quantum Electron.* **125**, 2272–2280 (1989)

[10.24] P. Zouganeneli, P.J. Stevens, et al.: Design trade-offs and evaluation of the performance attainable by GaAs-$Al_{0.3}Ga_{0.7}As$ asymmetric Fabry-Perot modulators, *IEEE J. Quantum Electron.*, **31**, 927–943 (1995)

[10.25] A. Tomita, Y. Kohga and A. Suzuki: 5:1 on-off contract InGaAs/InP multiple quantum well fabry-perot etalon modulator, *Appl. Phys. Lett.* **55**, 1817–1819 (1989)

[10.26] R.H. Yan, R.J. Simes and L.A. Coldren: Surface-normal electro-absorption reflection modulators using asymmetric Fabry-Perot structures, *IEEE J. Quantum Electron.* **27**, 1922–1931 (1991)

[10.27] C.H. Lin, J.M. Messe and Y.C. Chang: A low-voltage high-reflectance-change normally off refractive GaAs/$Al_{0.2}Ga_{0.8}As$ MQW reflection modulator, *IEEE J. Quantum Electron.* **30**, 1234–1240 (1994)

[10.28] K.K. Law, J.L. Merg and L.A. Coldren: Supperlattice surface-normal asymmetric Fabry-Perot reflection modulators: optical modulator and switching, *IEEE J. Quantum Electron.* **29**, 727–739 (1993)

[10.29] W.C.H. Choy and E.H. Li: The applications of an interdiffused quantum well in a normally on electroabsorptive Fabry-Perot reflection modulator, *IEEE J. Quantum Electorn.* **33**, 382–392 (1997)

[10.30] H. Yamamoto, M. Asada and Y. Suematsu: Theory of refractive index variation in quantum well structure and related intersectional optical switch, *J. Lightwave Technol.* **6**, 1931–1940 (1988)

[10.31] K. Wakita, Y. Kawamura, et al.: Long wavelength waveguide multiple quantum well optical modulators, *IEEE J. Quantum Electron.* **23**, 2210–2215 (1987)

[10.32] T. Ido, M. Koizumi, et al.: Polarization and wavelength insensitive MQW electroabsorption optical gates for WDM switch systems, *IEEE Photon. Tech. Lett.* **8**, 788–790 (1996)

[10.33] K.G. Ravikumar, T. Aizawa and R. Yamauchi: Polarization-independent field-induced absorption-coefficient variation spectrum in an InGaAs/InP tensile-strained quantum well. *IEEE Photon. Technol. Lett.* **5**, 310–312 (1993)

[10.34] K. Wakita, I. Kotaka, et al.: Polarization-independent electroabsorption modulators using strain-compensated InGaAs-InAlAs MQW structures, *IEEE Photon. Technol. Lett.* **7**, 1418–1420 (1995)

[10.35] T. Yamaguchi, T. Morimoto, et al.: Polarization-independent waveguide modulator using a novel quantum well with mass-dependent width, *IEEE Photon. Technol. Lett.* **6**, 1442–1444 (1994)

[10.36] M. Kato, K. Tada and Y. Nakano: Wide-wavelength polarization-independent optical modulator based on tensile-strained quantum well with mass-dependent width, *IEEE Photon. Technol. Lett.* **8**, 785–787 (1996)

[10.37] J.C. Zucker, K.L. Jones, et al.: Strained quantum well for polarization-independent electrooptic waveguide switches, *IEEE J. Lightwave Technol.* **10**, 1926–1930 (1992)

[10.38] W.C.H. Choy, E.H. Li and J. Micallef: Polarization-insensitive electro-absorptive modulation using interdiffused InGaAs(P)/InP quantum wells, *IEEE J. Quantum Electron.* **33**(August 1997)
[10.39] D.A.B. Miller, D.S. Chemla, et al. *Phys., Rev. B* **32**, 1043–1060 (1985)
[10.40] R.H. Yan, R.J. Simes and L.A. Coldren: *IEEE Photon. Technol.* **2**, 118–119 (1990)
[10.41] D.A.B. Miller, D.S. Chemla, et al.: *Phys. Rev. Lett.* **53**, 2713 (1984)
[10.42] For a comprehensive review of SEEDs, see D.A.B. Miller: *Opt. Quantum Electron.* **22**, S61 (1990)
[10.43] L.M.F. Chirovsky, M.W. Focht, et al.: *Proceedings of the Topical Meeting on Photonic Switching*, Opt. Soc. Amer., Washington, DC, p. 150 (1991)
[10.44] D.G. Deppe and N. Holonyak Jr.: *J. Appl. Phys.* **64**, R93 (1988)
[10.45] J. Werner, T.P. Lee, et al.: *Appl. Phys. Lett.* **57**, 810 (1990); T. Wolf, C.L. Shieh, et al.: *Appl. Phys. Lett.* **55**, 1412 (1989)
[10.46] O. Wada, A. Furuya and M. Makiuchi: *IEEE Photon. Tech. Lett.* **1**, 16 (1989)
[10.47] I. Suemune, T. Takeoka, et al.: *IEEE J. Quantum Electron.* **22**, 1900 (1986); S. Tarucha and H. Okamoto: *Appl. Phys. Lett.* **48**, 1
[10.48] E.H. Li, E.S. Koteles and J.H. Marsh: *Selected Topics in Quantum Electronics* 4, IEEE J. (1998)
[10.49] Jyh-Chwen Lee, T.E. Schlesinger and T.F. Kuech: *J. Vac. Sci. Technol.* **B5**, 1187 (1987)
[10.50] M. Kawabe, N. Matsuura, et al.: *Jpn. J. Appl. Phys.* **23**, L623 (1984)
[10.51] J.J. Coleman, P.D. Dapkus, et al.: *Appl. Phys. Lett.* **40**, 904 (1982)
[10.52] W.D. Laidig, N. Holonyak Jr., et al.: *Solid State Commun.* **38**, 301 (1981)
[10.53] J. Cibert, P.M. Petroff, et al.: *Appl. Phys. Lett.* **49**, 223 (1986)
[10.54] K. Kash, B. Tell, et al.: *J. Appl. Phys.* **63**, 190 (1988)
[10.55] H. Leier, A. Forchel, et al.: *J. Appl. Phys.* **67**, 1805 (1990)
[10.56] E.H. Li, J. Micallef and B.L. Weiss: *Jpn. J. Appl. Phys.* **31**(1A/B), L7–L10 (1992)
[10.57] J. Micallef, E.H. Li and B.L. Weiss: *Electron. Lett.* **28**(6), 526–528 (1992)
[10.58] E.H. Li and B.L. Weiss: *Appl. Phys. Lett.* **28**(9), 885–886 (1992)
[10.59] J. Micallef, E.H. Li and B.L. Weiss: *Appl. Phys. Lett.* (to be published)
[10.60] E.H. Li, B.L. Weiss and K.S. Chan: *Phys. Rev. B* (1992)
[10.61] E.H. Li, B.L. Weiss and J. Micallef: *Appl. Phys. Lett.* (1992)
[10.62] Li, E.H. and B.L. Weiss: *Proc. SPIE* **1675** (1992)
[10.63] J. Micallef, E.H. Li, et al.: *Proc. SPIE* **1675** (1992)
[10.64] J. Micallef, E.H. Li and B.L. Weiss: *Superlattice Microstruct.* (1992)
[10.65] E.H. Li and B.L. Weiss: *J. Appl. Phys.* **70**(2), 1054–1056 (1991)
[10.66] E.H. Li and B.L. Weiss: *IEEE Photon. Technol. Lett.* **3**(9), 787–789 (1991)
[10.67] E.H. Li and B.L. Weiss: *Appl. Phys. Lett.* **59**(25), 3312–3314 (1991)
[10.68] E.H. Li and B.L. Weiss: *IEEE J., Quantum Electron.* (1992)
[10.69] E.H. Li, B.L. Weiss and J. Micallef: *Opt. Quantum Electron.* **24**(1992)
[10.70] K.P. Homewood and D.J. Dunstan: *J. Appl. Phys.* **69**, 7581–7584 (1991)

[10.71] N. Holonyak Jr., W.D. Laidig, et al.: *Appl. Phys. Lett.* **39**, 102 (1981)
[10.72] D.G. Deppe and N. Holonyak, Jr.: *J. Appl. Phys.* **64**, R93 (1988)
[10.73] B.L. Weiss: *Opt. Quantum Electron.* **23**, S779 (1991)
[10.74] I.V. Bradley, B.L. Weiss and J.S. Roberts: *Opt. Quantum Electron.* **23**, S823 (1991)
[10.75] Y. Hirayama: *Jpn. J. Appl. Phys.* **28**, L162 (1989)
[10.76] B.L. Weiss, I.V. Bradley, et al. *J. Appl. Phys.* **71**, 5715 (1992)
[10.77] R.P. Bryan, J.J. Coleman, et al.: *Appl. Phys. Lett.* **55**, 94 (1989)
[10.78] O. Wada, A. Furuya and M. Makiuchi: *IEEE Photon. Tech. Lett.* **1**, 16 (1989)
[10.79] K. Meehan, K.C. Hsieh, et al.: *Appl. Phys. Lett.* **48**, 861 (1986)
[10.80] T. Fukuzawa, S. Semura, et al.: *Appl. Phys. Lett.* **45**, 1 (1984)
[10.81] J. Werner, E. Kapon, et al.: *Appl. Phys. Lett.* **55**, 540 (1989)
[10.82] F. Julien, P.S. Swanson, et al.: *Appl. Phys. Lett.* **50**, 855 (1987)
[10.83] E. Kapon, N.G. Sloffel, et al.: *Appl. Phys. Lett.* **52**, 351 (1988)
[10.84] B.L. Weiss, A.S. Wismayer and J.S. Roberts: *Electronic Lett.* **25**, 653 (1989)
[10.85] T. Wolf, C.L. Shieh, et al.: *Appl. Phys. Lett.* **55**, 1412 (1989)
[10.86] D.G. Deppe, L.J. Guido, et al.: *Appl. Phys. Lett.* **49**, 510 (1986)
[10.87] J.D. Ralston, S. O'Brien, et al.: *Appl. Phys. Lett.* **52**, 1511 (1988)
[10.88] J.D. Ralston, W.J. Schaff, et al.: *Appl. Phys. Lett.* **54**, 534 (1989)
[10.89] E.S. Koteles, B. Elman, et al.: *Superlattices Microstruc.* **5**, 321 (1989)
[10.90] E.S. Koteles, B. Elman, et al.: *J. Appl. Phys.* **66**, 5532 (1989)
[10.91] Y. Suzuki, H. Iwamura and O. Mikami: *Appl. Phys. Lett.* **56**, 19 (1990)
[10.92] M. Ghisoni, P.J. Stevens, et al.: *Opt. Quantum Electron.* **23**, S915 (1991)
[10.93] M. Ghisoni, G. Parry, et al.: *Jpn. J. Appl. Phys.* **30**, L1018 (1991)
[10.94] J.F. Gibbons: *Proc. IEEE* **56**, 295 (1968) and **60**, 1062 (1972)
[10.95] D.H. Lee and J.W. Mayer: *Proc. IEEE* **62**, 1241 (1974)
[10.96] G. Dearnaley, J.H. Freeman, et al.: *Ion Implantation*, North-Holland, Amsterdam (1973)
[10.97] K.A. Pickar. In: R. Wolfe (ed.): *Advances in Solid State Science*, Academic Press, N.Y., p. 151
[10.98] W.K. Hofker and J. Poloiteik: *Philips Tech. Rev.* **39**, 1 (1980)
[10.99] C.W. White and P.S. Peercy: *Laser and Electron Beam Processing of Materials*, Academic Press, N.Y. (1980)
[10.100] B.L. Weiss and A.C. Wismayer: Ion implanted strip optical waveguides in GaAs/GaAlAs MQW material, *Mater. Res. Soc. Proc.* **126**, 83–88 (1988)
[10.101] M. O'Neill, et al.: Multiple quantum well optical waveguides with large absorption edge blue shife produced by boron and fluorine impurity-induced disordering, *Appl. Phys. Lett.* **55**, 1373–1375 (1989)
[10.102] B.L. Weiss, I.V. Bradley and J.S. Roberts: Oxygen implantation induced disordering of AlGaAs/ GaAs quantum well structures (to be published)
[10.103] C. Jeynes: Ovel applications of ion implantation, *Vaccum* **39**, 1047–1056 (1989)
[10.104] J. Lindhard and M. Scharff: Kgl. Danske Videnskab. Selskab, *Mat.-Fys. Medd.* **27**, 15 (1952)

[10.105] J. Lindhard, M. Scharff and H.E. Schiott: Kgl. Danske Videnskab. Selskab., *Mat.-Fys. Medd.* **33**, 14 (1963)

[10.106] N. Bohr: *Mag. Phil.* **25**, 10 (1913)

[10.107] H. Runge: *Phys. Status Solidi A* **39**, 395 (1977)

[10.108] W.S. Johnson and J.F. Gibbons: *Projected Range Statistics in Semiconductors*, Stanford U.P., Stanford (1969)

[10.109] S. Furukawa, H. Matsumura and H. Ishiwara: *Japan. J. Appl. Phys.* **11**, 134 (1972)

[10.110] A. Barthel, J. Lorentz and H. Ryssel: Instrum. Nucl., *Methods Phys. Res.* **B37/3**, 312 (1989)

[10.111] H. Ryssel, J. Lorenz and K. Hoffmann: *Appl. Phys.* **A41**, 201 (1986)

[10.112] J.W. Mayer, L. Eriksson and J.A. Davies: *Ion Implantation in Semiconductors*, New York (1970)

[10.113] R.G. Wilson and G.R. Brewer: *Ion Beams*, New York (1973)

[10.114] W.D. Laidig, N. Holonyak, et al.: *J. Electron. Mater.* **11**, 1 (1982)

[10.115] W.D. Laidig, N. Holonyak, et al.: *Appl. Phys. Lett.* **38**, 776 (1981)

[10.116] W.D. Laidig, J.W. Lee, et al.: *J. Appl. Phys.* **54**, 6382 (1983)

[10.117] J.W. Lee and W.D. Laidig: *J. Electron. Mater.* **13**, 147 (1984)

[10.118] K. Ishida, T. Ohta, et al.: *Jap. J. Appl. Phys.* **24**, L620 (1985)

[10.119] M.D. Camras, N. Holonyak, et al.: *Appl. Phys. Lett.* **42**, 185 (1983)

[10.120] K. Meehan, J.M. Brown, et al.: *Appl. Phys. Lett.* **44**, 700 (1984)

[10.121] K. Meehan, N. Holonyak, et al.: *Appl. Phys. Lett.* **45**, 549 (1984)

[10.122] N. Iwata, Y. Matsumoto and T. Baba: *Jap. J. Appl. Phys.* **24**, L17 (1985)

[10.123] L. Gonsalez, J.B. Clegg, et al.: *Appl. Phys.* **A41**, 237 (1986)

[10.124] M. Kawabe, N. Shimizu, et al.: *Appl. Phys. Lett.* **46**, 849 (1985)

[10.125] J. Kobayashi, M. Nakajima, et al.: *Jap. J. Appl. Phys.* **25**, L736 (1986)

[10.126] M.E. Greiner and J.F. Gibbons: *Appl. Phys. Lett.* **44**, 750 (1984)

[10.127] E.V.K. Rao, H. Thibierge, et al.: *Appl. Phys. Lett.* **46**, 867 (1985)

[10.128] E.A. Dobisz, B. Tell, et al.: *J. Appl. Phys.* **60**, 4150 (1986)

[10.129] P. Gavrilovic, D.G. Deppe, et al.: *Appl. Phys. Lett.* **47**, 130 (1985)

[10.130] D.G. Deppe, G.S. Jackson, et al.: *Appl. Phys. Lett.* **50**, 632 (1987)

[10.131] D.G. Deppe, L.J. Guido, et al.: *Appl. Phys. Lett.* **49**, 510 (1986)

[10.132] T. Miyazawa, T. Kagawa, et al.: *Appl. Phys. Lett.* **55**, 828 (1989)

[10.133] J.R. Manning: *Diffusion Kinetics for Atoms in Crystals*, Van Nostrand, Princeton (1968)

[10.134] S.B. Zhang and J.E. Northrup: *Phys. Rev. Lett.* **67**, 2339 (1991)

[10.135] R.W. Jansen and O.F. Sankey: *Phys. Rev. B* **39**, 3192 (1989)

[10.136] M. Scheffler and J. Dabrowski: *Philos. Mag. A* **58**, 107 (1988)

[10.137] G.A. Baraff and M. Schlwer: *Phys. Rev. Lett.* **55**, 1327 (1985)

[10.138] J.F. Wager: *J. Appl. Phys.* **69**, 3022 (1991)

[10.139] J.A. Van Vechten and U. Schmid: *J. Vac. Sci. Technol. B* **7**, 827 (1989)

[10.140] K.B. Kahen, D.L. Peterson, et al.: *Appl. Phys. Lett.* **55**, 651 (1989)

[10.141] S.Y. Chiang and G.L. Pearson: *J. Appl. Phys.* **46**, 2986 (1975)

[10.142] J.D. Ralston, S. O'Brien, et al.: Room-temperature exciton transitions in partially intermixed GaAs/AlGaAs superlattices, *Appl. Phys. Lett.* **52**, 1511–1513 (1988)

[10.143] T.E. Schlesinger and T. Kuech: Determination of the interdiffusion of Al and Ga in undoped(Al,Ga)As/GaAs quantum well, *Appl. Phys. Lett.* **49**, 519–521 (1986)
[10.144] J.C. Lee, T.E. Schlesinger and T. Kuech: Interdiffusion of Al and Ga in(Al,Ga)As/GaAs superlattice, *J. Vac. Sci. Technol.* **B5**, 1187–1190 (1987)
[10.145] K. Kash, B. Tell, et al.: Aluminum ion-implantation enhanced intermixing of GaAs-AlGaAs quantum-well structures, *J. Appl. Phys.* **63**, 190–194 (1988)
[10.146] Y. Hirayama: Mechanism of Ga implantation-induced intermixing of GaAs-AlGaAs material, *Jpn. J. Appl. Phys.* **28**, L162–L165 (1989)
[10.147] K.P. Homewood and D.J. Dunstan: Optical characterization of thermal mixing in quantum wells and heterostructures using a Green's function method, *J. Appl. Phys.* **69**, 7581–7584 (1991)
[10.148] J. Crank: *The Mathematics of Diffusion*, 2nd ed., Oxford UP, Oxford, p. 15 (1975)
[10.149] M. Abramowitz and I.A. Stegun: *Handbook of Mathematical Functions*, National Bureau of Standards, Washington, D.C. (1964)
[10.150] D.E. Aspnes, S.M. Kelso, et al.: Optical properties of AlGaAs, *J. Appl. Phys.* **60**, 745–767 (1986)
[10.151] S. Adachi: GaAs, AlAs, and AlGaAs (material parameters for use in research and device applications, *J. Appl. Phys.* **58**, R1–R29 (1985)
[10.152] E.H. Li, J. Micallef and B.L. Weiss: Confinement subband and transitions in a diffusion induced InGaAs/GaAs non-square quantum well, *Jpn. J. Appl. Phys., Pt. 2* **31**(1A/B), L7–L10 (1992)
[10.153] E.H. Li and B.L. Weiss: The optical properties of AlGaAs/GaAs hyperbolic quantum well structures, *J. Appl. Phys.* **70**(2), 1054–1056 (1991)
[10.154] T.E. Schlesinger and T. Kuech: *Appl. Phys. Lett.* **49**, 519 (1986)
[10.155] K.P. Homewood and D.J. Dunstan: *J. Appl. Phys.* **69**, 7581 (1991)
[10.156] E.H. Li and B.L. Weiss: *J. Appl. Phys.* **10**, 1054 (1991)
[10.157] E.H. Li and B.L. Weiss: *IEEE J. Quantum Electron.* **QE–29**, 311 (1993)
[10.158] J.D. Ralston, S. O'Brien, et al.: *Appl. Phys. Lett.* **52**, 1511 (1988)
[10.159] A.T. Meney: *Superlatt. Microstruct.* **11**, 47 (1992)
[10.160] E.H. Li, J. Micallef and B.L. Weiss: *Jpn. J. Appl. Phys.* **31**, L7 (1992)
[10.161] S.L. Chuang, J. Schmitt-Rink, et al.: *Phys. Rev. B* **43**, 1500 (1992)
[10.162] T.Z.Y. Ting, E.T. Yu and T.C. McGill: *Phys. Rev. B* **45**, 3583 (1992)
[10.163] G. Neofotistos, H. Guo, et al.: *IEEE Trans. Electron. Devices* **ED–36**, 74 (1989)
[10.164] J.M. Luttinger and W. Kohn: *Phys. Rev.* **97**, 869 (1955)
[10.165] D.J. Ben-Daniel and C.B. Duke: *Phys. Rev.* **152**, 638 (1966)
[10.166] D.A. Broido and L.J. Sham: *Phys. Rev. B* **34**, 3917 (1986)
[10.167] K.S. Chan: *J. Phys. C* **19**, L125 (1986)
[10.168] I.J. Fritz, PJ. Gourley and L.R. Dawson: *Appl. Phys. Lett.* **51**, 1004 (1987)
[10.169] R. Dingle: *Festkoperprobleme* **XV**, 21–48 (1975)
[10.170] P.J. Price and F. Stern: *Surface Science* **132**, 577–593 (1983)
[10.171] J.B. Xia: *Physical Review B* **38**(12), 8358–8364 (1988)
[10.172] R. Houdré, C. Hermann, et al.: *Surface Science* **168**, 538–545 (1986)

[10.173] T. Katsuyama, M.A. Tischler, et al.: *Appl. Phys. Lett.* **51**, 539–531 (1987)
[10.174] X. Marie, J. Barrau, et al.: *Superlattices and Microstructures,* **10**(1), 95–98 (1991)
[10.175] R. Dingle, W. Wiegmann and C.H. Henry: *Physical Review Letters* **33**(14), 827–830 (1974)
[10.176] G. Bastard, J.A. Brum and R. Ferreira. In: H. Ehreneich and D. Turnbull (eds.): *Solid State Physics*, **44**, Academic Press, New York, pp. 229–415 (1991)
[10.177] R. Dingle, W. Wiegmann and C.H. Henry: *Phys. Rev. Lett.* **33**, 827 (1974)
[10.178] A. Harwitt and J.S. Harris Jr.: *Appl. Phys. Lett.* **50**, 685–687 (1987)
[10.179] L.C. West and S.J. Eglash: *Appl. Phys. Lett.* **46**, 1156–1158 (1985)
[10.180] B.F. Levine, R.J. Malik, et al.: *Appl. Phy. Lett.* **50**, 273–275 (1987)
[10.181] A. Ahn and S.L. Chuang: *IEEE J. Q-E* **23**, 2196 (1987)
[10.182] E.J. Roan and S.L. Chuang: *J. Appl. Phys. Col.* **69**, 3249 (1991)
[10.183] L.C. West and S.J. Eglash: *Appl. Phys. Lett.* **46**, 1156 (1985)
[10.184] B.F. Levine, R.J. Malik, et al.: *Appl. Phys. Lett.* **50**, 273 (1987)
[10.185] A.D. Wieck, J.C. Maan, et al.: *Phys. Rev. B* **35**, 4145 (1987)
[10.186] M. Nakayama, H. Kuwahara, et al.: *Appl. Phys. Lett.* **51**, 1741 (1987)
[10.187] P. von. Allmen, M. Berz, et al.: *Semicond. Sci. Technol.* **3**, 1211 (1988)
[10.188] E. Rosencher, E. Martinet, et al.: *Appl. Phys. Lett.* **59**, 3255 (1991)
[10.189] Y. Zhao, D.C. Tsui, et al.: *Appl. Phys. Lett.* **60**, 1510 (1992)
[10.190] J.D. Ralston, D.F.G. Gallaghter, et al.: *J. Appl. Phys.* **71**, 3562 (1992)
[10.191] V. Berger, N. Vodjdani, et al.: *Appl. Phys. Lett.* **60**, 1869 (1992)
[10.192] E.B. Dupont, D. Delacourt, et al.: *Appl. Phys. Lett.* **60**, 2121 (1992)
[10.193] D. Elbert, E. Ehrenfreund, et al.: *J. Appl. Phys.* **71**, 6198 (1992)
[10.194] B. Jogai, M.O. Manasreh, et al.: *Phys. Rev. B* **46**, 7208 (1992)
[10.195] J. Katz, Y. Zhang and W. I. Wang: *Appl. Phys. Lett.* **61**, 1697 (1992)
[10.196] K.K. Choi, B.F. Levine, et al.: *Appl. Phys. Lett.* **52**, 1979 (1988)
[10.197] K.K. Choi, B.F. Levine, et al.: *Appl. Phys. Lett.* **53**, 231 (1988)
[10.198] A. Pinczuk and G. Abstreiter. In: M. Cardona and G. Gutherodt (eds.): *Light Scattering in Solids,* **V**, Springer-Verlag, Berlln, p. 153 (1989)
[10.199] K. Bejema, R. Merlin, et al.: *Phys. Rev. B* **36**, 1300 (1987)
[10.200] A. Pinczuk, S. Schmitt-Rink, et al.: *Phys. Rev. Lett.* **63**, 1633 (1989)
[10.201] P. Yagos, P. Boucaud, et al.: *Phys. Rev. Lett.* **70**, 1018 (1992)
[10.202] S.Y. Yuen: *Appl. Phys. Lett.* **43**, 813 (1983)
[10.203] D. Ahn and S.L. Chuang: *J. Appl. Phys.* **62**, 3052 (1987) and *IEEE. J. Quantum Electron.* **QE–23**, 2196 (1987)
[10.204] D.S. Ahn, L. Chuangm and Y.-C. Chang: *J. Appl. Phys.* **64**, 4056 (1988)
[10.205] K.M.S.V. Bandara, D.D. Coon, et al.: *Appl. Phys. Lett.* **53**, 1931 (1988)
[10.206] D.H. Ehlers: *Phys Rev. B* **38**, 9706 (1988)
[10.207] D.J. Newson and A. Kurobe: *Appl. Phyts. Lett.* **53**, 2516 (1988)
[10.208] Z. Ikonic, V. Milanovic and D. Tjapkin: *Appl. Phys. Lett.* **54**, 247 (1989)
[10.209] J.-W. Wu: *Phys. Rev. B* **39**, 7992 (1989)
[10.210] Y.-C. Chang and R.B. James: *Phys. Rev. B* **39**, 12672 (1989)
[10.211] L.D.L. Brown, M. Jaros and D.C. Herbert: *Phys. Rev. B* **40**, 1616 (1989)

[10.212] Z. Ikonic, V. Milanovic and D. Tjapkin: *Solid State Commun.* **72**, 835 (1989)
[10.213] S. Nojima: *Phys. Rev. B* **41**, 10214 (1990)
[10.214] E.J. Roanand and S.L. Chuang: *J. Appl. Phys.* **69**, 3249 (1991)
[10.215] K.J. Kuhn, G.U. Iyengar and S.J. Yee: *Appl. Phys.* **70**, 5010 (1991)
[10.216] S.L. Chuang, M.S.C. Luo, et al.: *Phys. Rev. B* **46**, 1897 (1992)
[10.217] D. Teng, C. Lee and L.F. Eastman: *J. Appl. Phys.* **70**, 5010 (1991)
[10.218] H. Xie, J. Katz and W.I. Wang: *J. Appl. Phys.* **72**, 3681 (1992)
[10.219] H. Xie, J. Katz and W.I. Wang: *Appl. Phys. Lett.* **61**, 2694 (1992)
[10.220] P. Man and D.S. Pan: *Appl. Phys. Lett.* **61**, 2799 (1992)
[10.221] von Allmen: *P. Phys. Rev. B* **46**, 13351 (1992)
[10.222] W. Chen and T.G. Andersson: *Appl. Phys. Lett.* **60**, 1591 (1992), and the references cited therein
[10.223] V. Berger, E. Dupont, et al.: *Appl. Phys. Lett.* **61**, 2072 (1992), and the references cited therein
[10.224] F. Capasso: *J. Vac Sci. Technol. B.* **1**, 457 (1983)
[10.225] F. Capasso: *Science* **35**, 172 (1987)
[10.226] R. Dingle: Lightwave communication technology. In: R.K. Willardson and A.C. Beer (eds.): *Semiconductors and Semimetals*, **22,** Academic Press, Orlando, Fl. (1985)
[10.227] R. Dingle: Application of multiquantum wells, selective doping and superlattices. In: R.K. Willardson and A.C. Beer (eds.): *Semiconductor and Semimetals*, **24**, Academic Press, San Diego, Calif. (1987)
[10.228] G.C. Osbourn: *Semicond. Sci. Technol.* **5**, S5 (1990)
[10.229] B.F. Levine, K.K. Choi, et al.: *Appl. Phys. Lett.* **50**, 1092 (1987)
[10.230] B.F. Levine, G. Hasnain, et al.: *Appl. Phys. Lett.* **54**, 2704 (1989)
[10.231] B.F. Levine, G. Hasnain, et al.: *Appl. Phys. Lett.* **56**, 851 (1990)
[10.232] R.L. Whitney, F.W. Adams and K.F. Cuff: *Intersubband Transitions in Quantum Wells*, NATO ASI Series B, Physics **288**, p. 93
[10.233] S.A. Lyon: *Preliminary Reports, Memoranda and Technical Notes*, Mat. Res. Council Summer Conf., La Jolla, Sec C, July 1982, pp. 51–59 (available through Government Document Center)
[10.234] L.C. Chiu, J.S. Smith, et al.: *Appl. Phys. Lett.* **43**, 331 (1983)
[10.235] J.S. Smith, L.C. Chiu, et al.: *J. Vac. Sci. Technol. B* **1**, 376–378 (1983)
[10.236] B.F. Levine, K.K. Choi, et al.: *Appl. Phys. Lett.* **50**, 1092–1094 (1987)
[10.237] F. Capasso: *J. Vac Sci. Technol. B* **1**, 457 (1983)
[10.238] F. Capasso: *Science* **235**, 172 (1987)
[10.239] E.H. Li, Bernard L. Weiss and K.S. Chan: *Physical Review B.* **46**, 15181 (1992)
[10.240] B.L. Weiss (ed.): *Opt Quantum Electron* **23**, Specific issue, S779 (1991)
[10.241] B.J. Sealy: *Semicond. Sci. Technol.* **3**, 448–454 (1988)
[10.242] T.E. Schlesinger and T. Kuech: *Appl. Phys. Lett.* **49**, 519–521 (1986)
[10.243] E.S. Koteles, B. Elman, et al.: *Supperlatt. Microstruct.* **9**, 533 (1991)
[10.244] H.C. Casey Jr. and M.B. Panish: Academic Press, New York, p. 16 (1978)
[10.245] D. Moss, R.L. Williams, et al.: *App. Phys. Lett.* **57**, 3139 (1991)
[10.246] L.F. Lester, W.J. Schaff, et al.: *IEEE Photonics Technology Letters* **3**, 403 (1991)

[10.247] E. Yabolonvitch and E.O. Kane: *Journal of Lightwave Technology* **LT–4**, 504 (1986)

[10.248] S.W. Corzine, R.H. Yan and L.A. Coldren: *Appl. Phys. Lett.* **57**, 2835 (1990)

[10.249] L.C. West and S.J. Eglash: *Appl. Phys. Lett.* **46**, 1156 (1985)

[10.250] A. Zuzzman, B.F. Levine, et al.: *Appl. Phys.* 5101 (1991)

[10.251] Y. Arakawa and A. Yariv: Quantum well lasers — gain, spectra, dynamics, *IEEE J. Quantum Electron.* **QE-22**, 1887–1889 (1986)

[10.252] N. Peyghambarian and H.M. Gibbs: Optical nonlinearity, biastability, and signal processing in semiconductors, *J. Opt. Soc. Amer. B* **2**, 1215–1227 (1985)

[10.253] E.H. Li, B.L. Weiss and A. Laszcz: *Electronics Letters* 885 (1992)

[10.254] J.D. Ralston, M. Ramsteiner, et al.: *J. Appl. Phys.* **70**, 2195 (1991)

[10.255] A.G. Steele, M. Buchanan, et al.: *J. Appl. Phys.* 8234 (1994)

[10.256] D. Ahn and S.L. Chuang: *J. Appl. Phys.* **65**, 2882 (1989)

[10.257] D. Ahn and S.L. Chuang: *IEEE J. Quantum Electron* **QW-23**, 2196 (1987)

[10.258] E.H. Li and Bernard L. Weiss: *Phys. Rev. B.* **46**, 15181 (1992)

[10.259] Kelin J. Kuhn, Gita U. Iyengar and Sinclair Yee: *J. Appl. Phys.* **70**, 5010 (1991)

[10.260] K.J. Kuhn, G.U. Iyengar and S.J. Yee: *Appl. Phys.* **70**, 5010 (1991)

[10.261] D.J. Ben Daniel and C.B. Duke: *Phys. Rev.* **152**, 683 (1966)

[10.262] E.H. Li: *SPIE Proc.* **2886**(151), 263–329 (1996)

[10.263] E.J. Roan, and S.L. Chuang: *J. Appl. Phys.* **69**(5), 3249 (1991)

[10.264] E.H. Li and B.L. Weiss: *Phys. Rev. B.* **46**, 15181 (1992)

[10.265] E.H. Li, J. Micallef and B.L. Weiss: *Japan. J. Appl. Phys.* **31**, L7–L10 (1992)

[10.266] A. Kock, E. Gornik, et al.: *Appl. Phys. Lett.* **60**(16)(1992)

[10.267] T. Ando, A.B. Fowler and F. Stern: *Rev. Mod. Phys.* **54**, 437 (1982)

[10.268] N. Bloembergen: *Nonlinear Optics*, 4th ed., World Scientific, Singapore (1992)

[10.269] P.A. Franken, A.E. Hill, et al.: *Phys. Rev. Lett.* **7**, 118 (1961)

[10.270] M.J. Kelly: *Low Dimensional Semiconductors*, Oxford UP, Oxford (1995)

[10.271] B.F. Levine, K.K. Choi, et al.: *Appl. Phys. Lett.* **50**, 1092 (1987)

[10.272] D. Walrod, S.Y. Auyang, et al.: *Appl. Phys. Lett.* **59**, 2932 (1991)

[10.273] Y. Huang and C. Lien: *J. Appl. Phys.* **75**, 3223 (1994)

[10.274] S. Sen, F. Capasso, et al.: *Appl. Phys. Lett.* **51**, 1428 (1987)

[10.275] K. Tada, S. Nishimura and T. Ishikawa: *Appl. Phys. Lett.* **59**, 2778 (1991)

[10.276] B.L. Weiss (ed.): Special issue on quantum well mixing for optoelectronics, *Opt. Quantum Electron.* **23**, S799 (1991)

[10.277] E.H. Li, B.L. Weiss and A. Laszcz: *Electron. Lett.* **28**, 885 (1991)

[10.278] A.G. Steele, M. Buchanan, et al.: *J. Appl. Phys.* **75**, 8234 (1994)

[10.279] R.W. Boyd: *Nonlinear Optics*, Academic, New York, Chap. 3 (1992)

[10.280] E.H. Li, B.L. Weiss and K.S. Chan: *Phys. Rev. B* **46**, 15181 (1992)

11. Interdiffused Quantum Wells — Optical Properties and Device Applications Part II

11.1 Interband Optical Properties of DFQW

11.1.1 Interband Transitions

The diffusion QW (DFQW) structure is of considerable interest in its possibility to realize and improve performance of devices, such as lasers, waveguides, and modulators which use post-grown thermal-intermixing of semiconductor QW structures and the interband transition of the three different DFQW material systems: the AlGaAs/GaAs, the InGaAs/GaAs and the InGaAs/InP. The first material system is lattice-matched while the rest are developed strained structures in the DFQW.

11.1.1.1 AlGaAs/GaAs

The square of the normalized wavefunction (ψ^2) shown in Fig. 11.1 for an EQW (L_d = 40Å) demonstrates that the broadening effect also causes the wavefunction to spread out with a lager ι^2 magnitude (probability) away from the center of the well, which is similar to that reported for the parabolic [11.1] and trapezoidal [11.2] QWs. This variation of ψ^2 shape will affect the overlapping wavefunctions $\langle\psi_{Cl}|\psi_{Vl}\rangle$, such as those involved in the interband absorption coefficients. The upper levels of the EQW, such as C4 and H6, as shown in Fig. 11.1, clearly indicate that the cut-off rule may be applied effectively, since the contributions from these levels are minimal. The dipole transitions between the electron (level l_C) and hole (level l_V) subbands follow the selection rule of $\Delta l = l_C - l_V$, which represents the magnitude of overlap. For an infinitely high barrier QW ($\Delta E_g \rightarrow \infty$) the strongest "diagonal" rule ($\Delta l = 0$) applies since the ls between the two bands are completely orthogonal and deviations from this rule are often erroneously called "forbidden". A relaxed selection rule ($l_C + l_V$ = even) allows for a finite barrier square-QW although the "off-diagonal" ($l_C - l_V \neq 0$ and even) transitions are quite small. In the present EQW case, these off-diagonal transitions are enhanced, which is shown in Fig. 11.2 for a selection of overlapping square values, $|\langle\psi_{Cl}|\psi_{Vl}\rangle|^2$, for the default set as a function of L_d. It can be seen in Fig. 11.2(a) that for the diagonal transitions, the overlapping reduces during the initial interdiffusion which then increases to about its as-grown magnitude. On the other hand, as shown in Fig. 11.2(b), for the off-diagonal transitions, the overlapping initially increases and then either remains approximately constant or reduces at large L_d. The changes during the initial stages of interdiffusion are because of the nonlinearity of the well shape, which is magnified to $10Å \leq L_d \leq 30Å$. This is due to the relaxed orthogonality of the wave functions, which may be described analytically from a similarly shaped hyperbolic QW [11.3 and 11.4]. This effect of reversing the magnitude between the diagonal and off-diagonal transitions for the above range of values of L_d is similar to that observed and predicted using an external applied electric field [11.5 and 11.6]. The strength of the optical transitions is also governed in part by the value of this overlap and the trend is the same as in the hyperbolic QW case [11.4].

The HH related interband transition energy E'_{pq} and its strength f_{pq} for AlGaAs/GaAs QWs at the zone center are shown in Table 11.1. The

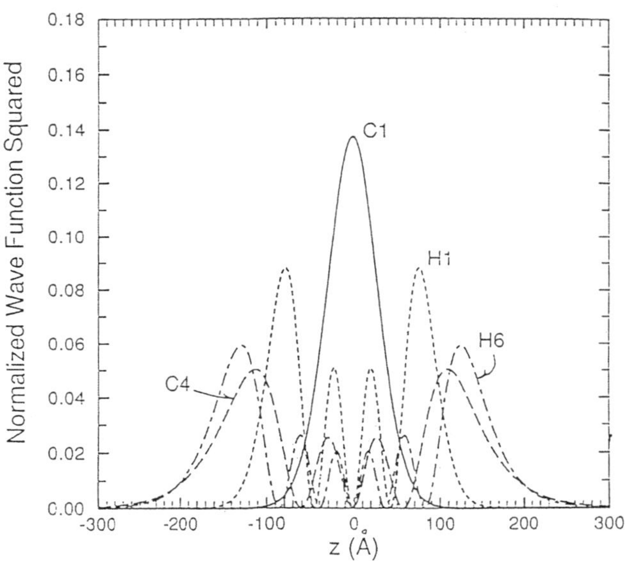

Fig. 11.1 The square of envelope wave functions at a selection of the subband edges for the EQW ($w_0 = 0.3$, $L_z = 100Å$, $L_d = 40Å$) [11.48].

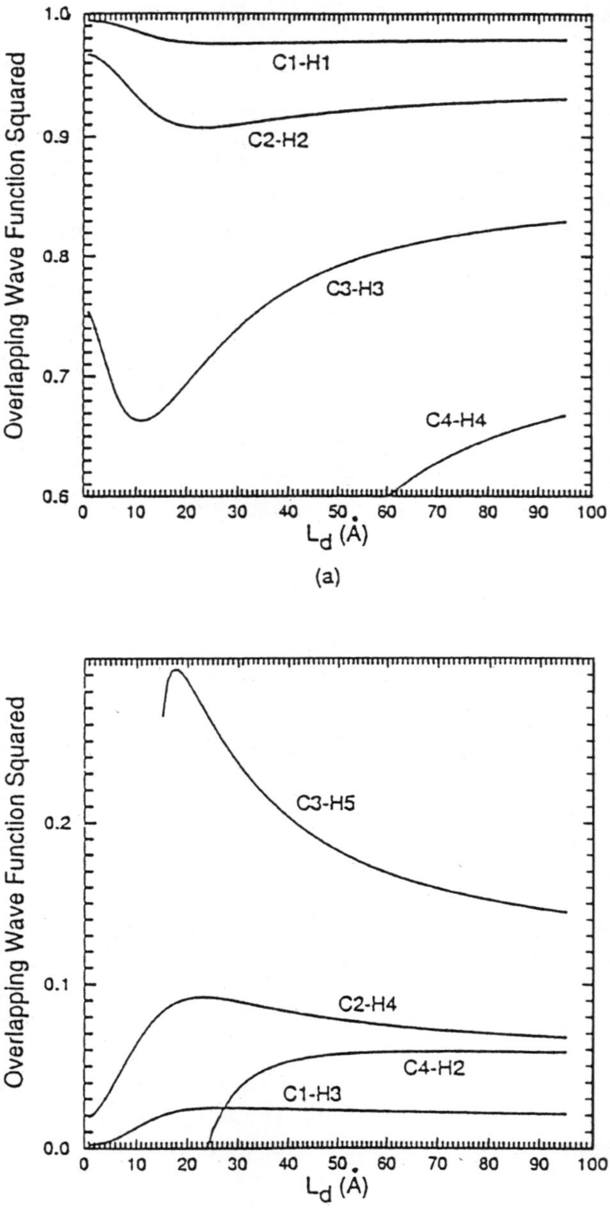

Fig. 11.2 a, b The square of the overlapping wave functions, $|\langle \psi_{C1} | \psi_{V1} \rangle|^2$, at the subband edges for an EQW ($w_0 = 0.3$, $L_z = 100$Å) with varying L_d. Only a selected number of overlapping are shown ($\Delta l = l_C - l_V = 0$) [11.29].

Table 11.1 Integrated Transition Energy and Strength (w_O = 0.3, L_Z = 100Å, Q_C = 0.6).

Selected Levels	Interband Transitions			
	E'_{pq} (eV) AG-SQW (L_d = 0Å)	E'_{pq} (eV) HQW (L_d = 40Å)	f_{pq} AG-SQW (L_d = 0Å)	f_{pq} HQW (L_d = 40Å)
C1-HH1	1.4624	1.6414	0.9333	0.8991
C1-HH3	1.5241	1.6794	0.0069	0.0497
C2-HH2	1.5756	1.7118	0.7632	0.7070
C2-HH4	1.6659	1.7444	0.0234	0.1529
C3-HH3	1.7395	1.7658	0.3525	0.4408
C3-HH1	1.6778	1.7278	0.0085	0.0346

dimensionless transition strength is defined here as the normalized wave function overlap squared divided by the respective normalized transition energies ($E_{pg}/\tilde{E}_g$) squared, i.e.,

$$f_{pq} = \left|\langle \psi_{Cp} \,|\, \psi_{Vq} \rangle\right|^2 (\tilde{E}_g \,/\, E'_{pq})^2 . \tag{11.1}$$

The enhancement of the off-diagonal overlapping wave functions (see Table 11.1) also demonstrates a relaxed selection rule for the non-square QW.

11.1.1.2 InGaAs/GaAs

The interband transitions energy at Γ, and the squared envelope part of the electron-hole overlapping wave function, $|\langle\psi_{Cp}|\psi_{Vq}\rangle|^2$ of an InGaAs/GaAs are depicted in Table 11.2 for various values of L_d. The overlapping wave function is an inner product, and represents the dipole transition selection rule. For a square-QW, the diagonal transitions are much larger than the off-diagonal transitions; and this corresponds to a reduction of the $|\langle\psi_{Cp}|\psi_{Vq}\rangle|^2$ by 0.025 in the C1 related transitions. The reduction of the off-diagonal transitions from the diagonal one is not as pronounced in the DFQW as in the square-QW. For instance, in the case of L_d = 40Å, from the C1-HH1 to the C1-HH3, a reduction factor of 0.09 results. This is more than three times that of the square-QW case. Also, the C1-HH3 overlapping value is increased three times from the SQW. It should be noted that the largest value of $|\langle\psi_{Cp}|\psi_{Vq}\rangle|^2$ in the DFQWs is due to the transition C3-HH5, and, in fact, it is larger than the diagonal C3-HH3 transition. The diagonal

Table 11.2 Interband transition energies and square of the overlapping wavefunctions.

	Interband Transitions							
	E_{pq} (eV)				$\|<\chi_p\|\chi_q>\|^2$			
	$L_d = 0Å$	$L_d = 10Å$	$L_d = 40Å$	$L_d = 80Å$	$L_d = 0Å$	$L_d = 10Å$	$L_d = 40Å$	$L_d = 80Å$
C1-HH1	1.2464	1.2542	1.3192	1.3625	0.9533	0.9301	0.9145	0.9232
C1-HH3	1.3041	1.3141	1.3491	1.3754	0.0264	0.0591	0.0817	0.0722
C1-HH5	–	–	1.3681	1.3853	–	–	0.0036	0.0045
C1-HH7	–	–	–	1.3914	–	–	–	0.0000
C2-HH2	1.3400	1.3506	1.3696	1.3845	0.6956	0.6338	0.6881	0.7470
C2-HH4	–	1.4075	1.3946	1.3959	–	0.3520	0.2793	0.2223
C2-HH6	–	–	–	1.4041	–	–	–	0.0304
C3-HH3	–	–	1.4015	1.4017	–	–	0.1408	0.4081
C3-HH1	–	–	1.3716	1.3888	–	–	0.0213	0.506
C3-HH5	–	–	1.4205	1.4116	–	–	0.4061	0.4009
C3-HH7	–	–	–	1.4177	–	–	–	0.1403
C4-HH4	–	–	–	1.4125	–	–	–	0.0254
C4-HH2	–	–	–	1.4011	–	–	–	0.0305
C4-HH6	–	–	–	1.4207	–	–	–	0.1673
C1-LH1	1.3276	1.3333	1.3685	1.3910	0.8496	0.8265	0.7728	0.7449
C3-LH1	–	–	–	1.4173	–	–	–	0.2230

$|\langle\psi_{Cp}|\psi_{Vq}\rangle|^2$ decreases then increases, and vice versa in the off-diagonal case, as the diffusion proceeds. This is due to the SQW-like characteristic in the smaller and larger L_d cases, and the DFQW characteristic in the medium L_d cases. These variations of $|\langle\psi_{Cp}|\psi_{Vq}\rangle|^2$ are similar to those reported in a square-QW under an external electric field. An optimal diffusion stage still needs to be found for a more favourable off-diagonal transition.

This enhancement of off-diagonal transitions can be explained by the orthogonality of the overlapping wave functions involved. In the finite barrier SQW, the wave functions are represented by standing waves, and therefore a rigid, although not strict, orthogonal relationship is obeyed, which leads to a weaker off-diagonal overlapping and a stronger selection rule (p – q = 0 transitions favourable). However, in the DFQWs, the wave functions are quite different due to a different confinement profile (see Fig. 11.3), in which the orthogonality is not so rigid and thus leads to the enhancement effect and a relaxation of the selection rule. In addition, the shifting of the subband levels due to diffusion modifies the individual wave function, which also contributes to $|\langle\psi_{Cp}|\psi_{Vq}\rangle|^2$.

Figure 11.4 shows the change in the in-plane strain (at the well center) with increasing L_d, the variation of the ground state transition energy for the electron-heavy hole (C1-HH1), and the electron-light hole (C1-LH1) transitions with increasing interdiffusion. The compressive strain in the well is maximum when L_d = 0 and decreases with increasing L_d. The ground state transition which dominates the C1-HH1 transition and is a result of the compressive nature of the strain. As L_d increases, the ground

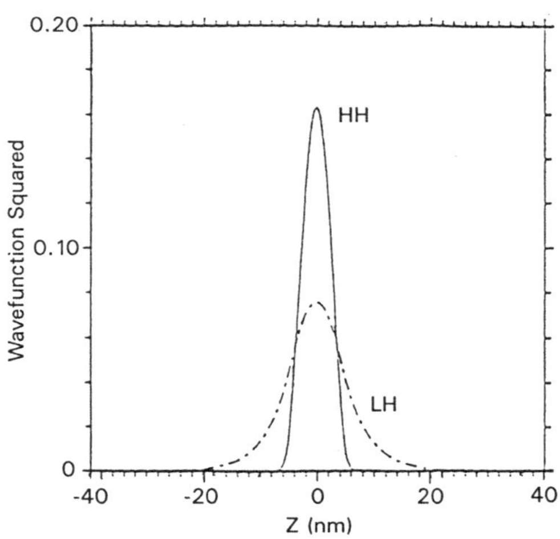

Fig. 11.3 Normalized wave function squared for the ground state HH and LH .

state transition energy increases, as evidenced by experimental results [11.7]. The change in C1-HH1 transition energy from $L_d = 0$ to $L_d = 10$nm corresponds to a blue shift in the wavelength of about 90nm. Thus the absorption edge of the intermixed QW structure is shifted to shorter wavelengths. From Fig. 11.4, it can be seen that the difference between C1-HH1 and C1-LH1 transition energies decrease with increasing L_d, with the effect that intermixing reduces the polarization sensitivity of the InGaAs/GaAs QW [11.8].

Figure 11.5 shows the variation of the C1-HH1 transition energy with both the well width and the indium content for various values of L_d. The C1-HH1 transition energy is more sensitive to the indium content variation than to the well width variation. The transition energy change from $L_d = 0$ to 8nm for $L_z = 5$nm is almost the same as that for $L_z = 10$nm (Fig. 11.5(a)). From Fig. 11.5(b), it can be seen that the higher the indium content, the larger is the possible transition energy variation with interdiffusion. These results show that the carrier confinement profile of intermixed InGaAs/GaAs QWs is more sensitive to the indium content than to the well width of the as-grown QW. This indicates that using intermixing to tailor the transition energy, it would be more favourable to start with an as-grown $In_wGa_{1nw}As/GaAs$ QW structure having as high a value of x as

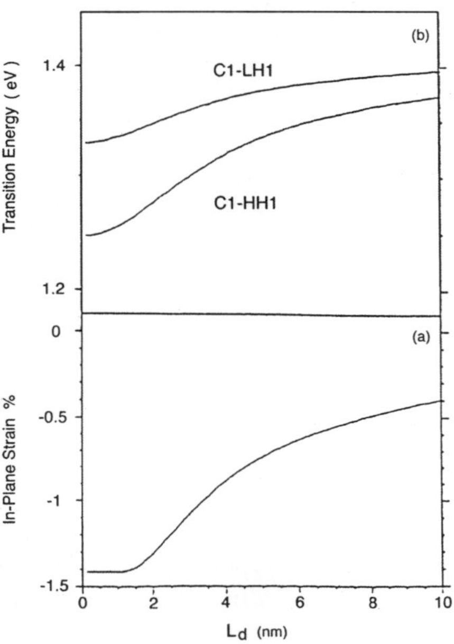

Fig. 11.4 a, b Variation of **(a)** the in-plane strain and **(b)** the ground state transition energy with L_d [11.14].

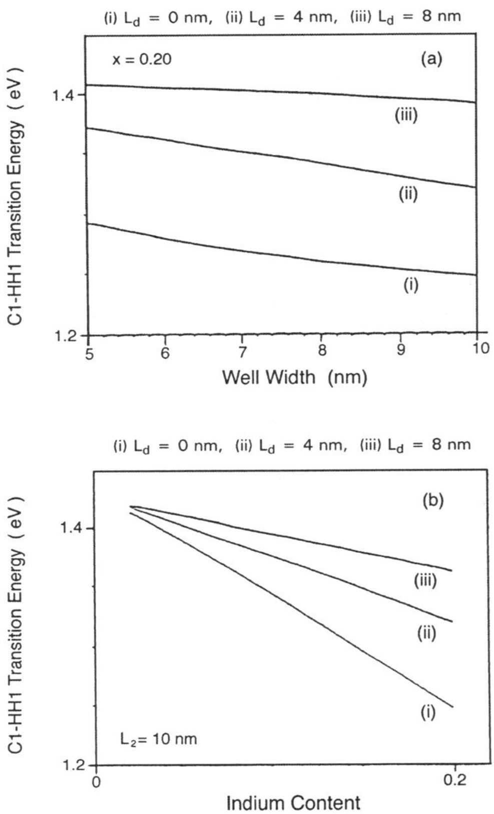

Fig. 11.5 a, b Variation of the C1-HH1 transition energy with **(a)** the well width, **(b)** the indium content for different L_d values [11.49].

possible. The penalty would be a smaller well width so as to maintain a coherently strained structure, although the ground state transition energy variation with intermixing is larger for a QW structure with a large x and small L_z in contrast to one with a small x and large L_z.

11.1.1.3 InGaAs/InP

As mentioned in Sect. 10.2, there are three different types of interdiffusion mechanism involved in the intermixing of the InGaAs/InP material system. They involved (1) group III interdiffusion only, (2) group III and group V interdiffusion, and (3) group V interdiffusion. Due to the different components involved in each of the mechanism of interdiffusion, the resulting features in the transition are distinguished from each other. The first two types of interdiffusion are discussed below.

(1) Group III interdiffusion only

The variation of C1-HH1 and C1-LH1 transition energy with interdiffusion is shown in Fig. 11.6. The QW bandgap energy is defined as the C1-HH1 transition, the red shift with interdiffusion reflecting the compressive nature of the strain induced in the QW by the intermixing process. The HH1 state occurs in the miniwells for small values of L_d. As the interdiffusion proceeds the bandgap energy decreases, corresponding to a shift to longer wavelengths, as shown by experimental results for Zn-induced intermixing [11.9–11.11]. The shift of the band-edge wavelength to longer wavelengths with intermixing contrasts with the experimental results for intermixed AlGaAs/GaAs [11.12] and InGaAs/GaAs [11.13], and for InGaAs/InP with a comparable extent of interdiffusion for the two sublattices, where the band-edge wavelength shifts to shorter wavelengths, as discussed below. For the interdiffusion length range in Fig. 11.6 the shift in bandgap energy corresponds to a shift in band-edge wavelength from about 1.4 to about 1.9μm. Another contrasting effect is the difference between the C1-HH1 and C1-LH1 transition energy. From Fig. 11.6, we see that intermixing under these conditions leads to an increasing separation between the C1-HH1 and C1-LH1 interband transition energy gap, while in intermixed InGaAs/GaAs QW structures this separation decreases [11.14].

(2) Identical interdiffusion rate for group-III and group-V sublattices

It is shown in Fig. 11.7 that the C1-HH1 and C1-LH1 transition energies increase as the intermixing increases. This results in a bandgap shift to

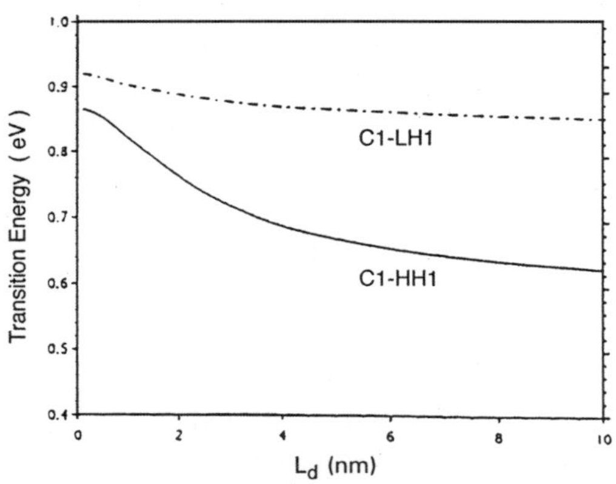

Fig. 11.6 Variation of the C1-HH1 and C1-LH1 transition energies with L_d [11.53].

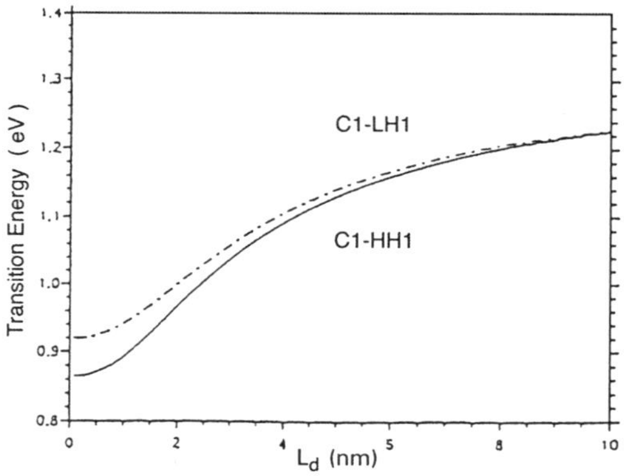

Fig. 11.7 Variation of the C1-HH1 and C1-LH1 transition energies with interdiffusion assuming identical interdiffusion rates for group III and group V sublattices.

shorter wavelengths (blue shift), as reported from measured-intermixed InGaAs/InP QWs using S diffusion [11.15]. Similar results occurred in intermixed AlGaAs/GaAs QWs [11.12], where the interdiffusion process is much better understood. The change in the C1-HH1 transition energy in going from $L_d = 0$ to 10nm corresponds to a shift to shorter wavelengths of about 400nm, from about 1.4 to about 1.0μm, which is much larger than that for the InGaAs/GaAs [11.14], and is a direct result of the compositional change brought about by the intermixing process. The C1-HH1 and C1-LH1 transition energies converge as L_d increases, so that in the intermixed InGaAs/InP structure, the separation between the two transition energies almost reduces to zero. Since the intermixed structure remains lattice-matched, the HH and LH confinement profiles still coincide, and the difference between the two transition energies is simply the result of the different quantization effects for the HH and LH due to their different effective masses. As interdiffusion proceeds, effectively wider well results, and the difference in the quantization effect for the HH and LH therefore decreases.

11.1.2 Electro-Optic Effect

11.1.2.1 Quantum-Confined Stark Effect

Among the electro-optic effects in QWs, the quantum-confined Stark effect (QCSE) is most notable, since strong exciton absorption occurs at room temperature, and it demonstrates that QWs are suitable for the realization of optical switching devices, such as the self-electro-optic effect devises

(SEEDs). Demonstrations using these devices have included arrays of SEEDs as large as 128×256 (32K elements). One drawback of these devices is their switching power, and at present the 32K array, requires about 1pJ to change state.

The extent of the Stark shift for a given applied field is dependent upon the QW structure (i.e. the well width, the barrier height, and the shape of the confinement profile) and is directly proportional to the effective mass of the carrier, which is dependent upon the subband structure. The diffused QW (DFQW) provides an enhanced Stark shift because the applied field lowers the subband edge in the tilted nonlinear confinement profile more than it does in the case of the square QW confinement profile. Also the average of the spatially dependent effective mass increases with increasing distance from the well center in the nonlinear profile.

(1) AlGaAs/GaAs

In the case of AlGaAs/GaAs QW structures, the QW intermixing process results in the interdiffusion of Al and Ga across the well barrier interface, which results in a nonlinear QW potential profile with modified confined states, transition energies, and wave functions. This process has already been used to produce optical confinement in lasers and wave-guides, and it has the potential to improve modulator performance by lowering the driving voltage and increasing the on/off ratio. The Stark shift energy ΔE for the C-HH transitions in different QW structures is shown in Fig. 11.8(a) below for applied fields (F) up to 50kV/cm. This includes an as-grown QW ($L_d = 0$), two DFQWs ($L_d = 20$ and 40Å), and an "equivalent" square QW (i.e. a square QW with the same well depth and zero field ground state transition energy as the DFQW ($L_d = 40$Å), but with a different Stark shift, which is obtained by adjusting the square well width to 86Å)(see Fig. 11.8(b)). It is important to consider the equivalent square QW since the depth and transition energy of the as-grown QW are very different from those of the DFQW at F = 0, and thus it is not suitable for an accurate comparison of the QCSE in these QW structures.

A similar method of comparison has also been used by others [11.16], where only the ground state transition energy is matched by considering an as-grown QW with a modified width as the equivalent QW, but the effect of the barrier height (well depth) is not considered. This demonstrates the improved Stark shift in a DFQW ($L_d = 40$Å) is greater than that in both the 100Å as-grown and 86Å equivalent square QWs. Also, it is shown that the variation of the non-square QW structure (in width, depth and shape) enables the Stark shift and absorption edge to be modified in a different way from that achieved by varying only the well width. The difference in the Stark shift of the as-grown and equivalent square wells is relatively small. This can be explained qualitatively by the compensation effect between a reduction of well width (ΔE increases) and barrier height (ΔE increases), and an increase of the carrier effective mass (ΔE increases) in the equivalent QW. The Stark shift maybe estimated by a variational approximation [11.17], whereΔE (of the equivalent QW) $\propto (0.026716)(86)^4$ and ΔE (of the as-grown QW, $L_d = 0$) $\propto (0.016662)(100)^4$. The ratio of their Stark shifts = 0.8771, which agrees well with our exact numerical calculation of 0.87.

The parameters used for the equivalent square well are C1 = 31.963meV, C2 = 120.143meV, HH1 = 8.552meV, HH2 = 33.232meV, $m_{C1} = 0.0781$, and for the as-grown QW, they can be found in [11.18] with $m_{C1} = 00688$, where the electron non-parabolicity has been included.

It can be seen from Fig. 11.8(a) that the Stark shift in the DFQW (L_d = 40Å) is larger than that for all the other QWs and at 50kV/cm it reached a

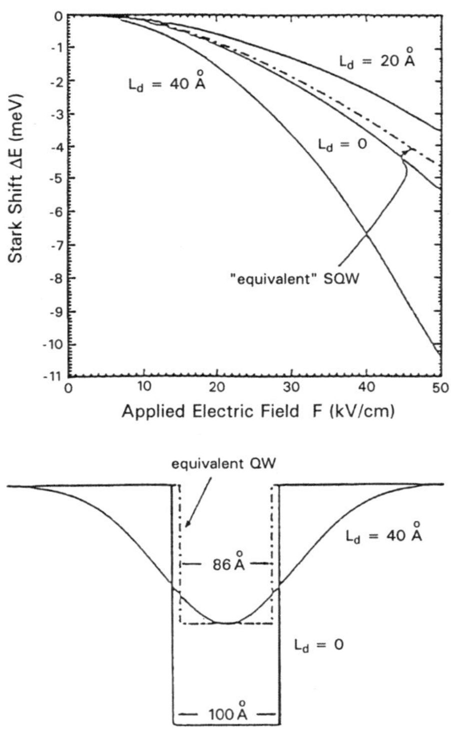

Fig. 11.8 a, b **(a)** The exciton Stark shift, $\Delta E = E_{exc}(F) - E_{exc}(F = 0)$, shown for applied fields (F) up to 50kV/cm for an as-grown square, two interdiffused (L_d = 2 and 4nm) $Al_{0.3}Ga_{0.7}As/GaAs$ QWs (L_z = 10nm), and an equivalent square QW (w = 0.3, L_z = 8.6nm, L_d = 4nm). **(b)** Schematic diagram for the QW confinement potentials (L_d = 0nm, 4nm, and the equivalent square well) as described in (a) [11.21].

two-fold enhancement (93%) over the equivalent QW. This confirms the enhancement effect for the Stark Shift in an interdiffusion-induced QW, which is also similar in trend to the results measured for a linear graded-gap QW and agrees with a twofold enhancement (80%) reported for impurity-free vacancy-diffused QWs [11.16 and 11.19]. These results indicate the advantage of using intermixed QWs for the realization of high speed optical modulators with low driving voltages. As L_d increases from zero, the Stark shift reduces initially (L_d = 20Å) and then it increases to a value larger than that found in the as-grown QW (L_d = 0). In other words, during the early stages of interdiffusion, a higher voltage will be needed to drive the modulator to produce the required Stark shift.

(2) InGaAs/GaAs

For the growth direction z along <001>, the lattice-mismatch in coherently-strained pseudomorphic InGaAs/GaAs QWs results in a biaxial compressive strain perpendicular to the growth direction, which increases the energy gap among the conduction band, both the degenerate valence bands and the spin-orbit split-off band, and a uniaxial component of the strain parallel to the growth direction, which lifts the degeneracy of the HH and LH bands at the Brillouin zone center Γ, with the HH band moving towards the conduction band-edge and the LH band moving further away from the conduction band-edge. The shear strain also results in mixing between the LH and split-off bands.

The intermixing process is assumed to result in the interdiffusion of In and Ga atoms during the thermal annealing. After the interdiffusion, the In concentration will have an error function distribution. This varying In compositional profile implies that the carrier effective mass, the unstrained bandgap, and the hydrostatic and shear strain, all vary along the growth direction.

Figure 11.9 shows the exciton Stark shift energy for the HH related transitions with the applied electric field, for various values of L_d. It can be seen that the Stark shift increases in magnitude with increasing strength of the applied field in all cases. At the corresponding values of the applied field, the Stark shift decreases slightly in the initial stages of interdiffusion (L_d = 1nm) and then increases to a value larger than the as-grown (L_d = 0nm), showing a significantly larger value for $L_d > 2$nm. This result shows that for a sufficiently large value of L_d the exciton Stark shift in the intermixed QW is greater than in the as-grown QW case. The applied field lowers the ground state subbands, and thus the ground state transition energy, to a greater extent in the intermixed QW ($L_d > 2$nm, in this case), and the average of the spatially dependent effective masses over the well region is larger in the intermixed QW, as a result of the increasing effective masses with increasing distance from the well center. The increase in Stark

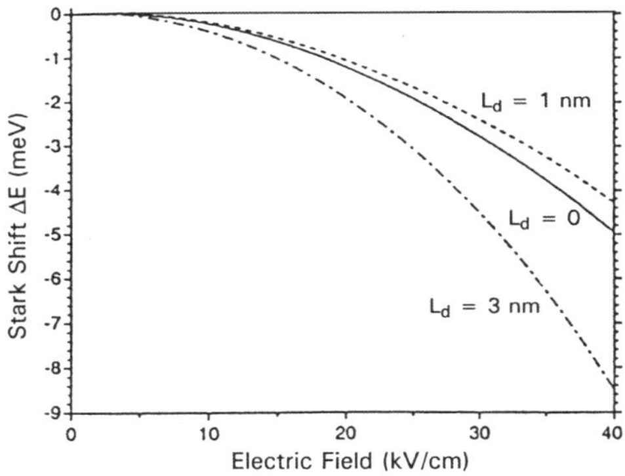

Fig. 11.9 Stark shift with applied electric field for an $In_{0.2}Ga_{0.8}As/GaAs$ single QW, with an as-grown L_z = 10nm, for various values of interdiffusion length L_d.

shift in the intermixed QW is similar in trend to the results calculated for a graded-gap QW structure. It can be seen that for L_d = 3nm, the Stark shift at 30kV/cm is enhanced by a factor of 65% over the as-grown QW case. This enhancement could be the basis of improved device performance in electroabsorption modulators which can be driven with a low operating voltage and possess a high on/off ratio. The Stark shift values for the as-grown QW compare favorably with the calculated values reported for an InGaAs/GaAs QW structure with similar barrier heights of 100meV [11.20]. The Stark shift values determined here for the strained, intermixed InGaAs/GaAs QW are comparable to the Stark shift values obtained when considering interdiffusion in an $Al_{0.3}Ga_{0.7}As/GaAs$ QW with L_z = 10nm [11.21].

11.1.2.2 Electroabsorption — AlGaAs/GaAs

The interband electroabsorption effect near the band-edge of DFQW structures has received considerable attention for modulator applications as it results in a much larger absorption change due to the QCSE than for the bulk material. The field-induced change of the absorption coefficient,

$$\Delta\alpha(F) = \alpha(F) - \alpha(F = 0)$$

(where α and F are the absorption coefficient and applied external electric field), is calculated for wavelengths around the bandedge of the DFQWs.

The DFQW is modeled by an error function confinement profile where the subbands are considered with the conduction band non-parabolicity and the valence band-mixing under an applied electric field. The subbands are then used to calculate the heavy-hole- (HH) and light-hole (LH)-related 1S exciton binding energies and wave functions by a perturbative-variational method as mentioned in Sect. 10.4. The incident light, which propagates along the plane of the QW layers, is TE and TM polarized, where its electric field vector is perpendicular and parallel to the growth axis of the QW layers respectively. The external electric field is applied perpendicular to the QW layers, such as a p-i-n structure.

The as-grown QW structure consists of a single GaAs well, of width L_z = 100Å, which is clad with 1000Å thick $AL_{0.3}Ga_{0.7}As$ barriers, and the DFQW modeled by an error function confinement profile. Figure 11.10 shows the $\Delta\alpha$ spectra for the TE and TM polarizations of an as-grown QW, a DFQW (L_d = 40Å) with fields up to F = 40kV/cm (TM polarization). The maximum F for the different polarization is due to the ground state HH and LH carriers, which is sensitive to the TE and TM polarization respectively. The values of $\Delta\alpha$ are determined with the well and barrier layers weighted equally with the relevant parameters and a broadening factor Γ_b taken to be 3meV and 6meV for the exciton and bound states respectively. As can be seen, two N-wave-like dispersions occur for the TE-polarized $\Delta\alpha$ spectrum: the one at the longer wavelengths has a larger amplitude due to the HH 1S-exciton. A similar single N-wave-like dispersion can also be found for the TM-polarized case, due to the contribution from the allowed LH 1S-exciton. The dispersion amplitude grows with increasing field for all cases of interdiffusion, and is greater for the TM-polarization, and shifts towards shorter wavelengths as L_d increases.

For the TE-polarization, $\Delta\alpha$ reduces and becomes negative as the interdiffusion proceeds beyond L_d = 30Å at the shorter wavelength dispersion peak for F = 50kV/cm (see Fig. 11.10(a)–(c)). This is due to the fact that the reductions of the absorption peaks with F are much more enhanced for the more extensively intermixed QW (i.e. L_d = 40Å). The HH and LH peaks should be merged for Stark shift to a different extent and therefore remain separated (see F = 30kV/cm in Fig. 11.10(c)) until the ground LH state becomes unbound (tunnels away) and leaves behind a weakly bound HH state, which results in a drastic reduction of the exciton absorption peak, as can be seen between F = 40kV/cm and 50kV/cm for L_d = 40Å. This results in an enhanced (larger $\Delta\alpha$) electro-absorption effect for the TE-polarization in the DFQWs.

The positive and negative dispersion peaks of the $\Delta\alpha$ spectra are larger for the TM- than TE-polarization (see Fig. 11.10(a)–(f)). These larger $\Delta\alpha$ peaks are due to the suppression of the HH transition at k = 0 for the TM-polarized light and the nonlinear confinement profile of the DFQW, which is more sensitive to the applied electric field. The effects of both of the above contribute to a lower and flatter absorption spectrum under an applied electric field (note that the maximum field for the TM case is 40kV/cm). This phenomenon of a larger $\Delta\alpha$ for the TM-polarized light is also demonstrated using a parabolic nonlinear QW profile, whose trend compares favourably with the measured waveguide absorption spectra of the intermixed QWs. Although the magnitude of $\Delta\alpha$ is greater for the TM-

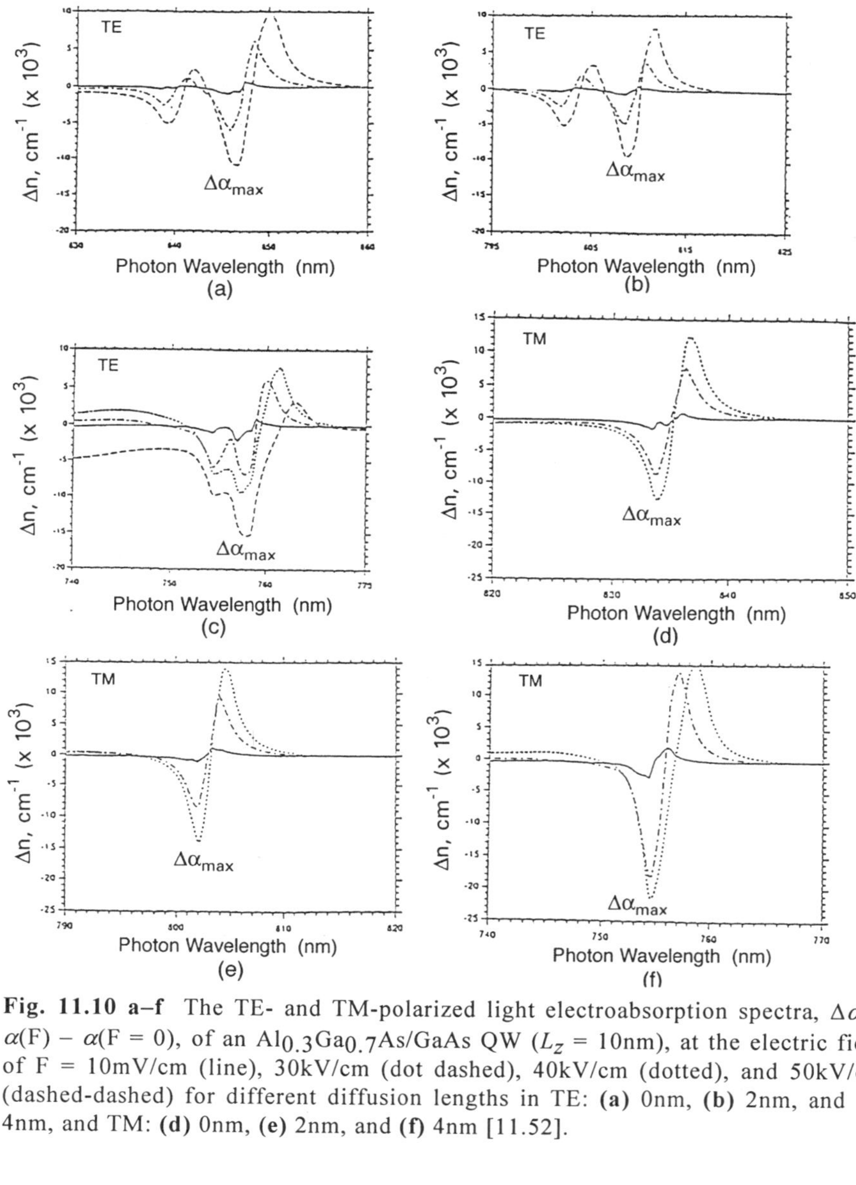

Fig. 11.10 a–f The TE- and TM-polarized light electroabsorption spectra, $\Delta\alpha = \alpha(F) - \alpha(F = 0)$, of an $Al_{0.3}Ga_{0.7}As/GaAs$ QW (L_z = 10nm), at the electric field of F = 10mV/cm (line), 30kV/cm (dot dashed), 40kV/cm (dotted), and 50kV/cm (dashed-dashed) for different diffusion lengths in TE: **(a)** 0nm, **(b)** 2nm, and **(c)** 4nm, and TM: **(d)** 0nm, **(e)** 2nm, and **(f)** 4nm [11.52].

polarized light, it must be noted that the enhanced forbidden transitions, which become allowed in the presence of the applied field, may broaden the lowest energy exciton transition (an effect which is not considered here).

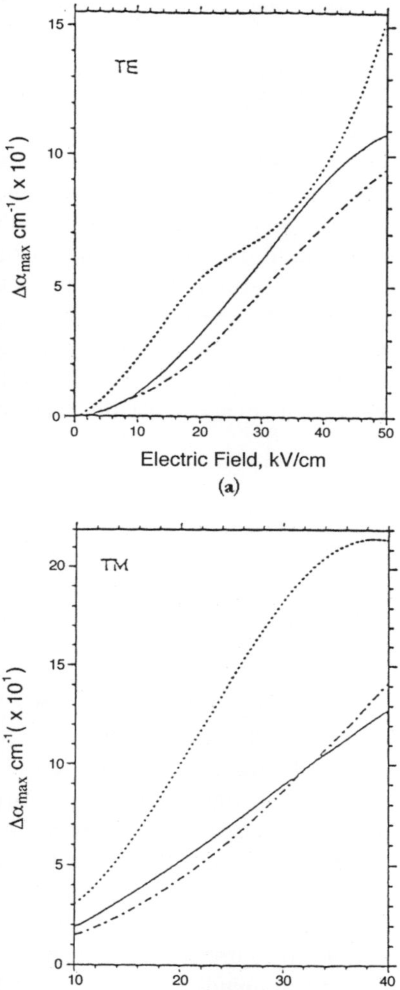

Fig. 11.11 a, b The maximum change of absorption, $\Delta\alpha_{max}$, as a function of the applied field F for diffusion length of L_d = 0nm (line), 2nm (dot-dashed), and 4nm (dotted) at their respective wavelengths. **(a)** TE-polarization and **(b)** TM-polarization [11.52].

The peak value of $\Delta\alpha$ ($\Delta\alpha_{max}$) at its respective wavelength is shown in Fig. 11.11 as a function of F for L_d = 0, 20, and 40Å. In general, this maximum change occurs for dispersion peaks with negative values and for approximately the same wavelengths for the respective L_ds (see Fig. 11.11). All of these curves are nonlinear for all values of L_d used and $\Delta\alpha_{max}$ for the DFQW (L_d = 40Å) is ≈ 45% and 65% greater than that of the as-grown square QW for TE-polarization at F = 50kV/cm and for TM-polarization at F = 40kV/cm, respectively. The $\Delta\alpha_{max}$ for L_d = 20Å is smaller than or comparable to that for the respective as-grown QW cases at the maximum Fs, which is due to an increased lowest subband energy level in the DFQW for a small L_d, as described in [11.53]. The larger $\Delta\alpha$ should provide a stronger electroabsorption effect for modulators operating in the normally OFF mode, for operation at wavelengths around the F = 0 peak. The change of the absorption coefficient has been determined, due to an applied electric field of up to 50kV/cm on various intermixed $Al_{0.3}Ga_{0.7}AS$/GaAs QWs with an as-grown well width of 100Å, where the incident light is TE- and TM-polarized. The results show an enhancement of the electro-absorption effect for the intermixed QW.

11.2 Device Applications

11.2.1 Delineated Waveguides

As mentioned earlier, the transmission methods for telecommunications during the last decade has been dominated by the evolution of opto-electronics. Electronic integrated circuits are being replaced by optical integrated circuits in order to enhance the performance of electronic integrated circuits and also to perform novel functions suited to optics. An essential feature of these optical components is a method of optically linking the various parts of the circuit together, and optical waveguides can be, therefore, considered as the connecting wires of the optical circuits.

An efficient way to realize waveguiding structure in optoelectronic integrate circuits is provided by the impurity-induced disordering (IID) technique. The masked ion implantation process produces a modification of the QW material which in turn modifies its refractive index. This creates a refractive index step between the implanted and non-implanted regions and produces a lateral confinement for photons in the lateral dimension. The work is to perform a simulation on a two-dimensional waveguiding structure fabricated using such technique.

The structure of the device to be considered consisted of layers of AlGaAs/GaAs QWs on a GaAs substrate, the schematic of the structure is shown in Fig. 11.12.

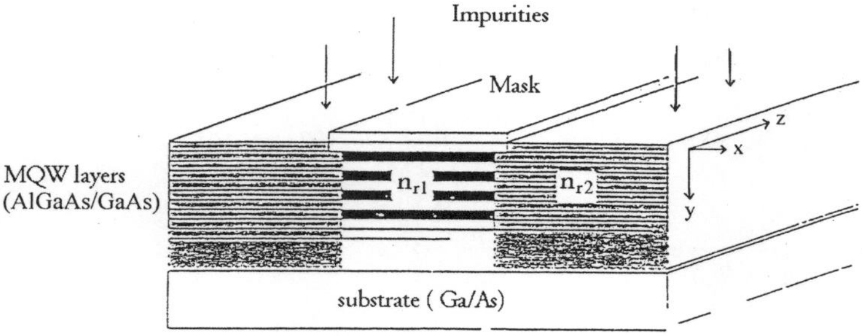

Fig. 11.12 Schematic diagram of the waveguiding structure.

11.2.1.1 Impurity Profile as Implanted, $N(x, y, t = 0)$

In this model, ion implantation is selected to be the way to inject the impurity to the structure. Before finding this concentration profile, we first consider the distribution of ions when only a single ion beam is implanted to a material. Based on [11.23], for one incident ion beam, the distribution function $f(x,y,z)$ is as follows:

$$f(x,y,z) = \frac{1}{(2\pi)^{3/2}\langle\Delta R_p\rangle\langle\Delta x\rangle\langle\Delta z\rangle} \times \exp\left\{-\frac{x^2}{2\langle\Delta x\rangle^2}-\frac{z^2}{2\langle\Delta z\rangle^2}-\frac{(y-\langle\Delta R_p\rangle)^2}{2\langle\Delta R_p\rangle^2}\right\} \tag{11.2}$$

where y is the implantation depth, x,y are the lateral coordinates, R_p is the projected range, ΔR_p is the longitudinal standard deviation, and Δx, Δz are the lateral standard deviation (all these three values depends on the implanting ions, implantation energy and the substrate material).

Thus the distribution $F(x,y,z)$ of ions implanted through a mask-window of size "2a" can be calculated based on the contributions of all incident ions (see Fig. 11.13):

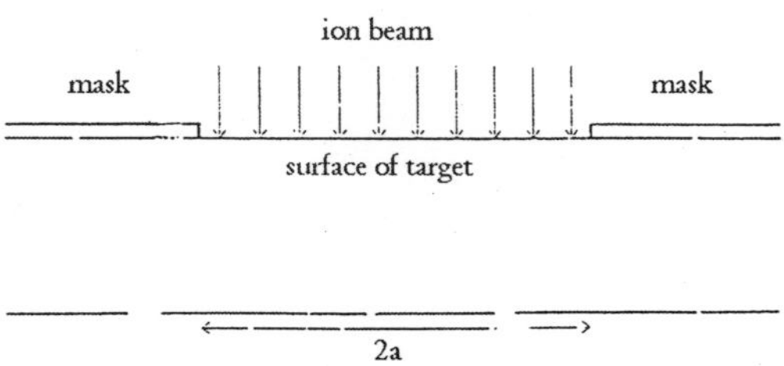

Fig. 11.13 Ion implantation of a target covered by a mask with a slit of width $2a$.

$$
\begin{aligned}
F(x,y,z) &= \frac{N_0}{(2\pi)^{3/2}\langle \Delta R_p\rangle\langle \Delta x\rangle\langle \Delta z\rangle}\exp\left\{-\frac{(y-\langle \Delta R_p\rangle)^2}{2\langle \Delta R_p\rangle^2}\right\} \\
&\quad\times \int_{-a}^{a}\int_{-\infty}^{\infty}\exp\left\{-\frac{(x-\varsigma)^2}{2\langle \Delta x\rangle^2}-\frac{(z-\eta)^2}{2\langle \Delta z\rangle^2}\right\}d\varsigma\, d\eta \\
&= \frac{N_0}{\sqrt{2\pi}\langle \Delta R_p\rangle}\exp\left\{-\frac{(y-\langle \Delta R_p\rangle)^2}{2\langle \Delta R_p\rangle^2}\right\} \\
&\quad\times\left\{\frac{1}{\sqrt{\pi}}\int_{\frac{x-a}{\sqrt{2}\langle \Delta x\rangle}}^{\frac{x+a}{\sqrt{2}\langle \Delta x\rangle}}\exp\left(-t^2\right)dt\right\}
\end{aligned}
\tag{11.3}
$$

where N_0 is the implantation dose. The equation is valid assuming the mask's edge is infinitely steep and the mask itself is thick enough to stop any impurity from penetrating.

In this equation the distribution function along x axis is shown in the last brace. In case when the width of the slit is much larger than $\langle \Delta x\rangle$ (i.e. $a >> \langle \Delta x\rangle$), the distribution function along x axis can be approximately expressed by the complementary error function for all positive x.

$$
\begin{aligned}
F(x,y) &= \frac{N_0}{\sqrt{2\pi}\langle \Delta R_p\rangle}\exp\left\{-\frac{(y-\langle R_p\rangle)^2}{2\langle \Delta R_p\rangle^2}\right\} \\
&\quad\times\left\{\frac{1}{\sqrt{\pi}}\mathrm{erfc}\left(\frac{x-a}{\sqrt{2}\langle \Delta x\rangle}\right)\right\}
\end{aligned}
\qquad (x \geq 0) \tag{11.4}
$$

where

$$
\mathrm{erfc}(t) = \int_t^{\infty}\exp(-u^2)\,du\ . \tag{11.5}
$$

In our proposed structure, we put a mask in the center and the schematic diagram of the structure to be modeled is shown in Fig. 11.14.

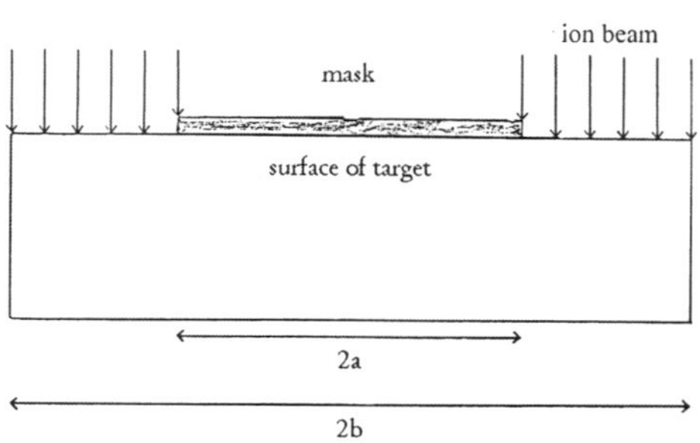

Fig. 11.14 Ion implantation on a target with a mask width of $2a$.

Again using similar calculations we can get the distribution function $F(x,y,z)$:

$$F(x,y,z)=\frac{N_0}{(2\pi)^{3/2}\langle\Delta R_p\rangle\langle\Delta x\rangle\langle\Delta z\rangle}\exp\left\{-\frac{(y-\langle R_p\rangle)^2}{2\langle\Delta R_p\rangle^2}\right\}$$
$$\times\left\{\int_{-b}^{-a}\int_{-\infty}^{\infty}\exp\left\{-\frac{(x-\varsigma)^2}{2\langle\Delta x\rangle^2}-\frac{(z-\eta)^2}{2\langle\Delta z\rangle^2}\right\}d\varsigma\, d\eta\right. \tag{11.6}$$
$$\left.+\int_{a}^{b}\int_{-\infty}^{\infty}\exp\left\{-\frac{(x-\varsigma)^2}{2\langle\Delta x\rangle^2}-\frac{(z-\eta)^2}{2\langle\Delta z\rangle^2}\right\}d\varsigma\, d\eta\right\}$$

where "$2b$" is the entire width of the structure to be considered. Therefore the impurity profile $N(x,y)$ injected to the structure is:

$$N(x,y)=F(x,y)=\frac{N_0}{\sqrt{2\pi}\,\langle\Delta R_p\rangle}\exp\left\{-\frac{(y-\langle R_p\rangle)^2}{2\langle\Delta R_p\rangle^2}\right\}$$
$$\times\left\{\frac{1}{\sqrt{\pi}}\left(\operatorname{erfc}\left(\frac{x-a}{\sqrt{2}\langle\Delta x\rangle}\right)-\operatorname{erfc}\left(\frac{x-b}{\sqrt{2}\langle\Delta x\rangle}\right)\right.\right. \tag{11.7}$$
$$\left.\left.+\operatorname{erfc}\left(\frac{x+b}{\sqrt{2}\langle\Delta x\rangle}\right)-\operatorname{erfc}\left(\frac{x+a}{\sqrt{2}\langle\Delta x\rangle}\right)\right)\right\}$$

Actually the impurity profile of many as-implanted structures, for example two masks or periods of masks (see Fig. 11.15), can be calculated by performing the appropriate integration. In addition, for masks with finite thickness (where impurities may penetrate through) and arbitrary mask edge (which will be a more realistic case, see Fig. 11.16), we can use models for implantation into multilayer targets [11.24]. The Numerical Range-Scaling Model is one example:

$$N(x)=N_1(x) \qquad \text{in layer 1} \tag{11.8}$$

$$N(x)=\alpha N_2\left(x-t\left(1-\frac{R_{p2}}{R_{p1}}\right)\right) \qquad \text{in layer 2.} \tag{11.9}$$

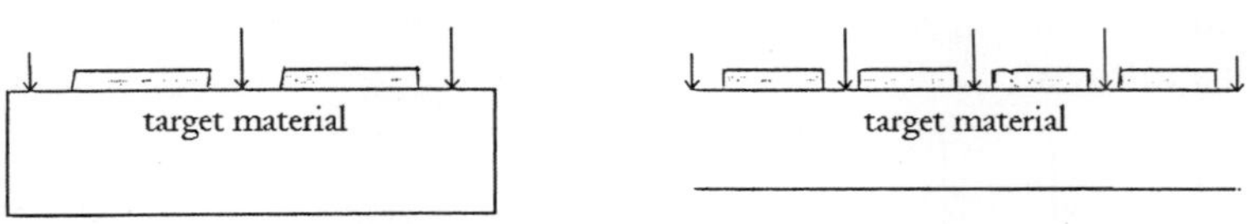

Fig. 11.15 Structure with two masks and two periods of masks.

Fig. 11.16 Masks with arbitrary shape.

Here, R_{p1} and R_{p2} are the projected ranges in layer one and layer two, respectively, t is layer one's thickness, N_1 and N_2 are the distributions in the semi-infinite layer one and two respectively, and α is calculated to yield the conservation of the implantation dose.

Using the above equation for a particular structure, we can then calculate the impurity profile $N(x,y)$ as the initial condition to proceed to further calculation.

11.2.1.2 Diffusion Coefficient and the Diffusion Length of the QW Layers

The diffusion coefficient of Al is proportional to the vacancy generated by the ion implantation. The local implanted impurity density determines the vacancy profile and hence determines the Al diffusion coefficient, D_{Al}. The Al concentration profile after thermal annealing $C(x,y,t)$ can now be described by

$$\frac{\partial C(x,y,t)}{\partial t} = \vec{\nabla}(D_{Al}(x,y,t)\vec{\nabla}C(x,y,t)) \qquad (11.10)$$

where t is the annealing time. The waveguide is discretized into cells of $\Delta x \Delta y$. Each of the Δys consists of one period of the QW structure, i.e. $\left|-\frac{1}{2}L_b\right| + L_z + \left|\frac{1}{2}L_b\right|$, where L_z is the well width and L_b is the barrier thickness, and the center of the ith cell coincides with the center of the QW. Once the Al concentration profile is obtained for a specific annealing time $t = \tau$, each of the ith cell is associated with a diffusion length, $L_d(x, y, \tau)$, upon the knowledge of which n_r can be found. The association is determined by the total diffusion of Al concentration from the barrier into the well in one cell. This is described by

$$f(L_d(x_i, y_i, \tau)) = \int_{y_i - L_z/2}^{y_i + L_z/2} C(x_i, \xi) d\xi \qquad (11.11)$$

where the total diffusion $f(L_d)$ can be expressed as a function of L_d which is defined as follows:

$$f(L_d) = \int_{-L_z/2}^{L_z/2} w_0 \left(1 - \frac{1}{2}\left[erf\left(\frac{L_z + 2\varsigma}{4L_d}\right) + erf\left(\frac{L_z - 2\varsigma}{4L_d}\right)\right]\right) d\varsigma \qquad (11.12)$$

the diffusion length L_d can be found by solving (11.4) and (11.5). The 2-D refractive index profile n_r can be readily determined [11.25] by using this $L_d(x,y,\tau)$ for each cell.

11.2.1.3 Refractive Index Profile, β and Φ

In the diffusion length profile, the number of grid points in the y direction is much greater than the number of QWs, and we use the mean value to represent one diffusion length value of every QW.

Adopting a simple model [11.26], we then calculate the refractive index profile $n_r(x,y)$ for a certain· propagating wavelength. After getting the refractive index profile, we can then solve the wave equation:

$$\nabla^2\Phi(x,y)+(n_r^{\,2}(x,y)k^2-\beta^2)\Phi=0 \qquad (11.13)$$

where Φ represent the intensity of the major field component, β is the waveguide propagation coefficient, and k is the wave number. The wave-guiding properties of the IID-produced 2-D AlGaAs/GaAs QW waveguide is studied based on this Helmholtz equation.

Before the Helmholtz equation is solved using some numerical method, the guiding conditions was estimated using the effective index method [11.27]. In the effective index method, five regions with different refractive indices are used to represent the refractive index profile in the waveguiding structure (see Fig. 11.13). The technique consists of calculating the propagation constant of the slab-guide obtained by letting the long dimension of the original rectangular guide approach infinity. This propagation constant is then used to define an effective dielectric constant. A second slab-guide, filled with material of the previously calculated dielectric constant, is obtained by allowing the short dimension of the rectangular guide to approach infinity. The propagation constant of this second slab is taken to represent the original rectangular waveguide.

To illustrate the application of the effective index method on our proposed structure, consider the rectangular dielectric guide of Fig. 11.17.

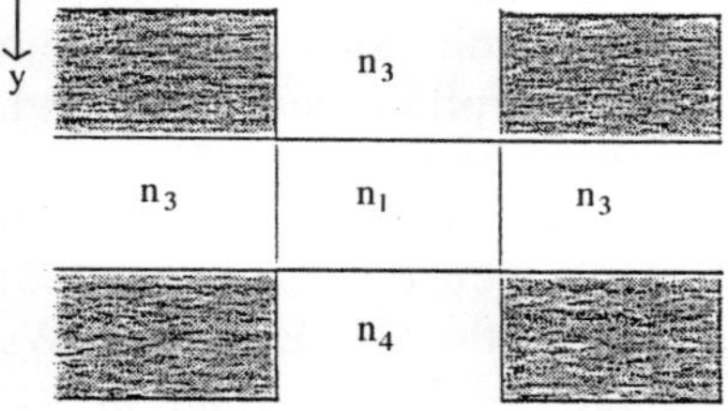

Fig. 11.17 Configuration of the rectangular dielectric waveguide.

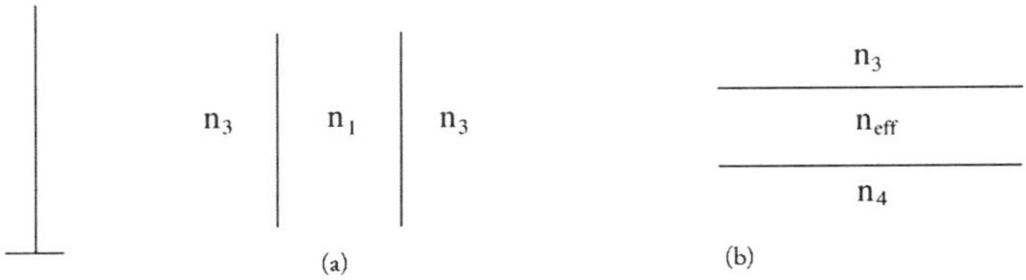

Fig. 11.18 a, b **(a)** A 1-D guide with equivalent confinement in the *x* direction. **(b)** An equivalent 1-D guide with confinement in the *y* direction. n_{eff} is the effective index of the mode in guide (a).

We first neglect the indices in the shaded areas. Working first in the *x* direction we calculate the effective index n_{eff} of the mode of interest of the 1-D guide in Fig. 11.18(a). The addition of confinement in the *y* direction is now represented by the equivalent 1-D guide of Fig. 11.18(b), with a guiding layer of index n_{eff} and cladding layer indices n_3 and n_4. The effective index n_{eff} of the mode of this equivalent planer guide is taken to describe the mode of the original 2-D structure. This procedure is particularly simple when one index difference $(n_1 - n_2)$ is large and the other index differences $(n_1 - n_3; n_1 - n_4)$ are small, as in the case of a IID waveguide.

11.2.2 Modulators

Optical communication systems often require modulating elements which change the intensity, phase or polarization of the optical signal. This makes the research of optical modulator become more significant in opto-electronics development: the waveguide type electroabsorptive modulator, the vertical cavity Fabry-Perot modulator with the application of a novel group III-V semiconductor material system and the diffusion induced non-square QW due to post-grown thermal annealing.

The electric field-induced refractive index and absorption coefficient change for DFQW systems, such as the $Al_{0.3}Ga_{0.7}As/GaAs$ in polarization-dependent incident light, for small and medium interdiffusion lengths, with different applied fields, can be used for electroabsorption modulators with reduced chirping in a wide range of operating wavelengths. The large change in the electric field-induced refractive index variation [11.28] ($\Delta n/n \approx 1\%$), as compared to the bulk group III-V semiconductor ($\Delta n/n \approx 0.01\%$) and $LiNbO_3$ ($\Delta n/n \approx 0.1\%$) in particular, is due to the QCSE in the QWs around the excitonic absorption edge. A small Δn is important in the realization of high speed absorption modulators to eliminate the nuisance of the inherent chirping, whereas a large Δn is required for the realization of directional couplers. The flexibility of the DFQW in achieving this condition is very attractive. The various Δn achieved by changing the interdiffusion length L_d with application of different field strengths induce chirping in the DFQW, which can also be utilized to realizing phase modulation devices. The phase modulation due to n is derived by considering the ideal situation for the phase difference per unit length, where $\Delta\phi = 2\pi\Delta n/\lambda$, in a waveguide type modulator. The distinctive $\Delta\phi$ can also serve as a multiplexer type of application and also be used as a wide-bandwidth directional coupler modulator and detector.

11.2.2.1 Waveguide Type Modulator

In the conventional waveguide modulator structure, in order to reduce the parasitic capacitance and the fringe field effects, the active region is isolated by coating it with a polymide (Fig. 11.19). The use of the DFQW as the active region will improve the electroabsorptive properties and chirping parameter of the modulator. The ion-implanted diffusion (IID) technology is used for the confinement in the cavity region. By attaching metal contact to the electrodes to apply an electric field across the QW (Fig. 11.20), the change in the electroabsorptive effect and the electro-optic effect can be employed to modulate the incident light. The material involved in the active guiding region of these modulators are the DFQW. The active cavity can be further modified by thermal annealing. The wavelength of the modulator can be tuned to short wavelength since the interdiffusion will increase the transition energy of the emission in the confined well.

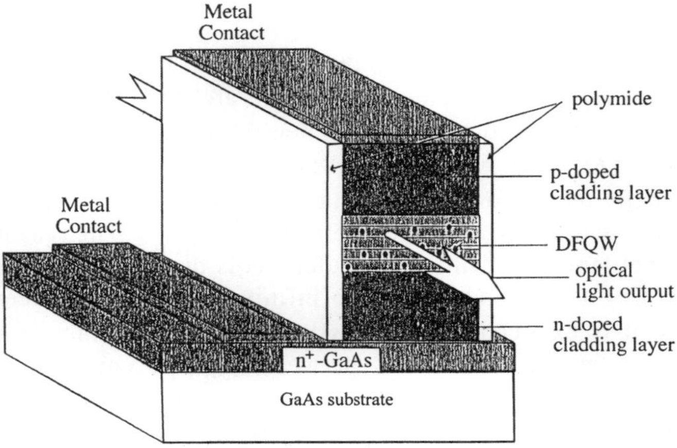

Fig. 11.19 The 3-D schematical structure of the LPM modulator.

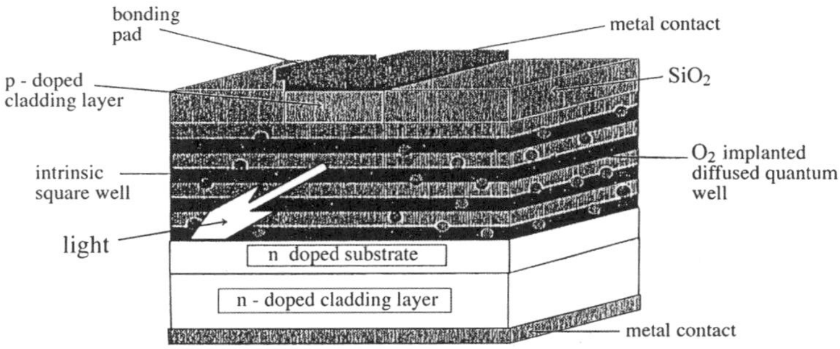

Fig. 11.20 The 3-D schematical structure of a waveguide type modulator.

In the interest of high-speed intensity waveguide type modulators, the effect of the DFQW on the chirping problem of electroabsorptive modulation is important, since this will affect the frequency broadening of the output intensity spectrum of the device. The degree of chirping can be indicated by a linewidth enhancement factor β_{mod} [11.29], which is due to an electric field applied in the modulator as in the case of the charge carrier injected into the laser device. The relation is defined as:

$$\beta_{\text{mod}} = \Delta n(\lambda)/\Delta k(\lambda) \tag{11.14}$$

where Δk is the change of the extinction coefficient that corresponds to the imaginary part of the complex refractive index. Physically, a change in the absorption coefficient implies a change in the extinction coefficient, since they are related by $\Delta k = \lambda\Delta\alpha/(4\pi)$, and through the Kramers-Krqnig relations, this change in Δk leads to a change in Δn. So, for a while, the electro-optic effect is in principle related to electroabsorption while the factor β_{mod} can be considered as a ratio between the two effects.

Figure 11.21 shows the chirping effect in the DFQW for an applied field F = 0, 100kV/cm with L_d = 30Å. It shows that there are three dispersible singularities in the β_{mod} spectrum, which implies that the spectrum of Δk has crossed the zero point ($\Delta k = \Delta\alpha = 0$) three times [11.28]. At these points, β_{mod} should theoretically tend to ∞ since the material mainly contributes to the electro-optic effect without any change in electroabsorption. The reason the singularities do not tend to infinity in calculations is due to finesse. So far the infinite chirping (wavelength chirp to infinity) have not yet been observed. However, it is expected that β_{mod} is large but finite rather than infinite in magnitude. On the other hand, between any two consecutive singular points, β_{mod} must cross zero. This denotes that $\Delta n = 0$ while $\Delta\alpha$ remains finite and non-zero, and the material now attributes mainly to electroabsorption with no contribution to the electrooptic effect. The sign

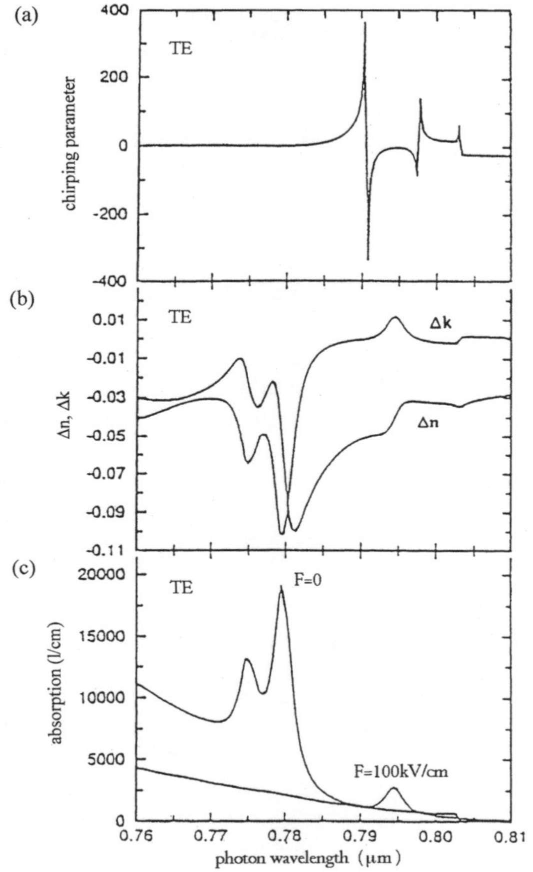

Fig. 11.21 a–c **(a)** The chirping coefficient β_{mod} for $Al_{0.3}Ga_{0.7}As$/ GaAs with L_z = 10nm, L_d = 3nm and F = 100kV/cm. There are three dispersible singularities in the spectrum between 785 and 805nm. **(b)** Δn (solid line) and Δk (dotted line) for the same structure as in (a). **(c)** The absorption spectra of the case L_d = 3nm with F = 0 (solid line) and F = 100kV/cm (dotted line) [11.28].

of β_{mod} depends on that of Δn and $\Delta\alpha$ and there are two possible cases to be considered. First, for $\beta_{mod} > 0$, which occurs in certain wavelength ranges, both parameters (Δn and $\Delta\alpha$) are negative . When Δn is negative, the phase modulation is negative and since $\Delta\alpha$ is also negative, it improves the transparency so that more light can be transmitted out. Therefore, this implies a relatively larger intensity of the output light with a phase reduction. Second, for $\beta_{mod} < 0$, Δn and $\Delta\alpha$ are in opposite signs with $\Delta n < 0$ and $\Delta\alpha > 0$. This results in a reduction of the output light intensity with a phase increment. In addition, the sign of β_{mod} changes as the zero points are crossed, since $\Delta\alpha$ changes sign while Δn keeps the same sign. An interesting feature to note is that at around Δn_{min}, the slope of Δn changes sign from negative to positive in the range from λ = 780–782.5nm. The light signal with an increased positive slope of n will blue-shift in energy and that with a decreasing negative slope of Δn will red-shift in energy. An input light pulse, centered at the above minimum point of Δn and spread within the respective wavelength range, will be compressed, and this can be used in pulse compression applications. Figure 11.21(c) shows the residence loss as a reference. It is also seen that the heavy- and light-hole exciton peaks mix together and undergo a red shift when F = 100kV/cm. It brings a large $\Delta\alpha$ at 779.5nm for application in light intensity modulations. Using the DFQW as cavity in waveguide type modulator systems, such as the $Al_{0.3}Ga_{0.7}As$/GaAs with small and medium interdiffusion lengths (L_d = 10 to 20Å) under fields of 100kV/cm, improved chirping and electro-absorption can be obtained. Furthermore, in a selected set of interdiffusion lengths and fields, the material can be used as an electroabsorption modulator with reduced chirping in a wide range of operating wavelengths (758–874nm).

11.2.2.2 Vertical Cavity Fabry-Perot (FP) Modulator

The Vertical Cavity Fabry-Perot (FP) Modulator, grown epitaxially by the Molecular Beam Epitaxy (MBE) or Metalorganic Chemical Vapour Deposition (MOCVD), reduces the complexity for coupling in devices as compared to the waveguide type modulators. This integratablilty with the recent vertical-cavity surface-emitting laser technology minimizes the chirping effect and releases one more degree of freedom (the vertical direction) from the conventional two-dimensional optical integration. Therefore a three-dimensional optical integration can be developed. The vertical cavity Fabry-Perot (FP) Modulators consist of a two asymmetric quarter-wavelength Bragg reflectors (stack in alternation layers with different indices), which are employed as the top and bottom part of the QW cavity structure. The asymmetric FP structure will be generated so as to strengthen the electroabsorptive properties of the whole structure in order to produce a higher contrast ratio (CR) and minimize the power consumption.

However, since the surface of normal modulators have much shorter optical interaction length (typically about a few micrometer, and thus limit the magnitude of modulation) while that of the waveguide type can have up to about 100μm. Normally, the on-off (contrast) ratio (CR) of a transverse modulator is about 10:1 [11.30], because the maximum change in the excitonic absorption is limited by the line broadening [11.31]. One of the ways to enhance the modulation is to increase the layers of the QW structure, but the applied bias voltage has to increase simultaneously. However, the thicker the fabricated layers, the higher the inhomogeneous of the field in the QW structure. Moreover, generally a high voltage ~ 1V is needed for the operation of the longitudinal modulators. Since a few volts are supplied in incorporating the parallel arrays on the Si-bases LSI electronics to form high-bandwidth optical interconnections [11.32], alternative ways are needed. The three main difficulties we need to overcome are: (a) reducing the operating voltage swing and thus minimizing the amount of the required applied voltage, (b) reducing the operating increase of the effective path for expected signal modulations, and (c) creating sharp resonance characteristics which are very sensitive to the change of refractive index [11.33]. Two quarter-wavelength identical Bragg reflectors stack of alternating layers with different indexes, which sometimes are called grating mirrors with different reflectivities, is an asymmetric FP configuration. This configuration has advantages over the symmetric FP configuration that the electro-absorptive properties of the whole structure will be further strengthened sufficiently to perform higher CR [11.34 and 11.35]. It has been reported that the CR of the transverse QW optical modulation has been improved considerably to 100:1 by applying an asymmetric FP configuration [11.36], and even higher CR (> 100:1) can be achieved by replacing the QW cavity with a superlative [11.34]. The former is associated with the QCSE while the later is based on the effect of the Wannier-Stark localization in superlative, both due to an applied electric field [11.36–11.39].

Conventional vertical cavity FP modulators in QW systems have been constructed with material such as the AlGaAs/GaAs and InGaAs/InP [11.40]. The best performance (CR > 100:1) has been achieved using the superlative in the cavity region. The modified optical properties of the $Al_{0.3}Ga_{0.7}As/GaAs$ DFQW with the QCSE is generally enhanced when different strengths of fields are applied on the vertical asymmetric FP reflection modulator. The modified DFQW modulator can have optical properties. such as high change in the reflectance (ΔR), low resonance reflectance on the OFF-state, low power consumption, precise tunable operating wavelength, and simple fabrication technology. The method used in tunable wavelength operation is achieved by incorporating a transmission modulator and corner-cube retroreflector as a means of matching the wavelength. A non-permanent method is to utilize the QCSE to tune to the operating wavelength.

There is a challenge for both symmetric FP and asymmetric FP structures in terms of a precise control of the accuracy in thickness and composition of the layers, so that it can operate at the expected wavelength [11.41]. There are several ways of modifying the operation wavelength λ_{op}, either permanently or non-permanently. The permanent method, for the reflective modulator only, is to incorporate a transmission modulator and corner-cube retroreflector. This provides a means of fine tuning the λ_{op} to match the source [11.35]. The non-permanent method is by ion-implantation. An electric field is applied to the QCSE in the QW structure for Weinner-Stark localization of SL where the λ_{op} can be modified, and the top and bottom Bragg reflectors of the FP modulator should be preliminary doped with one n-type and the other p-type (p-i-n structure) [11.34 and 11.36]. For QW cavity structures, the QCSE can be employed to modify the reflectivity at the operation wavelength of the FP modulator.

Here an annealing DFQW is introduced as an intrinsic material of the FB cavity, so that the λ_{op} of the modulator can be fine-tuned by intermixing under control the regular as-growth QW, because the intermixing of the QW structure will vary the bandage energy of the well and therefore amend the λ_{op} of the FP modulator. It is expected that this structure can serve as a high performance FP modulator by improving the change of reflectivity of the modulator as well as reducing its residence loss in OFF-state. Since a simple thermal annealing is required for the intermixing the QW structure, the technology can be simplified to produce a FP modulator operating at a precise λ_{op}. It will have high change of reflectivity and low residence loss, and may make the FP-DFQW modulator become a significant candidate in vertical optoelectronic integrated technology. Together with the development of the same-plane monolithic integration, a 3-D optoelectronic integrated circuit (OEIC) may be realized in the future.

11.2.2.3 Modeling

Consider a simple DFQW resonant-etalon reflection modulator structure, as shown in Fig. 11.22, where the inner intrinsic DFQW consists of layers of $Al_{0.3}Ga_{0.7}As/GaAs$. The extent of diffusion is characterized by the diffusion length ($L_d = 0$). The front and back quarter-wavelength Bragg reflectors are alternative AlAs and AlGaAs layers with reflective indices taken from [11.42] which are p- and n- doped as in a p-i-n structure. The bottom reflectance R_B is set at about 0.97, the top one R_T is calculated from the relation $R_T = R_B\exp(-2\alpha_c L_c)$ where α_c and L_c is calculated from a model without the polarization sensitivity. The relation of R_T is the impedance-matching condition for zero total reflectivity at resonance, so that the contract ratio CR of the modulation can become infinity. The reflectance of the FP modulator can be calculated by using the scattering matrices method. The arrow indicates the direction of propagation of the field, where F (forward) and B (backward) are the electric field of the light. The total reflectance is defined as $R_{tot} = (B_1/F_{1)})^2 = |(S_{21} + R_{B5}{}^{1/2} S_{22})/(S_{11} - R_{1T}{}^{1/2}S_{12})|^2$, where R_{B5} and R_{1T} are the reflectance of the interface between the reflectors, the negative sign of R_{1T} indicates that the refractive index of medium T is higher than that of air ($n_{air} = 1$), and S_{ij} are the matrix elements of a 2×2 scattering matrix S defined as $M_T \times M_{QW} \times M_B$, where M_j is the

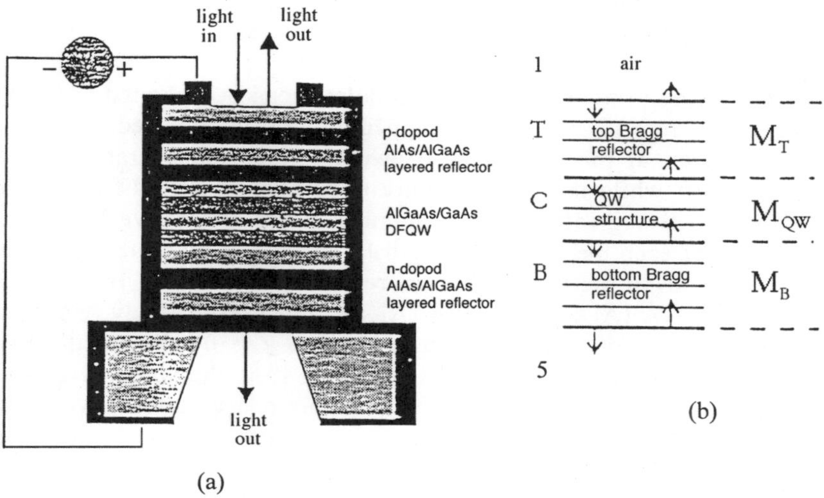

Fig. 11.22 a, b (a) Schematic cross-section of the three-section epitaxial Fabry-Perot reflection modulator. **(b)** Illustration of the scattering matrices in deriving at the design equations [11.51].

scattering matrix of region j for $F_j = m_{j11}F_{j+1} + m_{j12}B_{j+1}$ and $B_j = m_{j21}F_{j+1} + m_{j22}B_{j+1}$, where m_{jab} are elements of M_j, and the subscripts a and b are integers.

The acquired 2×2 matrices S is defined as $\begin{pmatrix} S_{11} & S_{12} \\ S_{21} & S_{22} \end{pmatrix}$ and can be evaluated so that the transmission T and the reflectance R can be calculated as transmission $T = \left| \frac{1}{S_{21}} \right|^2$ and reflectance $R = \left| \frac{S_{11}}{S_{21}} \right|$ respectively. The resonance reflectance in FP mode will became zero while the impedance condition holds. This condition is valid when the cavity phase is a multiple of 2π.

The difficulties with designing DFQW FP modulators are the preset impedance conditions R_T and R_B are the reflectance at the top and bottom Bragg reflectors respectively; α_c and L_c are the absorption coefficients and thickness of the cavity. The phase matching condition for the operating wavelength will not hold for a new operation wavelength. There are two possible solutions: The first is to fix the cavity thickness in different interdiffusion cases while the phase deviation is compensated by rearranging the Bragg reflectors. This means that not only the total number of periods of the alternative layers have to be adjusted, but also the thickness of each layer have to be altered for the quarter-wavelength condition. The second is to adjust both the thickness of the cavity and the distance of the Bragg reflector plates.

It is interesting to note that the FP modulator can be operated at the FP mode for cases from $L_d = 0$ to 40Å, stepped by 10Å, without redesigning of the active region of the modulator. However, since the operation wavelength will shift while the QW structure undergoes an interdiffusion, the top and bottom Bragg reflectors, the number of the alternative layers and the thickness of each quarter-wavelength layer have to be rearranged. The number of periods of the top and bottom reflectors are 8.5 and 16.5 [11.43] of alternative 734Å (AlAs) and 639Å ($Al_{0.3}Ga_{0.7}As$) respectively, when the cavity structure is an as-grown square well ($L_d = 0$). While the QW structure undergoes an interdiffusion to L_d = 40Å, the top and bottom reflectors will be amended to 7.5 and 16.5 periods of alternative 633Å (AlAs) and 548Å ($Al_{0.3}Ga_{0.7}As$) respectively. As a whole, the total thickness of the modulator will be reduced for different DFQWs' cavities, Table 10.5 shows the modulation properties of the DFQWs modulators with an identical thickness of cavity. For the cases from $L_d = 0$ to 20Å, the change of reflectance (ΔR_{tot}) has roughly the same value. Moreover, there are two possible λ_{op}s: the first λ_{op} of $L_d = 10$ will have a relative low residence loss but a high minimum value of R_{tot}, with R_{tot}(min) at the FP mode, while the second λ_{op} is the lowest (1.0418×10^{-5}) with a low residence (0.100) when comparing with other L_ds' cases, and the figure of the merit $\Delta R_{tot}/R_{tot}$ is extremely high, about 9600. If the interdiffusion is further increased to L_d = 40Å, the λ_{op} will further blue-shift [11.43] to 0.7645μm. Also the applied electric fields F (kvVcm) will reduce to 50–80kV/cm, though the R(min) becomes high. The same as-grown QW cavity structure can operate between 0.8747 and 0.794μm under a different interdiffusion L_d from 0 to 30Å. If the high R_{tot}(min) (0.8747–0.7645μm) is not an obstacle in an operation, the wavelength tuning range will then increase to about 110nm.

In fact, the whole modulation will further be improved by adjusting the cavity thickness L_c for $n\pi$ phase variations of the field with different L_ds. In order to compare the result with that of an as-grown square QW modulator, the adjusted L_cs of the different L_ds are set nearest to the L_c of the square QW modulator. Figs. 11.23(a) and 11.23(b) show the reflectance spectra of the different applied fields of the cases $L_d = 0$ (square QW) and L_d = 40Å respectively.

For the case $L_d = 0$, the OFF-state operates at λ_{op} = 0.875μm with field F = 130kV/cm while the ON-state is with field F = 0. It is found that the λ_{op} of the FP mode of the higher field operation will shift to shorter wavelengths because the change of the reflective index Δn of the DFQW cavity is under reverse bias. The Δn is due to the QCSE and can be calculated from the spectrum of the change of absorption coefficient by the

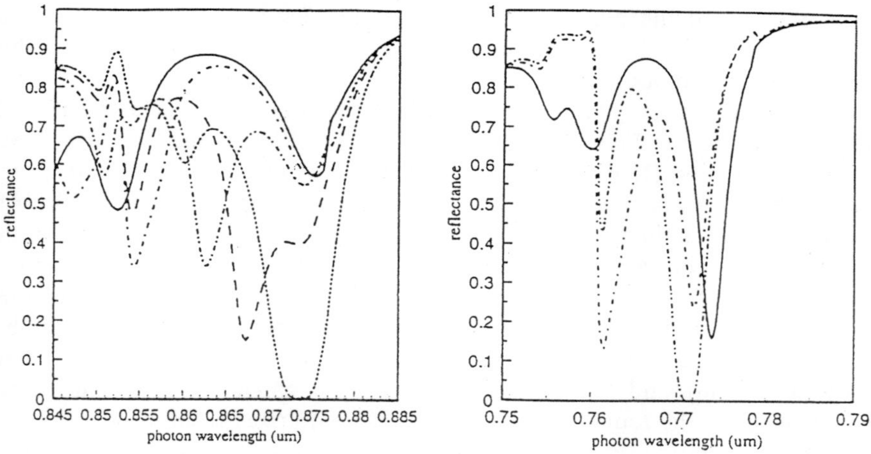

Fig. 11.23 a, b **(a)** The reflectance spectra of L_d = 0 with field F = 0 (solid), 50 (dash-dot), 90 (dash-double dot), 110 (long-dash) and 130kV/cm (dot). **(b)** The reflectance spectra of L_d = 4nm with field F = 0 (solid), 50 (dash-dot), and 80 (dash-double dot) [11.51].

Kramer-Kronig relation. Moreover, for the spectra of the case F = 0, 50, and 90kV/cm, the reflectance are close together at the λ_{op}. This can be explained by the packing of the absorption spectra at the λ_{op}, as shown in [11.44]. The R_T is designed for zero R_{tot} at F = 130kV/cm (off-state), so from Fig. 11.23, the modulation is operated at the ON-OFF state with $\Delta R_{tot}/R_{tot}$ = 0.6427/ 3.313×10^{-3} = 194. In Fig. 11.23(b), the case of L_d = 40Å, the λ_{op} blue shifts to 0.7727μm. The FP resonance with zero reflectance is operated at F = 80kV/cm, the ΔR_{tot} of the ON (F = 0) and OFF (F = 80kV/cm) states is 0.6872 with an extremely low R_{tot}(min) value of 4.7633×10^{-5}. Thus the $\Delta R_{tot}/R_{tot}$ is as high as 15000. Moreover, the applied field required for operation also reduces from 130kV/cm at L_d = 0 to 80kV/cm in this case. This can therefore minimize the voltage swing. The results of other applied fields are shown in Table 10.6. The ΔR_{tot} increases from L_d = 0 to 20Å and then reduce to 0.4873 at L_d = 40Å. While the R_{tot}(min) remains very low and within only an acceptable fluctuation, the residence loss of the different cases of L_ds is just between 550 and 860cm^{-1}, except the second possible λ_{op} which can operate at a low R_{tot}(min) with L_d = 10Å and the residence loss of 1200cm^{-1}. In two outstanding cases, the first with L_d =30Å, the residence loss and the R_{tot}(min) reduce to the lowest at 350cm^{-1} and 2.327×10^{-4} respectively. For the second case, L_d = 40Å and the applied fields can be reduced to 80kV/cm (lower power consumption) with an extremely high $\Delta R_{tot}/R_{tot}$ of 15000, as mentioned above. Together with the distinct feature of the blue shift of λ_{op} and the simple thermal annealing process for producing the DFQW, this FP reflection modulator will become competitive among the similar kinds of modulators.

In the further development of the intermixing of the GaAs/AlGaAs MQW asymmetric Fabry-Perot modulators (AFPM) using impurity-free vacancy diffusion (IFVD) by rapid thermal annealing, a blue shift of up to 52meV has been achieved [11.45]. While maintaining the resolvability of the heavy and light hole excitons, which enable the production of both normally-off (contrast > 10dB for –5V bias) and normally-on (reflection change > 30% for –8V) AFPMs from the same wafer, we can show how the IFVD can be used to tailor the optoelectronic properties after growth. Proton-implanted p-i-n $GaAs/Al_wGa_{1nw}As$ MQW modulators have been realised [11.46] and the measurement of the carrier lifetime in these materials show the reduction in the photocurrent arises from a reduction in lifetime due to the implant-induced damage. Also the reduced lifetime decreases the optically-excited QW carrier population, leading to an increase in the saturation intensity.

11.3 Summary

The application of the DFQW structures to improve the optical properties of the optical and photonic devices is of much interests to the use of interdiffused QWs. In particular these structures have shown that the optical properties of the QWs, such as the spectral absorption coefficient and the refractive index which are important for the design of efficient photonic and optoelectronic devices, may be varied from that of the as-grown square QWs by interdiffusing (i.e. interdiffusion across the well barrier interface). In particular, during the initial stages of interdiffusion, the width of the well at the top is significantly greater than at the bottom, and this effect results in the introduction of additional bound states in the QW, which in turn leads to a modified absorption coefficient and refractive index. Also, during this initial diffusion stage, the depth of the QW does not change significantly so that the band-edge shift is minimal. As the interdiffusion proceeds, the well depth decreases and the newly-created bound states that were created at the top of the well moves into the continuum, with the changes in the absorption and refractive index that would normally be expected for interdiffusing. The exciton absorption coefficient with no applied field for the TE-polarization shows an enhanced oscillator strength in the more extensively-interdiffused QWs, which is due to the increasing density of states and the merging of the heavy and light hole excitons resonance. Whilst the change of refractive index is primarily concerned with the variation of the density of the bound states in the QW, this variation due to the continuum states also has to be included to produce an accurate picture of these refractive indices, especially since their absolute changes are small. For the interdiffused QW at small L_ds, the birefringence is small at long wavelengths (>1mm) and relatively larger at wavelengths around the absorption-edge. This will provide interesting features for waveguiding and modulation applications.

The quantum-confined Stark effect is also analyzed for the interdiffused QW, which includes the effect of exciton absorption for both TE- and TM-polarizations. Results show an enhanced Stark shift in the interdiffused QW, which is twice of that produced by an equivalent square QW. Also the fundamental exciton absorption peak is reduced much more than that in the square QW case, because of the heavy holes becoming unbound. This is due to the earlier vanishing of the confinement effect of the barrier in the interdiffused QW by the same applied field, thus demonstrating the advantage of using an interdiffused QW structure. These results are important since they provide improved characteristics for the optical properties of the QWs. Tunneling is another important consequence of the interdiffused QW, since the barrier of a interdiffused QW is not a sharp abrupt wall (as in the case of square QWs), but is in fact a nonlinear profile which is smooth and flatter. Consequently the tunneling properties differ from those of an abrupt barrier, as described by classical quantum mechanics. This gives a unique characteristic to the interdiffused QWs due to the field-induced tunnelling, which cannot be achieved using square QWs and may be especially applicable to devices utilizing tunnelling. For far-IR wavelengths up to ≈ 20mm, the nonlinear optical absorption due to the intersubband transitions results in an approximately constant absorption peak magnitude and width in the different stages of interdiffusion. The peak position of the intersubband transitions varies from 12 to 50mm with increasing interdiffusion, and this gives a huge tuning range of approximately 40mm. The effect of interdiffusion on the built-in compressive strain of an InGaAs/GaAs QW results in a reduction of the overall interband transition energies and the bandgap of the QW, which provides tuning of the absorption-edge at wavelengths around 1mm. Although the effect of interdiffusion will increase the bandgap, this structure still provides a strong photon lateral confinement at these wavelengths.

In response to the needs of optoelectronic integrated circuits (OEICs), devices fabricated by using selectively interdiffused QW structures allow a planar technology to be used for the fabrication of such OEICs. However, it may also provide a means for improving the device performance and the development of novel optical devices based on the properties of the non-square QWs. The topics presented here are only concerned with the optical properties of the interdiffused QWs for applications in photonic and opto-electronic devices, such as waveguides and modulators, which are the fundamental optical components used in OEICs.

In order to provide optical lateral confinement in a planar QW environment, a significant refractive index step between the optical guiding region (lightly-interdiffused QW with a larger refractive index) and the confining region (extensively-interdiffused QW with a smaller refractive index) is required. It is also important to be able to predict how this step will vary with the interdiffusion as the values of the diffusion coefficient for the two regions are different. A simple model uses the Moss' rule, which states the inversely proportional relation between the bandgap the the refractive index of a material. For the AlGaAs/GaAs and InGaAs/GaAs material systems, interdiffusion (interdiffusing) always produces an increased bandgap in the well and a decreased bandgap in the barrier, thus the required refractive index step is produced by this process. The analysis from the model developed here shows that the realistic situation is much more complicated and subtle; for instance, this refractive index step depends on the wavelength and polarization of the optical beam, the extent of interdiffusion and the waveguide structure used.

For the case of a planar waveguide geometry, where light is propagating along the layer of the QWs, the polarization sensitivity is as strong as with the conventional square QW structures. The advantage of using this geometry is to provide a longer interacting length of the optical signal and the QWs. The effect of interdiffusion will reduce this polarization factor, and guiding will be found for both polarizations as expected. The results indicate the refractive indices for the TE- and TM-polarization vary differently with interdiffusion, because the contributions from the heavy and light holes for the two polarizations are different, due to a different polarization factor for the two type of holes. The electroabsorption effect of the interdiffused QWs under this waveguide-modulator structure is enhanced over the square QWs and will therefore reduce the interaction length that is required for the square QWs with a prescribed ON/OFF ratio. This makes the interdiffused structures attractive for the integration of small size devices. The QCSE is enhanced in this geometry using the more interdiffused QWs and gives a larger absorption-edge shift, which may be used for the design of modulators with a lower driving voltage. These attractive characteristics can be utilized with strong photon lateral confinement for both the TE- and TM-polarizations, if a large difference in the extent of interdiffusing is used.

For light propagating perpendicular to the QW layers as in the case of absorption (reflection) modulators, such as SEEDs, the polarization is less sensitive than that for parallel light propagation, because of the different selection rules for the allowed transitions. The results demonstrate that strong photon confinement can be obtained at the longer wavelengths (> 1 μm) for both polarizations. However, for wavelengths around the band-edge where modulators operate, the refractive index step may become negative, resulting in antiguiding. This is because the absorption peaks, due to the different bound states, create a crossing of the refractive indices for the different degrees of interdiffusion (especially during the initial stages of interdiffusion) and modifie the trend of refractive index steps at these wavelengths. Similar effects may also occur at longer wavelengths for certain structures. Therefore in designing the absorption modulators for the normal incident input beam and interdiffused QWs, care should be taken to have the appropriate well width and the extent of interdiffusion. A 100Å-wide well with a small amount of interdiffusion appears to provide both requirements, although the structures have not yet been optimized.

The results presented here show that interdiffusing may be used to modify the optical properties of selected areas of samples in a controlled manner for the optimization of device characteristics and the realization of new devices.

Interdiffuse delineated stripe optical waveguides in InGaAs/InP QW structures have also been investigated, together with the computer modeling of the propagation characteristics of the waveguides with non-square refractive index variations across the guiding layer. The implantation of P into the InGaAs/InP was used to produce significant QW interdiffusing, which was used to fabricate the low loss interdiffuse delineated stripe optical waveguides.

References

[11.1] C.L. Yang and Q. Yang: *Phys. Rev. B* **37**, 1364 (1988)
[11.2] J.N. Schulman *J. Vac. Sci. Technol. B* **1**, 644 (1983)
[11.3] E.H. Li, B.L. Weiss and J. Micallef (to be published)
[11.4] E.H. Li and B.L. Weiss: *IEEE J. Quantum Electron* (to be published)
[11.5] R.C. Miller, D.A. Kleinman, et al.: *Appl. Phys. Lett.* **39**, 1 (1981)
[11.6] G.D. Sanders and K.K. Bajaj: *Phys. Rev. B* **35**, 2308 (1987)
[11.7] G.P. Khotiyal and P. Bhattacharya: *Journal of Applied Physics* **63**, 2760 (1988)
[11.8] J. Micallef, E.H. Li and B.L. Weiss: *Electonics Letters* **28**, 526 (1992)
[11.9] I.J. Pape, P. Kam-wa Li, et al.: *Electron Lett.* **24**, 910 (1988)
[11.10] K. Nakashima, Y. Kawaguchi, et al.: *Appl. Phys. Lett.* **52**, 1383 (1988)
[11.11] S.A. Schwarz, P. Mei, et al.: *Appl. Phys. Lett.* **53**, 1051 (1988)
[11.12] T.E. Schlesinger and T. Kuech: *Appl. Phys. Lett.* **49**, 519 (1986)
[11.13] E.S. Koteles, B. Elman, et al.: *Superlattices and Microstructures*, **9**, 533 (1991)
[11.14] J. Micallef, E.H. Li, et al.: *Proc. SPIE* **1675**, 211 (1992)
[11.15] I.J. Pape, P. Kam-wa Li, et al.: *Electron Lett.* **24**, 1217 (1988)
[11.16] M. Ghisoni, P.J. Stevens, et al.: *Opt. Quantum Electron.* **23**, S914 (1991)
[11.17] G. Bastard, E.E. Mendez, et al.: *Phys. Rev. B* **28**, 3241 (1984)
[11.18] W. Bloss: *J. Appl. Phy.* **65**, 4789 (1989)
[11.19] J.D. Ralston, W.J. Schaff, et al.: *Appl. Phys. Lett.* **54**, 534 (1989)

[11.20] D. Coffey: *J. Appl. Phys.* **63**, 4626 (1988)
[11.21] E.H. Li, K.S. Chan, et al.: *Appl. Phys. Lett.* **63**, 533 (1993)
[11.22] E.H. Li, B.L. Weiss and K.S. Chan: *Phys. Rev. B* **46**, 15181 (1992)
[11.23] S. Furukawa, H. Matsumura and H. Ishiwara: Theoretical considerations on lateral spread of implanted ions, *Japan. J. appl. Phys.* **11**, 134–142 (1972)
[11.24] H. Ryssel, J. Lorenz and K. Hoffmann: Models for implantation into multilayer targets, *Appl. Phys. A* **41**, 201–207 (1986)
[11.25] E.H. Li: *JAP* **82**, 6251 (1997)
[11.26] E.H. Li, B.L. Weiss, et al.: Polarization dependent refractive index of an interdiffusion induced AlGaAs/GaAs quantum well, *Appl. Phys. Lett.* **62**, 550–552 (1992)
[11.27] H. Kogelnik: *Topics in Applied Physics* **7**(2), Springer-Verlag, New York
[11.28] E.H. Li and W.C.H. Choy: *IEEE Photonics Technology Letters* **7**(8)(August 1995)
[11.29] E.H. Li, B.L. Weiss and K.S. Chan: *Phys. Rev. B* **46**(23), 15181–15192 (1993)
[11.30] H. Shen and M. Wrabeck: *Appl. Phys. Lett.* **62**(23), 2908–2910 (1993)
[11.31] K.W. Jelly, R.W.H. Engelman, et al.: *Appl. Phys. Lett.* **55**, 70 (1989)
[11.32] M. Whitehead, G. Parry and P. Wheatly: *IEEE Proceedings* **136**, Pt. J, No. 1, 52–58 (1989)
[11.33] A. Tomiita, Y. Kohga and A. Suzuki: *Appl. Phys, Lett.* **55**(18), 1817–1819 (1989)
[11.34] K.K. Law, R.H. Yan, et al.: *Appl. Phys. Lett.* **57**(13), 1345–1347 (1990)
[11.35] K.K. Law, L.A. Coldren and J.L. Merg: *IEEE Photon. Tech. Lett.* **3**(4) (1991)
[11.36] M. Whitehead, A. Rivers and G. Parry: *Electronic Lett.* **3**(1991)
[11.37] J. Bleuse, G. Bastrad and P. Voisin: *Phys. Rev. Lett.* **60**(3), 220–230 (1988)
[11.38] E.E. Mindezm, F.A. Rueda and J.M. Hong: *Phys, Rev. Lett.* **60(**3), 220–230 (1988)
[11.39] I. Bar-Joseph, K.W. Goossen, et al.: *Appl. Phys. Lett.* **55**, 340–342 (1989)
[11.40] A. Tomrea: *Appl. Phy. Lett.* **55**, 1817 (1989)
[11.41] J. Fritz, T.M. Breman, et al.: *Appl. Phys. Lett.* **63**(4), 494–496 (1993)
[11.42] B. Jensen and W.D. Jensen: *IEEE J. of Quantum Electron.* **27**, 40–44 (1991)
[11.43] E.H. Li and W.C.H. Choy: *Jap. J. Appl. Phys.* **35**, L496–L498 (1996)
[11.44] W.C.H. Choy and E.H. Li: *IEEE, J. of Quan. Elect.* (to be published)
[11.45] M. Ghisoni, G. Parry, et al.: *Jap. J. Appl. Phy.* **30**(6A), L1018–L1020 (June 1991)
[11.46] T.K. Woodward, W.H. Knox, et al.: *IEEE J. of Quantum. Electron.* **30(**12) (1994)
[11.47] E.H. Li: *IEEE-JQE* **38** (1996)
[11.48] K.M. Lo and E.H. Li: First Optoelectronics and Communications Conference (OECC'96), *Technical Digest*, 18P-5 (1996)
[11.49] J. Micallef, E.H. Li and B.L. Weiss: *Superlattices and Microstructures* **13**(1), 125–132 (1993)

[11.50] J. Micallef, E.H. Li and B.L. Weiss: *Appl. Phys. Lett.* **61**(4), 435–437 (1992)

[11.51] W.C.H. Choy and E.H. Li: The application of an interdiffused quantum well in a normally on electroabsorptive Fabry-Perot reflection modulator, *IEEE J. Quantum Electron.* **33**(3), 382–392 (1997)

[11.52] E.H. Li and B.L. Weiss: *IEEE Photonics Technology Letters* **4**(4), 445 (1993)

[11.53] J. Micallef, E.H. Li and B.L. Weiss: *J. Appl. Phys.* **73**(11), 7524–7532 (1993)

Appendices

Appendix 1. Materials Parameters of $Al_xGa_{1-x}As$

	Unit	*Temp*	$Al_xGa_{1-x}As$
E_g(eV)	eV	RT [LT]	$1.424 + 1.594x + x(1-x)(0.127 - 1.31x)$ $1.5194 + 1.36x + 0.22x^2$
Δ_0	eV	RT [LT]	$0.34 - 0.065x$ $0.341 - 0.066x$
m_c	m_0	RT LT	$0.0632 + 0.0856x + 0.0231x^2$ $0.0665 + 0.1006x + 0.0137x^2$
m_{hh}	m_0	RT LT	$0.50 + 0.2x(x < 0.45)$ $0.34 + 0.16x(x < 0.45)$
m_{lh}	m_0	RT LT	$0.088 + 0.0372x + 0.0162x^2$ $0.0951 + 0.0441x + 0.0097x^2$
c_{11}	$\times 10^{11}$dynes/cm^2	RT LT	$11.90 + 0.12x$ $12.23 + 0.27x$
c_{12}	$\times 10^{11}$dynes/cm^2	RT LT	$5.38 - 0.08x$ $5.71 - 0.37x$
a_0	Å	RT LT	5.6533 + 0.0078i $5.6525 + 0.0075x$
dE_g/dP	eV/bar	RT LT	$11.5 - 1.3x$
ε (static)	ε_0	RT LT	$13.18 - 3.12x$ $12.40 - 1.8x$
g_1 g_2 g_3	M M M	RT LT	$6.68 -2.52x$ $2.34 - 1.73x$ $7.65 - 3.89x$ $2.41 - 1.73x$ $3.28 - 1.86x$

Appendix 1 (continued)

b	eV	RT LT	$-1.7 + 0.2x$
$m_{lh_{\parallel}}$	m_0	RT LT	$0.23 + 0.13x$ $0.21 + 0.11x$
$m_{hh_{\parallel}}$	m_0	RT LT	$0.11 + 0.02x$ $0.12 + 0.06x$

RT = 300°K, LT ≤ 10°K; $m_0 = 9.11\times10^{-31}$kg, $\varepsilon_0 = 8.85$pF/m

Appendix 2. Materials Parameters of $In_xGa_{1-x}As_yP_{1-y}$.

	Unit	*Temp*	$In_xGa_{1-x}As_yP_{1-y}$
E_g	eV	RT	$1.35 - 1.17y + 0.668(1-x) - 0.069y(1-x) + 0.18y^2 + 0.03(1-x)y^2 + 0.758(1-x)^2 - 0.322y(1-x)^2$
		LT	$0.418 + 1.10(1-x) + 1.005(1-y) - 0.493x(1-x) - 0.180y(1-y) - 0.265x(1-x)(1-y) - 0.030(1-x)y(1-y) + 0.347(1-x)(1-y) + 0.875xy(1-x)(1-y)$
SOS (Δ_0)	eV	RT	$0.34(1-x)y + 0.41xy + 0.08)1-x)(1-y) + 0.11x(1-y)$
		LT	$0.341(1-x)y + 0.38xy + 0.082(1-x)(1-y) + 0.108x(1-y)$
m_e	m_0	RT	$0.0632(1-x)y + 0.0213xy + 0.17(1-x)(1-y) + 0.077x(1-y)$
		LT	$0.0665(1-x)y + 0.023xy + 0.17(1-x)(1-y) + 0.08x(1-y)$
m_{lh}	m_0	RT	$0.088(1-x)y + 0.024xy + 0.16(1-x)(1-y) + 0.12x(1-y)$
		LT	$0.0951(1-x)y + 0.027xy + 0.14(1-x)(1-y) + 0.089x(1-y)$
m_{hh}	m_0	RT	$0.5(1-x)y + 0.41xy + 0.54(1-x)(1-y) + 0.6x(1-y)$
		LT	$0.34(1-x)y + 0.3413xy + 0.67(1-x)(1-y) + 0.45x(1-y)$
c_{11}	$\times 10^{11}$ dynes/cm^2	RT	$11.9(1-x)y + 8.329xy + 14.05(1-x)(1-y) + 10.11x(1-y)$
		LT	$12.23(1-x)y + 8.33xy + 14.39(1-x)(1-y) + 10.22x(1-y)$
c_{12}	$\times 10^{11}$ dynes/cm^2	RT	$5.38(1-x)y + 4.526xy + 6.203(1-x)(1-y) + 5.61x(1-y)$
		LT	$5.71(1-x)y + 4.53xy + 6.52(1-x)(1-y) + 5.76x(1-y)$
a_0	Å	RT	$5.6533(1-x)y + 6.0583xy + 5.4505(1-x)(1-y) + 5.8687x(1-y)$
		LT	$5.6525(1-x)y + 6.0584xy + 5.4512(1-x)(1-y) + 5.8688x(1-y)$

Appendix 2 (continued)

	Unit	Temp	$In_xGa_{1-x}As_yP_{1-y}$
ε (static)	ε_0	RT	$13.18(1-x)y + 14.55xy + 11.1(1-x)(1-y) + 12.35x(1-y)$
		LT	$12.4(1-x)y + 14.55xy + 10.86(1-x)(1-y) + 11.77x(1-y)$
γ_1	m	RT	$6.68(1-x)y + 18.9xy + 4.05(1-x)(1-y) + 5.04x(1-y)$
		LT	$7.65(1-x)y + 19.67xy + 4.2(1-x)(1-y) + 6.28x(1-y)$
γ_2	m	RT	$2.34(1-x)y + 8.37xy + 1.10(1-x)(1-y) + 1.56x(1-y)$
		LT	$2.41(1-x)y + 8.37xy + 0.98(1-x)(1-y) + 2.09x(1-y)$
γ_3	m	RT	$3.28(1-x)y + 9.29xy + 1.66(1-x)(1-y) + 2.76x(1-y)$
		LT	
b	eV		$-1.7(1-x)y + (-1.8)xy + (-1.8)(1-x)(1-y) + (-2.0)x(1-y)$
$m_{lh_{11}}$	m_0	RT	$0.23(1-x)y + 0.082xy + 0.34(1-x)(1-y) + 0.29x(1-y)$
		LT	$0.21(1-x)y + 0.087xy + 0.32(1-x)(1-y) + 0.22x(1-y)$
$m_{hh_{11}}$	m_0	RT	$0.11(1-x)y + 0.031xy + 0.19(1-x)(1-y) + 0.15x(1-y)$
		LT	$0.12(1-x)y + 0.035xy + 0.20(1-x)(1-y) + 0.11x(1-y)$

RT = 300°K, LT ≤ 10°K, $m_0 = 9.11\times10^{-31}$kg, ε_0 = 8.85pF/m

Problems

Chapter 2

2.1 (1) Prove that the damping coefficient h is just the half-magnitude half-width of the absorption curve.

(2) Draw the scaled curves of $\chi'(\omega)$ and $\chi''(\omega)$.

(3) Prove that $\chi'(\omega) = \frac{1}{2}\chi''_{\max}(\omega)$ at the point h is the half-magnitude half-width of the absorption curve.

2.2 Write the physical meaning of each term in the second-order nonlinear susceptibility tensor element $\chi^{(2)}_{\mu\alpha\beta}(\omega)$ as in (2.76).

2.3 Prove that all the even-order susceptibilities $\chi^{(2r)}(\omega) = 0$ for the centrosymmetric media.

2.4 Write the representations of the third-order susceptibility by considering the intermolecular interactions $\chi^{(3)}{}_{\mu\alpha\beta\gamma}(-\omega_4;\omega_1,\omega_2,\omega_3)$.

2.5 Introduce the fourth-order nonlinear susceptibility formula with Feynman diagrams.

2.6 Prove the Kramers-Kronig Relations:

$$\chi'(\omega) = \frac{1}{\pi} P.V. \int_{-\infty}^{\infty} \frac{\chi''(\omega')}{\omega' - \omega} d\omega'$$

$$\chi''(\omega) = -\frac{1}{\pi} P.V. \int_{-\infty}^{\infty} \frac{\chi'(\omega')}{\omega' - \omega} d\omega'.$$

Chapter 3

3.1 Determine the refractive index n, distinction κ and absorption α by $\varepsilon_1 = n^2 - k^2$ and $\varepsilon_2 = 2n\kappa$. Find the approximate expressions of Δn, $\Delta\kappa$ and $\Delta\alpha$ when $n >> \kappa$, if $\Delta\varepsilon = \Delta\varepsilon + i\Delta\varepsilon^2$.

3.2 Discuss that the conditions for the quasi-monochromatic planar wave approximation are continuous waves or pulse waves with pulse widths longer than 100fs.

3.3 Prove that the condition of slowly-varying amplitude approximation (3.30) is

$$\left|\frac{\partial^2}{\partial z^2}\widetilde{E}(\omega,z)\right| << \left|k\frac{\partial}{\partial z}\widetilde{E}(\omega,z)\right| << \left|k^2\widetilde{E}(\omega,z)\right|.$$

3.4 Discuss the pulse-width > 0.1ps condition for (3.68)

$$\frac{\partial A}{\partial z}+\beta_1\frac{\partial A}{\partial t}+\frac{i}{2}\beta_2\frac{\partial^2 A}{\partial t^2}+\frac{\alpha}{2}A=i\gamma|A|^2\cdot A.$$

3.5 Indicate the physical meaning of L_{eff} and consider the cascading cases of EDFAs.

3.6 Calculate, tabulate, and compare the results of the following data:
(1) Find L_{D} and L_{NL} if T_0 is 1ps, 100ps, or 1ns, and if P_0 is 1mW, 10mW, 100mW, or 1W.
(2) Find L_{NL}, using λ = 1.55μm, $|\beta_2|$ = 20ps²/km, and γ = 20/W·km.
(3) Find L_{eff}, if α is 0.17, 0.2 or 0.32dB/km, and if L_{fiber} is 20km or ≅ ∞km.

Chapter 4

4.1 Discuss the physical patterns of the Manley-Rowe relations in the SHG, the sum-frequency, and the difference-frequency processes.

4.2 Discuss the physical patterns of the parametric nonlinear effects between the multiwave processes. Draw the interactions of the light waves with energy levels in the medium, as well as the spatial patterns of the wave vectors of the four-wave mixing.

4.3 (1) Derive the angle phase-matching conditions of positive-uniaxial and negative-uniaxial crystals in the type-I (or parallel) phase-matching.
(2) Find the phase-matching angle of the KDP crystal in the type-I phase-matching condition, if n_0^{ω} = 1.502, $n_0^{2\omega}$ = 1.530, $n_e^{2\omega}$ = 1.488.

4.4 Prove that the far-infrared difference-frequency of two noncollinear laser beams is generated only in the case of anomalous dispersion in crystals.

4.5 Explain the benefits of using $LiNbO_3$ as an SHG crystal (hint: use temperature phase-matching).

4.6 (1) Find the result of $\Lambda = (\lambda/\pi)\cdot(n_{2\omega} - n_{\omega})$, the periodical domain inversion of nonlinear coefficient in the quasi-phase matching.
(2) From the point of material preparation and the SHG efficiency, discuss the structure coefficient $\Lambda \cong (1\sim n)\lambda \cong (1\sim n)$μm.

Chapter 5

5.1 Analyze the factors that affect a perfect phase-matching SHG: the injection beam angle, the beam linewidth and the crystal temperature.

5.2 Analyze some of the differences between the OPO and laser oscillators.

5.3 Find the gain coefficient $G_{OPA}(\text{cm}^{-1}) = 2\cdot(\omega_1\omega_2/n_1n_2)^{1/2}\cdot(\mu_0\varepsilon_0)^{1/2}\cdot\chi c^{(2)}\cdot E_3$ (MKS) of the traveling wave parametric amplifier, if the signal wave amplified $\lambda_{1\text{-signal}} = 0.98\mu\text{m}$, the idle wave $\lambda_{2\text{-idle}} = 1.15\mu\text{m}$, the pump wave $\lambda_{3\text{-pump}} = 0.53\mu\text{m}$, the nonlinear coefficient $d = \varepsilon_0\chi c^{(2)}/2 = 5\times10^{-23}$ (MKS), the refractive indices $n_1 = n_2 = 2.2$, $n_3 = 2.4$, and the pump power $P_{3\text{-pump}} = 4\times10^6\text{W/cm}^2$, where $\varepsilon_0 = 0.0885\times10^{-10}$ (F/m), and $\mu_0 = 1.257\times10^{-6}$ (H/m).

5.4 Design an optical system to measure $\chi^{(3)}_{xyyx}$ of a transparent medium by using DFWM. It is known here that $\eta_{\text{WFM-OPC}} = tg^2(|\zeta|L) = 8.5$, and $I = P/S = cn(|E|^2)/8\pi$.

Also, take the YAG-SHG pump wave $\lambda_{3\ \text{pump}} = 0.53\mu\text{m}$, output energy $\hat{U} = 50\text{mJ}$ (linear polarization), pulse-width 8ns, medium length $L = 10\text{cm}$, its refractive index $n = 1.62$, with beam diameter 2mm. Calculate the value of $\chi^{(3)}_{xyyx} = (cn|\zeta|)/(2\pi\omega|E|^2)$.

5.5 Prove the SBS threshold pump power density $I_{\text{SBS threshold}}$ in bulk silica is approximately 10^6–10^7W/cm^2. The input laser wavelength $\lambda_{1\text{-B}} = \lambda_{2\text{pump}} = 1.55\mu\text{m}$, $(1/\alpha) = 5\times10^{10}\text{nt/m}^3$ (α is the elastic coefficient of the medium), $\delta_s = 20\text{cm}^{-1}$, $\delta = 0.02\text{cm}^{-1}$, $\varepsilon_{1r\text{-SBS}} = \varepsilon_{2r\text{-Pump}} = 2.2$, the electrostriction coefficient of the medium $S \cong \varepsilon_0 = 10^{-11}$(MKS), and $n_{1\text{-B}} \cong n_{2\text{-P}} = 1.5$ are the refractive indices.

5.6 Compare the features of SRS and SBS in the following table:

	SRS	SBS
Transmission Directions		
Quantity of Stokes Shift		
Threshold Pump Power		
Causes in Physics		

Chapter 6

6.1 The free-exciton in a 3D-bulk is described by the hydrogen-like model, its bonding-energy can be solved from wave equation

$$[-(h^2/2\mu)\nabla - e^2/\varepsilon r]\cdot\varphi(r) = E_n\varphi(r)$$

and the solution is

$$E_n = -13.6\times(m_{eh}^*/m)\cdot(\varepsilon_0/\varepsilon)^2\cdot(1/n^2) = Ry^*/n^2$$

where the Rydberg constant $Ry^* = -13.6\times(m_{eh}^*/m)\cdot(\varepsilon_0/\varepsilon)^2$ (n = 1,2,3, ...), m is the free-electron mass, m_e^* and m_h^* are the effective mass of electron and hole respectively, and $m_{eh}^* = m_e^* m_h^* / (m_e^* + m_h^*)$ is the electron-hole reduced mass.

The effective Bohr's radius $a_{\text{B-3D}}$ of a free-exciton in a semiconductor is

$$a_{\text{B-3D}} = \varepsilon h^2/(m_{eh}^* e^2)(nm).$$

If the known terms in the bulk GaAs are: $m_e^* = 0.067\text{m}$, $m_h^* = 0.45\text{m}$, $\varepsilon = 13.7\varepsilon_0$, and $m_{eh}^* = 0.058\text{m}$, find the solution of:

(1) the ground-state bonding-energy of the 3D-excition E_1,

(2) the effective Bohr's diameter of a free-exciton in a semiconductor $2a_{\text{B-3D}}$,

(3) the location of the ground-state 3D-exciton absorption peak $E_{a\text{-peak1}}$ (from the lower side of the absorption-edge).

6.2 Introduce the exciton energy E_n and effective Bohr radius $a_{\text{B-3D}}$ in semiconductors by a hydrogen-like model. If n = 1, compare these with kT = 26mV at room temperature and with the Bohr radius of hydrogen

$a_{\text{B-hydrogen}} = (\varepsilon m_0/m_r)\cdot a_{\text{B-3D}}$.

6.3 From (6.17), $\alpha(\omega,0) = (c/\omega)(\omega - \omega_g)^{1/2}$ for $\omega > \omega_g$. Calculate the material constant C ($= 2.3\times10^{12}\text{cm}^{-1}\text{s}^{-1/2}$ for $\lambda = 0.85\mu\text{m}$).

6.4 Explain the causes of the uniform broadening Γ_{SC} and the nonuniform broadening Γ_{QW} of the 2D-exciton spectrum linewidth.

If $\Gamma_{\text{SC}} = 3\times10^{-22}$J and $\Gamma_{\text{QW}} = 3.46\times10^{-22}$J, find the exciton lifetime τ for the Gauss and Lorentz line shapes. Compare these with the measured lifetime of τ = 280fs, which approximation is more accurate?

6.5 Explain the origins of the further broadening of the 2D-exciton spectrum linewidth and the shortening of lifetime in the case of an applied extra field.

6.6 From Fig. 6.19, analyze the possibilities of using the 2D-QW to realize the phase modulators and intensity modulators.

6.7 The QCSE-induced ground-state energy change can be found by the second-order perturbation:

$$\Delta E_{\text{1-SHIFT}} = E_1 - E_1^{(0)} = -2.2\times10^{-3}\,(m_e^* e^2/h^2)U^2 L_{wz}^4.$$

The larger the well width L_{WZ}, the stronger the QCSE and hence the larger the $\Delta E_{1\text{-SHIFT}}$. Is there a limit on the maximum well-width in the QCSE?

Chapter 7

7.1 Compare the following two systems with respect to the GVD, the limit on bit-rate and L, if the initial chirp and higher ordes of the GVD are neglected: One in which there is considerable linewidth $\delta\omega$ of the light sources, e.g. $\delta\omega T_0 >> 1$, and in which the modulation-induced linewidth can be omitted; and the other in which there is considerable pulse-modulation induced $\delta\omega$, and in which the linewidth of the light sources can be omitted, as in a dynamic single-mode laser [7.2 and 7.16].

7.2 If $\beta_2 = 20\text{ps}^2/\text{km}$, $\gamma = 20\text{W/km}$, and L =50 km, choose the appropriate initial pulsewidth T_0 and injection peak power P_0, in the cases that: (1) both the GVD and nonlinear effects are not important, (2) the GVD is dominant, (3) the SPM is dominant, and (4) both the GVD and nonlinear effects are important.

7.3 Calculate the maximum frequency shift due to the SPM-induced spectrum broadening for m = 1 and m = 3, if the maximum phase-shift φ_{SPMMAX} is 2.5π, and the initial pulsewidth T_0 = 100ps.

7.4 Find the peak power $P_{1\text{st}}$ needed to produce the fundamental (N = 1) soliton in the cases: (1) T_0 = 1ps, (2) T_0 = 10ps, (3) $\beta_2 = -2\text{ps}^2/\text{km}$ (dispersion-shifted fiber), if $\gamma = 20\text{W/km}$. For the first two cases, $\beta_2 = 20\text{ps}^2/\text{km}$.

7.5 Find the peak power $P_{1\text{st}}$, $P_{2\text{nd}}$, and $P_{3\text{rd}}$ for the first-, second- and third-order solitons, if the GVD coefficient of the DSF at 1.55μm, D_{DSF} = 2ps/nm.km and the GVD coefficient of the conventional fiber at 1.55μm, $DCON$ = 17ps/nm.km, $A_{\text{eff}} = 50\mu\text{m}^2$, $n_2 = 3.2\times10^{-16}\text{cm}^2/\text{W}$, and T_{FWHM} = 20ps. Discuss the means to lower the required peak power.

7.6 Analyze the causes of the Gordon-Haus limit and how to bypass it by using filters in the EDFA cascading communications.

7.7 Find the soliton transmission distance limit from the Gordon-Haus effect, both with and without filters. Use $\lambda = 1.55\mu\text{m}$, D = 1ps/nm.km (DSF) and D = 17ps/nm.km (conventional fiber), $\gamma = 0.055\text{km}^{-1}$, corresponding to 0.25dB/km, T_{FWHM} = 20ps, $L_{\text{P-P}}$ = 30km, B = 10Gbit/s, $A_{\text{eff}} = 35\mu\text{m}^2$, $n_2 = 3.2\times10^{-16}\text{cm}^2/\text{W}$, and Ω_f = 723GHz (filter frequency shift bandwidth $\Delta ff \cong$ 115GHz).

Chapter 8

8.1 In Fig.8.11(a), from the permitted power penalty ($P_{\text{Dispersion}}$ = 1dB or 3dB), find the dispersion factor $F_{\text{Dispersion}}$.
Then from the definition of

$$F_{\text{Dispersion}} = \frac{1}{\pi} \cdot B^2 L \cdot D(\lambda) \cdot \frac{\lambda^2}{c},$$

the transmission product B^2L limited by GVD could be exposed. Accordingly, if the CSMF is at λ = 1.55μm, get the L from B. (Of course, there are some other ways to understand it, such as from Fig.8.11(b).)

(a)

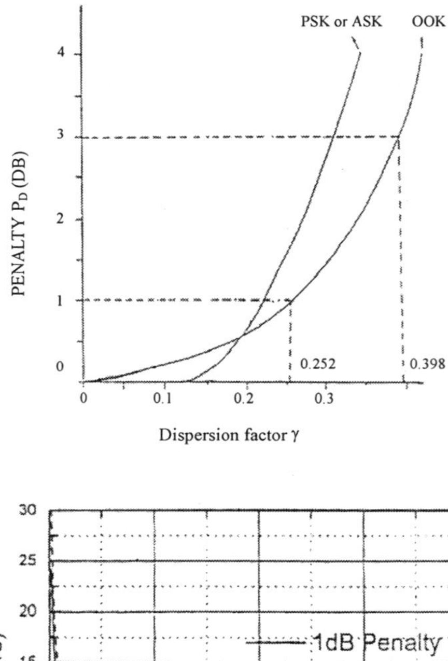

(b)

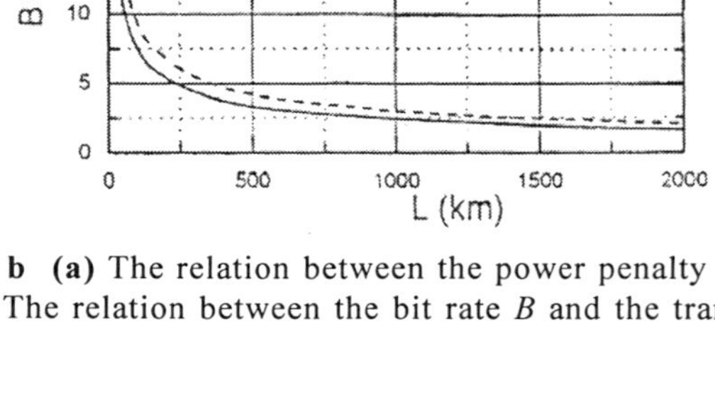

Fig. A1 a, b **(a)** The relation between the power penalty and the dispersion factor. **(b)** The relation between the bit rate *B* and the transmission distance *L*.

8.2 Many approaches of the dispersion compensation or management have been proposed in a 1.55μm single wavelength (SW) or WDM system, such as (a) transmitted at zero dispersion wavelength of the DSF, (b) trunk fiber at $D = +17$ps/nm·km of the CSMF and compensated by a fiber section, (c) trunk fiber at $D = +2.0$ps/nm·km and managed by a fiber segment. Analyze and compare three of them.

8.3 Try to prove the severe FWM will occur if one wavelength of the two is rightly located at the zero-dispersion point in the WDM systems (refer to the FWM in Chapters 4 and 5, and the detail discussion in [8.7]).

8.4 Analyze the FWM suppression by the channel spacing nonuniformity

8.5 Acquire and discuss the performances of the GVD compensation transmission by the midway spectrum inversion (Optical Phase-conjugation, OPC).

8.6 Find the permitted channel power $P_{\text{ch-permitted}}$ for a power penalty of 1dB due to the SRS. If $A_{\text{eff}} = 90\mu\text{m}^2$, $G_{\text{Raman}} = 6.3 \cdot 10^{-14}$m/W, and

(1) the channel number $N = 16$, and $\Delta f_{\text{spacing}} = 100$GHz, 1000km with 10 amplified sections,

(2) the channel number $N = 64$, and $\Delta f_{\text{spacing}} = 50$GHz, 300km with 3 amplified sections.

8.7 Discuss the purposes, characteristics and perspectives of the optical network (ON) or the all optical network (AON).

About the Authors

Professor Charles K. Kao

Professor Kao is internationally known as the "Father of Fibre Optics". He pioneered the research and was involved in the subsequent development of optical fibre communication transmission systems, which are now the fixed transmission backbone of all major communication trunk routes in the world, and will be extended to cover all local routes.

Professor Kao received a B.Sc. degree in 1957 in Electrical Engineering from the University of London and a Ph.D. degree in Electrical Engineering in 1965, also from the University of London. His research experience included theoretical and experimental studies of electromagnetic waves at microwave and optical frequencies, as well as telecommunications systems techniques and applications.

As the first executive scientist and subsequently the corporate director of research at ITT, he launched in early 1980s a program addressing the high frequency limits of signal processing known as "terabit technology" to assist his R&D managers to improve the effectiveness of materials and device research.

From 1987-1996, he was the Vice-Chancellor of The Chinese University of Hong Kong. He has edited a number of books and is the author of two books on optical fibers and one on the business of high technology.

Professor Kao is a Member of National Academy of Engineering (USA), Fellow of the Academia Sinica (Taiwan), Foreign Member of the Royal Swedish Academy of Engineering Sciences IVA (Sweden), Fellow of the Royal Academy of Engineering (UK), Fellow of the Royal Society (UK), Fellow of IEEE (USA), Fellow of IEE (UK), Fellow of HKIE (Hong Kong) and Academician of the Chinese Academy of Sciences. Among his many international honours and awards are the Stewart Ballantine Medal, Rank Prize, L.M. Ericsson International Prize, Alexander Graham Bell Medal, Marconi International Fellowship, Faraday Medal of IEE, the Japan Prize, and the Draper Prize (USA).

Professor Yili Guo

Professor Guo, was born in Shanghai, China, in 1934 and has been teaching at the Department of Electronic Engineering, Tsinghua University, Beijing, China since 1959.

He has been engaged in research on microwave electronics since 1950s, and on lasers, laser techniques, information optoelectronics and nonlinear photonics since 1970s.

His major fields of interest include four wave mixing and frequency conversion, stimulated Raman scattering, wide-band tunable high power lasers, semiconductor optical bi-stability and optical gates, photonic switching and ATM optical switching systems. Since 1996, he focused his research on the high capacity broadband DWDM optical networks, smart data optical networking, OXC, OADM and the optical nonlinearities in the optical communications and optical networks. Professor Guo taught a graduate course on nonlinear photonics at Tsinghua University from 1981 to 1999.

Professor Guo was a visiting scholar at Tokyo Institute of Technology, Tokyo, Japan in 1987, at University College London (on a fellowship sponsored by The Royal Society, UK) in 1990–91, and at The Chinese University of Hong Kong in 1995.

He has published over 100 papers in professional journals and is the coauthor of the book, *Optoelectronics and Lightwave Technology* (John Wiely, 1992).

Professor Guo received a State Award of Advanced Science and Technology (Second Grade) from the State Science and Technology Commission of China for his contributions on the wide-band tunable nonlinear high power laser.

Professor Eddie Herbert Li (1957-2001)

Professor Li received his BS degree in Physics from the Western Washington State College, USA, in 1979, his M.Phil. degree in Applied Mathematics and Electronic Engineering from the University of Hong Kong in 1990, and his Ph.D. degree in Electronic Engineering from the University of Surrey, UK, in 1992.

Between 1979 and 1988, Professor Li had worked for the Kirsten Aeronautical Laboratory in Seattle, and K.K. Engineering Company and Micro Systems in Hong Kong. He joined the Department of Electrical and Electronic Engineering, The University of Hong Kong in 1994 and was the Associate Dean of Engineering in 2000.

He was actively involved in the research of III-V compound semiconductor materials and devices, and his other research interests included optoelectronic device fabrication, modelling and characterization, acoustic-optic devices for slow-rate switching applications, photovoltaic devices and the development of blue band-gap materials.

He was an Honorary Life Advisor of Pi-Mu-Epsilon, American Mathematical Society, USA., and received in 1990 the *Distinguished Leadership Award* from the International Directory of Distinguished Leadership, ABI Inc., USA. He was the co-recipient of the 1992 *J. Langham Thompson Premium Prize*, awarded by the Institution of Electrical Engineers, UK. He had also been a *Distinguished Lecturer* for the Institute of Electrical and Electronics Engineers (IEEE), Electronic Device Society (EDS), USA since 1997. Dr. Li had published over 250 articles and technical papers in journals and monographs, and had written and edited several books.

He was a Fellow of the Institute of Physics (UK), the Royal Statistical Society (UK), the British Society of Commerce (UK), the Institute of Management Specialists (UK), and the Institute of Mathematics and Its Applications (UK), and a Senior Member of IEEE (US).

Professor Kin-Seng Chiang

Professor Chiang received his B.Eng. and Ph.D. degrees in electrical engineering from the University of New South Wales, Australia, in 1982 and 1986, respectively.

He first worked for six months in the Department of Mathematics, Australian Defence Force Academy, Canberra, where he developed mathematical models for optical fiber fused tapered couplers, then from 1986 to 1993 in the Division of Applied Physics, Commonwealth Scientific and Industrial Research Organization (CSIRO), Australia, as a Research Scientist/Senior Research Scientist where he established a fiber-optics laboratory and initiated research in the areas of optical fiber sensors and nonlinear fiber optics.

He received a Japanese Government Research Award for Foreign Specialist through which he visited in 1987–1988 the Electrotechnical Laboratory in Tsukuba, Japan to work on an optical fiber interferometric sensor. From 1992 to 1993, he worked also for the Optical Fibre Technology Centre, University of Sydney, on a project on grating-based fiber sensors.

He joined the City University of Hong Kong in 1993, and is at present the Chair Professor of Electronic Engineering. In 1999, he was awarded the Croucher Senior Research Fellowship for 2000-2001.

He holds two patents and has published over 200 papers in international journals and conference proceedings, as well as chapters in several books. His research interests include optical waveguide theory, numerical methods, integrated and fiber optics, fiber sensors and devices, and nonlinear optics.

Index